Materials in Mechanical Extremes

Fundamentals and Applications

This unified guide brings together the underlying principles, and predictable material responses, that connect metals, polymers, brittle solids and energetic materials as they respond to extreme external stresses.

Previously disparate scientific principles, concepts and terminology are combined within a single theoretical framework, across different materials and scales, to provide the tools necessary to understand, and calculate, the responses of materials and structures to extreme static and dynamic loading. Real-world examples illustrate how material behaviours produce a component response, enabling recognition – and avoidance – of the deformation mechanisms that contribute to mechanical failure. A final synoptic chapter presents a case study of extreme conditions brought about by the infamous Chicxulub impact event.

Bringing together simple concepts from diverse fields into a single, accessible, rigorous text, this is an indispensable reference for all researchers and practitioners in materials science, mechanical engineering, physics, physical chemistry and geophysics.

Neil Bourne has been studying materials under extreme conditions for his entire career. He is a former Chair of the American Physical Society's Topical Group on Shock Compression of Condensed Matter, holds fellowships from the American Physical Society and the Institute of Physics, and obtained his Ph.D. and Sc.D. from the University of Cambridge. He has held appointments at the Universities of Manchester, Cambridge, Cranfield, and Imperial College, London, and as a Distinguished Scientist at the Atomic Weapons Establishment (AWE).

Materials in Mechanical Extremes

Fundamentals and Applications

NEIL BOURNE

CAMBRIDGE
UNIVERSITY PRESS

CAMBRIDGE UNIVERSITY PRESS
Cambridge, New York, Melbourne, Madrid, Cape Town,
Singapore, São Paulo, Delhi, Mexico City

Cambridge University Press
The Edinburgh Building, Cambridge CB2 8RU, UK

Published in the United States of America by Cambridge University Press, New York

www.cambridge.org
Information on this title: www.cambridge.org/9781107023758

First published 2013

Printed and bound in the United Kingdom by the MPG Books Group

A catalogue record for this publication is available from the British Library

Library of Congress Cataloguing in Publication data
Bourne, Neil, 1964–
Materials in mechanical extremes: fundamentals and applications / Neil Bourne,
Atomic Weapons Establishment, UK.
    pages   cm
Includes bibliographical references and index.
ISBN 978-1-107-02375-8
1. Materials – Mechanical properties.   2. Mechanics, Applied.   I. Title.
TA405.B6755   2013
620.1′1292 – dc23   2012039024

ISBN 978-1-107-02375-8 Hardback

Additional resources for this publication at www.cambridge.org/mechanicalextremes

Contents

Preface

I cannot explain my curiosity about extreme phenomena in nature; nevertheless I have been drawn to the science that surrounds them – from those occurring on the scale of solar systems to those at work at the smallest regimes within matter. Extreme forces surround us; they govern our weather, the cores of planets, components of engineering structures and the ordering of particles within atoms. At the scales of interest in this book they are either gravitational or electrostatic in origin. Forces drive mechanical routes to impose change and materials are forced to respond to these pressures in non-linear, counter-intuitive and utterly fascinating manners; frequently more quickly than not only the senses, but the recording media that exist today can track. Nothing that changes does so instantaneously; every mechanism takes some time, however small. This mean that the integrated response follows a delicate framework of competing pathways that reorder as the driver for the forces changes. As with many processes, one can only see patterns apparent in retrospect. Furthermore, the difficulties encountered achieving these states mean that there are many untracked routes that matter can take to respond about which we know little. Thus despite the years this book has taken to come to this point, it can only provide a snapshot of behaviour as I see it.

Nevertheless, matter allows the nature of its bonding to be probed by subjecting it to load and the reader will learn to appreciate the variety of materials behaviours and their causes that allow the design of structures or even new materials to withstand the environments considered. The behaviours observed are complex and seemingly counter-intuitive, and quantifying them has frequently filled books in the past with extended solid mechanics. This has made texts rich in analysis and specialised in application and required the reader to be expert in the mathematics of non-linear behaviour. However, it seemed that a reader with an appreciation of the physical sciences and elementary algebra required an open text to emphasise behaviours not analytical subtleties. Thus this book unites principles covering a broad canvas at a level accessible to graduate students. Further, it addresses the regime in which the strength of matter may be described with extensions of solid mechanics at the continuum rather than extrapolation of atomic theory and quantum mechanics at the atomic scale.

It is common in academic life to classify problems and approaches by discipline: physics, chemistry, materials science, engineering, geophysics, cosmology. Each has its own unique history and this development has ensured a rich vocabulary of terminology within each field. However, the cross-cutting themes discussed here have a common root which applies across length scales and amplitudes and describes the consequences of strength under loading in each of the areas. At some pressure, atoms are forced so closely together that inner electron states become perturbed and the nature of strength itself changes too; this book does not consider this regime. However, below this threshold and from scales within nano-crystals to those of planets within solar systems, a common description, *akrology*, can be applied to the subject.

In 1990 a visitor to the Cavendish laboratory asked me if I knew what a shock wave was. His name was Zvi Rosenberg and I said I did. Over the next 20 years I have tried to justify that statement and attempted to understand and describe what such a front means to a solid material, and to him I owe a debt of gratitude. Within a few years of that time a launcher was fashioned to load materials and a course was developed at the University of Cambridge. There is much exploration and analysis distilled into this volume that has its origins in those times. The field has developed from its roots within national laboratories across the globe and spread over the last decades to infuse university and industry too. This has left gaps in the coverage offered by other texts and the time was thus right for a wider volume encompassing the range of topics covered by this field, focused on the materials themselves not upon the applications that use (or abuse) them.

This book was written as a single discourse working from an introduction in Chapter 1 and ending with a more detailed description of an asteroid impact on Earth in Chapter 9 using the concepts developed in the text. A series of tools are described as the reader works through the book. Chapter 2 gives an analytical framework on which to hang the discussion of what follows. Chapters 3 and 4 describe the platforms and diagnostics typically used to investigate the mechanisms occurring. The meat of the text, in Chapters 5–8, covers the response of metals, brittle solids, polymers and plastics and energetic materials. Finally, Chapter 9 summarises the features of the response of all classes of matter under intense loading in a manner that indicates their possible applications in extreme environments. Since solids and their structure are at the core of this volume, an appendix which summarises materials science, for the benefit of those of us with different backgrounds and training, is included as a reference. It is written from my perspective as a physicist but I hope that the many simplifications I have made will not detract from making it useful to readers from other backgrounds. Although the text contains references to specific work by various authors they and other significant works are collected in the Bibliography at the end of the text to preserve flow.

I am deeply grateful to my colleagues (who are also my friends) who have supported me in the preparation of this book. Rusty Gray, Zvi Rosenberg and Marc Meyers gave constant encouragement from the onset and as always I appreciated sound advice from N. S. Brar, Dennis Grady, Yogi Gupta and Ken Vecchio. Thanks in particular to those who have given me their time reading and commenting on various sections; the work would not be complete without their assistance. These include Jeremy Millett, Rusty Gray, Eric Brown, Marcus Knudsen, Peter Dickson, David Funk, Philip Rae and Rade Vignjevic. I must thank friends across the national laboratories with whom I have worked over the years including Billy Buttler, Kurt Bronkhorst, Carl Cady, Ellen Cerreta, Bob Cauble, Datta Dandekar, Rob Hixson, Neil Holmes, Jim Johnson, Veronica Livescu, Paul Maudlin and Anna Zurek. In the UK, friends and colleagues in universities including David Clary, Bill Clyne, Bill Clegg, John Dear, Lindsay Greer, Stefan Hiermaier,

Ian Jones, Phil Martin, Alec Milne, John Ockenden, Steve Reid and Phil Withers. Also within AWE including Graham Ball, Stephen Goveas, Hugh James, Brian Lambourn, Simon McCleod, Nigel Park, Steve Rothman, Glenn Whiteman, and within DSTL Richard Jones, Bryn James and Ian Pickup. Much interaction and inspiration has come from working with students and staff within my groups in Cambridge, Cranfield and Manchester including Patty Blench, Gary Cooper, Lucy Forde, Simon Galbraith, Robert Havercroft, Dave Johnson, Yann Meziere, Natalie Murray, Gary Stevens, Stephen Walley and finally to John Field who allowed me free rein to start a new area of work all those years ago. Finally, it is inevitable that I have failed to properly acknowledge someone or some organisation within this book that has contributed to its content and I offer my apologies in advance for the omission.

None of this would have been possible without the love and support of my close family; thanks to my parents, to Heather and to my children Freya and Oliver for everything. I dedicate this book to them.

Neil Bourne, 2012

Extreme positions are not succeeded by moderate ones, but by contrary extreme positions.
Friedrich Nietzsche

1   Natural extremes

1.1   Akrology

1.1.1   Extremes

The dynamic processes operating around us are often treated as transients that are not important when compared with the fixed states they precede. However, an ever-increasing knowledge base has illuminated this view of the operating physics and confirmed that extreme regimes can be accessed for engineering materials and structures. Matter is ever-changing, its form developing in a series of nested processes which complete on the timescales on which mechanisms operate; processes that occur on ever smaller timescales as length scales decrease. This book is concerned with the response that occurs when loads exceed the elastic limit. This affects behaviour in the regime beyond yield which encompasses a range of amplitudes and responses. However, it concerns condensed materials and loading, eventually taking them to a state where they bond in a different manner such that strength is not defined; this limit represents the highest amplitude of loading considered here. Nonetheless the driving forces are vast and awe-inspiring, while the different rates of change observed in operating processes are on scales that span many orders of magnitude. The following pages will highlight prime examples from the physical world and then provide a set of tools that classify mechanisms in order to analyse significant effects of these processes on the materials involved. The wide range of observations and applications create simple but powerful principles that are outlined in what follows.

Materials are central to the technologies required for future needs. Such platforms will place increasing demands on component performance in a range of extremes: stress, strain, temperature, pressure, chemical reactivity, photon or radiation flux, and electric or magnetic fields. For example, future vehicles will demand lighter-weight parts with increased strength and damage tolerance and next-generation fission reactors will require materials capable of withstanding higher temperatures and radiation fluxes. To counter security threats, defence agencies must protect their populations against terrorist attack and design critical facilities and buildings against atmospheric extremes. Finally, exploitation of new deep sea or space environments requires technologies capable of withstanding the range of operational conditions found in these hostile locations. The range of conditions under this umbrella spans high-energy fluxes, severe states and intense electromagnetic loading, but in what follows thermomechanical extremes on condensed matter will be considered. To advance in all of these areas requires a greater understanding of new behaviours and an ability to model the controlling mechanisms.

There is benefit in investing significant effort to map out these nested chronologies, as advances in understanding of materials and processes reveal that dangers and rewards come from embracing new modes of thinking. A key requirement is to consider timescales and length scales that operate outside the regimes in which intuition can operate; regimes in which glass stops an incoming projectile when a metal plate will not, even though the former exists as a pile of dust after the event whilst the latter retains much of its original form. It is the aim of this book to present a simple guide to key methodologies used to define incoming impulses and to track the material response to the extreme loading transmitted. The intention is to build an understanding that may be applied to new, more extreme regimes of loading and response based upon experience gained in the ambient environment.

1.1.2   Extreme material physics

The condensed phase (for most materials) defines a pressure and temperature range of interest which may be approximately fixed at less than 1 TPa and less than 10 000 K respectively. Pressure has one of the largest ranges of all physical parameters in the universe (pressure in a neutron star is $c.$ 10^{33} Pa), so that most of the materials in the universe exist under conditions that are very different from the ambient state on the surface of the Earth. Compression induces changes in bonding properties at the atomic scale, synthesising new compounds and causing otherwise inert atoms or molecules to combine. Integrated thermal and mechanical loading creates new structures within matter. These compressions (reducing interatomic spacings by up to a factor of two and increasing densities by over an order of magnitude) result in changes in the electronic structure that begin to shift notions of chemical interaction and atomic bonding. For example, electrons surrounding nuclei or ions become delocalised, changing insulators into metals, and eventually adopting new, correlated electronic states. Instantaneous application of an impulse provides a pump to drive the deformation of materials, and varying its amplitude and duration allows a window into the operative mechanisms that lead to plasticity and damage evolution within them. Under dynamic loading these extreme conditions may be exploited to explore the balance between mechanical ($P\Delta V$) and thermal ($T\Delta S$) energies by examining how this dichotomy governs physical and chemical phenomena in the condensed state. Additionally, shortened loading periods provide a filter to select governing mechanisms according to operating kinetics. The system may then attain a final metastable state that lies beyond equilibrium thermodynamic constraints.

If the fundamental mechanisms can be understood, exciting opportunities to use such extreme thermomechanical conditions to design and manufacture new classes of materials will open up. Such advances may allow limits such as the theoretical strength (the stress needed to shear atomic planes of an ideal crystal across one another) to be attained. In this manner materials may be designed for application and clearly the key property of interest is the strength of a material statically and during flow. Empirical discovery techniques can achieve incremental advances. Steel-making, for example, dates back over 30 centuries, but the strengths of most present-day commercial alloys are less than a factor of two higher than the steel used in swords during the medieval era. The strengths of commercially available steels are of the order of 1 to 5% of the ideal strength limit and this is typical for currently available bulk materials. This is principally linked to the misconception that the properties of materials can be derived on the basis of atomic structure alone, whilst in reality most engineering properties are dominated by defects within the microstructure. Thus a shift in perception and boundary conditions is necessary to bridge the gap between materials today and the theoretically achievable and this requires critical questions to be answered. What are the most important length scales and defect distributions that control deformation and fracture? What are the ultimate strength and temperature performance limits for structural materials and what is their development after yield? Finally, can dynamic processes be harnessed to capture and maintain theoretical limits for some operating period if not permanently?

Materials found across natural environments experience mechanical extremes of pressure, temperature and strain rate or survive electromagnetic loads of great violence. Reaching an understanding of such states and the response of materials subject to them is key in fully describing operating deformation mechanisms. With the knowledge gained it would be possible to contemplate designing not only structures to operate within such environments, but also to adopt strategies to engineer materials with optimised properties to survive there, should physically based models become available. These extreme material states may exist in nature in inaccessible repositories such as at high temperature and pressure at the centre of planets. Alternatively, they may represent the results of one of the two principal dynamic inputs that may reach these states in short times. These two driving stimuli come from forces generated during explosion or impact, and whilst both of these may occur as a result of some natural process, they may also be harnessed to engineer particular effects such as welding or cutting in a controlled manner.

The mechanical response of objects under load is taught to scientists from an early age through Newton's laws. These, in their simplest form, treat a body as a finite mass concentrated at the object's centre and, with the application of conservation of mass, momentum and energy, Newtonian mechanics describes the macroscopic world. A simple illustration of the utility of this treatment is that of the impact of spheres in the once-popular executive toy, the Newton's cradle. Here, equal masses, suspended from a common stationary framework, are allowed to sequentially impact one upon the other (Figure 1.1). When the first impacts a second, momentum is transferred to it and, if it is free, it may travel onward at the same velocity as the impactor. If there are several balls of equal mass, then the force is transferred through the stack to accelerate the last in the sequence to a velocity consistent with its need to match the same momentum to the first. This is a simple and effective illustration of the common experience of impacting bodies, understood using the assumption that the mass acts from a point at the centre of the body.

To move out of the time and length scales that human perception can respond to, requires description of the processes by which momentum is imparted from one sphere to the next. The deceleration of one ball on a face of a central sphere must transmit out a wave front. The contact point is decelerated whilst the first planes of atoms in the

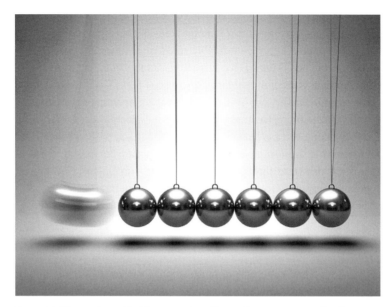

Figure 1.1 Newton's cradle illustrates the effects of waves propagating in a system where all the collisions are elastic. Higher stress impacts can deform materials and produce counter-intuitive physics at the shock front. Source: image copyright shutterstock.com/FreshPaint.

target start to accelerate. Since the two are touching they must travel at the same speed and a wave front travels forward into the target and back into the projectile, accelerating the one and decelerating the other. When the returning wave reflects at the free rear surface it releases stresses in the impactor to zero and accelerates the material ahead of the returning front to the initial impact speed whilst stopping the material behind. Momentum is conserved and Newton mechanics correctly describes the response: forces may act from the centre of mass of the moving objects in the time frame of the office in which the toy sits.

This process has a short high-pressure (or more correctly, high-stress) phase, and after some time, equilibration (a key concept of this book) has occurred; this governs the processes of inelastic flow and chemical reaction described in what follows. The impact state exists until the stress has been relieved within the spheres and the appropriate masses accelerated to their steady speed. In the case of elastic waves the approximate time to reach equilibration, t_{equil}, is given by

$$t_{equil} = \frac{2d}{c_L}, \tag{1.1}$$

where d is the diameter of the sphere and c_L is the wave speed in the metal. Whether the mass is in equilibrium or not depends upon the moment at which an observer chooses to sample the state of the system. If the waves have not released the stresses the state is still equilibrating; if they have, it is in equilibrium and Newtonian mechanics applies. Rheology defines a dimensionless parameter, the Deborah number, to represent the state of fluidity of a material and this is used in glaciology to describe how morphology develops over time – 'the mountains flow before the lord' as the prophet Deborah foretold in the Old Testament (Judges 5:5). Whereas the Deborah number is qualitative, another criterion is defined here to include the impulse properties and kinetics of the operating equilibration or localisation mechanisms. In this picture, all processes are regarded as dynamic and, moving into Norse mythology, the Freya number, F, is defined to reflect dynamic transformation and liberation of energy within materials under load since 'it is Freya who makes fire with steel and flintstone' (Schön, 2004).

The Freya number determines the totality of the deformation mechanism and flags the achievement of a stress-equilibrated state. It is defined to represent the extent to which a mechanism is driven to completion by a stress or temperature excursion during the time for which the impulse is active in the following manner:

$$F = \frac{t_{\text{relax}}}{t_{\text{obsvn}}}, \tag{1.2}$$

where t_{relax} refers to the characteristic relaxation time for the step in the rate-limiting process and t_{obsvn} is the length of the observation period on the system. If there is time for compression and release to decelerate the incident sphere and accelerate the impacted one, then stress equilibration has occurred. If this is the case F will be small and the state will be defined. Conversely, if one observes whilst the waves are travelling then F will be large and the state observed will be transient and contain some material in the initial state of the material at the start of the process. Thus F is a means to define thresholds for the equilibration of states in dynamic processes and will be used through the book to define the observed response. From the office frame the stress-state equilibration within the impacting sphere takes a few microseconds whilst the seated executive has seconds to observe the impact. In this case F is very small so that the details of the process can be neglected in describing the cradle. If the same sphere were fired at a police officer in a protective vest the failure time of the ceramic insert would be of the same order as the slowing of the bullet on impact – in this case F must be greater than 1 if the officer is to survive.

Of course the impulses considered in Newton's cradle only induce elastic responses. Our interest is in inelastic behaviour where the amplitude of the stress will overcome the strength of the material and deform it irreversibly. As a result the processes will not be periodic like the cradle but rather unique, increasing the entropy of the systems on which they act. It is important to consider the effects of densification on the metals' microstructure and thus the rise of the pulse, and the time and length scale over which the rising pressure will act, will have an important effect. Further, the period for which the loading is maintained will determine whether mechanisms can act to equalise these compressed states over the body. Finally, the induced states will not only be those of high pressure and velocity within the material (which increase strain and change its volume) but also high induced temperatures since shock waves may be applied for a time so brief that there is insufficient time for heat to conduct away.

The characterisation of condensed matter under extremes of loading is best bounded within a region of common physics described not by the pressure applied but by the

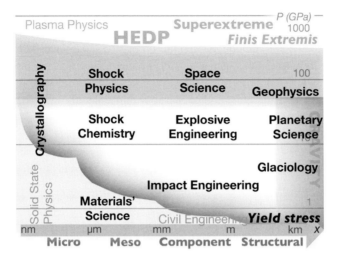

Figure 1.2 The range of disciplines accessed across a notional *P–x* space. At the smallest length scales atomic physics dominates, with the emphasis on energy levels and structure, at the largest planetary science describes the development of structures in solar systems. The world of engineering occupies the continuum scale. These disciplines span matter across its extreme states and define akrology, the science of extremes. The details and terms used in the diagram will return through the text and be explained as they arise.

strength of the microstructure it adopts. This region spans the elastic limit at the lower boundary and at the upper boundary the electronic states correlate so that valence electrons no longer determine bonding and delocalisation occurs. Beyond this point the strength of the compressed solid has a different nature to that in the regimes considered in this book and in what follows this is called the super-extreme state. As pressure increases at all scales, the response homogenises. At a particular pressure a threshold called the *finis extremis* is reached, which represents the upper limit for valence electrons bonding. Above this limit, high-energy-density physics (HEDP) describes the homogeneous state that exists. Within the bounds between this threshold and the yield stress of condensed matter, the range of subject interests extends from atomic physics excited by laser impulses, to integrated systems of behaviour found in planetary science. The variety of subjects amassed across this slice of pressure–volume space is illustrated in the cartoon of Figure 1.2.

Between the yield surface and the *finis extremis*, observed behaviour is defined by the pressure applied and the strength observed. The former is a thermodynamic variable, defined at scales beyond the unit cell and the latter is a function of the volume element sampled by the experiment or phenomenon of interest. The interpretation of measurement in the laboratory thus depends on this sampled volume (a function of the defect population contained with the material under load), the length of the impulse applied and the resolution limit of the detector. Thus the impulse defines the state observed at the length scale of interest and the physics that results is contained within a region where a common terminology and common physics defines a common space which may be described by the developed methodologies of solid mechanics.

Akrology: The science of the mechanisms operating and the responses resulting in condensed matter loaded to an extreme, mechanical state. (Origin: Greek, *akros*, 'extreme, highest, topmost'.)

This book will explain the principal mechanisms that reorder a material's structure and change the mechanical and physical properties it exhibits. Of course there are many extremes and loading by intense radiation pulses for instance will not be considered here. Such materials may be metallic, polymeric, brittle or even reactive. Connecting understanding of the mechanisms operating with mathematical description of the behaviours observed permits the principles of material response to be classified to allow the engineer to design structures.

The application of such principles additionally provides understanding of observed events on Earth and in the cosmos. These extreme events may be subdivided into those found in nature and those man-made. There are miscellanies of man-made dynamic events of note, in both the civil and defence fields, including terrorist devices that provide a pressure pulse by explosion or cutting devices that use metal jets. Even transportation exceeds the sound speed of the air surrounding us, requiring knowledge of supersonic flows (although the fluid mechanics of shock waves in gases will not be considered in this book). The result of the detonation of a reactive material or the acceleration of a metal jet requires the generation and propagation of a shock wave and it is the shock, a pressure impulse of rapid rise travelling through matter, that represents the limit in rate for dynamic loading.

1.2   Natural extremes

In the natural world the forces of nature frequently reach extreme states and, as the globe warms and kinetic energy within the atmosphere increases, provide increasing dynamic loading upon structures or even populations. The scale of planets and stars, and the laws of gravitational attraction, inevitably mean that at their centre, pressures and temperatures are extremely high. At the centre of the Earth, for example, the pressure is believed to be of the order of 360 GPa and the temperature of the order of 7000 °C. Furthermore, these static extremes mean that the core is primarily iron and behaves as if a single crystal in the anisotropic hexagonal close-packed (HCP) phase. In the centre of stars the mechanical states quickly exceed these and elements exist here as condensed plasmas. In these regimes the electronic states have homogenised; indeed the centre of the Earth is in this state and analogies with single crystals under ambient states may not be useful. This book will consider states of matter in which concepts of strength are extended to the point at which the energy density in the impulse exceeds the bond energy of the material (which is *c.* 300 GPa for an element like tungsten). Beyond this point the term strength as it is understood from solid mechanics has a different meaning.

On planets, impact has shaped their topography, and indeed one theory has our Moon as part of a Mars-sized object *Theia* (around 10% of the Earth's mass) that impacted the young Earth around 4.5 billion years ago (4.5 Ga). Smaller impactors have continued to pepper the Earth's surface as the solar system cleared rocky bodies such as asteroids into stable orbits and, at much higher impact speeds, so have comets from outside the solar system. Today there are over 175 confirmed impact structures on Earth and the number is growing as surface topography is better resolved. One in particular has caught public imagination since there is now general agreement that 65 million years ago (65 Ma) a bolide impacted the Earth and caused sufficient disruption to the ecology that all non-avian dinosaurs became extinct (this will be considered further in the last chapter). But such impacts are thankfully not common. More often found are the smaller-scale dynamic effects that result, for example, from mass movement of snow in avalanches, soil in landslides or water in tsunamis. When a process leads to explosion, wave propagation, with travelling mass and elevated pressure, is the necessary progenitor for shock wave formation. Such waves may also result from a series of natural phenomena such as earthquake, or volcanic eruptions.

To belittle such phenomena as small scale belies the devastating effect these natural processes can have on populations. Of the top ten deadliest natural disasters, half can be attributed to earthquake and tsunami, whilst of the remainder, the Baqiao Dam disaster (1975) occurred by catastrophic material failure under extreme loading under torrential rainfall from the super typhoon Nina. Earthquake, tsunami, volcanic eruption or impact provide an input stimulus which defines the initial state and the development of conditions that a material or structure must withstand. To acquire insight into how matter itself responds requires the deformation to be tracked for the time over which the pulse develops. In the text below these methodologies will be followed to show common links in the operating phenomena that underpin extreme loading events.

1.2.1   Volcanoes

Many Europeans encounter active volcanoes in Italy at either Etna on Sicily or Vesuvius on the bay of Naples; the latter is famed for the devastation and destruction of two Roman cities (Pompeii and Herculaneum) and was documented by the historian Pliny. The macabre secrets of those events are still emerging as more archaeological evidence is uncovered, but the key features of the event are written in the deposits around the volcano. The eruption occurred in two phases in the late summer of 79 AD. Small earthquakes started taking place, building in intensity over four days to failure of the dome that left a vertical plume of ash hovering in the sky above the volcano, which then collapsed down the mountain's west side in a torrential, pyroclastic flow that buried all in its path. This black cloud descended the slopes of Mount Vesuvius and flowed down to the Roman resort of Herculaneum. The first Plinian eruption vented the upper magma chambers and launched a column of ash and pumice 30 km into the sky. It has been estimated that the mountain was pumping out mass at a rate of 150 000 tons per second and that, in all, 2.6 km^3 of rock were exhausted out of the mountain. The first effects on Pompeii consisted of a rain of ash and pumice onto the roofs of the city until, bowed by the ever-increasing load, many collapsed under the weight, killing inhabitants sheltering in misperceived safety (Figure 1.3).

When the volcano's upper chambers were exhausted of their stores of gasified lava, a second source, deeper in the mountain was ejected. Changes in the type, temperature

Figure 1.3 Eruptions at Vesuvius. A reconstruction of the AD 79 event (top: from the Discovery Channel's *Pompeii*, courtesy of Crew Creative Ltd) and as seen by Athanasius Kircher in 1638 (bottom: from *Mundus Subterraneus*, 1664).

and rate of expulsion of this mass led to the collapse of the towering column of pumice above the mountain and the launch of six devastating flows. These blasts are termed pyroclastic surges and flows and they account for the most destructive and lethal aspects of an eruption in a stratovolcano (a volcano built up of alternate layers of lava and ash) such as Vesuvius. In 1902, surges were observed (and photographed) travelling down Mt Pelee on Martinique at speeds of 30 m s^{-1}. The flow engulfed St Pierre and destroyed the city, claiming 26 000 lives. A pyroclastic flow is triggered by an explosion of hot gas, expelled from the volcano as the summit collapses, and is preceded by a surge of a mixture of gases from deep within entraining rocks, and dust. The following incandescent flow is a composite of the volcanic debris and ephemera entrained by the rapidly moving and heated gases. The temperature within it is between 200 and 800 °C, capable of carbonising the timbers of doors in its path and, as observers at Mt St Helens discovered, it can travel at up to 150 m s^{-1}. This lethal wave accounts for around half of volcano fatalities, killing by asphyxiation under the hot ash, burning and boiling under blow-torch temperatures, or impact from the entrained rocks propelled at high velocity in the flow. Six of these waves accounted for Pompeii and Herculaneum and for 20 000 inhabitants in a few hours.

The wave is propelled by the pressures and the blast wave from the rupture of the volcanic summit and driven forward by the potential energy converted by the descent of the entrained mass. In addition, the high gas flow velocities and elevated temperatures can fluidise this moving bed of material to allow high-speed flow. In the case of Vesuvius, the flow terminated as the hot magma intruded into and mixed with sea water, causing vaporisation, explosion and ejection of lapilli, which encased victims in a rock crust tableau that maintained their form long after their bodies decomposed, providing striking evidence of their unfortunate fate. In the case of Mt St Helens (1980), sufficient water was vaporised from Crater Lake to drive a shock front into the upper atmosphere where condensed water droplets formed a cloud layer at the boundary with the troposphere. The largest eruption in recent history occurred at Tambora (Indonesia) in 1815 (the year of Waterloo). The air shock travelled to Sumatra (c. 2000 km) and heavy volcanic ash falls were observed a similar distance away. The skies were dark for two days following the eruption and the eventual total death toll was at least 71 000. The year 1816 was the 'year without a summer', with snow in New England in July and August, and crop failure across Europe causing riots and famine. Yet it is still the eruptions that levelled Pompeii and Herculaneum that grasp the imagination since they occurred within a continent with a developing civilisation and have a written account by Pliny. The events of those two days echo down the millenia and tell not only of historic events, but also warn of what might happen today; the next eruption of Vesuvius is overdue with the last major event in 1944. Towns near Naples are rising and falling at a high rate – at one a shift of two metres in three days was measured, showing that, beneath the quiet monster, lava is gathering and stretching the crust above it.

Stratovolcanoes like Vesuvius, Etna or Pelee, placed at convergent plate boundaries, represent a group where viscous magma leads to explosive eruptions. This high-viscosity melt does not allow escape of volcanic gases, which thus build and pressurise the sealed volcano until eventually failure of the cone occurs. Dynamic dome fracture triggers an ejection of magma and this pressurised flow can jet high into the sky or blast sideways as a fluidised bed – the pyroclastic flow. The potential energy stored under these structures is clearly high: an estimate of the total energy in the AD79 eruption gives an equivalent to 100 000 nuclear explosions on the scale of Hiroshima ('Little Boy': 13–16 ktons).

In May 1980, the eruption of the volcano Mt St Helens was performed to a gallery of eager observers from across the world since it occurred at a time when it could be observed with ease by advanced equipment. The same may be said of other sites such as Etna on Sicily, which provides a show every two years or so. Yet Mt St Helens, like Vesuvius, displays a different nature to these quieter companions. Rather than ejecting molten oceanic crust, it mixes its diet with the lighter continental material, forming a brew that sticks in the channels the volcano creates to release the pressures that grow beneath. All along the Pacific coast the Juan de Fuca plate is sinking beneath the North American plate, forming the Cascade Mountains with active and dormant volcanoes such as Mt Rainier, Mt Adams and Mt Baker. As a consequence of the continental crust being subducted beneath the North American plate, the eruptions are violent involving a considerable number of dynamic processes and explosive events. By late March of 1980 there were signs of imminent activity. On 20 March earthquakes were detected around the area, whilst a week later on 27 March explosions were reported. Then on 18 May there was a significant (magnitude 5+) earthquake and the scene was set for the events to come. At this time the mountain formed a regular cone rising high above the placid Spirit Lake below. It reached a height of 9677 ft at daybreak on 18 May, but by the evening it would only be 8364 ft high. At 08.32 the pressure developed within the dome at the peak of the mountain caused the northern part of it to flow in two massive sections down into Spirit Lake. The release of gases, entraining rocks as they passed, sent a pyroclastic flow travelling at $c.$ 100 m s^{-1} down the mountain, leaving felled trees in its wake.

The effects of the blast remain visible today with forests uprooted and boulders, trees and ice blocks carried tens of kilometres from the epicentre. Another notable feature of the flow produced was that for a large proportion of the surrounding observers, the events were observed in silence. This resulted from the particular geography of the land coupled with the failure of the cliff face and led to a strong channelling of the sound waves in a northerly direction. Although dramatic, the flow was travelling much slower than a shock wave would have in the same medium. In fact, the hot material, flowing north down into Spirit Lake, vaporised it and drove an air shock into the upper atmosphere that condensed a layer of cloud above. This lay next to one formed earlier above the site of the eruption itself and so directly above the cone. The measurement stations provide some detail on the physical conditions of the flow. One kilometre from the eruption, the ambient pressure was 20% above normal, dropping to 5% above normal 5 km away; the temperatures in the hot flow reached 600 K.

Composite volcanoes deform at the large scale with the same broad range of deformation mechanisms found at the microscale. The loading impulse sets in train a structural failure followed by a steady flow phase until the impulse is exhausted and the remaining crust can lock. These three phases, *localisation*, *flow* and *interaction*, can be described knowing the amplitude of the impulse and the duration of the event and may be tied to the equations for transport that describe the flows observed over the times they act. The responses of different classes of material reflect their atomic structure and this book will consider those for metals, polymers, brittle materials and explosives in what follows.

1.2.2   Tsunami

Dynamic impulses may come from the land or the sea. At close to 08.00 local time on 26 December 2004, a large earthquake occurred along a length of the plate boundary beneath the Pacific Ocean. The Indian Ocean Sumatra–Andaman earthquake was one of the largest ever recorded (9.1–9.3 moment magnitude). A section of the India plate ruptured in two phases spaced over a period of a few minutes, and slipped around 15 m beneath the Burma plate along a length of subduction zone around 1600 km in extent. The rupture travelled at up to 3 m s^{-1} along the zone boundary, displacing the seabed laterally and upward. Although the motion was principally in its plane, there was a vertical shift of several metres, displacing a huge volume of water in a short time. This shift triggered a tsunami which ranks (at the time of writing) as the sixth deadliest natural disaster in history, with around 300 000 lives lost worldwide. The upward shift of the ocean started a surface wave, recorded 2 hours after the first quake and in deep water about 60 cm in amplitude, travelling at up to 300 m s^{-1} away from the line source on the seabed, 30 km below. As the wave travelled towards land, the front slowed and rose to heights of 30 m as it reached shallow water and ploughed inland. Indonesia, Sri Lanka, India and Thailand all experienced waves of this prodigious height and even South Africa, 8500 km distant, witnessed a wave rising 1.5 m.

Seven years later, at 14.36 on 11 March 2011, an earthquake shook Japan. A 5 mm upthrust on a 180 km wide seabed drove a quake of magnitude 9.0 from an epicentre 70 km to the east of Tohoku, Japan. The whole of the island of Honshu was moved 2.4 m east on that day and shifted the Earth on its axis as the planet redistributed its mass to change its moment of inertia; the length of the day reduced by 1.8 μs as a consequence. The shift of the seabed drove a tsunami with waves up to 40 m in height that struck the eastern Japanese coast line, in some cases travelling 10 km inland, stripping towns and cities in its path. The tsunami propagated outward across the Pacific, with waves of 0.5 m in the Philippines, and surges along the Pacific coast of South America: Chile and the Galapagos Islands witnessed waves up to 3 m and Peru felt a pulse of 1.5 m. The tsunami even fractured icebergs off Antarctica. The shock triggered a series of major emergencies, the most serious being the meltdowns at the Fukushima nuclear plant. The earthquake left 16 000 dead with a further 3000 missing or injured.

The energy of these events fuelled wave propagation over enormous distances, along with the aftershock and reflections that continued for a considerable time. The energy of the Japanese quake has been estimated at *c*. 50 PJ, whilst that of the Indian Ocean event was *c*. 20 PJ even though the latter earthquake was more powerful. Rayleigh (surface) waves were excited, raising the land surface worldwide by an amplitude of around 1 cm.

Thus, the delivery of such monumental quantities of water at these prodigious rates, in response to tectonic activity underwater, provides another means by which a dynamic impulse may be delivered to coastal targets. The death tolls were vastly different from the 230 000 people dead from the Indian Ocean event, showing that new understanding and warning of these dynamic slip events can save lives and economies, particularly in the developing world.

1.2.3   Krakatoa

The eruption that rolls all of these geophysical cataclysms, including pyroclastic flows and immense tsunamis, into one stupendous drama, occurred in 1883 in Indonesia. This island chain has over 130 active volcanoes, and one member of this group, seemingly insignificant to the occupying population, is named Krakatoa (or Krakatau). Its eruption, however, combined a series of dynamic volcanic effects like none other in recent times. The volcano shows evidence of periodic activity from at least 60 000 years ago, with more recent signs of activity in *c.* 535 AD and 1680. Pre-eruption, the site consisted of three coalesced volcanoes: Perboewatan (122 m), Danan (445 m) and Rakata (823 m). In the eruption, the first two, along with half of Rakata, collapsed leaving a submarine caldera around 250 m beneath the sea. On 20 May, the first mild explosions were heard from Perboewatan. The process accelerated to include activity on Danan by mid July. On 26 August the plume above Krakatoa had extended to 25 km, and over the hours of that Sunday climbed further to 36 km. The process culminated on Monday, 27 August with four eruptions starting at 05.30 followed by others at 06.42, 08.20 and 10.02 local time. That final explosion was heard 4700 km away and remains the loudest sound in recorded history, logging levels of 180 dB at 160 km from the site. Every barograph in the world recorded the air shock as it reverberated around the globe for five days after the explosion.

The mountain in its death throes exploded in a series of intense pyroclastic flows which generated tsunamis as they entered the sea. These flows had a range of effects. On local islands all vegetation was stripped off and the 3000 people living in these areas were washed into the sea. Burning accounted for two thousand corpses close to the site with more as far away as Sumatra, 40 km distant. The cause was over-pressurisation of a magma chamber that contained lava with dissolved gas; injection of new, hotter magma may have expelled this gas, bursting the rock above in a massive explosion. All facets of the behaviour noted in the events at Vesuvius were present and, again, connected reservoirs containing a multiphase magma at high temperature and pressure fractured at the surface by further incremental injection. The expanding viscous mixture launched a jet in a fast, outward flow which propelled a tsunami out over a large distance. Vaporisation of the water drove an air shock which was heard at great distances.

These flows, a hot mix of gases and molten and solid rock, were propelled under pressure and gravity down the sides of the mountain, out of the volcano and into the sea.

The prodigious distances reached must mean that the driving pressure was augmented by the vaporisation of the water surface to enhance the flow sped. The front pushed the sea surface up into a wave front 40 m high which was driven out from the volcano. This tsunami devastated the coasts it touched. The most graphic account was from the captain of the *Loudan*, who climbed the wave in his ship to survive its passage into land and record the events in his log. The propagating sea wave was observed at points around the region and even in Aden, 3800 nautical miles from the mountain.

Although these rapid dynamic events are the main subject of interest here, other effects had an equally devastating effect on the surroundings. The violent blast generated vast quantities of tephera, estimated to be 20 km^3, or 20 times that of Mt St Helens in 1980. This ash fell at distances up to 2500 km downwind in the days following eruption with coatings on buildings in Singapore (840 km away) and Cocoa Island (1155 km south west). The ascending cloud contained SO_2, which combined with vapour in the air to create acidic aerosols and volcanic dust. Propelled above 20 miles, this material was circulated in the stratosphere around the world, taking 13.5 days to circle the equator. This acted as an atmospheric shield to reflect sunlight away from the planet and gave rise to a series of effects. Global temperatures dropped by around two degrees for five years after the eruption (Tamboro, the largest eruption in the last 10 000 years gave rise to the 'year with no summer' (1815) when snowfalls and frost occurred in June, July and August in New England and red skies were seen around the world (Figure 1.4). However, later eruptions showed the global temperature to be more sensitive to the SO_2 concentration than to effects from other ejecta. Even though smaller, the volcano El Chichon lowered temperature 40 times more than Mt St Helens because of increased SO_2 emissions despite the greater volume of ash in the latter. Colourful sunsets were seen for the next three years. In fact it has been speculated that Munch's apocalyptic 1893 painting *The Scream* (Der Schrei der Natur [*The Scream of Nature*], Edvard Munch, 1863–1944), with its writhing red/green sky and reflection in a swirling harbour, were at least inspired by the climatic effects from Krakatoa.

The contemporary Dutch colonial authorities recorded 36 417 people killed. Of these there were no survivors of the 3000 people living on the island of Sebesi, about 13 km from Krakatoa and the pyroclastic flows killed another thousand people at Ketimbang, 40 km north on the coast of Sumatra. Only about one third of the island remained above sea level after the summer eruption. The energy liberated by the explosions has been estimated at 200 megatons of TNT, equivalent to approximately 13 000 times the yield of the Little Boy bomb which devastated Hiroshima, Japan. Even the largest bomb ever exploded, the Tsar Bomba (57 megatons), only had an explosive power one quarter that of Krakatoa. Man cannot approach the effects when nature unleashes explosive power.

On 29 December 1927, fisherman in the area saw steam rising from beneath the sea, and by 26 January 1928 the rim of a new cone, Anak (the son of) Krakatau, had appeared. Since the 1950s, the island has grown at an average rate of 13 cm per week with periodic gentle outbursts from the cone. It is a matter of time before the mountain speaks again.

Figure 1.4 Paintings by J. M. W. Turner (23 April 1775–19 December 1851). Top, *Eruption of Vesuvius* (1817); bottom, *Chichester Canal* (*c*. 1828). Turner's paintings frequently include vivid red skies connected with contemporaneous volcanic activity around the globe (see reproductions on the web for full colour).

1.2.4 Supervolcanoes: Yellowstone

Whereas the volcanoes discussed above lie generally on subduction zones where the magma produced necessarily induces explosive eruptions, the largest structure considered here, the supervolcano at Yellowstone National Park, lies instead on a mid-continent hot spot. Here a rising magma stream fills a massive chamber which is 60 km long and up to 40 km wide with a volume of some 15 000 km^3. The magma within the chamber is partially solid and clearly pressurised so that it too contains gases kept in solution whilst the volcano remains quiescent. However, on a regular timeframe, the volcano has erupted with gargantuan explosions; the last three of these events were 2.1 million, 1.3 million and 640 000 years ago. It has frequently been noted that the times between the first two of these eruptions, 800 000 a and 660 000 a (*a* is an abbreviation for *anni* [years]), might imply that the volcano may erupt again at any time. These three major eruptions were on a scale not yet seen in modern times and deposited ash across the whole of the continental USA. There is visible evidence for smaller-scale events, the period of these being *c*. 20 000 a on average, with the last, 13 000 a ago, leaving a 5 km diameter crater.

The operating mechanism is similar to that described above for Krakatoa and Vesuvius, but on a different scale, and again sees expansion of a pressurised, heavily gasified magma with release by the failure of a thin surface crust leading to a large gas expansion at ground level.

The Yellowstone volcano is many thousands of times more powerful than Mt St Helens and even than Krakatoa, and its next major eruption will have cataclysmic consequences that will change the course of civilisation on Earth. The expected consequences will be threefold. Firstly, in the area 100 km from the epicentre, pyroclastic flows will engulf anything living in the area. Major effects will considerably extend the affected zone since ash deposition could severely affect a region of 1000 km radius. The heavy ash fall would deposit mass onto the roofs of houses, which would likely collapse and could kill or severely injure people, echoing the smaller-scale but still observable effects at Pompeii in 79 AD. A disaster in the modern age will have further, more subtle effects: ash will intrude into electronics and machinery and cripple communication, transport and emergency response. Global effects from any such eruption would be crippling: those at Eyjafjallajökull in 2010 and Grimsvötn in 2011 closed air transport in Europe as a result of eruptions on scales that are insignificant in comparison with the eruption of a supervolcano. That at Laki in 1783–1784 was a basaltic eruption from an opened fissure in Iceland and it had several other effects. Large quantities of SO_2 were emitted, leading to many deaths across Europe, but another effect was fluorine emission, which caused dental and skeletal fluorosis. Half the livestock and a quarter of the population of Iceland perished in the event. Fluorine and fluorides might also be emitted on a future eruption of Yellowstone, along with sulphur and nitrogen dioxide in prodigious quantities.

The final global effect relates to the distribution and conversion of the SO_2 into an aerosol which would ascend into the stratosphere and form a cloud that would be relatively stable at such an altitude for many years – the Earth could, in a few tens of hours, be entirely surrounded in a cloud thick enough to exclude sunlight from the surface for years. This would cause a global temperature drop of several degrees, perhaps more in critical areas. Climatic effects such as monsoon rains could be curtailed, constraining agriculture. Coupled with acid rains, these would critically strangle food production and give rise to potential global famine and disease. The severity of the effects depends

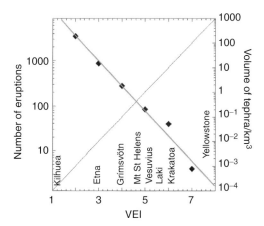

Figure 1.5 VEI. Number of eruptions and volume of tephera for eruptions over the last 10 000 years. The Volcanic Explosivity Index (VEI) ranks volcanic events.

upon the quantity and the speed of venting of the gases from the magma chamber, and that is critically determined by the physical properties of the rock initially sealing the structure. Of course, a myriad other effects might accompany the volcanism since tipping one component of a multivariable system by such a large amount is likely to move one metastable state to another. Whether life could adapt fast enough to match the speed of the change is a critical concern; again the speed of the response to an impulse determines the ability of a system to respond. Can the system adapt and reach a new equilibrium before the impulse has ceased to stress the environment?

A measure of the explosiveness of eruptions has been suggested which ranks such quantities as plume height and ejecta volume and uses them to place volcanoes on a scale from 0 (non-explosive) to 8 (mega-colossal). This quantity is the Volcanic Explosivity Index (VEI, see Figure 1.5). On this scale the category 8 volcano releases a volume of tephera of greater than 1000 km^3 but occurs on average less frequently than 10 000 years. These volcanoes are the bringers of catastrophe and Yellowstone is one of the largest of these still active. In the last 36 million years there have been 42 identified as magnitude 8 or greater. The VEI categories for other eruptions mentioned here were, for comparison, Krakatoa (6), Laki (6), Pelee (4), Vesuvius (5) and Mt St Helens (4).

The most recent historical VEI 8 super-eruption was the Lake Toba event of 71 500 a. It occurred at a time when humanity was evolving towards its modern state, so some effect may be inferred upon human development. It was 40 times larger than anything witnessed in the last two centuries and expelled an estimated emitted volume of 2 800 km^3 during the eruption. The severity of the event is believed to have caused a human population bottleneck by reducing the breeding population; other primates show similar, but less obvious effects. Evidence in the Greenland ice cores gives a precise picture of the chemical state of the atmosphere at the time. Over a six year period, more sulphur was deposited in the ice than in the last 100 000, reflecting a concentrated aerosol of sulphuric acid cloaking the planet, reflecting sunlight and cooling climate for that period with temperatures likely dropping over 10 °C in the northern hemisphere. The cumulative effects of all of this gave six years of volcanic winter followed by a millennium of the coldest, driest climate seen up to the present day.

Today the Yellowstone Plateau is rising and falling at an average of 1.5 cm across the caldera each year, but local areas are moving around 10 cm. Although an eruption may be imminent, the period between them is uncertain as the sample number for past events is so small. A more reasonable statement would be that an eruption in Yellowstone is to be expected in the next 10 000 years. For an idea of frequency, the record of volcanic eruptions in the *past* 10 000 years maintained by the Global Volcanism Program of the Smithsonian Institution shows that there have been four eruptions with a VEI of 7, 39 of VEI 6, 84 of VEI 5, 278 of VEI 4, 868 of VEI 3 and 3477 explosive eruptions of VEI 2. In the interim the more knowledge that can be accumulated of the effects of such eruptions in the past, and the physical state of the rock and magma in the chamber today, the better the planning will be to mediate for an event that may impose another bottleneck for humanity when it occurs.

1.2.5   Volcanism and impact within the solar system

The Earth is not alone in venting the heat from its core into eruptions of molten rock at its surface. The Moon (as a close neighbour) shows evidence of ancient volcanic activity in its dark, basaltic *mare* observable from Earth when skies are clear. Other surface features have volcanic origin but there is no active volcanism now. Mars possesses huge volcanoes: Olympus Mons is 300 km in diameter and 25 km high, the largest known in the solar system. These structures are inactive too although there may be more recent activity confirmed in future exploration. Venus has been through a phase where there were huge amounts of volcanism with flood basalt planes visible across the whole surface. It is likely that volcanic activity continues there and this may be as violent as that on Earth.

The greatest presently observed activity in the solar system is found on Io, a moon of Jupiter, that is kneaded by the gravitational forces proximal to the largest planet. Other volcanic activity on colder, deeper space objects appears to involve different ejecta but a similar operational principle. Cryovolcanism describes the eruption of a liquid or vapour of water, ammonia or methane which condenses again as it returns back to the frozen surface. Again it is a thermal cycle driving a melted phase extruded, potentially explosively, through a flow constriction at a surface hot spot where it forms a jet which subsequently condenses on the cold surface.

The largest scale of impact structures seen routinely are the craters that cover the planets and their satellites in the solar system. There have been hundreds identified on Earth and some are associated with dramatic events in evolution such as the mass extinctions that occurred at the boundary (literally between layers of rock in some places) between the Cretaceous and Paleogene (the beginning of the Tertiary) periods. This particular division marks the level above which the large dinosaurs met their demise. The margin is also marked by an iridium layer around the globe which suggests that a massive impact was responsible for the climatic change which gave rise to this event.

The site of impact, Chicxulub, in the northern part of the Yucatan peninsula of Mexico, has also been identified by several indicators including high-pressure mineral phases such as stishovite (high-pressure phase of silica) identified around the site – a 'smoking gun' from a large impact event. In such events, kilometre sized bolides (a catch-all term that could describe an asteroid or comet) impact the earth at speeds of tens of kilometres per second. At the impact point there is vaporisation and the pressures induced are TPa. Kilometres from the site, the pressures still exceed the yield stress of the rocks and beyond even this boundary, faults within give huge fractured volumes around the crater triggered by the dynamic tensile hoop stresses in the expanding front. Such an event causes the ejection of a plume into the atmosphere, shielding the earth's surface from sunlight until it settles again. A year of cold followed by a thousand of intense heat followed Chicxulub; too fast a change in conditions for massive dinosaurs to adapt so that equilibrium in their population could not be reestablished. The plume was one mechanical effect that must occur but there was also a large electromagnetic pulse that resulted from the impact that had little mechanical effect on the systems of the day. Of course, other geological phenomena such as volcanism on the other side of the globe at a similar time, exacerbated the effects and hastened the demise of the large dinosaurs. However, it will be useful to return to the Chicxulub impact in the last chapter to analyse the conditions that existed armed with the greater understanding of material response under dynamic loads presented in the meat of this book.

In all cases a shock must be driven either by an impacting body or by explosive expansion. Any object orbiting the sun has experienced the full gamut of impactors, from grains weighing microgrammes to bolides weighing some proportion of the mass of the earth. All must be considered as potential threats, particularly now that humanity designs structures launched into space for communication or transport. Every planetary object surveyed shows cratering, so that over long enough timescales a catastrophic impact must occur. Micrometeorites may be travelling at speeds ranging from 5 to 75 kilometres per second. Clearly at the higher end of this range, there is little one can do to protect from the threat. But equally one must ask whether the size of the impact makes the impulse delivered so short that the threat is not significant. For smaller particles at up to ten kilometres per second, there is some chance of protecting satellites with spaced armour shields which act to diffuse the momentum from an incoming particle over a region where it may be absorbed.

To understand the large craters observed on planets, experiments with controlled impactors and accurate imaging were designed. In the Deep Impact experiment a projectile was launched at hypervelocity to impact upon a comet, not only to understand the processes resulting (such as cratering of the surface), but also to interrogate ejecta in order to identify the materials from which the comet formed. A penetrating body may uniquely sample not only surface conditions, but also an interior formed at times when the solar system was in its infancy. Ernst Tempel was an itinerant astronomer who started his career as a lithographer and he travelled from his native Germany through France and Italy. He discovered 13 comets and two minor planets, and one of these, 9P/Tempel 1, an ellipsoidal comet around 14 km at its maximum dimension, and 4 km at its minimum, was chosen as a target for a NASA impact mission. A spacecraft was launched which

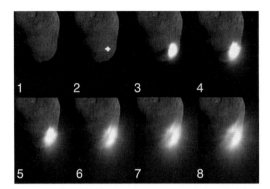

Figure 1.6 An image sequence of the impact onto comet Tempel I during the Deep Impact experiment. Source: image courtesy of NASA/JPL-Caltech/UMD.

converged with the orbit of the comet. Twenty-four hours before the planned encounter, a 370 kg, copper-encased cone was jettisoned into its path, and the observation craft manoeuvred into a position from which it flew close by, and observed the results. The relative velocity of the impactor and target ensured that the copper projectile and Tempel 1 struck obliquely at 10.1 km s^{-1} at 05.52 GMT on 4 July 2005 (Figure 1.6). At that moment, the projectile had a kinetic energy of 18 GJ, equivalent to detonating four tons of the high explosive TNT. If the target were water this would have generated a pressure shock of 125 GPa (a million times atmospheric pressure), and double that if it were rock. However, the comet was lower density still, around half that of water, and the pressures generated were consequently less. Nonetheless, the temperatures of thousands of Kelvins existing at the impact site vaporised the copper impactor and the upper surface of the target comet into a fireball of high-pressure and high-temperature vapour and the bright flash was observed and its spectrum analysed from the space-craft and the observatories watching from Earth.

The formation of the crater ejected a huge plume of material that dominated the later observations. Imaging at various wavelengths gave some clue to the geometry of the impact site, whilst spectroscopy allowed identification of the material from the surface and from the interior of the comet excavated by the crater. What was seen showed that the comet consisted primarily of water, methanol and carbon dioxide, held loosely by gravity at a low density. The size of the plume masked the crater beneath, believed to be over 100 metres in diameter. The comet was of such low density that, rather than forming a shallow crater as it might have done on Earth, it burrowed much deeper into the core.

The momentum transferred in such a collision was still insufficient to shift the orbit of the body. However, if a threat was ever identified for Earth, and there was time for the orbit to shift, a more direct collision might move a potential threat an Earth diameter to avoid catastrophe. The mission has revealed many of the features of the comet's structure. Its composition contains less ice and more dust than expected, although its penetration into the core was not conclusive. Certainly the results have shown the value of impact missions in probing the composition and the history of planetary bodies, and already the Deep Impact flyby craft has been repositioned to observe comets in the future.

The comet that was the target in this mission was, is, and will be one of the impactors that may career towards Earth one day. The exploration and developments in imaging technology that have freed observations from the Earth's surface into voyaging probes have provided detailed evidence that every body in the solar system has been subject to myriad impacts. Craters similarly cover the Earth with over 175 identified and more added each year. Thus impact has been a dominant geological process starting with the collision of a Mars-sized object obliquely impacting the proto-earth and creating the planet that we inhabit today with its orbiting moon.

1.2.6 Natural triggers

A volcano on the surface of the Earth is the thin, visible upper wall of a series of high-pressure reservoirs filled from zones beneath by a porous mixture of magma and gases, principally sulphur and nitrogen dioxides. It is also unconstrained by surrounding material. The initial conditions within this reservoir are c. 0.5 GPa of pressure with a temperature of around 800 °C. Further injection of magma from below fails the rock wall at its weakest point and explosively vents the upper chambers. The gaseous suspension of solid and gas, at high temperature and above ambient density, form first a convectively driven jet upwards through the vent, and later, as the upper chamber exhausts, a wave front spreading out under gravity and driven by the expansion of magma lower in the connected reservoir system beneath the volcano. If this multiphase flow intrudes into the sea, it vaporises the water and drives an air shock out from its head which can travel great distance from the volcano; across the sea, it may plane over the waves at high speeds despite its density, as vaporisation of the water beneath helps to drive it forward. Ahead of this a wave travels out as a tsunami, driven and strengthened by the expanding flow behind it.

The size of the reservoir and the properties of the rock surrounding it determine the magnitude of the event. The pressure within it is controlled by its feed from the mantle. Thus, a volcano may simplistically be regarded as a pressurised vessel, with a bursting disc atop a vent at the highest point, ready to convert the potential energy of compressed gas within into work in propelling flow at the surface. As such the classification of event magnitude should rank as the pressure contained in the reservoir multiplied by the volume of the confined chamber in the first instance. Amongst the volcanoes chosen for description above, Yellowstone holds the greatest threat for the future.

To understand and classify these phenomena requires a simple view of the driving mechanical forces and an understanding of the resulting deformation within the range of materials encountered under load. The build up of gas and the failure of a confining disc vents a high-pressure drive that propels pressure waves and entrained materials from the source. The analogy with gas-driven launchers is clear and their operation will be described in Chapter 3. Similarly the imposition of a step change in pressure (by material failure) or driving particles to move with a velocity at a boundary by impact, propagates waves that can be simply described. In the following chapters these devices and behaviours will be expanded to provide a toolset to describe these problems capable of use in engineering structures and devices for the future.

1.3   Key concepts in dynamic loading

To quantify the effects of the phenomena discussed, a range of experiments and techniques to study materials will be described here. The mechanical quantities of interest and the timescales of high-speed events will be quantified so that the reader can attack a problem with sufficient rigour that its essence will be clear and numerical quantities may be derived to define states of behaviour that will allow design in appropriate regimes. Several years ago, observers on Earth were privileged to observe the break-up of the Shoemaker–Levy comet. In its path around the Sun it disintegrated into a series of icy spheres which impacted the surface of Jupiter observed by a suite of instruments from Earth. The spheres were up to five kilometres in diameter and impacted at up to 50 km s^{-1}. Consider the impact of these 20 spheres. The volume of spheres of 1 km diameter is $V = (4/3)\pi r^3 = 5.23 \times 10^8$ m^3 and their mass is $V_\rho = 5.23 \times 10^{11}$ kg.

The spheres have a kinetic energy of $20 \times 1/2\, mv^2 = 1.3 \times 10^{22}$ J. The largest nuclear blast, conducted by the former USSR, was equivalent to 58 Mtons and the chemical energy contained was equivalent to 5 MJ/kg. Thus, the energy of the blast was 2.9×10^{17} J, equivalent to 44 800 nuclear blasts. Clearly natural events dwarf human engineering devices but nevertheless such phenomena have common features that must be clarified here. The overview of planetary phenomena has indicated some features of materials behaviour in natural extremes and has described the loading impulses, mapped material response and tracked system behaviours thereafter. These examples have shown that two important progenitors to dynamic material response are pressure loading from explosion and impact. Observation of their effect in deforming targets, and disassembling natural phenomena into operating mechanisms, are key to understanding and predicting their consequences. Such observation must conclude that most of these processes are propagated by driven waves travelling out from the source of the impulse, and that the most destructive of these are shock waves.

Common usage has used the terms explosion, shock and impact in wider contexts. However, science has defined the terms by which the shock may be described technically. Stokes (1851) described a process in which 'a surface of discontinuity is formed in passing across which there is an abrupt change of density and velocity', and this provided the foundation in later years for Rankine (1870) and Hugoniot (1887a, 1887b, 1889) to formulate an initial attempt at the underlying theory in the form recognised today. These works represent the genesis of descriptions of phenomena that this book aims to describe.

A shock may be simply described thus:

A travelling wave front across which a discontinuous, adiabatic jump in state variables takes place.

There are three phenomena that provide the descriptive framework to drive the shock compression considered. One is the rapid release of gas resulting from the chemical transformation of a solid explosive molecule to its component parts. The rapidly expanding gas cloud drives a discontinuity (shock) through the explosive fuelled by a chemical reaction to its rear. Alternatively, an energy source vaporises a material and the hot plasma cloud expands and loads a surface, driving shock waves into the material. A final cause is the motion of an object through a medium at faster than its wave speed on the one hand, or where the impact stress exceeds the elastic strength on the other. When the material into which the shock propagates has no strength, the processes observed are described as hydrodynamic, and this approximation is also applied to the case of loading to such a pressure that the strength is regarded as negligible. Thus the progenitors of the compressive impulses that this book will describe are explosion, plasma and impact. The classes of material, inert and reactive, and their response to these stimuli form the subject matter of what follows. Of course, the stimuli may be insufficient to result in detonation on the one hand or a shock front on the other. Less energetic results may ensue: either burning where chemical reaction might occur or the propagation of an elastic wave front in an inert material. The property that determines which of these occurs is the material's strength, the threshold above which interatomic bonds are broken. Thus the behaviour of different classes of material and their internal defect population becomes an issue, particularly since the defects sampled are a function of the duration of the blow and how quickly it is applied. The magnitude of the shock also determines the state that the material adopts, and, in particular, this may result in a structural rearrangement in some to give a new phase. Although much may be gained from observation, a theoretical backbone must unite these observations. This book aims to apply basic mathematics to the problems attacked, whilst indicating the breadth of further, more rigorous approaches that can be found in other texts.

1.3.1 Explosions and explosives

The common perception of a dynamic event involves concepts that owe their origins to childhood experiences of fireworks. In many peoples' minds, an explosion is merely an event that involves the emission of light and a sharp noise. Yet the rapid events that the term 'explosion' encompasses range from the energy liberated by a blasting cap (of the order of 10^3 J), through the largest man-made events (atomic events at $c.$ 10^{15} J), up to the much larger natural events such as Krakatoa (at $c.$ 10^{23} J) and culminating at $c.$ 10^{50} J with supernovae.

In considering explosions fashioned by man, Berthelot (1883) wrote:

An explosion results from the 'Sudden expansion of gases in a volume much greater than their initial one, accompanied by noise and violent mechanical effects.'

Explosion may be further divided into the following classes: physical, chemical or nuclear. A physical explosion involves the rapid application of some force that results from an applied stimulus. Rapid application of heat to a substance may lead to instantaneous vaporisation. Lasers use this method to create a plasma which expands and applies load to compress targets in experiment. This also happened in the Mt St Helen's eruption in 1980 when hot material entered Spirit Lake; the consequent vaporisation drove a shock wave that travelled into the upper atmosphere. Of course, the mechanical loading of a pressure vessel may cause it to fail, allowing the rapid expansion of the included, pressurised fluid. This may extend from a bursting balloon at low pressures to the rupture of a gas cylinder at high pressures. In the one case, fragments of rubber may unzip, allowing the sound of a report to go forth. In the other, a blast wave will ensue, and if the cylinder fragments in a brittle manner, metal shards may be propelled into the surrounding area. In the case of a nuclear event, the physical processes of fission or fusion release vast quantities of energy in very short times. There is rapid expansion of the surrounding air, and vaporisation of material. All of these events are lumped under the term 'explosion' in what follows. It is chemical explosion that will be discussed in this book since one of the classes of materials considered is reactive. Here a substance (or substances) undergoes particular chemical reactions that produce both gas and heat and occur over a short time. It is necessary to find a suitable unstable chemical (or chemicals) to undergo such a reaction. There are many unstable substances, but most are not useful. The UK Explosives Act (1875) is more explicit, defining an explosive to be 'a substance used or manufactured with a view to produce a practical effect by explosion'. But clearly such a material must have other desirable characteristics. These will include the need for the substance to be inert to materials placed in contact with it and for it to be stable when stored. Indeed, an ideal explosive ought to be completely inert for most of its life, but very easy to set off when required. Clearly this is a requirement if devices using the substance are to be both reproducible and safe to use. When the time comes, the material must produce the hot, high-pressure gas required for doing significant work on its surroundings. An explosive then is:

A single or mixture of substances which when suitably initiated decomposes explosively with the evolution of heat and gas. (after Bailey and Murray, 1989)

For example, 1 kg of high explosive (HE) initiated in a bore hole produces a power of 5 GW over the duration of the reaction (microseconds) and occupies less than 1 litre before use. Furthermore, it may be stored for years providing suitable care is taken to prevent degradation of its performance during storage.

1.3.2   Blast waves

Air shocks driven by an explosion propagate a front travelling out into the atmosphere. They must continue to be driven from the rear to maintain their intensity, given that in most cases they are continually increasing their contained volume. This is never possible to maintain indefinitely as sources are generally finite, and so the amplitude will decrease with distance, and the positive pressure pulse will be matched by a release phase where the pressure drops below ambient. The gas flow in the pressure phase of

Table 1.1 Effects of impact at various velocities (v)

$v < 50$ m s^{-1}	Structural response: elastic
$50 < v < 200$ m s^{-1}	Local indentation and penetration. Plastic response and structural deformation
$500 < v < 3$ km s^{-1}	Pressures > strength, hydrodynamic, strength important
$3 < v < 12$ km s^{-1}	Hydrodynamic behaviour, compressibility can't be ignored
$v > 12$ km s^{-1}	Explosive impact, solids vaporised

the shock drives any material it encounters in the direction of its propagation, attaining a particle velocity at which the gas moves. The flow in the release phase is in the opposite direction. Some of the most intriguing blast sequences from the nuclear trials of the 1950s show buildings flung first one way by the propagating wave and then ripped apart in the reverse direction as the release arrives. If nothing else this illustrates that buildings are sturdier in compression than tension. The study of gas shocks forms a field of study of its own, yet their effects are important to consider in quantifying phenomena that range from the sonic boom from a supersonic airliner at one extreme to the damage that may result from an exploding chemical storage tank at the other. Ultimately, wave interaction with a structure results in momentum transfer and localisation that may fail the object it meets. More momentum may be passed when the object has significant mass and so it is impact that applies the greater forces, inevitably concentrated over smaller areas.

1.3.3   Impact

The effects of impact range over several fields of study. Differences in classification and interpretation relate to geometrical effects resulting from the mode of the loading. For instance, in one dimension consider a situation where loading is in mode I so that only uniaxial compression and tension result. The chief variable factors then become the amplitude and duration of the pulse applied. On the other hand, in the general case there are multimode loadings and these may occur from low impact speeds where the material may not reach its yield strength and thus responds elastically, to the high-velocity case where the loading induces pressures that are so high that the strength is vastly exceeded and the material behaves as if it were a fluid. This is known as the hydrodynamic response, as at high pressures strength may be regarded to a first approximation as irrelevant. Such assumptions may be simplistic in certain cases, as will be seen in later sections. When impact speeds exceed 3 km s^{-1}, the loading is termed hypervelocity, although the term has a variable threshold. Table 1.1 gives a brief introduction to the ranges of impact velocity and the effects that result.

For instance, when a rod impacts a target a series of physical phenomena ensue which this book will describe in more detail in later sections (see Figure 1.7; the reader may also look to other specialist texts on ballistics). Firstly, at the impact site a high contact

Figure 1.7 An X-ray image showing a steel rod 10 mm in diameter hitting a block of glass at a velocity of about 500 m s^{-1}. The first image was taken 3 μs after impact, the second 30 μs later. By the third frame, the rod has eroded a pit in the glass and the block begins to fail. In the final frame, 80 μs after impact, a disc of metal sheared off by the glass can be seen stationary at the base of the eroded crater. Source: reprinted from Bourne, N. K. (2005a) On the impact and penetration of soda-lime glass, *Int. J. Impact Eng.*, **32**: 65–79, Copyright 2005 with permission from Elsevier.

pressure is generated in both impactor and target. Waves travel forward and back from the impact site and transmit information through the rod and target. The waves will be of several forms. Compression waves (elastic and shock fronts) will propagate, but shear waves, commencing with the lateral flow to the metal, must follow. All of this induces a material response, which ranges from dislocation generation and storage or twin formation in metals, to fracture in brittle materials. Later chapters will track the inelastic response of different classes of material. After these fronts have passed, reflections release the state of compression, frequently inducing local tension in various zones. This dynamic tensile failure is termed spallation. In some metals and polymers, shear banding occurs as softening overcomes lateral heat conduction in localised deformation regions. These zones allow slip of large volumes of material with consequent plug formation if the geometry allows. Plastic flow at the impact point may allow the penetrator to enter the target. Alternatively, parts of a structure may be allowed to bow under the impact load and absorb the forces induced. Designed failure is in some cases a goal, for example in the quest to improve the crashworthiness of cars and improve safety. At the highest velocities, bumper shields, protecting communications satellites, must be capable of preventing relatively small particles of debris reaching the vital working parts of the device. Here the tactic is to allow penetration, followed by vaporisation and the production of a cloud of smaller particles which exit from the rear of the shield. The cloud is of smaller particles and spreads over a larger area than the original impact, which allows a second shield to catch the resultant debris and protect the structure it covers.

1.3.4   Strain rate

The Newton's cradle described earlier is an example illustrating wave behaviour in impacting solids. Spheres are suspended on a steel frame and impacted to amuse bored executives. If, for instance, the spheres on the cradle have a diameter of 10 mm and are made of a hard steel, then an elastic wave, travelling at around 5000 m s^{-1}, crosses the ball in $c.$ 2 µs. Typically, for laboratory-sized items, waves travel in millionths of a second across objects measured in millimetres. Conveniently 1 km s^{-1} and 1 mm µs^{-1} are the same unit so that, armed with a few simple concepts, it is quick and easy to compute distance and time. The SI unit for pressure (in the typical scenarios considered here) is GPa, which is approximately ten thousand times atmospheric pressure. For example, a copper plate impacting another with a velocity of 500 m s^{-1} induces 10 GPa (which is 20 times higher than the elastic limit of the material). This means that the material compresses to a higher density behind the shock front.

All of these concepts define the speed, or the rate at which a material is loaded. The amplitude of the pulse, in terms of the maximum stress applied, or the maximum strain through which it may extend a target, is one measure, but clearly if the loading takes millennia in one case and microseconds in another, different results might be expected and are generally observed. The quantity that is used as an indicator of this behaviour is the strain rate of the pulse and is simply the strain attained divided by the time it takes to do so. In the case of a shock wave, the time required to rise to the maximum strain applied is determined by the speeds at which dislocations can move in the crystals loaded. For the example of the shock in copper quoted above, 5% strain in 50 ns is equivalent to a strain rate of 10^6 s^{-1}, which is a typical figure in shock loading. However, if a cylinder is loaded by a pulse, it takes roughly three reverberations of the wave, to and fro, to equilibrate it to a uniform stress state from which it might deform uniformly. It reaches a strain of a few per cent in 10 µs, which is a rate of 10^3 s^{-1}. It is important to understand that strain rate is an averaging process. When impacting the cylinder, the strain rate on the impact face is infinite. The stress state equilibrates after several bounces of the wave, but nevertheless the material has still responded to processes that occur in the initial stages. Understanding precisely to what it applies is crucial since it is the choice of the user as to what increment of strain is chosen to describe it and whether or not the state is steady at that time. Thus because the term is an averaging quantity one must be careful not to define it for times when deformation has completed or when a state is steady. Secondly it must be applied to a particular volume of material and again the whole of this must have attained an equilibrated stress state before a global measure may be defined. Thus strain rate is a measure of the strain achieved and the time taken to do so of an unsteady region that results in a volume of material after the pulse which is in stress-state equilibrium.

More will be said on the use of this term throughout the book and with the concepts developed a fuller discussion of these results will occupy Section 9.7. Nevertheless strain rate is a useful indicator of the rapidity of loading, and as such has been introduced as a parameter into engineering models of continuum behaviour under rapid loading when applied to laboratory length scales. However, it must be interpreted with caution when it is applied to small volume elements of material over short times and not used to apply to bulk conditions in the surroundings.

The best-known result of varying strain rate in an impulse is that generally a material shows a greater strength the faster the load is applied, since one enters a realm where the loading of the pulse is on the same timescale as the dissipation processes that define the inelastic behaviour of whatever form. The jump in yield stress at strain rate of around 10^4 (depending on the material) marks the transition from continuum loading, where the stress is in equilibrium, to wave dominated loading, where it is equilibrating. In this case dynamic and static yields have different boundary conditions and should be expected to be different. In some materials, however, hardening or softening behaviour occurs as different mechanisms are induced as loading is applied more quickly or more slowly. Conventionally, a simple mathematical description that may be applied to a material to describe this shows how its yield strength varies with the strain rate applied. Such a relationship is called a constitutive description and recent advances in the subject have resulted in new forms for such equations to fit different regimes of behaviour.

There is a range of material motions that stem from impact or explosion and result from the transient loading pulse within them. These drivers propel waves that travel through solids, liquids and gases and place the material they have swept through into a state of compression, tension or shear. One goal of experiments is to populate simple mathematical descriptions of the phenomena to describe the response of materials. The basic laws of conservation of mass, momentum and energy, and classical mechanics drive the descriptions of the thermodynamic states. To these will be added the concepts of elastic and inelastic (in metals, plastic) deformation bounded by a yield surface. To focus on material response, it is generally the simplest loading that is applied experimentally, and these techniques and the platforms that apply them will be explored in the text below.

At each point in a solid, a stress and strain can be defined. Stress is interpreted as internal tractions that act on a defined datum plane. The principal directions are particularly important where the principal stresses act normal to these internal planes. Engineering strains are defined in a global sense as the ratio of elongation with respect to the original length. These definitions are used here where the loading is compressive and the displacements are small. However, the engineering stress–strain curve does not give a true indication of the deformation characteristics at large strains in tension, for instance, since variations in geometry distort interpretation of the results. These issues will be touched on and the reader is referred to the engineering texts for full discussion of these effects.

1.3.5   Length scales

The remainder of this book will discuss physics and chemistry operating at a range of length scales from the atomic to that in which structures operate; however, the response that results will almost always be observed at the laboratory scale and it is here that effort must be concentrated to understand the integrated effect of operating mechanisms at lower scales. Processes occurring on small length scales take short times to complete and the length scales and timescales used here will be connected by rate constants of one sort or another. The most important determinant of kinetics will be the velocity of particles in the flow since this controls communication of state from one location to another. The scales at which humanity operates are those of the macroscale. This has a characteristic length scale of a metre but such units come from antiquity as a measure of human dimensions (regal strides or distances between nose and thumb). In this text the macroscale is divided into component and structural scales. A structure is a composite of components held together by joints that are either introduced by man or by nature at the ambient scale. The component scale is that of one of those elements and the division between them is the defect that exists at the interface. Within a component the material is composed of assemblages that are held together at another boundary. In a metal these constituents are called grains and these may be of one element or a number of them according to the material. Grains of a material with inclusions of a different type constitute a microstructure and materials science has developed tools to investigate these, which are typically at micron dimensions at the mesoscale. Yet an individual grain has its own structure determined by the packing of atoms in repeated cells. This scale is called the microscale and its components are atoms and the defects that accompany this packing. The lowest length scale that is important here is the atomic scale, which defines the Bohr radius of atoms that are repeated within the microscale structure.

Thus a typical element at the structural scale contains components, which at the component scale contain grains that consitutute the mesoscale. These grains contain atoms with particular packing at the microscale, and unit cells of repeating structure at the microscale. Each scale has a boundary: the faces of a component that are joined to make a structure; the grain boundaries between elements of a microstructure that make up a component; and the stacking of cells that together form a grain. With boundaries come defects since no stacking is perfect in reality, and as this book will illustrate it is these defects that seed the behaviours in extreme loading that will be discussed in what follows.

1.4 Final remarks

The goal of the following pages is to describe the response of materials and structures as they respond to the extreme states found throughout nature. Gravitational forces apply constant, often extreme compression, but there is need to understand the transient states set up by the loading applied. The variables of interest will, as ever, be the pressure, the volume and the temperature or relatives, but in all cases their development as a function of time is a key feature to be described. Clearly the final thermodynamic state of the matter under load should be independent of the path taken to pass from the initial conditions. However, equilibrium thermodynamics applies when the state has equilibrated, and equally waves can be described in this manner only when they are steady. In all of these considerations strength will prove a barometer reflecting the state, the volume over which it is measured and the time at which the measurement is taken. Over the longest length scales and timescales it becomes very small and hence the objects observed through the universe are spherical. However, the means by which the loading interacts with strength of jointing in stuctures, elements within the microstructure or of individual bonds at the microscale, are the mechanisms by which deformation and change occurs and these form the rich tapestry that this book will aim to illustrate.

Thus an approach will be taken which considers the response at a series of length scales. At the continuum, the relations between physical variables are set by the large-scale boundary conditions imposed by the dimensions of the problem and material thresholds such as the yield strength of the material averaged over the whole target. Constructing a framework based on mechanics at this scale, it becomes clear that details of the response, minor for some classes of material, are important in others. Further response is generally dependent upon microstructure at the meso- or micro-scale. In the descriptions of dynamic materials, such behaviours are not fully appreciated and therefore not embedded into theoretical treatments at the continuum. Thus in later chapters, details of the microstructure and defect assemblages are described which determine the form of developments necessary to add onto the existing constitutive laws. These subscale models have been variously described but their incorporation into continuum descriptions represents one of the key developments of the last decade.

Whereas physically based descriptions exist for quasi-static loading, this is much less the case for the dynamic case. This is in part because there has been less work done in this area, but also because there is still incomplete information on operating mechanisms under these fast loadings. Thus it is necessary to highlight the need to make measurements at small length scales and to image the development of assemblages of defects or structures responding dynamically. This means that the measurements will therefore have to be made at faster rates than are presently typical with devices that do not yet exist. Thus the challenge for diagnostics at the present time is to image at greater resolution, and also much faster, than has been required in previous developments in the field, in order that the theoretical descriptions are more complete and become properly physically based. With greater resolution, and smaller length scales modelled, there is a need to look at interactions of defect assemblies in three dimensions. Furthermore, once descriptions have been constructed, the response at these scales needs to be mapped to the continuum. The devices and techniques to accomplish the former, and the frameworks for constructing theory to deliver the latter, are the challenge for the next decade. Great advances are being made, but the prize will be fully physically based dynamic material descriptions. These, with increased computer power, offer the promise of virtual design in extreme environments and this will drive great advance in a series of sectors from high-pressure energy generation to structures in space.

This book attempts to describe this regime with a common language gleaned from the training of physical science. However, each field across this landscape has its own terminology and results. One key contributor is materials science and an appendix is included to highlight key concepts that the reader should refer to when progressing through the text.

1.5   A note on units

In SI units, the pressure will be measured in pascals but the values of pressure in the shock will correspond to gigapascals. Atmospheric pressure, 1 bar, is 10^5 Pa so the pressure in a 1 GPa shock corresponds to 10 kbar. Higher pressures are measured in megabars: 100 GPa = 1 Mbar. Another variable that recurs in what follows is that of velocity which in SI is measured in m s^{-1}, or, under these conditions, km s^{-1}. There is a simple equivalence between the velocity measured in the laboratory frame, and that measured in the faster time frames considered here, i.e. 1 km s^{-1} = 1 mm μs^{-1} = 1 μm ns^{-1}.

1.6   Selected reading

Benson, D. J. (1992) Computational methods in Lagrangian and Eulerian hydrocodes, *Comput. Meth. Appl. Mech. Eng.,* 99: 235–394.

Bailey, A. and Murray, S. G. (1989) *Explosives, Propellants and Pyrotechnics*. London: Brassey's.

Bourne, N. K. (2009) Shock and awe, *Physics World*, 22(1): 26–29.

Fortov, V. E. (2011) *Extreme States of Matter on Earth and in the Cosmos*, Berlin: Springer.

French, B. M. (1998) *Traces of Catastrophe: A Handbook of Shock-Metamorphic Effects in Terrestrial Meteorite Impact Structures*. LPI Contribution No. 954. Houston, TX: Lunar and Planetary Institute.

Meyers, M. A. (1994) *Dynamic Behavior of Materials*. Chichester: Wiley.

2 A basic analytical framework

In the previous chapter, a series of examples was given to illustrate a range of material responses that stem from impact or explosion and result from a transient loading pulse within the material. These drivers propel waves travelling through solids, liquids and gases and place the material they have swept through into a state of compression, tension or shear. This chapter will describe these disturbances in more detail and attempt to give simple mathematical descriptions of the phenomena and the material's response. This basic approach is really a development of solid (a branch of continuum) mechanics to embrace additional features of loading at higher speeds and amplitudes; there are many, more complete texts available on the basics of solid mechanics that the reader may consult. The strategy here is to keep the derivations as simple as possible; again there are texts that derive relations with more generality than here but it is vital that the reader realises the assumptions, and more importantly their limits, in what follows. Particularly, it should be noted that solid mechanics assumes material behaviour based upon observations made in ambient states. Electronic bonding itself changes nature at around 300 GPa, so it is unrealistic to expect theory extended from the elastic state to apply in these regimes. Thus assumptions made and their limitations in the loading states the reader wishes to consider must be fully understood before using the formulae below. The basic laws of conservation of mass, momentum and energy, and classical mechanics will drive the descriptions of the thermodynamic states. To that will be added the concepts of elastic and inelastic (in metals, plastic) deformation bounded by a yield surface. To focus on material response, it is generally the simplest loading that is applied experimentally. Thus these states will be mentioned below to highlight particular relations to which the text will return.

At each point in a solid, a stress and strain can be defined. Stress is interpreted as integrated internal forces and moments that act on a defined internal plane within the volume of interest for a problem. The principal directions where the principal stresses act normal to these defined planes are of importance here. Engineering strains are defined in a global sense as the ratio of elongation with respect to the original length. These definitions are used here where the loading is compressive and the displacements are small. However, the engineering stress–strain curve does not give a true indication of the deformation characteristics at large strains (in tension for instance) since variations in geometry distort the interpretation of the results. With a few notable exceptions, the pulse lengths in dynamic loading are small so that strains are generally small. The reader is referred to the engineering texts for a full discussion of these effects when they are not.

Pulses rise to peak stress at different rates within materials, and a feature of the dynamic loading of materials is that, in general, a material has a greater strength the faster a load is applied. One reason for this behaviour is that the rise of the loading pulse is on the same timescale as the dissipation processes that define inelastic behaviour of whatever form. The standard mathematical description that may be applied to a material thus relates its yield strength to the variation of the strain rate applied. Such a relationship is called a *constitutive description*. However, although advanced solid mechanics is a vital tool for these problems, the reader must always remember that continuum response is the integration of material deformation processes that can only proceed on defined timescales and thus defined length scales. If the response time is sufficiently short, then a series of mechanisms with slower kinetics are filtered out, leading to a different end state. Thus operating deformation mechanisms and the time (and thus the position) of the material station of interest, must always be kept in mind. This may mean that a result or analytical equation determined from solid mechanics (but derived using quasi-static data) may not apply when the input is a transient loading pulse (such as may be supplied by a rapidly pulsed laser, for example).

2.1   Loading states

In the final released state, returned to ambient conditions, wave propagation beyond the elastic limit transforms the material into a different shape and into a multiaxial state defined by the boundary conditions of the loading and the transit of waves defined within it. In the dynamic loading of materials, impact and explosion always generate compression states in the first instance and so it is these that the following sections will address. The simplest geometry to describe is one in which a material is compressed between rigid anvils down one axis and where it is said to be in a state of uniaxial stress. If that axis is the x direction, then $\sigma_x =$ constant, and $\sigma_y = 0$, $\sigma_z = 0$. Here a target, loaded quasi-statically, is in a state where it compresses (shortens) down the impact axis and expands normal to this in lateral directions. In the elastic region these strains are connected by Poisson's ratio. Although this picture is familiar, it does not account for the time taken for material in a central region of a target to lose confinement around it from surrounding material. This may occur, but only after a wave has travelled from the periphery to this region allowing lateral expansion.

During the past two centuries the study of material response has defined terminology for various limits derived from measurements of engineering thresholds. A few of these are defined below with reference to the schematic in Figure 2.1.

2.1.1   Yield strength

The *yield strength* (or the *yield point*) is the stress that a material can undergo before moving from elastic to inelastic (plastic) deformation. This point is well defined in

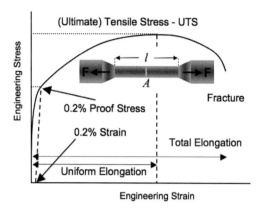

Figure 2.1 Engineering stress–strain curve determined in a tensile test on a metal.

hardened and tempered or annealed metals or glasses, but can be ill defined in others that have a composite nature. Thus alternative thresholds are defined.

2.1.2   Proof stress

The stress that will cause a specified small, permanent extension of a tensile test piece is known as the *proof stress*. Commonly the stress to produce 0.2% extension is quoted. This value is used to represent the yield stress in materials not exhibiting a definite yield point in composites and alloys and applies to ductile materials where the compressive and tensile quadrants of the yield surface are similar.

2.1.3   Ultimate tensile strength

The *ultimate tensile strength* (UTS) of a material is the limit stress at which it actually breaks, with a sudden release of the stored elastic energy.

In tension, a rigid solid cylinder of length l and area A_0, drawn down a deformation axis by a uniform force F, deforms by stretching down its length (and conserving volume) so that the radius of the deforming section contracts. The stress–strain path taken by such a sample is shown in Figure 2.1 and a schematic of the deforming target with some dimensions is shown there too. If the dimensions of the deformed section are A and l then the stress in the deformed section is $\sigma = F/A_0$ and the strain is $\varepsilon = [\Delta l/l_0]$, where σ and ε are known as the engineering stress and the engineering strain, respectively. The engineering stress–strain curve does not give a true indication of the deformation of a sample when strain becomes large (greater than c. 5%) because it is based upon the original dimensions of the specimen, which change continuously during the test. Also, a ductile metal, for instance, which is pulled in tension, becomes unstable and necks. Thus the scale on which the deformation is considered determines the validity of the definition used.

In contrast, the *true stress*, based on the actual cross-sectional area of the specimen, yields a stress–strain curve that increases continuously up to fracture. If the strain measurement is also based on instantaneous measurements, the curve that is obtained is known as a true stress–true strain curve, also known as a *flow curve*.

Since volume is conserved ($A_0 l_0 = Al$), then taking a definition of instantaneous strain the true strain, e, and the true stress, s, may be defined as

$$e = \ln \left[\frac{l}{l_0} \right] = \ln \left[\frac{A_0}{A} \right] \quad \text{and} \quad s = (1 + \varepsilon)\sigma \tag{2.1}$$

In what follows, stress and strain will be used as shorthand for engineering stress (σ) and engineering strain (ε) when applied to shock, since deformation is generally small.

These states of uniaxial stress must be contrasted with those of uniaxial strain, where the material is compressed down a single direction and there is no time for further correction to allow for a triaxial load. It is this state that is important for the type of problem considered here since dynamic loading may only progress on a timescale commensurate with that taken for waves to reach areas of interest from boundaries. Thus initially planar loading remains so until waves arrive from the periphery to load the material triaxially. Since compacting a material on an impact surface compresses all particles in the vicinity in one direction, that is the direction of the incoming impactor.

2.1.4 Residual strain due to shock loading

In order to understand the effects of material deformation and damage, whether by manufacturing routes or by loading in the environment in which they are used, it is common to compare a target before and after the working has occurred. Recovery techniques for shock are a subject in themselves and this discussion will be deferred until a later chapter. If comparison of the effects of loading is to be useful, some account of the residual strain imparted by the deformation process must be included to offset the reload curve. An estimate of this may be recovered from the total shear strain undergone,

$$\varepsilon_{\text{res}} = \frac{\sqrt{2}}{3} [(\varepsilon_x - \varepsilon_y)^2 + (\varepsilon_x - \varepsilon_z)^2 + (\varepsilon_y - \varepsilon_z)^2]^{\frac{1}{2}}. \tag{2.2}$$

In the case of shock loading the strain is one-dimensional and $\varepsilon_x = \varepsilon_y = 0$, leaving only ε_z. This simplifies the equation for the strain and recovers ε_{res} for the shock state. Clearly before retest the material has also been released and, assuming the two processes give the same strain, this gives a total effective strain of

$$\varepsilon_{\text{res}} = \frac{4}{3} \ln \left(\frac{v}{v_0} \right) = \frac{4}{3} \ln \left(\frac{U_s}{U_s - u_p} \right). \tag{2.3}$$

where U_s and u_p are the shock velocity and the velocity of particles in the following flow, and v and v_0 are the specific volumes after and before compression.

Comparisons of annealed materials with materials recovered and then reloaded for body-centred cubic (BCC) and face-centred cubic (FCC) metals show different

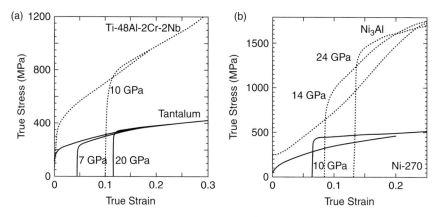

Figure 2.2 Reload stress–strain behaviour at 0.001 s^{-1} for annealed and pre-shocked material. (a) TiAl and Ta showing no hardening behaviour, (b) Ni and Ni$_3$Al showing hardening on reload from the shocked state. Source: reproduced from Bourne, N. K., Millett, J. C. F. and Gray III, G. T. (2009) On the shock compression of polycrystalline metals, *J. Mat. Sci.*, **44**(13): 3319–3343. With kind permission from Springer Science and Business Media.

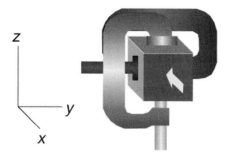

Figure 2.3 Uniaxial strain loading. Lateral motion is constrained with loading purely longitudinally.

behaviours. Figure 2.2 shows the behaviour for four cubic materials: FCC nickel; the ordered FCC intermetallic Ni$_3$Al; the BCC metal tantalum; and two alloys based on the intermetallic phase TiAl. In Figure 2.2(a) the behaviour for the BCC metals shows two curves for TiAl and three for tantalum. One pair are for the unshocked materials whilst the others are displaced by varying amounts along the strain axis corresponding to the effective plastic strains given in Eq. (2.3). An expression for the volumetric strain, V/V_0 can be recovered from the conservation relations for shock discussed later in this chapter. These metals have plastic behaviour which shows no effect from the shock loading excursion that the material has undergone, while the FCC metals in Figure 2.2(b) show that the material has strengthened as a result of the shock impulse.

Imagine compressing a material in an impact for which two of the directions perpendicular to one another are fixed so that there is no motion as it compacts ($\varepsilon_x \neq \varepsilon_y = \varepsilon_z$, $\sigma_x \neq \sigma_y = \sigma_z$) as shown in Figure 2.3. There must then be a compressive force applied down the third to compress the block down this, and only this axis. In the elastic range one may express the strain down all the directions in the following manner using Hooke's law:

$$\varepsilon_x = \frac{\sigma_x}{E} - v\frac{\sigma_y}{E} - v\frac{\sigma_z}{E} = \frac{1}{E}[\sigma_x - v(\sigma_y + \sigma_z)]$$

$$\varepsilon_y = \frac{1}{E}[\sigma_y - v(\sigma_x + \sigma_z)]$$

$$\varepsilon_z = \frac{1}{E}[\sigma_z - v(\sigma_x + \sigma_y)].$$

Now $\varepsilon_y = \varepsilon_z = 0$, so $\sigma_y = v(\sigma_x + \sigma_z)$ and $\sigma_y = v(\sigma_x + \sigma_z)$ and therefore

$$\sigma_y = \sigma_z = \left[\frac{v}{1-v}\right]\sigma_x. \tag{2.4}$$

This relates the three orthogonal stress components in a uniaxially strained material. Typical values for Poisson's ratio, v, are generally close to one third for metals, whereas the value is closer to a quarter for brittle materials. These values are not universal for these classes of material – for instance, there are metals (such as titanium aluminide, Al_3Ti) with a Poisson's ratio of 0.18. However, using these generic values reveals that the lateral stress is about half the longitudinal stress for a metal, but only around one third of this for a brittle material, in the elastic range.

Continuing on from above, and substituting in for the lateral stress components,

$$\varepsilon_x = \frac{1}{E}[\sigma_x - v(\sigma_y + \sigma_z)] = \frac{\sigma_x}{E}\left[\frac{(1+v)(1-2v)}{1-v}\right] \text{ so that}$$

$$\sigma_x = \left[\frac{1-v}{(1+v)(1-2v)}\right]E\varepsilon_x, \tag{2.5}$$

which shows that to achieve a fixed strain when loading in a uniaxial strain state requires augmented values compared to that required for a uniaxial stress state, since the factor in the squared brackets is always greater than one (since $v < 1/2$). Thus a material will be measured to have a higher yield stress if it is confined such that uniaxial strain conditions are maintained for the measurement time. To give an idea of how different that number will be, consider a few cases for different classes of material. For tantalum, a metal with a Poisson's ratio of 0.34, the factor by which uniaxial strain exceeds uniaxial stress is 1.55. Thus for metals and polymers with Poisson's ratio of around a third, the Hugoniot elastic limit (the limit in 1D strain) will be c. 50% higher than the yield stress. For hard materials like TiB_2, with a Poisson's ratio of 0.09, the uniaxial stress and strain strengths will be comparable.

Relating the conditions of the loading to other quantities, engineering definitions may also be used to derive an expression for the shear stress τ, thus

$$\tau = \frac{1}{2}(\sigma_x - \sigma_y). \tag{2.6}$$

At stresses up to the elastic limit, substitution of quantities allows rearrangement to give

$$\sigma_y = \left[\frac{\nu}{1-\nu}\right]\sigma_x,$$

then

$$\tau = \frac{1}{2}\left(1 - \frac{\nu}{1-\nu}\right)\sigma_x \Rightarrow \tau = \frac{1}{2}\left(\frac{1-2\nu}{1-\nu}\right)\sigma_x. \tag{2.7}$$

Introducing the hydrodynamic pressure, p, defined as

$$p = \frac{1}{3}(\sigma_x + \sigma_y + \sigma_z), \tag{2.8}$$

substitution leads to

$$\sigma_y = \sigma_z \Rightarrow 3p = \sigma_x + 2\sigma_y = \sigma_x + 2\sigma_x - 4\tau,$$

which can be rearranged to give

$$p = \sigma_x - \frac{4}{3}\tau \quad \text{or} \quad \sigma_x = p + \frac{4}{3}\tau. \tag{2.9}$$

Equation (2.9) shows the separation of the hydrostatic and deviatoric components of the stress. Readers familiar with solid mechanics will appreciate that stress tensors are broken down into components that act to change the volume of a body and those which act to distort it. Hydrostatic stresses do not act to distort the volume and the pressure is simply the mean of the principal stresses. The components that cause the volume to deviate from its original form are combined into the shear term. The hydrostatic response is defined by the material's equation of state (of which more later). A note of caution should be made here that one assumption of the above is that the material is isotropic, i.e. that it deforms similarly along all directions. This is really a special case as most real materials are anisotropic, deforming more or less easily along preferred directions. It is important to remember this, as what follows will assume isotropy in most cases.

The phenomena of interest here will take a material above its yield stress and in general cause it to fail as a wave travels. It will thus be useful to work out when such failure will occur and how it relates to other states. The classical view of a yield criterion is that a material behaves elastically until it starts to deform at the yield stress, Y, and this is generally determined under conditions of one-dimensional stress. Similarly, the criterion is easier to interpret in uniaxial strain loading. However, in general the loading is multiaxial, and under these conditions the criterion is more difficult to interpret. This has been implicitly assumed in some of the discussions above, but it will be useful to discuss simple representations of yield before proceeding.

One criterion for yield that is certainly reasonable given the behaviour of ductile materials, which deform by shear along crystal planes, is the maximum shear stress criteria formulated by Tresca. Clearly plastic deformation is not possible if all stress components are equal (which defines a hydrostatic pressure). Tresca suggested a simple yield criterion to define the limit of elastic behaviour, which occurs when the greatest difference between the principal stresses exceeds the yield strength, Y. Thus, in simple tension,

$$\sigma_x - \sigma_y = Y. \tag{2.10}$$

This defines a surface in three-dimensional stress space defined by the axes σ_x, σ_y and σ_z, which form a sphere about the origin in which the locus of points obeying the criteria lies at a constant distance from it. Such a surface in stress space is known as a yield surface.

Another commonly encountered criterion is that formulated by von Mises. The von Mises criterion assumes that failure occurs when the energy expended in distortion shearing the body reaches the same energy for yield or failure, which can be expressed as

$$(\sigma_x - \sigma_y)^2 + (\sigma_x - \sigma_z)^2 + (\sigma_y - \sigma_z)^2 = \text{constant}. \tag{2.11}$$

Now, in uniaxial stress, $\sigma_x = Y$ and $\sigma_y = \sigma_z = 0$ so $2Y^2 = \text{constant}$. Whereas in uniaxial strain

$$\sigma_x = Y, \; \sigma_y = \sigma_z \text{ so } \sigma_x - \sigma_y = Y. \tag{2.12}$$

Thus, in uniaxial strain at least, the Tresca and von Mises criteria reduce to the same relation. These criteria may be represented as a surface with axes in the direction of the principal stresses. Within the surface, the behaviour is elastic, yet when a particular loading path reaches it, the material yields and transits from elastic to plastic (in general, termed inelastic) behaviour. The final process causing failure in materials is brittle failure of rocks, glasses or polycrystalline materials. There are several classes of criteria that attempt to describe such processes. The first employ empirical attempts to describe data whilst a second adopts a physical model of the failure process. This approach is epitomised by the application of the Griffith theory of brittle fracture to describe a criterion for failure. In this case the general form for inelastic failure is

$$(\sigma_x - \sigma_y)^2 = Y(\sigma_x + \sigma_y). \tag{2.13}$$

Thus, deformation constrains the loading path to the yield surface for increased stresses. Many such criteria are possible and some of these will be mentioned in subsequent chapters. Experimentally, the initial requirement for a new material of interest is testing to map the yield surface for a variety of stress and strain states. Dynamic experiments must additionally show the evolution of the yield surface with time and thus it must be defined for each rate of loading since deformation mechanism thresholds will be exceeded at different times with different impulse histories and the integrated effects will be different in each case. This is the physical reason for the catch-all term 'strain rate dependence,' that is used in engineering to describe this behaviour. Thus a material is found to behave as if stronger when time for deformation is short, relative to longer time flow, when they can operate for longer. It is a challenge for present research to evaluate tests to define mechanisms operating and then others to populate the constitutive models that describe them. An overview of constitutive models will be given later.

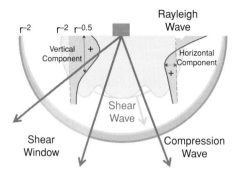

Figure 2.4 Two groups of waves found in a continua: surface and bulk waves. Principal amongst these are longitudinal (P) and shear (S) waves indicated in the figure. Source: after Meyers (1994).

Historically, understanding of observed phenomena began with consideration of elastic waves. A summary of the range of those operating at the continuum scale is shown in Figure 2.4. There are two classes evident, each with different boundary conditions: one describes surface waves, the second concerns those within the bulk. Surface waves explain a variety of phenomena, including geophysical applications and tsunamis where various elastic and non-linear waves have been identified. The best known of these is the Rayleigh wave but there are others with distinct regimes of behaviour that may be coupled with various ground motions. However, this work will be concerned primarily with those travelling through the bulk of a material and here elastic and plastic longitudinal and shear waves (P and S waves in geological terminology) are key. Derivations of the properties of these elastic waves appear in many texts so that only a summary will be presented here.

2.2   Elastic waves in solids and on bars

The simple derivation of wave speed shows the velocity to be the root of modulus, divided by the density of the medium through which they propagate. This is true of bulk waves travelling in a medium with modulus K, where the wave speed c_0 is given by

$$c_0 = \sqrt{\frac{K}{\rho}},$$
(2.14)

and ρ is the density. In a medium contained within a semi-infinite half-space, the longitudinal wave speed c_L reaches a higher value than this:

$$c_L = \sqrt{\frac{K + 4/3\mu}{\rho}},$$
(2.15)

where μ is the shear modulus. Finally, shear waves travel with speed c_S, depending on the shear modulus and density:

$$c_S = \sqrt{\frac{\mu}{\rho}}. \tag{2.16}$$

Clearly these quantities can be combined so that

$$c_L^2 = c_0^2 + {}^4\!/\!_3\, c_S^2. \tag{2.17}$$

It will have not escaped the reader that the longitudinal stress was connected to pressure and shear stress in a similar manner (Eq. (2.9)). All of these are manifestations of the underlying framework of solid mechanics on which this representation sits.

These waves travel in semi-infinite media, yet when adjacent free surfaces are present, the boundary conditions on the waves alter and their velocities are changed accordingly. For instance, an elastic wave may be induced by striking one end of a long bar. The wave travels down it at a speed c_{Rod}, given by

$$c_{Rod} = \sqrt{\frac{E}{\rho}}, \tag{2.18}$$

where E is the Young's modulus of the bar material when it is in a uniaxial stress state. A group wave speed is a more appropriate term for such a loading pulse since each frequency travels at a different velocity such that a step edge disperses with distance. The full analysis of this phenomenon is due to Pochhammer (1876) and Chree (1889), but Rayleigh produced a simpler solution where he defined a group velocity for such waves given as

$$c_g = \left(1 - 3\nu^2\pi^2 \left[{}^a\!/\!_\Lambda\right]^2\right) c_{Rod}, \tag{2.19}$$

where a is the radius of the bar, and Λ is the length of the pulse.

Waves on the surface of materials may take various forms depending upon the trajectory of particles entrained into them. The best known of these is the Rayleigh wave (a special case of the Stonely wave), where the particles take elliptical trajectories under the surface of a solid. The Rayleigh wave has a form that is complex and may be approximated as

$$c_R = \left(\frac{0.863 + 1.14\nu}{1 + \nu}\right) c_S, \tag{2.20}$$

where ν is the Poisson's ratio of the medium. The value of the Rayleigh wave speed is always close to the shear wave speed, e.g. for steel it is about 90% of it.

This chapter began by describing the basis of what follows as a subset of solid mechanics. All of this can be described from general assumptions relating materials properties through bulk formulations. Having presented the group of equations and shown wave speeds as functions of the various moduli and density, it is useful for completeness to mention that there is an elegant derivation of all of this from the wave equation using two arbitrary constants (Lamé's constants). It is a simple result that in the elastic range the stress and strain within a uniform, isotropic material are interconnected by Young's modulus. Yet for an arbitrary material the connection between the two follows the tensor relation

$$\sigma_{ij} = C_{ijkl}\varepsilon_{kl},$$

where C_{ijkl} is a tensor of elastic constants and represents a 9×9 matrix (thus containing 81 elements). These elements are not independent and the 81 may be reduced in the general case to 21. However, crystal symmetry reduces the number still further until, for an isotropic solid, there are only two arbitrary parameters that describe the material. These are λ and μ, known as Lamé's constants. In the simplest case these can be written in a shorthand notation,

$$\sigma_{ij} = C_{ijkl}\varepsilon_{kl},$$
$$\sigma_{ij} = 2\mu\,\varepsilon_{ij} + \lambda\delta_{ij}\varepsilon_{kk}, \text{ where } \varepsilon_{kk} = \varepsilon_{11} + \varepsilon_{22} + \varepsilon_{33}, \text{ the dilation, } \Delta. \quad (2.21)$$

Here, the suffices i, j, k and l are summed over the integers 1 to 3. Thus, the first equation is the generalised form of Hooke's Law discussed above. But assumptions of isotropy yield the second that is the shortened form. It is perhaps sobering to note that very few materials are isotropic at ambient pressures, yet the framework provided by understanding the form of the description for those that are is important, since materials homogenise and become more isotropic as pressure rises to the limits below which strength is defined.

2.3   Shock loading

Whereas elastic waves displace material that subsequently returns to its original position and density after the front has passed, the shock permanently and irreversibly deforms it into a higher density state. The process occurs so quickly that it is not possible for heat transfer to occur in the time taken for the front to pass. This is simply revealed by a flux calculation. Heat flow equilibrates a metal body of dimension R over a time R^2/α, where α is the thermal diffusivity of the object. A stress wave equilibrates the stress state in several wave transits over a time of the order of R/c, where c is a wave speed. The ratio α/c is of order 10^{-7} for a metal so that unless objects reduce in dimension to scale with this quantity, the process can be regarded as purely adiabatic and, on release of the pressure where the structure is unperturbed, isentropic as well. When loading times drop below 10 ns the assumptions of Fourier's Law and continuum heat conduction fail to describe processes operating in microscopic volumes, and so these states are different from those found at larger scales and equilibrium thermodynamics in this form must be used with caution.

Thus, the shock is adiabatic and is also accompanied by a step increase in entropy. This concept was not quickly embraced by academia. As early as 1851 Stokes postulated that:

. . . a surface of discontinuity is formed in passing across which there is an abrupt change of density and velocity . . .

which addressed several of the salient concepts, yet influential figures like Lord Rayleigh maintained that it was not possible (Rayleigh, 1877).

Nevertheless, it can be shown that in media where the sound speed increases as the density does, ramped compression fronts steepen until they eventually form a shock. This is key in demonstrating that a shock front is a consequence of the properties of most solids. Given their existence, the next stage of interest is in the relations connecting thermodynamic quantities in the shocked state. There are two routes to derive the relations of interest. The first is to consider the front as resulting from a particular type of flow. The necessary steepening and formation of a shock may be derived for compressible fluids using the isentropic flow equations. Alternatively, it is possible to merely assume that there is a discontinuity in the flow and then derive a series of relations to describe the behaviour from conservation of quantities across the jump. This route was the one adopted by Rankine (1870) and Hugoniot (1887a, 1889) independently over a century ago in the derivation of the relations that bear their name.

2.3.1   Rankine–Hugoniot conservation relations

Visualisation of a fluid compressing as it is driven forward by some stimulus illustrates the nature of a shock wave and allows derivation of basic quantities that define its state. A convenient analogy may be drawn from a snow-plough entering a fresh fall of snow, for the sake of argument at velocity u (Figure 2.5). In time t the plough's blade advances a distance ut through the snow but it piles up on the front of the plough and as the truck moves forward, more snow is added, so the compressed region of influence extends for each advance so that the front is moving at a higher velocity, S, than the plough that drives it. Compressed snow on the front of the plough is clearly of greater density than that ahead, and further, a ball picked up by the flow travels on the compressed mass, moving (with the plough) at u. The front is an analogue of a shock wave. A densification front travels through a medium, imparting momentum to the fluid in which it travels and across which density, particle velocity and pressure change discontinuously. The flow is made up of waves that may be represented in one dimension by a schematic which indicates flow as a function of time. In such a diagram distance and time are the axes, with distance conventionally the abscissa, and a wave travelling with constant speed represented by a straight line with slope proportional to the velocity. The process of the snow-plough picking up the ball is shown in the diagram. The motion of the front of the plough, the accruing snow front (the shock), and the path of a particle (the ball) first entrained in the flow and then travelling with the plough speed, are shown (Figure 2.5).

Clearly the analogy drawn is far from perfect (as none of those that have been used over the years ever have been) since the timescales, length scales and processes are not observable directly. However, representations serve their purpose although the snow compresses and travels ahead of the driving plough, but is not confined so that its pressure does not increase. Neither does it travel in the same direction as the plough for all its length since it may escape around the edges of the blade. Indeed, a wave driven

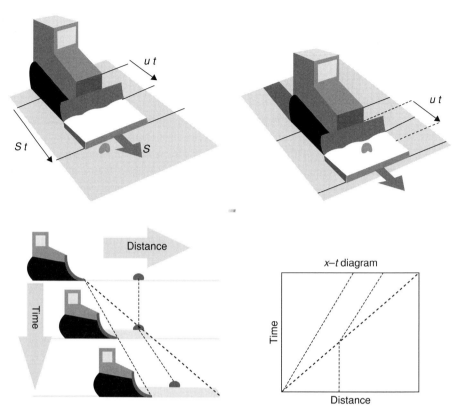

Figure 2.5 A plough compacts snow ahead of it, travelling at speed u into an uncleared field. The front moves at speed S and the plough and snow at u. Below, the paths of the blade and snow front as a function of time can be shown in an x–t diagram.

into a finite body does not obey these assumptions either since any material at the free surfaces can flow outward, so releasing the pressure built in the target. A release front travels at the bulk wave speed in the compressed material which is always higher than the speed at which the shock front can travel into virgin material. Thus shocks must be driven to supply the work needed to compress the material through which they run, decaying eventually to an elastic wave. If the plough stops so does the shock front; it cannot continue without a drive. It is because materials at higher stress are non-linear (such that higher stress levels travel faster than lower) that waves steepen and shock waves can form. Correspondingly, the releases at higher stresses travel faster than those at lower stresses so that whilst compression fronts sharpen into shock discontinuities, release fronts disperse with distance.

2.3.2   Hugoniots

As mentioned above, in the mid-nineteenth century opinion was divided as to the existence of discontinuous waves, with luminaries like Stokes supporting and Rayleigh

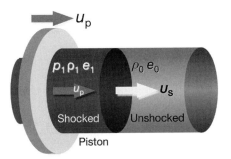

Figure 2.6 A piston, travelling at velocity u_p from the left, compresses a confined fluid and drives a shock at U_s into the undisturbed medium. In general, states ahead of the shock are given subscript 0 and behind 1, but pressure and particle velocity ahead of the front are assumed to be zero in what follows for simplicity. P, ρ and e describe pressure, density and specific internal energy, respectively.

dismissing the existence of shocks. It was Rankine (1870) and Hugoniot (1887a, 1887b, 1889) who eventually derived the relations for a travelling wave front across which a discontinuous, adiabatic jump in state variables takes place.

To describe this jump it is necessary to define the mechanical state in fluid either side of the shock. To do so assumes several features of the material and flow; firstly that the wave travels into a fluid that has no strength (i.e. that the shear modulus is zero so that *stress* and *pressure* are the same); secondly, that body forces (such as gravity, etc.) are negligible. Clearly, the only physical change permitted is densification so that the material is not considered to change phase or chemically react as a result of the shock. Equally, the flow is assumed to be steady and energy is not dissipated by this means. Finally, the analysis can only proceed if the initial and final states are in mechanical equilibrium. Energy and volume are defined to be those that apply per unit mass of material and are lower case in the following text. By this means the conservation laws can be generalised so that $\rho v = 1$, making conversion between these two quantities easy.

In what follows assumptions and simplifications will be made so that the physics is not masked by mathematical complexity. For instance, the shock will be assumed to be travelling into matter at rest. Furthermore, the pressure behind a shock is generally very much greater than the ambient state and so the initial pressure is assumed to be zero. A vital and often forgotten point is that the relations that will be derived below apply to steady waves: waves for which the rise time is small compared with the pulse length and which do not evolve with travel as they progress through the medium. Waves will be assumed to have equilibrated the material state behind their front through a travel distance in the solid. Having set the scene, assume that a planar wave sweeps down a rigid tube, driven, like the snow by the plough, by a piston at velocity u_p (Figure 2.6). Material either side of the front that travels ahead of the piston is either at rest or at a velocity u_p, yet it travels at the greater speed, U_s. There is conventional terminology for these quantities; u_p is known as the particle velocity and U_s the shock velocity in the medium.

From the point of view of a surfer travelling with the boundary, each unit area of the front sweeps in a mass $\rho_0 U_s$ and loses $\rho_1(U_s - u_p)$ flowing out behind the front, which must equate. Thus

$$\rho_0 U_s = \rho_1(U_s - u_p), \text{ conservation of mass.} \tag{2.22}$$

For the same unit area, the force across the front, p, must (by Newton's second law) be balanced by the change in momentum across it, which is mass $\rho_0 U_s$ travelling at u_p so that

$$p = \rho_0 U_s u_p, \text{ conservation of momentum.} \tag{2.23}$$

Finally, the change in internal energy across the front may be equated to the pressure and volume as follows:

$$e_1 - e_0 = \frac{1}{2} p(v_0 - v_1) \text{ conservation of energy.} \tag{2.24}$$

Substituting in density and the two other relations it is clear that the change in internal energy is the change in kinetic energy of material across the front.

But, of course, the fluid itself has an equation of state (EoS) connecting its internal energy with the pressure and volume states

$$e = e(p, v) \text{ equation of state.} \tag{2.25}$$

Thus, the descriptions of the flow constitute a system of four equations with five unknowns (p, v, e, U_s, u_p). The relations between the variables are frequently described graphically in the $p-v$, $p-u_p$ and U_s-u_p planes. These curves represent the loci of all possible final states attainable by a single shock from a given initial state and the curve centred on a particular initial state at standard conditions is known as the *principal Hugoniot* for the material. The final piece in this jigsaw is an experimental relation. For most materials tested in the regime below the *finis extremis* (that don't change phase; see Chapter 1) it transpires that the relationship between the shock velocity achieved and the driving velocity input, is a linear relation, i.e. U_s is related via two constants to u_p thus

$$U_s = c_0 + S u_p. \tag{2.26}$$

The constants c_0 and S are generally determined by experiment. However, the first is related to the bulk sound speed (the limit of wave velocity as the input falls to zero) and for pure materials can be shown to be so. The second is a number normally between 1 and 2 (see Appendix D for some typical values for various materials). Armed with all this, a Hugoniot curve may be simply constructed for an arbitrary material using (2.23) and (2.26), which define all the possible final shock states that might be achieved when a material is accelerated to a particular velocity, u, and this is called the shock equation of state (Eq. (2.31) below). Shock behaviour is derived by first obtaining the experimentally determined constants c_0 and S for a particular material and then constructing expressions for other variables of interest that define the compressed state. Of course the primary assumption is that a single set of constants is applicable to the full range of states accessed. This is not the case when a material changes phase because a new pair of constants is appropriate for the new phase, reflecting the changed microstructure. This is one example where the Hugoniot is not a simple quadratic and there are others where at low particle velocities the response of the material is not hydrodynamic. Polymethyl-methacrylate (PMMA), for example, has a convex curved region at low values of u_p. In these cases the more complex behaviour has often been fitted with a polynomial form for (2.26) with terms in u_p^2 and higher powers added as required to represent the data more accurately.

Nevertheless, the linear relation holds well for most materials and a simple approach (assuming a representation called a Mie–Gruneisen equation of state) provides an indication of the relationships between the constants c_0 and S. The bulk modulus K, is related to the bulk sound speed and the density as shown above (2.14) and so $K = \rho c_0^2$. Further, considering the isentropic limit when waves are acoustic and $TdS = 0$, the second law of thermodynamics may be expressed as $e = -\int p \, dv$, which (with the supplied EoS) gives an expression that equates the constant S such that

$$\frac{\partial K}{\partial P} = 4S - 1. \tag{2.27}$$

Thus c_0 and S are both functions of the bulk modulus of the material in the hydrodynamic limit.

If the Hugoniot has an initial state at an arbitrary pressure, particle speed and velocity, it is said to be *centred* on that state. Further, the suite of possible Hugoniots includes one that has an initial state centred at standard conditions of ambient pressure, and initially at rest, known as the *principal Hugoniot*, and release states represented by isentropes have a *principal isentrope* through this point. It is clear from the above that the Hugoniot is then not a thermodynamic path followed as a material loads up, but represents the locus of end states. Thus a shock jumps discontinuously from the point around which it is centred to the final state achieved and follows a line drawn between the points in the plane in which the curve is displayed. Having discussed the ambient state as at zero pressure and at rest, initial state pressures and volumes will be reintroduced to illustrate the generality of this relation in what follows. Solving (2.22) and (2.23) together, it may be seen that

$$\mathscr{R} = \rho_0^2 U_s^2 - \frac{(p - p_0)}{(v - v_0)} = 0, \tag{2.28}$$

which describes a line in the pressure–volume plane with slope proportional to the shock velocity U_s. Here the line in question has a slope $\rho_0^2 U_s^2 = Z_s^2$, where Z_s is the dynamic *shock impedance* of the material. Further, using (2.24)

$$\mathscr{H} = (e - e_0) - \frac{1}{2}(p + p_0)(v_0 - v), \tag{2.29}$$

which represents a rectangular hyperbola about the initial state in the p–v plane (Figure 2.7(a)).

The curve may also be represented in the pressure–particle velocity plane (Figure 2.7(b)). A target at rest is at atmospheric pressure and stationary so the principal Hugoniot passes through the origin. It is readily apparent that in either pressure–volume

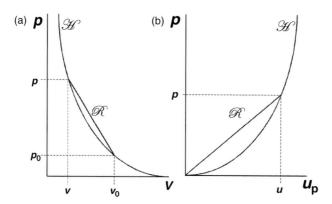

Figure 2.7 Hugoniots in (a) pressure–volume space and (b) pressure–particle velocity space. The Hugoniot, $\mathscr{H}$, defines the locus of shock states and the Rayleigh line $\mathscr{R}$ connects the initial and final states.

or pressure–particle velocity space the Hugoniot is concave-up, which implies that the Rayleigh line, whose slope is proportional to U_s, always increases with P. The loading path follows the Rayleigh line, reaching the final state in a single discontinuous jump.

The Hugoniots, presented in this fashion, provide a useful means of quantifying energy balance through shock processes. From (2.23) and (2.24),

$$u_p^2 = p(v_0 - v),$$

so that the specific kinetic energy (KE) is

$$\mathrm{KE} = \frac{1}{2}p(v_0 - v).$$

Now from (2.24),

$$e - e_0 = \frac{1}{2}p(v_0 - v),$$

so that the specific internal energy (IE) is

$$\mathrm{IE} = \frac{1}{2}\rho(v_0 - v),$$

Therefore,

$$\text{gain in IE} = \text{gain in KE.} \tag{2.30}$$

The terms may be visualised by plotting the Hugoniot, $\mathscr{H}$, and the Rayleigh line, $\mathscr{R}$, and noting that the energy is represented by the shaded area beneath the Hugoniot curve between v and v_0. The triangular region above p_0 corresponds to the change in specific kinetic energy, whilst the whole shaded area corresponds to that in specific internal energy (Figure 2.8).

More practically, problems in shock are frequently best understood using a frame of reference in which states are represented in terms of pressure–particle velocity rather than pressure–volume. This is simply because surfaces connected to one another need to

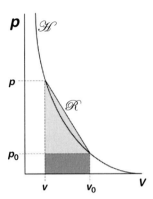

Figure 2.8 Energy balance in the shock process. The shaded area beneath the Rayleigh line R between v and v_0 corresponds to the change in internal energy during the shock process.

be travelling at the same speed and to be at the same pressure or stress. If an interface does not obey these constraints then waves must travel from it until either these conditions are achieved or the materials at the interface separate. Combination of the conservation of momentum (2.23) with the expression for the shock velocity U_s (2.26) gives

$$p = \rho_0(c_0 + Su_p)u_p, \tag{2.31}$$

which is an expression for the Hugoniot in the P–u_p plane. This is sometimes called the shock equation of state, but here will be referred to as a hydrodynamic curve since it represents the locus of shock states (achievable under adiabatic conditions), neglecting any strength the material might have.

The volumetric engineering strain induced in a solid by the shock process is simply given by the relative change in the volume of a compressed element $\Delta V / V$. The specific volume of an element is the volume per unit mass since the volume, v, is simply the reciprocal of the density, ρ. In this notation the engineering strain, ε, is given as follows:

$$\varepsilon = \frac{v_1 - v_0}{v_0} = \frac{\rho_0}{\rho_1} - 1, \tag{2.32}$$

which, using the conservation of mass (2.22), becomes

$$\rho_0 U_s = \rho_1(u_p - U_s) \Rightarrow \varepsilon = \frac{u_p}{U_s}. \tag{2.33}$$

Equation (2.26) (which defines the key experimental relation $U_s = c_0 + Su_p$) can now be rearranged to give an expression for the shock velocity in terms of strain,

$$U_s = \frac{c_0}{1 - S\varepsilon},$$

which, put back into the expression for the conservation of momentum (2.23), gives

$$p = \rho_0 U_s u_p = \rho_0 U_s^2 \varepsilon,$$

or

$$p = \frac{\rho_0 c_0^2 \varepsilon}{(1 - S\varepsilon)^2}. \tag{2.34}$$

The above relation shows the dependence of pressure with volumetric strain along the Hugoniot for the material. Since the Hugoniot and isentrope are close in the weak shock regime (see later), the pressure dependence of a quantity such as the bulk modulus can be estimated using (2.33) and the expression for the bulk sound speed in (2.14) thus

$$\frac{K_1}{K_0} = \frac{(1 - \varepsilon)(1 + S\varepsilon)}{(1 - S\varepsilon)^3}, \text{ where } K_0 = \rho_0 c_0^2. \tag{2.35}$$

Other moduli scale in this manner in this range. These relations connect the framework of solid mechanics to the specific loadings encountered in shock, and further linkages are explored below. Yet the derivations gloss over the fact that material has a finite strength and it is this which must be tackled to use these relations for the materials seen in most applications, since it will be apparent that strength has an effect even in the most extreme of loadings when the amplitude of the input pulse is many times that of the yield stress of the material. In the discussion above, the constants c_0 and S in Eq. (2.26) have been related to the bulk modulus and its pressure derivative, respectively. These relations can be combined to show variation in these quantities subject to the *caveats* relating to the assumptions made in deriving them.

2.3.3   Temperature

The discussion above has identified a range of pressure and volume relations but to complete a full thermodynamic description requires temperature. Conservation requires the total energy of the system to equate to the work done by the forces. The energy will be divided into two parts, internal and kinetic, and it is the internal energy that will manifest itself in temperature rise through the specific heat. Rapid compression of a material causes an increase in internal energy taken from that depositing it. In reality some of this will be converted to heat whilst some will be associated with microstructural rearrangements of atoms within molecules or crystals, and the energy of chemical bonds broken or formed. It may be assumed for simplicity that all of the internal energy change is associated with changes in the thermal state of materials and this will be followed below.

It is a simple matter to track the work done per unit time over the area A considered. If the matter is initially moving at speed u_0 then account must be taken of this term but the main component will amount to the force behind the shock at the characteristic particle velocity u_p. Thus

$$\text{work done} = p_1 A u_p - p_0 A u_0. \tag{2.36}$$

In what follows all relations will assume that relations are derived per unit area and unit time and this won't be explicitly stated for the sake of clarity. The material is initially at rest at low ambient pressure. As the shock travels forward at speed U_s it will sweep a mass of $\rho_1 A(U_s - u_p)$, which amounts to a change in kinetic energy of

$$\frac{1}{2}\rho_1(U_s - u_p)u_p^2. \tag{2.37}$$

This mass will be raised in temperature and assuming all internal energy is stored as heat and that the specific heat capacity under ambient conditions describes the state at pressure p_1, the change in internal energy (per unit area and time) may be written as

$$\rho_1(U_s - u_p)c_v T_1 - \rho_0 U_s c_v T_0, \tag{2.38}$$

where c_v is the specific heat capacity for the material (J kg^{-1} K^{-1}) and T_1 and T_0 are the temperatures ahead and behind the shock. Equating the work done to the kinetic and thermal energy gained

$$P_1 u_p = \rho_1(U_s - u_p)c_v T_1 - \rho_0 U_s c_v T_0 + \rho_1(U_s - u_p)u_p^2, \tag{2.39}$$

which simplifies using the conservation of mass and momentum relations derived earlier $(\rho_0 U_s = \rho_1(U_s - u_p); P_1 = \rho_0 U_s u_p)$ to

$$\rho_0 U_s u_p^2 = \rho_0 U_s c_v(T_1 - T_0) + \rho_0 U_s u_p^2$$

and leads to a temperature rise across the shock of

$$\rho_0 U_s u_p^2 = \rho_0 U_s c_v(T_1 - T_0) + \frac{1}{2}\rho_0 U_s u_p^2 \tag{2.40}$$

Clearly all these assumptions assume that bonding is such that the storage of heat in the solid at pressure is the same as that observed under ambient conditions. This may be the case near to the yield point (in the weak shock regime up to around ten times the elastic limit), but at greater compressions the bonding changes significantly and these assumptions will break down. The above also assumes all of the work done is shared by heat and kinetic energy. Of course in some circumstances this may not be the case; for example, changes of phase will change electronic bonding states as well as introducing the possibility of chemical reactions in energetics.

The previous sections have reviewed the analytical framework that exists to describe high-pressure states and processes. The jump conditions define a locus of these for a material, resulting in Hugoniot curves plotted in a space chosen to best match the conditions. In many cases this is with pressure generally along the ordinate and either volume or particle velocity along the abscissa. In situations where chemical reaction or phase transformation occurs and volume change is significant it is generally more useful to choose p–v space, whereas when calculations involving interactions of materials or waves are considered (such as impact or spallation, for example) then it more usual to choose p–u space.

2.4   The response of materials with strength

Once condensed matter possesses strength, compression takes on a more complex form and one must then correctly define a stress tensor for the loading that occurs. In the discussion above, the states across a shock should have been described as stresses not pressures if talking in the most general terms. In what follows the distinction will be made more explicit and the reader may substitute stress for pressure where necessary to understand the expressions derived. This is especially true at higher impact stresses when the strength is small in comparison with pressure. This is then the hydrodynamic approximation discussed earlier. A shock will be considered travelling down the x-axis whilst a coordinate system with two lateral components, y and z, will accompany it.

2.4.1   Elastic-perfectly plastic solid

The simplest assumption for the behaviour of a condensed-phase solid that has strength is that it loads elastically to a defined yield point and thereafter deforms perfectly plastically. Such a material obeys the Tresca or von Mises yield condition and relations for such behaviour have thus been presented below (2.45 and surrounding). Such behaviour is termed elastic-perfectly plastic or sometimes elasto-plastic. It is the former which will be used here. For completeness, the governing equations for this simple material will be summarised as a source on which to build more complicated behaviours that will follow. Real material deformation mechanisms do not in general allow such a simple description yet it serves to assemble and understand the following as a baseline on which to build.

The basic definitions of terms and behaviours must start with the variables: pressure, p, strain (engineering strain) $\varepsilon = u_p/U_s$, and principal stresses in a uniaxial strain state such that $\sigma_x \neq \sigma_y = \sigma_z$, $\varepsilon_x \neq \varepsilon_y = \varepsilon_z = 0$. The shear stress in an isotropic material is given by the difference between the longitudinal and lateral stress levels

$$\tau = \frac{1}{2}(\sigma_x - \sigma_y). \tag{2.41}$$

2.4.2   Elastic

In the elastic region, one-dimensional strain constrains the states and stresses are connected through Poisson's ratio, v. In this region

$$\sigma_x = \frac{v}{1 - v}\sigma_y. \tag{2.42}$$

2.4.3   Yield

At yield $2\tau = Y = (\sigma_x - \sigma_y)$, and the longitudinal stress takes the value known as the Hugoniot elastic limit, σ_{HEL}. This value is related to the yield stress Y in a one-dimensional state through Poisson's ratio, v, thus $\sigma_{HEL} = \dfrac{1 - v}{1 - 2v}Y$, which for metals (where $v = 1/3$) means $\sigma_{\mathrm{HEL}} = 2Y$, i.e. twice the strength simply due to inertial confinement. This state is the initial state for the plastic flow regime. In particular, the hydrodynamic and deviatoric stresses at the HEL are connected thus

$$\sigma_x = P_{\mathrm{HEL}} + \frac{2}{3}Y. \tag{2.43}$$

2.4.4 Plastic region

In the plastic region, the hydrodynamic pressure is defined to be

$$P = \frac{1}{3}(\sigma_x + \sigma_y + \sigma_x), \text{ which at the HEL} = \frac{1}{3}\left(\frac{1+v}{1-v}\right)\sigma_{\mathrm{HEL}}. \tag{2.44}$$

The longitudinal stress across the shock front is made up of the pressure component added to the deviatoric stress to deliver

$$\sigma_x = P + \frac{4}{3}\tau. \tag{2.45}$$

The pressure state and the longitudinal stress are thus related by $\sigma_x = P + \frac{2}{3}Y$. Under higher pressure states, the longitudinal and lateral stresses are also increased but the yield surface confines their difference. Thus the result of this is that the Hugoniot is displaced from the hydrodynamic curve by a constant offset for an elastic-perfectly plastic solid.

This simple model is the baseline description for material response in compression. All other behaviours build upon this, and thus it is important to understand its form. The individual mechanisms that allow dynamic inelastic flow are the key inputs that determine how such a description performs. If the kinetics of deformation are relatively slow with respect to the pulse length, then this description is insufficient; if plasticity mechanisms operate instantaneously then it does well. For instance, on one hand, pure FCC metals of low strength are represented well whilst complex alloys and most non-metals, on the other hand, are not. A major effect that requires further description is that of a change of phase in the material when the relation between pressure and volume changes markedly. Chemical reaction might be so described, and therefore reactive materials come into this class.

Figure 2.9 shows the Hugoniot for an elastic-perfectly plastic material. Such a solid has a stress–strain curve in a one-dimensional stress state that rises elastically to a limit and then deforms plastically on the yield surface. In the stress–particle velocity plane $(\sigma_x - u_p)$ the Hugoniot therefore shows a linear elastic path to the Hugoniot elastic limit (HEL) and then a non-linear rise at higher stresses. The hydrodynamic analogue of such a material (Eq. (2.31)) describes the Hugoniot of a material with the same density and shock constants but no strength. The states upon it are all at elevated temperature which can be estimated as shown above. The hydrostat for the material maps the hydrostatic part of the equation of state. It lies through the origin beneath the Hugoniot and beneath the hydrodynamic curve since the latter is at a higher (shock) temperature. In an elastic-perfectly plastic solid the Hugoniot is thus displaced upwards by the yield strength according to Eq. (2.45).

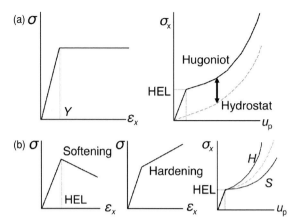

Figure 2.9 Hugoniot for an elastic-perfectly plastic material. In (a) the strength for an elastic-perfectly plastic solid material deforming when placed in a one-dimensional stress state is illustrated to the left and in shock under one-dimensional strain to the right. In general, material response may involve softening (S) or hardening (H) as illustrated in (b).

2.5   Impact states

Having derived the Hugoniot states from the conservation relations, a relation between the pressure behind the shock and the volume or particle velocity in the compressed material has been presented. Further, when a material exhibits strength the longitudinal stress in the shock propagation direction, σ_x, was shown to be related to the hydrodynamic pressure through $\sigma_x = P + \frac{4}{3}\tau$ (2.45), where P is the pressure and τ the shear strength of the solid. Together, these relations give a means of describing the experimentally derived Hugoniot curves for materials in shock experiments.

It will be necessary to define the shock states in a material subject to different initial conditions than those considered so far to explain the conditions induced by impact. The impact will be considered to be one-dimensional for the initial discussions to establish a simple case. To track these states it is more useful to consider events in the stress–particle velocity space since solids must achieve the same values of these quantities if they are to remain connected. To achieve appropriate conditions the Hugoniots may be translated according to the initial velocity or stress state. For instance, within a shock travelling in a direction opposite to the positive axis u_p, the states lie on a Hugoniot reflected through the initial state specified (Figure 2.10). Using these observations allows simple graphical solutions for one-dimensional impact problems to be attempted using experimentally determined Hugoniots. In the following section, four examples will be given. These include the impact of two bodies, reflection of a shock from a boundary, interaction of two shocks or two release waves and finally the ingress of a Taylor wave from an explosive detonating in contact with a metal.

To introduce these examples, consider the impact of a moving rigid object on a target material. Assume that a rigid piston, travelling at velocity v, impacts a stationary sample.

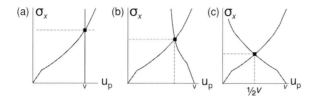

Figure 2.10 (a) Rigid piston moving at v impacts solid and induces stress where curves cross. (b) Impact of two dissimilar materials. (c) Symmetrical impact of two identical materials.

The condition that the piston is rigid in this sense means that it cannot compress and so moves at the same velocity regardless of the stress state induced. If some compression of the piston were to occur, then it would be decelerated but of course its rigidity precludes this. Thus at any shock stress, the piston will have particle velocity v, which gives it a particularly simple Hugoniot – a vertical line. In reality, high-impedance materials approach this behaviour. The shock state at impact is determined by plotting the Hugoniots of impactor and target in stress–particle velocity space. In this case the target is at rest so its Hugoniot passes through the origin. The rigid moving piston has a vertical Hugoniot through v so that where the two intersect is simply the stress corresponding to this particle velocity. This is to be expected since the rigid piston is driving the target at the velocity it travels and of course cannot be compressed. The Hugoniots and their intersections are illustrated in Figure 2.10(a).

   For a piston made from a real material, its Hugoniot must be reflected about the stress axis and have its ambient state fixed at the impact speed v on the u_p axis since a reflected Hugoniot represents a shock travelling in an opposite direction. The two cross again at the impact state achieved, and since the Hugoniot now has curvature, the intersection will be at a lower stress and reduced particle velocity than the state acquired by the rigid piston (Figure 2.10(b)). If the piston and target were to be made from the same material, then the symmetry of the Hugoniot curves would force their intersection to lie at precisely half the impact speed ($v/2$). It is sometimes useful to note that in a symmetrical case the impact is entirely rigid and thus the target is stationary in the centre of mass frame (i.e. a frame of reference moving at $v/2$). This state induces a flow where the forward and reverse conditions are the same; this is known as simple centred flow in the target (Figure 2.10(c)).

2.6   Distance–time diagrams

The challenge for all dynamic experiments is realising not just what states exist in a process but also determining when they occurred. Any means of ordering time is a vital window on processes that occur so fast that intuition is often misleading. One tool in aiding observation is given by a graphical representation of where waves (and by inference pressure fields) have reached spatially as a function of time. These plots are called distance–time, or X–t or x–t diagrams. The capital and lower case x in the titles refers to whether the plots represent a Lagrangian or Eulerian frame of reference,

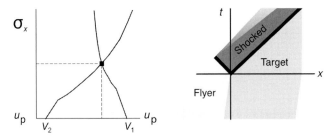

Figure 2.11 A flyer travelling at speed v_1 impacts a target travelling at a speed v_2; stress states to the left and x–t diagram to the right.

i.e. one tied to the material or one in the laboratory rest-frame. Figure 2.11 shows the state achieved when an impactor moving at speed v_1 impacts a target moving at speed v_2. Note that the target Hugoniot is simply shifted so that the zero stress point (ambient pressure) is 10^{-4} GPa, which will always be negligible compared with induced stresses. A plate (in 1D a line) moving to the right at speed v_1, impacts at $t = 0$ onto a more slowly moving target. A shock runs into the target, accelerating it to the induced particle velocity, whilst another shock travels back into the impactor plate, slowing it. The shock reflects from the free rear surface of the thinner flyer plate, changes phase and returns as a release reducing the pressure in both the target and flyer plates.

The representation considers a single spatial axis and shows waves as trajectories as time runs up the ordinate of the plot. Figure 2.12 shows to the left a Lagrangian and to the right an Eulerian representation of the symmetrical one-dimensional plate impact of a 5 mm copper flyer onto a 10 mm copper target at 500 m s^{-1}. This impact induces a stress of close to 10 GPa and as the impact is symmetrical the velocity of material behind the shock front is 250 m s^{-1}. The reader will observe that the Lagrangian plot is simple to construct and focuses upon wave transmission and subsequent interaction. Elastic waves (dashed) precede the plastic fronts. The Eulerian plot is more complex but shows the behaviour of the interfaces. The flyer comes in at 500 m s^{-1} and accelerates the target and decelerates itself immediately to half the impact speed as fronts travel back into the moving flyer decelerating it and forward into the target accelerating it. As the shock reaches the rear of the flyer it reflects and as the release propagates back it decelerates the flyer to rest. Its effect on the target is to accelerate its velocity further up to the full impact speed whilst of course returning the stress in the target back to its initial ambient value.

The shock in the target accelerates the material and compresses it up to the impact stress. As it reaches the rear surface it accelerates that to the impact velocity since it is unconfined and the release travelling back acts similarly, accelerating the material to the impact speed and returning the stress back to ambient. If the impactor and target had been the same dimensions then one can see from the symmetry of the geometry that the releases would have met at the impact face and the impactor would have stopped whilst the target would have accelerated off at the impact speed – the wave explanation

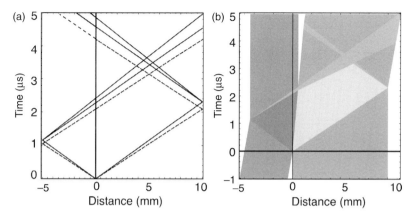

Figure 2.12 (a) Lagrangian (X–t) and (b) Eulerian (x–t) distance–time plots for the impact of a 5 mm copper plate hitting a 10 mm copper target at 500 m s^{-1}. Elastic fronts are dashed.

of Newton's cradle described earlier. In this case, however, the target is twice the size and the releases meet at its centre. A later section will address this and show the effect of such an interaction.

What should be obvious from this description is that different parts of the target material experience the shock stress for different lengths of time. The regions near the boundaries, although accelerated, are near free surfaces where the stress is always zero. On the impact face the material experiences the impact stress for a time determined by the transit time of the shock and release in the flyer plate. Figure 2.13 shows a schematic of the pressure histories at different locations in the target. The stress history on the impact face is viewed directly so that the rise and release of the pulse can be clearly seen. As one travels through the target the pulses remain of similar length; in fact they shorten as the head of the release catches the shock as will be seen below. Beyond the point where the releases interact, the material has seen a very different history. The pulse durations in this region rapidly decrease so that the impulse seen by the target reduces to zero at the rear surface. If the material deformation mechanism is time dependent, and the bulk hardens or softens with exposure to this impulse, then it will have been conditioned by the shock process to exhibit different strengths in different positions. Shock processing is already used to strengthen surfaces using these effects. For example, shot peening uses the impact of round metallic, glass or ceramic particles onto a surface to plastically deform and create a layer of compressive residual stress at the surface, and a tensile stress in the bulk on release. The surface layer in compression resists metal fatigue and corrosion and in some cases the lifetime of a component can be increased by as much as 15%.

When a shock interacts with a boundary, similar principles apply. Figure 2.14 shows an x–t diagram. A shock travelling into material A (which is at rest) encounters an interface with a second material B (also at rest) and the shock partially transmits and partially reflects from the boundary since B has a higher impedance than A. The interaction accelerates the interface (shown as a dotted line). Calculation of the stress–particle

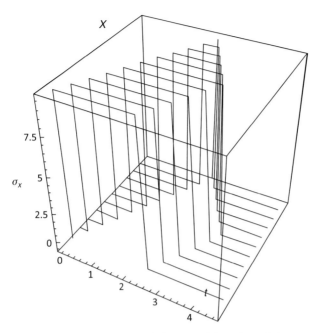

Figure 2.13 Stress histories at different positions through the target for the impact described in the text. (Stress in GPa, time in μs.)

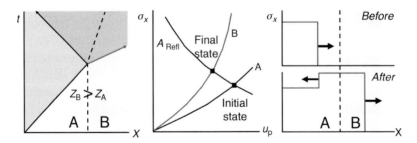

Figure 2.14 Distance–time (x–t) diagram for the interaction of a shock in material A with a higher impedance material B. Shock reflects back into A and transmits into B. States are indicated in the Hugoniot in the centre. Snapshots of stress histories before and after the interaction are shown to the right.

velocity states uses three Hugoniots, for the stationary materials A and B and one to describe the reflected wave from the interface. The initial shock has a state imposed in some previous event and this will be assumed to be at the initial state indicated. When it interacts with the higher impedance material, a shock must be reflected and transmitted so that the final state must lie on the reflected Hugoniot of material A, reflected through the initial state, and the Hugoniot of B, the higher impedance material. The reflected Hugoniot intersects at the final state as shown. To the right, the initial shock is shown entering from the left. After the reflection, the increased stress increment can be seen

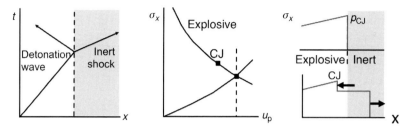

Figure 2.15 Distance–time (*x–t*) diagram for the interaction of a detonation wave in an explosive with an inert material. Shock reflects from the inert material and a wave transmits back into it. States are indicated in the Hugoniot in the centre and snapshots of stress distributions are shown to the right.

propagating forward and back into the media, alternating between the stress state and the particle velocity in the material.

A further example describes a means to calculate the states achieved when a detonation wave encounters an inert material. In many engineering applications explosives are used to drive metal – for instance, to produce jets used for cutting materials or structures. The discussion proceeds with reference to Figure 2.15. As will be seen, a detonating wave has a defined pressure and velocity due to the physical and chemical properties of the explosive. In particular, the wave has a stable pressure, the Chapman–Jouget (CJ) pressure, P_{CJ}, which such a wave will present at the inert boundary (see later discussion). The explosive products will have a Hugoniot (which will obviously contain the CJ point) that intersects the stationary inert adiabat at some pressure and particle velocity generally less than those of the explosive. That intersection defines the state of shock travelling forward into the inert target and back into the explosive products, so (as the schematic far right of the figure suggests) the wave coming back into the products relieves the pressure for some distance of travel back. The shock in the inert adiabat will eventually encounter a surface from which it will reflect, accelerating the material, and subsequent bounces of the wave between the inner pressurised wall and the free outer surface will continue to increase that velocity sequentially but with decreasing increment until a steady velocity is achieved. This velocity is the Gurney velocity (see Chapter 8), which serves as an engineering design equation for explosive acceleration of metals (in particular).

Having described how Hugoniots may be used graphically to solve one-dimensional impact problems, shock interactions with boundaries and detonation waves loading inert targets, the final examples cover how shocks interact with other shocks and, equally, how release waves behave when they meet. Release states are described by an isentrope discussed in more detail below.

Figure 2.16 shows the interaction of forward-moving shock 1 with a rearward-moving shock 2. The result is predicted using the methods used for impact above. The key observation is that after interaction the shocks reflect with their states mapped by the reflected Hugoniots plotted through the initial state on each curve (1 and 2). Thus the two reflected curves through the initial states intersect as shown at point 3, which

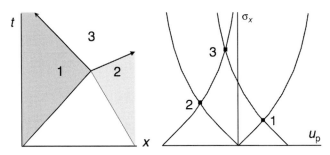

Figure 2.16 Interaction of two shock waves.

is the new state that results. If the shocks and the material were the same, then the interaction would be such that the particle velocities behind each shock would be zero, but the material at that point would be raised to a stress as expected.

Finally, imagine that two release fronts meet in the material. The schematic pulse of Figure 2.17(a) shows two such waves approaching each other. The ramps in each front may be represented as a series of shock wavelets, as shown in the sketch to the right. Calculating the effect of any interaction is now a matter of adopting the methods sketched out above. Each of the wavelets is represented in the distance–time plot of Figure 2.17(b) as a line of increasing slope (decreasing velocity). The initial state in the shock is determined by crossing the Hugoniots for impactor and target, giving the peak stress behind the shock, σ, which is shaded darkest in the x–t diagram. Subsequent states are determined by crossing the relevant isentropes across the Hugoniot at suitable intervals. For instance, the wavelet travelling into the shocked material from the left takes the state from the initial value to state 1, then to state 2, and finally to state 0, travelling down the isentrope for waves travelling to the right. Similarly, the wave travelling right to left moves to states on a backward-pointing curve through the initial state, but this time only travelling through state 3 to reach the unrelieved state $0'$.

A release from state 1 also occurs through interaction with a leftward travelling wave which takes the state down an isentrope through state 4 eventually to state 6. This final state is defined by the isentrope for the tail of the release propagating leftward that had initially relieved the shock to $0'$, and that isentrope takes the stress into tension. Subsequent steps take the material in a central region of the target down sequentially further into tension. The tails of the releases propagating to the left and right finally intersect to mark the maximum tension reached, which is $-\sigma$ as expected. All of the arguments above assume that the material can of course withstand such a tensile force, and the case where the bonds break and failure occurs is treated next.

In all of the examples here a series of techniques have been adopted using the intersections of the Hugoniot curves and isentropes for the materials to determine states. The slope of the Rayleigh line to the Hugoniot at any point is the dynamic impedance of the shock, and so these graphical solutions are known collectively as impedance-matching techniques to determine the shock state.

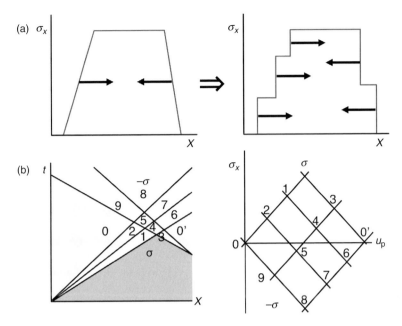

Figure 2.17 Interaction of two release fronts.

2.7   The release of dynamic compression

Clearly an impact puts the region traversed by the shock wave into compression and increases the target and impactor materials' density. Equally, processes may occur that release flyer and target back to the ambient state from whence they came. Rarefaction is a term used to describe the reduction of a medium's density (thus the opposite of compression) and since the process in this case is dynamic, the term rarefaction wave is used to describe the propagating region across which the process occurs. The mathematical description of rarefaction waves is often associated with gas dynamics and, for this reason, since this text concentrates on condensed-phase materials, the term release wave will be used to describe the process here.

Whereas the shock jump is sufficiently rapid that it is adiabatic (since there is no time for heat flow), expansion waves are not jumps but are continuous and the release process is additionally isentropic. Thus the waves disperse and the head of the wave will have velocity given by the bulk wave speed defined in Eq. (2.14). The material into which they travel is compressed, however, which means that the wave speed will be a function of the pressure at which it is held. An approach to represent this for the modulus has already been presented above (2.35). It is found that most materials have the property that the sound speed increases with density. On the one hand, this results in a ramp compression front driven into a material that shows an increasingly rapid rise time on the front until a shock is formed. If the front is decomposed (in a Huygen's manner) into a sum of wavelets, then since high-pressure components travel faster than lower-pressure ones, the peak of the wave will catch the base. This process is described as shocking up. Conversely, the same property results in an initially rapid jump in pressure (e.g. on the interface of a target as a shock front initially arrives) propagating back into the compressed material with the higher-pressure wavelets travelling faster than the lower. Thus the release front comprises a head, travelling at speed determined by the sound speed in the compressed material, and a slower tail travelling at that in the bulk. It thus disperses with time as it travels through the target. Of course the flow behind a shock has a particle velocity associated with it so that, in the laboratory frame, the speed of the head of the front would be the sound speed at pressure plus the particle velocity, whilst that of the tail would be the ambient sound speed. This difference in the speeds of the head and tail of the front results in it being sometimes termed a release fan by analogy with its appearance in a distance–time diagram. A further vital observation is that although the ambient sound speed is less than the shock velocity into uncompressed material, the head of a release, travelling at the sound speed at pressure plus the particle velocity behind the shock, will not be so. Thus release fronts will always catch a shock. This is perhaps an obvious observation since practical experience dictates that shock fronts have to be driven in some manner to remain propagating. Sonic booms occur only whilst the plane within the Mach cone travels supersonically; an explosion must create products to drive a blast front; and an impacting ball must remain in contact to accelerate a target that it hits. All of these are consequences of conserving momentum, and the release front is the means by which the end of the impulse is defined.

One exception to the dispersive nature of release occurs in the case of a material that has undergone a pressure-induced phase transition. Here the target has attained a higher density as a result of microstructural reorganisation. In this case, the head and tail of the release are travelling in materials of different density and a condition occurs where the head has a lower sound speed at higher pressure than the tail, resulting in a discontinuous jump as the material releases from the higher pressure phase to the lower one. This is a shock process occurring instantaneously and is termed a *rarefaction shock*.

Whereas the states acquired in a shock jump are found on the Hugoniot, those reached on release must follow a curve where the state is isentropic. The isentrope is thus the locus of release states and is followed back to ambient conditions after material has reached the shocked state.

The states represented by the isentrope thus differ from those plotted on the Hugoniot. At low pressures, where the shock temperature is low, the Hugoniot and isentrope are so close that they may be regarded as coincident. It is only at the highest pressures that the two diverge, and for engineering purposes (even with explosive loading) the two can be assumed to be the same in the weak shock region. However, it is stressed that even though these two curves may be close they describe different processes, since the Hugoniot is the locus of adiabatic end states which a solid adopts when rapidly compressed, whilst the isentrope is a thermodynamic path taken by the material as it unloads. That the curves are coincident at low pressures is a matter of convenience only.

To illustrate this further, consider a simple equation of state for a solid that expresses internal energy, e, in terms of pressure and volume using the bulk modulus, K, and the Grüneisen gamma, Γ, to which the text will return later. The quantity Γ can be measured and K deduced from the measured bulk sound speed. However, these values are only used as input variables here to illustrate a principle, since the equation itself is not accurate, but merely illustrative. The equation of state is given by

$$e = \left[\frac{p+K}{\Gamma}\right] v,$$

and the jump conditions for energy conservation gives

$$e - e_0 = \frac{1}{2}p(v_0 - v).$$

Substituting gives the Hugoniot

$$p = \frac{2K(v_0 - v)}{2v - \Gamma(v_0 - v)}. \tag{2.46}$$

The same equation of state may be used to derive an expression for the isentrope since the second law gives $de = TdS - pdv$, where $dS = 0$ (since the process is isentropic). Thus

$$e = \left|\frac{p+K}{\Gamma}\right| v,$$

and now

$$de = -pdv = \frac{v}{\Gamma}dp + \frac{(p+K)}{\Gamma}dv,$$

or

$$\int_0^p \frac{dp}{p(1+\Gamma)+K} + \int_{v_0}^v \frac{dv}{v} = 0,$$

which implies

$$p = \frac{K}{(\Gamma+1)}\left[\left(\frac{v_0}{v}\right)^{\Gamma+1} - 1\right], \tag{2.47}$$

which is the analytical form for the isentrope for this EoS.

These two expressions for Hugoniot and isentrope are used to derive the two corresponding curves for copper and the pressure is plotted as a function of the compression ratio (V_0/V or ρ/ρ_0) in Figure 2.18. The Hugoniot is uppermost and the isentrope below, reflecting the temperature state achieved in the two processes. The solid points are Hugoniot data in the compendium by Marsh (1980) to show data over this pressure range. The two curves do diverge on this scale but these pressures should be put in perspective with the regimes where they will occur. For copper projectiles impacting copper targets the velocity required to reach the pressure under consideration here will be around 3500 m s^{-1} – hypervelocity conditions. The Chapman–Jouget (CJ pressure; see below) of a detonating high explosive is around 40 GPa, and that induced in an adjacent copper plate would be less. An insert in the figure gives an indication of how much divergence occurs between curves over the pressure range up to 100 GPa. The percentage by which the Hugoniot exceeds the isentrope is given as a function of the pressure: deviation by 4% at 40 GPa, by 1% at 20 GPa and by $\frac{1}{3}$% at 10 GPa. The latter corresponds to a more

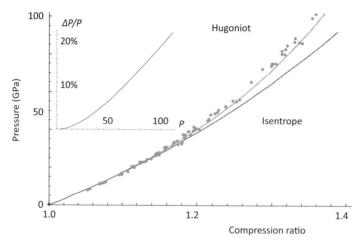

Figure 2.18 The isentrope and Hugoniot for copper. Source: data from Marsh (1980).

accessable impact state – one where a copper projectile hits a copper target at 500 m s^{-1}, a condition accessed frequently in the laboratory. Thus it is a good approximation to assume that the Hugoniot and the isentrope are closely positioned although describing different processes.

2.8   The compaction of porous materials

There is always interest in the development of materials capable of plastically deforming to absorb the effects of shock or impact. Most materials contain some degree of inhomogeneity and may be regarded as composites at the mesoscale. However, when the main bulk of the material contains a significant fraction of voids, with even larger variations in impedance, the nature of its response is markedly different. As the porosity of the bulk increases, the crushing of pore space becomes an ever-more efficient mechanism for absorbing shock energy and the dissipated work results in higher post-shock material temperatures than are observed in impacts into fully dense solids of the same substance. Thus one practical use of such behaviour is in the attenuation of shock waves in order to protect personnel or structures.

Solid foams form an important class of lightweight, cellular engineering materials to meet these threats. The foams themselves may be divided into two classes according to their morphology. The first group contains open-cell structured materials containing pores that are connected to one another forming an interconnected network (see Figure 2.19). The second class do not have interconnected pores and are termed closed-cell foams, and a special class of these, syntactic foams, contain hollow particles embedded in a matrix material.

When such foams are compressed, work is done bending and compressing the cell walls. As yet there is no detailed and universally applicable understanding of how the topological arrangement of the cell structure in the foam and the material behaviour of

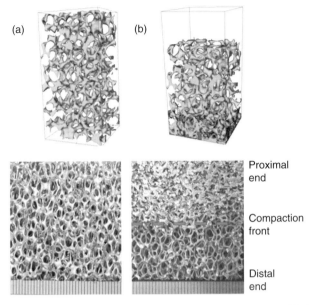

(a)   (b)

Proximal
end

Compaction
front

Distal
end

Figure 2.19 (a) Undeformed and (b) shock-compressed metallic foam. The upper two images show simulation of a microstructure taken from a tomographic scan of the material subsequently input as a mesh into an Eulerian hydrocode. The second image shows compacted foam as the cell is driven onto a rigid boundary from the top. The lower two images of the actual foam show the situation before impact and to the right, a partially compressed target. The scale is in mm.

the solid phase, relates to its strength properties. Similarly the macroscopic response at different loading rates can only be defined in general terms since porous solids cover such a wide range of geometries and microstructures. A simulation of the response of a metallic foam under compression is shown in Figure 2.20, which shows successive frames in the collapse of the aluminium foam shown earlier under impact from a piston accelerated from above. The structure crushes to high density against a rigid anvil where it locks and collapses to higher density before a high-pressure pulse returns through the compressed material. Examination of its response to successive increments of strain, provides a qualitative picture of the response of a porous material to rapid compaction. This can further aim to capture pertinent features of the mechanisms operating to put into a continuum description. Acceleration of a downstream surface within the material starts collapse once the surrounding structural strength has been exceeded. The structure may support considerable load and the relevant porosity will determine the threshold at which crushing begins. The collapse of the foam beyond this point is an irreversible process that takes it to another state with greater density, higher sound speeds and different strength. Thus the Hugoniot of the target develops with time (hardens) and is dependent upon the amplitude of the incoming impulse in the low-amplitude case before the structure collapses to full density. The threat becomes so severe in this limit that the target crushes until it becomes the fully dense matrix material as porosity is successively

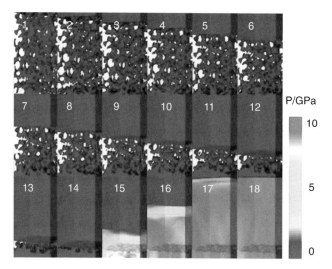

Figure 2.20 Three-dimensional mesoscale simulation of a collapse at 1013 m s^{-1}. Foam descends onto a rigid wall at the base at 1000 m s^{-1} and a compaction front is visible by frame 15. Frames in the simulation are 100 ns apart.

removed from the bulk, save that its temperature has been raised by the work done in compressing the pores during the collapse.

The effect of such compression is to attenuate the pulse entering the foam. The increase in density occurs over a greater time than the impulse rise since the collapse is structural with the failure of pillars at the microscale, which integrate over the volume of the foam to show a homogeneous ramp in pressure through the bulk as the matrix collapses. To treat such materials in a continuum manner suitable for using as a model within a code requires a simple description that captures the shift of the equation of state through PVT space. A widely used simple construct to do this, implemented in most hydrocodes, is Hermann's $P–\alpha$ (P for porous) model. It is both simple and computationally efficient and captures features that describe the crushing behaviour of voids in a porous material. The distension parameter, α, relates the macroscopic material density, ρ, to the density of the matrix, ρ_M:

$$\alpha = \frac{\rho}{\rho_M}. \tag{2.48}$$

There are a series of assumptions made to simplify the treatment. The most important of these is that voids do not reopen in the timescales of interest in stress-wave propagation. While there are strong experimental indications that this is true for ductile porous metals there may be conditions, particularly at low stress amplitudes, where this may not be the case. A further assumption is that the shear strength is negligible throughout the loading. There may also be a need to include shear strength when the matrix has significant strength. Finally, there is not an accurate treatment of temperature and the subsequent resulting phase changes such as melting and vaporisation at the highest stresses, and this limits the pressure range over which the model might be expected to

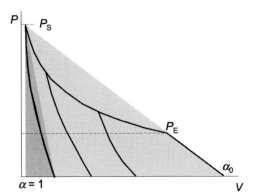

Figure 2.21 Shock response of a foam. The foam has an initial porosity α_0 and then crushes through an elastic region ($P < P_E$) where it can return to its initial state on unloading. Beyond this point it compresses until reaching the solid density of the matrix ($\alpha = 1$) at P_S. The shading represents the energy behind the shock in the compression of the foam (lighter) and the solid (darker grey).

be applicable. All that having been said, it has had great success in capturing a deal of the pertinent physics in porous crushing.

The EoS for the porous material is related to that of the matrix through

$$P(\rho, T, \alpha) = \frac{P(\rho_M, T)}{\alpha} \tag{2.49}$$

and the crushing through a relation for $\alpha = \alpha(P, E)$. Thus crushing behaviour is determined by the initial densities of the target, ρ_0, and the fully dense matrix material, ρ_S, respectively (see Figure 2.21). On loading the first motion results in elastic compression where crushing is reversible, followed by compaction where the deformation is irreversible. Thus, since compaction occurs only on compression, it is irreversible and there are several thresholds for different operating mechanisms in the loading that must be exceeded. There is a defined pressure for complete compaction to a solid, P_S, at which all voids are crushed out ($\alpha = 1$) where the material achieves solid density. Further, there is a lower limit for elastic crush within which recovery is possible. The P–α model does not consider deviatoric behaviour so that since shear strength is neglected, the only elastic sound speed admitted in the theory is the bulk sound speed. Approximate treatment of temperature rise suggests that a porous metal like aluminium reaches incipient melting after isentropic release from shock compression to a stress in the neighbourhood of only about 5 GPa. Thus a potential warning is that high internal energies and temperatures present in porous compaction must be adequately treated if accurate results are to be obtained from such calculations and the user must be careful of predictions obtained in this case.

To illustrate the application of the model, simulations are shown in Figure 2.22 which show the relative response of the aluminium alloy 6061 to a triangular shock and a porous material with 6061 matrix in which the P–α model is used to simulate behaviour. The shock in the matrix decays since release attenuates the impulse applied to the target. The

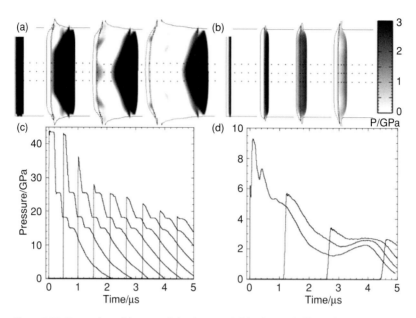

Figure 2.22 Examples of the use of the P–α model in the modelling of shock attenuation in foam. A composite flyer (designed to induce a triangular pulse) is impacted onto (a) Al6061 and (b) porous Al6061 of density around a third of fully dense aluminium ($\alpha = 0.37$) at 2500 m s^{-1}. Above snapshots of pressure field at 1.5 µs interframe times. Note the pressure scale in (b) is one quarter that in (a). Parts (c) and (d) show pressure histories for the stations indicated as dots in the simulations of (a) and (b).

figure shows snapshots taken of the pressure field at regular intervals as well as a series of measurement stations which track the pressure behind the shock. The simulations last for 5 µs in each case and have the same flyer and impact conditions. The porous aluminium absorbs the impulse and traps the momentum. As can be seen from the two examples, the wave propagates through the solid material at a rapid rate with the pulse slowing only because of the pressure decrease due to the release. On the other hand, the porous solid localises deformation at the impact face and very rapidly attenuates pressure through the target. Note that the scales for the pressure impulses shown differ, with that in Figure 2.22(a) four times that in Figure 2.22(b). A foam is thus an efficient and low-weight addition to structures that can result in significant attenuation and protection of personnel, vehicles or buildings from impulse such as blast so long as the impulse delivered does not reach the *interaction* phase where the void collapses and the material responds at full density. The modern world can use such effects to design safety into products as a matter of course and use of simple models in computer codes is a step down this road.

To summarise, the P–α model is both simple and robust, including as it does the most basic features needed to simulate the behaviour of porous materials in hydrocodes. However, the model is top level and attempts to cover a wide range of material types and morphologies without discriminating details of the responses of different material classes. Thus in its simplest form it cannot match material behaviour in detail, but then it does not attempt to. Further, it only addresses hydrostatic response of the material even though the degree of porosity clearly affects material strength parameters, which are important in some low-porosity materials. Finally, there is limited temperature dependence in the model, which must be addressed once phase transformation has occurred in the material. Clearly there is a large difference between the work done compressing a foam and that of an equivalent compression of a target made from the fully dense matrix material. This is illustrated in the schematic Hugoniot of Figure 2.21. The lighter shaded region represents the work done compressing to the pressure P_S, which represents the point at which the distension $\alpha = 1$. Approximating the isentrope of the solid phase to the Hugoniot at low pressures, the darker shaded region represents the work done compressing the solid phase. Clearly the areas are very different and this is reflected in the large temperature difference between the compression of a foam to that pressure compared with that of the matrix material.

2.9   The shape of Hugoniots and the wave structures that result

Consideration of the jump relations for a fully dense fluid with no strength has shown the locus of shock states (the Hugoniot) to have a parabolic form through the origin (Figure 2.7(b)). In shock experiments, one of these states is sampled, fixed by the impact speed or the explosive or plasma transmitting the impulse and the target material into which it runs. The experimental tools available today allow the measurement of a shock speed between two points or the use of a sensor embedded in the flow to track the particle velocity or stress history at a point. The form of such a history reflects the loading path that a material takes to reach the final state and the simplest form is that for a material that has no strength and flows as a fluid since these are the assumptions used to derive the jump conditions.

A simple general recipe to illustrate shock states using the Hugoniot to deduce a wave structure employs a graphical method using the flavour of the curve in a particular space to plot the end states in terms of the required variables. Assume that the object is to deduce a pressure history. The Hugoniot is a parabola and in the pressure–particle velocity plane the initial condition for a material at rest is at the origin. To reach a final shock state, one plots the initial point onto the curve and joins the two with a Rayleigh line as seen above. For a shock to be stable, one can show that, at the initial state, the Rayleigh line must be steeper than Hugoniot. Similarly, it can be shown that the Hugoniot must be steeper than the Rayleigh line at the final state. These requirements fix it to lie above and to the right of the Hugoniot in pressure–volume space, and above and to the left of it in pressure–particle velocity coordinates. This is the picture shown in Figure 2.23(a). For the hydrodynamic fluid, the shock is a step jump in pressure to the final state as a function of time or distance travelled through the target. In liquids, or materials at many times their elastic limit, this behaviour approximates what is observed. In reality, the shock cannot rise to the final state discontinuously. The rise is always finite as a result of the time required to overcome intermolecular forces. This finite rise is frequently described as viscosity continuing the hydrodynamic theme.

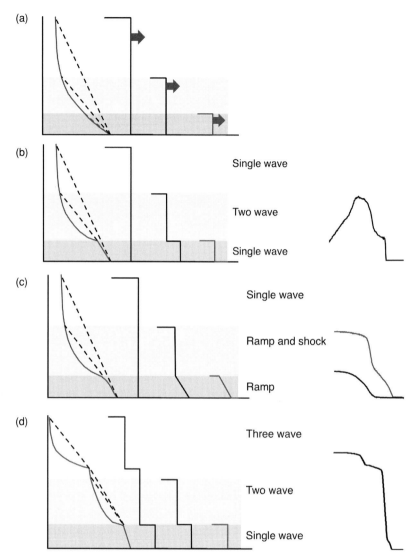

Figure 2.23 The evolution of Lagrangian stress profiles in solids: (a) hydrodynamic loading; (b) elasto-plastic behaviour; (c) porous behaviour; (d) phase transformation.

Although such a result describes the form of a shock seen in a fluid, solids have microstructure that must be overcome before the hydrodynamic state can be reached. The Hugoniot now has an elastic and a plastic branch (as described in Figure 2.9). The form of the curve is different since there are two loading regimes to consider. Both have different regimes of behaviour, which manifest themselves as alternate branches of the Hugoniot connected by a discontinuity at the Hugoniot elastic limit. To view the form of the shock, imagine shocking to three levels as shown in Figure 2.23(b). Although in the hydrodynamic case the shock merely steps up to higher and higher values of stress, there are three distinct regimes of behaviour for the elastic-perfectly plastic case. Within the elastic range, the stress can jump to values up to the elastic limit. Yet above the HEL, a new situation arises since this point is a cusp on the Hugoniot. There are two cases to consider. At the highest velocities, the final state will lie at such a stress that the Rayleigh line will bypass the cusp at the HEL and a single wave will result. However, at lower final stresses there will come a transition state where the Rayleigh line must split. Recall that it must lie (in the pressure–volume sketches of the figure) above and to the right of the Hugoniot. Thus a line can rise up to the cusp with one slope, and connect to the final state from there with a second line with a lower one. The slopes of the Rayleigh line are proportional to wave speed; thus the two lines imply that the shock has split into two waves. The lower stresses up to the elastic limit travel with the elastic wave speed whilst a second, slower shock travels behind at the shock speed. This structure is characteristic of observed profiles and is known as a two-wave structure. The transition from two-wave to single-wave shock structure represents a shift in regimes of behaviour. The stress range where two-wave structure exists is conventionally called the *weak shock* regime; that where it is overdriven the *strong shock* regime. Typically the transition point is at around ten times the Hugoniot elastic limit of the material but this value varies from solid to solid. This value represents the point at which the defect density is saturated as discussed below and in what follows the transition stress will be dubbed the Weak Shock Limit (WSL).

To the far right of the figure is a real stress history from impact on a steel in the weak shock regime. The wave at the impact face is a step jump, but after some travel into the target the wave has split showing elastic and plastic waves travelling at different speeds to the right. The wave also shows some structure after the elastic jump on the rise up to the peak stress. This results from the kinetics of this material's deformation mechanisms and reflects the limits of this construction which will only yield the position of the elastic wave and the head of the shock front.

In porous materials as described in the previous section, there is free space which can be removed at lower stresses before the matrix phase itself starts to compress. The Hugoniot for such materials frequently displays three sections; an elastic part, a convex region over the compaction stress range, and then the expected concave region in the hydrodynamic phase (Figure 2.23(c)). A convex region has interesting implications in trying to construct a Rayleigh line since a shock does not form in this space. The slopes of tangents to the curve describe the velocity of wavelets travelling at those stress levels. However, each of these is travelling slower as the stress level increases. The result of this is that a step at the impact face disperses to a ramp which broadens in a self-similar manner as the wave travels through the target. However, there comes a point at which there is a point of inflection on the Hugoniot and it curves upward in a concave manner once more. A tangent from the convex region can be drawn which intersects the Hugoniot at some final state and so a Rayleigh line can then be defined with an associated shock velocity. Thus in this intermediate regime, the shock will be composed of a ramp followed by a shock at higher stresses. The final stress level reached will determine where the Rayleigh line intersects the Hugoniot. Thus the stress break at the end of the ramp is not some material threshold but rather merely defines the jump off point from the convex Hugoniot. As ever, the highest shock state has the simplest wave profile. In the overdriven case, the convex region can simply be surpassed and a single hydrodynamic shock will exist to the highest pressure state indicative of the magnitude of the pulse and the ease with which it overcomes the material's bonding. To the right of the figure are two wave profiles from an open-structured borosilicate glass. The lowest profile shows the ramping behaviour up to a stress value indicative of collapse of the amorphous silica network. The second shows a ramping rise at lower stress levels, followed by a more rapid shock to the peak stress. These two histories illustrate the two lower regions of the behaviour described above.

The existence of regions in pressure–volume space where microstructural transformation leads to changes in moduli will always result in an altered form to the Hugoniot. At thresholds where such changes occur there will be cusps and shock wave splitting will result. One such change is a phase transformation and this results in a Hugoniot with cusps as shown in Figure 2.23(d). The schematic in the figure shows two thresholds for material behaviour. The first is the HEL showing the elastic-plastic transition, the second opens new hydrodynamic behaviour since the material has transformed phase. Such transformations may be thermodynamic (solid, liquid, vapour, gas) or displacive (often called martensitic by analogy with behaviour in steel in the shock literature). In a martensitic transformation, crystallographic change occurs when the unit cell transforms to a closer packed structure across the whole material at great speed. Clearly, an individual material compressed under shock will potentially open a number of these thermodynamic pathways and a Hugoniot covering a large range of states will have several cusps corresponding to these thresholds. In the example given here, only one is considered and the example chosen will be such a martensitic transition. In the lower region, the presence of a HEL and a plastic response in the deformed first phase leads to the two-wave structure noted in the initial example for elastic-plastic behaviour. The second cusp opens up the possibility of a further wave separating as it travels. The elastic and the shock velocities for the stress at the transition point will be two defined parameters for the material under investigation. These two velocities will describe the velocities of the first and second waves. The speed of the shock in the transformed region will be determined by the Rayleigh line from the second cusp to the final stress reached. Thus three waves can be separated one from the other to allow values for the dynamic transition stress to be read off for comparison with static high-pressure work (using the diamond anvil cell, for example) or quasi-isentropic loading using Z pinch platforms. The final solution will occur for the hydrodynamic case where the stress is so high that a Rayleigh line from the initial state can bypass the two cusps in the Hugoniot. To the right of the figure is a profile recorded by an embedded sensor in shocked, polycrystalline KCl showing an elastic limit and a transition between the B1 (NaCl) and the B2 (CsCl) structure which occurs at a modest stress threshold.

Thus the physical states a material assumes under load are intimately linked to the magnitude and transport of fronts equilibrating stress through it. A simple step measured on an impact face is transmuted into a complex series of steps and curves mirroring the evolution of features at greater depths within the microstructure. Although X-ray diffraction techniques are developing to track changes in the unit cell in real time, the behaviour of a bulk material is governed by the kinetics of assemblies of crystals with the

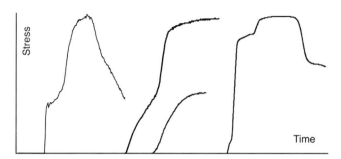

Figure 2.24 Lagrangian waveprofiles for (from left to right) a steel, borosilicate glass, polycrystalline potassium chloride.

defects that all matter possesses. At present, the measurement of wave profiles provides the most reliable window on bulk processes in a deforming material. The histories to the right of the figure above were recorded with stationary sensors embedded in the target. As the wave arrived they moved and travelling with the flow tracked the stresses rising and then falling. The sensors recorded stress as a function of time and the histories recorded are shown in Figure 2.24. They contain information on the physics of the dynamic deformation of the materials under load and will be returned to later in the text. Yet in themselves, the histories displayed come to hold a familiarity and a beauty that transcends the physics of their origins. It is in this manner that they hold their fascination for workers in the field.

Wave profiles, like the setting sun or the splash of a rain drop, are an interesting phenomenon in the natural world. They are rich in features which stimulate the curiosity toward the underlying responsible physics. Thus, like the sun and the rain, wave profiles in condensed matter deserve, in and of themselves, a satisfactory scientific explanation. It is for this reason that the shock wave profile has come to personify the field of shock wave physics. (Grady (2007))

2.10 Phase transformations

In previous sections the interplay of dynamic hydrodynamic compression with additional shear imposed by the lattice has been seen operating and has been classified for simple condensed materials. The individual responses of metals and non-metals will be covered in later chapters. At the macroscale uniform hydrodynamic compression at the continuum obeys the conservation laws of classical physics, yet at the atomic scale the motions of atomic planes and local defects are governed by multidimensional loading states. There are means of accommodating compression in amorphous materials as pressure is increased, principally by imposing an ordered microstructure on the material as will be seen in subsequent chapters. One means to accommodate plasticity in polycrystalline solids by atomic rearrangement is twinning that is seen widely in materials where atomic bond strength makes slip difficult. Another related phenomenon involves a solid–solid (Martensitic) phase transition where stacking changes to a closer-packed form by an atomic shuffle. Thus both twinning and solid–solid phase transitions are closely related, as the following examples will illustrate, and these transitions are classified as displacive or diffusionless and characterised by a collective movement of atoms across distances that are typically smaller than one nearest-neighbour spacing.

Elevated temperatures and pressures found in dynamic compressional loading are capable of inducing a material to change its thermodynamic state abruptly at a threshold and adopt a new phase. The development of physics from the nineteenth century provided the thermodynamic basis to describe the energy changes in transformations from one material state to another and the work that might be done on the system surrounding. However, phase transformation within a material refers to more than just transitions from solid to liquid to gas or to plasma. There are other available pathways that a shock can initiate; in particular the application of high pressure or temperature, and the presence of shear behind the front, can lead to a transition between different molecular packings. It was the German metallurgist Martens in the late nineteenth century who first explained how this displacive transformation might operate. His study of structural phase transitions in carbon steels identified changes in the crystallographic structure such as that between ferrite and austenite phases in iron. Since then, this explanation of transformations in steels, but also in other materials, has become known as martensitic transformation. These transitions occur since the structure may reorder and adopt a thermodynamically lower energy configuration under the influence of either temperature or pressure. Of course, the energy barrier between the two phases will determine whether the transition is induced by thermal effects (homogeneous nucleation) or from pre-existing defects (heterogeneous nucleation). An important recent application is the class of shape memory alloys that have surprising mechanical properties which can be explained by an analogy to martensite, since in the nickel–titanium system there are alloys in which the martensitic phase is thermodynamically stable. Such transformations may be between two ordered lattices or between crystalline and amorphous structures.

The shock process is sufficiently fast that it may be possible to achieve a transformation that has not the time to complete by slow diffusional processes. At high pressures, with interatomic distances reduced, diffusion through structures is not favoured. Thus shock favours displacive transformations such as those described above, but melting, solidification, vaporisation or condensation, or order–disorder transitions, have all been observed in shocked processes. In all cases, the change is governed by the thermodynamics and kinetics of the loading pathway accessed.

The importance of shock loading in identifying phase change in metals was first highlighted by the discovery by Minshall in 1954 of the shock-induced phase change in iron at 13 GPa (Minshall, 1955). It was initially assumed that it had reached the FCC phase but subsequently this martensitic transformation from BCC to the HCP phase was confirmed statically; it has been the most studied of all shock-induced transitions. Since then there have been a series of other notable highlights including carbon to diamond by De Carli in 1961 (De Carli and Jamieson, 1961), the extensive work on a range of geological materials by Ahrens, Grady and coworkers, and the transformation to metallic hydrogen in 1996 by a Livermore team (Weir *et al.*, 1996).

In what follows throughout this book it will not be possible to follow all of the investigations of a range of materials that have been conducted over the last 50 years. Rather the approach adopted earlier will be taken in choosing examples of different microstructure in order to show operating behaviours. The treatment in Chapter 5 will cover martensitic transformations, melting and solidification to define the bounds of the condensed state in metals. It will not extend to the work on shock-induced synthesis or to details of electromagnetic transitions except to note effects in two of the selected materials. There are many excellent texts on phase transitions and the reader is referred to those (see Duvall and Graham (1977) and the references therein). It will be important here to add specific notes to allow interpretation of the presented explanatory material so that details of the response can be interpreted. In this chapter, a basic physics framework will be sketched to set the stage for what follows.

2.10.1   Physical basis

In considering shock and release in a material that is transforming its structure, it will be necessary to derive conditions that constrain the shocked states achievable. This will involve derivation of the Hugoniot for the material, but this time encompassing elastic, plastic and now a transformed locus of states for the metal. Equally, on release, the challenge will be to have an equivalent appreciation of the isentrope in this space. In all of the transformations considered, it will be necessary to determine the thermodynamics of the system in order to predict the state it will adopt. In all cases the phase stability is going to be governed by the Gibbs free energy. The state that has the lowest value for this will have the most stable form.

Thermodynamic descriptions of transitions described them according to their energetic behaviours. This classification system (due to Ehrenfest) has been modernised over recent years since its basis does not hold in some cases. Under this scheme first-order phase transitions exhibited a discontinuity in the first derivative of the free energy with a thermodynamic variable, whilst second-order phase transitions are continuous in the first derivative but exhibit discontinuity in a second derivative of the free energy. Under the modern scheme, first-order phase transformations are defined to be those that involve a latent heat so that, during such a transition, a system either absorbs or releases a fixed amount of energy whilst its temperature stays constant. Thus the martensitic transition is first order, as are melting and solidification, whilst magnetic and order-disorder transitions are second order. Since energy cannot be transferred instantaneously, first-order transitions display regimes in which some parts of the system have completed the transformation whilst others have not, leaving the material in a mixed phase.

Classical thermodynamics describes the conversion of energy between thermal and mechanical modes and it is useful to sketch the continuum, state description of the process before embarking upon the microstructural features observed. States are described using macroscopic variables such as temperature, pressure and volume to define continuum conditions in dynamic loading. There are particular terms which will be usefully revisited for completeness here.

Internal energy (U)

The internal energy of a system is the sum of the potential energy and the kinetic energy. For many applications it is necessary to consider a small change in the internal energy, dU, of a system:

$$dU = dQ + dw = c_v dT - PdV, \qquad (2.50)$$

where $dQ =$ the heat supplied to a system, $dw =$ the work performed on the system, $c_v =$ heat capacity, $dT =$ change in temperature, $P =$ pressure and $dV =$ change in volume.

Enthalpy (H)

A potential accounting for both internal energy and PV work in fluids. The enthalpy is given by

$$H = U + PV,$$

and thus

$$dH = dU = PdV + VdP = TdS + PdV. \qquad (2.51)$$

Entropy (S)

Entropy is a measure of the disorder of a system. It reflects configurational disorder where different atoms occupy identical sites and the thermal vibrations of the atoms. A change in entropy is defined to be

$$dS \geq \frac{dq}{T}. \qquad (2.52)$$

Gibbs free energy (G)

The Gibbs free energy (G) can be used to define the equilibrium state of a system. It considers only the properties of the system and describes a measure of the energy which is available in the system to do useful work:

$$G = H - TS = U + PV - TS,$$

thus

$$dG = dH - TdS - SdT = VdP - SdT + (dQ - TdS). \qquad (2.53)$$

For changes occurring at constant pressure and temperature,

$$dG = dQ - TdS,$$

which requires $dG < 0$ for irreversible changes. Thus G is a minimum at equilibrium. This is particularly important in determining the most stable phase when a material may exist in several and that with the lowest free energy at a given temperature will be the most stable.

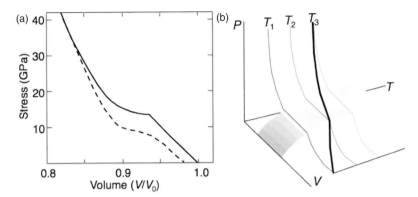

Figure 2.25 (a) Hugoniot and release isentrope for iron. (b) Form of the Hugoniot tracked by a family of isotherms. Source: after Duvall, G. E. and Graham, R. A. (1977) Phase transitions under shock wave loading, *Rev. Mod. Phys.*, **49**: 523–579. Copyright 1977 by The American Physical Society.

The *Helmholtz free energy*, F, is the equivalent of G for changes at constant volume and is defined as

$$F = U - TS. \tag{2.54}$$

2.10.2 Martensitic transitions

The experimental Hugoniot for iron can be seen in Figure 2.25(a). Its form can be understood by tracking a family of isotherms at increasing temperature shown in Figure 2.25(b). If the material compresses and moves to a state of lower entropy then at a fixed temperature the volume jumps under compression from its initial higher value to a lower one. This is shown as the dashed line in Figure 2.25(a). When a shocked state occurs, the temperature rises as well and this defines a locus for the final states. This may be understood by considering a family of isotherms (three are shown in Figure 2.25(b) to conceptually construct the Hugoniot. This must follow a path from the initial state to the final across a family of isotherms of increasing initial temperature. An analogous argument places the release isentrope for the shock process below the isotherm. These constructions explain the solid curves of Figure 2.25(a).

The form of the Hugoniot and the isentropes in Figure 2.25(a) determine the structure of the wave front that propagates when the material changes phase. As is the case for elastic–plastic transitions, propagating shocks are defined in the P–V plane by Rayleigh lines whose slope determines the shock's speeds. Further, stability of the shock fronts requires them to lie above the Hugoniot and release fronts to lie below the isentrope. The schematic Hugoniot of Figure 2.26(a) shows a schematic of the compression process in the P–V plane. The initial state of the material is at (P_0, V_0) and this is compressed in the normal manner to a point on the Hugoniot where the transition takes place (P_1, V_1). This is the same as that occurring in the normal shock process and the material shows an elastic and a plastic region in the same manner as that seen earlier. Behind the front the

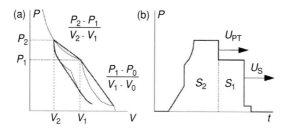

Figure 2.26 Schematic of the compression process in the P–V plane for a material with a phase transformation. (a) The Hugoniot and the Rayleigh lines in the P–V plane. (b) The wave profile and the pressure history. The dotted line represents the transformation front.

material is in a shocked plastic state S_1 and travels with a shock speed U_s which is related to the slope of the Rayleigh line connecting the two states. This is shown schematically in Figure 2.26(a), where slopes of the lines are indicated (and the elastic precursor state is assumed to be negligible for simplicity).

Beyond pressure P_1, the material is assumed to change phase and if shocked to the state at P_2, the compression front is composed of three travelling waves: the elastic, plastic and phase transformation waves, with each following after the other and the transition front separating plastically deformed material in state S_1 from transformed material in state S_2. If the amplitude of the shock introduced becomes greater, the phase transformation wave is driven faster and faster until the transition eventually becomes overdriven and a single wave results.

As seen earlier, the release path lies below the isentrope. Whereas Rayleigh lines represent single jumps that achieve the shocked state in one pass, release follows the isentrope with wavelets having speeds corresponding to tangents to that curve, each with lesser slope as the pressure drops back to ambient. With the isentrope for release in a transformed material, the situation is more complex. A section of such a curve is sketched in the figure. Initially the release follows the concave fan as happens under normal conditions. However, at some point the converging release returns to the cusp which marks the reverse transition and since the wavelets cannot assume multiple values, a component of the release must jump from the upper to the lower cusp of the Hugoniot down a shock trajectory shown. Note that there is hysteresis in a shock transition and that the state lies at a lower position than that at the transition. Once on the lower cusp, the release forms a divergent release of the form seen under normal circumstances. This jump then defines a shock in the release process which is termed a rarefaction shock that forms to bridge the region where wavelets speed becomes multivalued. The reverse transition stress itself is masked and cannot be directly read at a transition in a wave profile.

Thus the simple three-wave structure for a shock in a material with a displacive phase transformation appears schematically as shown in Figure 2.26(b). The release component of the pulse shows two dispersive parts separated by a rarefaction shock. Most of the work in the literature has mapped solid–solid martensitic transitions to try and understand complexities in the picture presented above. One of the key questions concerns the shear stresses and strains that are present in both phases. In particular, there is no clear view of the evolution of the defect population across the phase boundary, which would be key in constructing a physically based model for a martensitic transition. A further issue will include the micromechanics of the melting and solidification process. Clearly, information concerning the operating mechanisms by which the metal reaches the point at which it starts to lose strength, and the processes that occur over this mixed phase region, are key to understanding the transition. In this and in many other features of physical transformation, information is limited. Understanding such transitions is an important future goal for the subject.

2.11   Spallation

Spallation is the technical term used to describe the dynamic tensile failure of a solid. The quantity is of importance in a variety of instances where dynamic loading occurs. In nature, the failure and fragmentation of rock is key to understanding the effects of volcanic activity and bolide impact structures discussed in Chapter 1. In man-made devices, the shattering of brittle components such as car windscreens or protection against impact using armours of one type or another requires knowledge of fracture and fragmentation. Even within the body, the effect of blast on internal organs is of concern for the protection of personnel in hazardous environments. The nature of the observed response depends upon the material's strength and particularly upon the defects it contains. A key observation, dating back to the beginning of the last century, is that real materials can never attain the theoretical strength that one might calculate assuming only that one needs to overcome an intermolecular potential under ambient conditions. If one were to approximate the interatomic force with a simple sine function then the stress might be represented in the following manner:

$$\sigma = \sigma_T \sin \left[\frac{2\pi x}{a} \right] = \sigma_T \sin \left[2\pi \varepsilon \right], \tag{2.55}$$

where σ_T is the tensile strength and ε is the strain (displacement x divided by the interatomic spacing a). When ε is small, the sine term can be expanded in a Taylor series, and rearrangement gives

$$\sigma_T = \frac{E}{2\pi} \approx 0.16E, \tag{2.56}$$

where E is the Young's modulus. For a metal like pure copper, for example, the Young's modulus is $c.$ 130 GPa, which gives a value for the theoretical tensile strength of $c.$ 20 GPa. In dynamic experiments on the same metal (using plate impact) a value of $c.$ 1.5 GPa has been measured. This value is about 3% of the theoretical strength. For a brittle solid like a high-purity alumina the Young's modulus is $c.$ 400 GPa, which gives a value for the theoretical strength of $c.$ 64 GPa. In experiments on alumina using plate impact a value of $c.$ 0.4 GPa has been measured. This value is about 0.6% of the theoretical strength. It should be noted that the potential adopted is crude and that much better forms exist which will reduce the theoretical strength by up to a factor of two or so. Further, in some derivations the shear modulus, μ, is substituted for the bulk modulus, E. However, in all cases the measured value is still very much less than the theoretical one so that clearly other features determine tensile strength. In general terms, the theoretical strength of metals is around ten times greater, and in brittle materials one hundred times greater than the observed values.

The reason for the observed difference is that materials are not perfect at scales where failure processes occur. Thus the material is controlled by defects within crystals or bulk defects at grain boundaries, or within amorphous solids by inclusions or vacancies. At the mesoscale, where these can be viewed in the microstructure, the shock wave front is not steady and local inhomogeneous stress fields will exist in compression and in tension during the unloading process. It is the resulting growth of these defects through cavitation in ductile materials and fracture in brittle ones that gives rise to the creation of new free surfaces and spallation of the material. The onset of growing failure points from these defect sites occurs at a threshold stress at which damage sites can nucleate. Later stages of the process involve the spread of these by operating mechanisms with corresponding kinetics so that failure is not instantaneous. Thus more sophisticated modern treatments of the problem define three stages of spall failure. The first is nucleation of damage at defects within the material, the second growth of damage outward from this population of flaws, and the final stage the coalescence of these damaged zones to form a failed region within the material.

The construction of models to describe these processes is an active area of research. As yet, however, since operating mechanisms are only qualitatively known, spallation is poorly probed by the present range of experiments. This is because there is no means of constructing a history of any state variable in a spalling region using present techniques. Hydrocode simulation and measurement techniques assume a homogeneous continuum response across the spall region. The difficulty comes since they act to describe waves equilibrating the compressive stress state, which is normally the case, yet in tension the material localises deformation and materials in the region where electronic states remain inhomogeneous at the scales at which nucleation and growth of damage occur. A promising method under development tracks incipient spall sites using tomographic techniques (Figure 2.27(a)) but only after the event has occurred. This gives 3D density distributions showing failed zones that have occurred. Note that smaller impulses at the same stress level give very different damage fields in the steel (Figure 2.27(b) and (c)). It shows spatial information that sections lack (Figure 2.27(b) and (c)). One challenge, however, is that there is currently no time-resolved information from volumes failing that can be imaged non-invasively. Non-invasive imaging of the development of incipient damage to show development of the spallation region would be a tractable and important advance, as synopses of platform developments in later chapters will support.

The workhorse used at present to assess spall strength uses a remote, time integrated measurement of a wave emitted from the opening inner surface at an external one some distance away. To determine what is happening in the bulk, a simplistic analysis is adopted to back out conditions within from a surface measurement. The following section gives a synopsis of such a treatment.

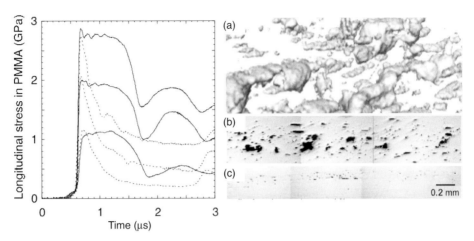

Figure 2.27 Ductile void growth in a stainless steel. Stress histories at three locations (left). Dotted lines show response from triangular pulses. Right: (a) X-ray tomographic image of ductile spall voids in a recovered metal target; (b) equivalent section of the same void distribution; (c) effects from a triangular pulse at the same stress magnitude on the void distribution.

2.11.1 Measuring spall strength in plate impact experiments

A standard means of determining dynamic tensile or spall strength to input into hydrocode failure models is to make a measurement in one dimension and extend it by assuming the failure surface is of simple von Mises type. One of the simplest geometries used to determine this quantity is the plate impact configuration, in which a flyer plate impacts in a planar manner onto a flat target. Such an experiment requires the geometry to be fixed so that release waves, originating at the rear of the flyer and at the back surface of the sample, interact within the target, producing dynamic tension over a thin zone. A release propagates from an interface at which there has been a drop in impedance, allowing the interface to move at speed greater than the particle velocity in the target. This impedance jump is greatest when the surface is free. Experiments designed to measure strength adopt free surfaces, or on occasion surfaces backed with windows. In what follows stress states will be derived for confined surfaces (the general state) and the reader can simply put the window impedance to zero to recover the free-surface state. An Eulerian representation of such behaviour is shown in Figure 2.28.

The flyer, F, enters from the left at a velocity of $2u$ and impacts, sending a shock wave into the target, T, and back into the flyer, reducing the particle velocity within the target to u but increasing the stress state to that of state 1. For illustration and simplicity the following will assume that the processes being tracked are all elastic, but the reader may simply extend these concepts to the general non-linear case. Assuming that the impedances of target and flyer are the same, and that both behave in an elastic manner, then the Hugoniots of both can be represented as straight lines in the P–u plane of slope Z_T. The state 1 can be found using impedance-matching techniques by reflecting the flyer Hugoniot through the point $(0, 2u)$ and determining its intersection with the

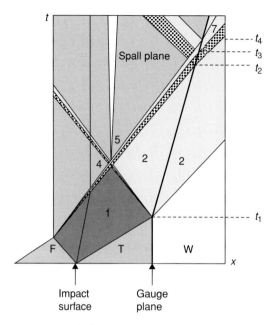

Figure 2.28 Eulerian representation of the impact of a flyer onto a target with a rear-surface window in which overlapping release waves lead to the formation of a spall plane.

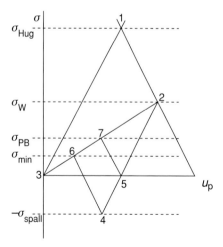

Figure 2.29 Hugoniot representation of states in the impact of flyer and target with subsequent release and spall plane formation.

target Hugoniot as shown in Figure 2.29. The shock wave travelling within the target next encounters a window material attached to its rear surface. Such a window is always required if a foil stress gauge is be inserted at this interface, since the gauge cannot function unless supported with a material of significant impedance. On the other hand, if surface velocity measurements are to be used as the principal diagnostic tool then the window may or may not be present, leading to variations in the form of the recovered

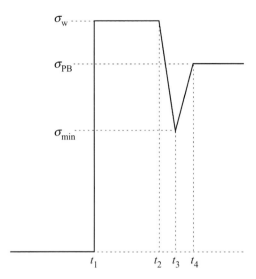

Figure 2.30 Schematic axial stress history at the target–window boundary.

signal. In the following discussion, to illustrate the general case, it will be assumed that there is a window material of low impedance in order to develop the derivation of the form of the signal. The extension to that in which the window is removed can be seen by geometry to result from the lowering of the window Hugoniot to the $\sigma = 0$ axis and is left as an exercise for the reader. Similarly, the longitudinal stress state in the window material will be followed here whilst leaving the related particle velocity signal as a further exercise.

As the wave encounters the low-impedance window interface it will be seen that the stress state at this position is reduced, propagating a release wave back into the target and a shock wave forward into the window. This puts material around the interface into state 2, which must thus lie on its Hugoniot and be connected to state 1 by the isentrope of the target material (the isentrope and Hugoniot follow the same locus when elastic behaviour is assumed in the weak shock regime). This state, whilst of lower stress than initially, accelerates target material to a higher particle velocity as a result of the low-impedance window to its rear. At time, t_1, the gauge first reacts to the loading pulse and rises to a stress σ_W, as seen in the schematic pulse shown in Figure 2.30. Although the states at the window interface have been considered here, the events occurring at the rear of the flyer plate (which has, in this example, a true free rear surface) must also be tracked. The shock wave decelerating the plate, and raising its stress to state 1, is released at the rear surface back to ambient stress and the plate material itself, which was travelling at a particle velocity of u, is decelerated fully to rest. This state 3 is represented by the origin in Figure 2.29 and corresponds to the construction of the target isentrope from state 1 back down to the $\sigma = 0$ axis.

The effect on the target of the interaction of the release waves from the flyer and the target rear surfaces is the key ingredient of the analysis. Each reduces the stress in its immediate vicinity back in the former case to ambient and in the latter to σ_W. Along the plane where the two waves meet the stress will be reduced from the original σ_{Hug} by $(\sigma_{\text{W}} + \sigma_{\text{Hug}})$, putting the target into net tension by a stress of magnitude σ_{W}.

If the stress exceeds the dynamic tensile stress of the material, σ_{Spall}, than it fractures or voids will open at the plane of interaction leading to the creation of a new free surface within the target. Once created, this free surface traps part of the release in the two sections of the tile created. For instance, a portion of the release from the rear surface can travel unhindered to the gauge plane since it lowers the stress to values supported by the material's strength. At a critical point in the unloading, state 4, the material opens up a new surface and the rest of the release is reflected back towards the flyer. Conversely, the release travelling from the window travels back towards the target until rupture occurs and then the remaining part of the wave reflects as a compression, state 5, from the new free surface travelling back to the gauge plane. It will thus be seen that the waves arriving at the gauge plane and giving rise to states 6 and 7 come from two separate interfaces within the material and convey first a relieving stress and then, as a result of reflection, a reloading. The release from σ_{W} to σ_{min} occurs from t_2 to t_3 whilst the reloading to σ_{PB} occurs from t_3 to t_4 as can be seen in Figure 2.30.

The relation of the magnitude of the drop in stress, $\sigma_{\text{W}} - \sigma_{\text{min}}$, and the reloading amplitude, $\sigma_{\text{PB}} - \sigma_{\text{min}}$, to the spall strength is interesting to pursue since it provides a route for the determination of the spall strength from the experiment. The construction can be easiest understood by reference to Figure 2.29. Assume that the bounding characteristic in the release that just allows the material to remain intact is given by state 4 and carries a tensile stress of σ_{spall}. This characteristic lies upon the isentrope for the target material and this state must have originated from state 2 at the window. This defines the stress and particle velocity at state 4. To determine the subsequent interaction with the window one must consider the reflection of the release as a compression and the transmission of a release here. This leads to state 6 lying on the isentrope through state 2 of the window and the Hugoniot for the target through state 4. This defines σ_{min} and its attendant particle velocity. Finally, consider the end of the release from the window which meets a new free surface within the target. At state 5 the release lies on the $\sigma = 0$ axis, representing the zero stress condition on the spall plane as well as, clearly, the isentrope through the window state 2. Partial transmission and reflection of the reflected compression at the interface leads to state 7 lying on the window Hugoniot at a stress of σ_{PB}. The signal now reverberates in the part of the target in which it is trapped, leading to further attenuation in amplitude and a train of similar pulses.

In order to derive the relation in stress amplitude Figure 2.29 is redrawn with more detail in the region bound by states 2, 3 and 4 (Figure 2.31). The figure has the original states but has dropped perpendiculars to the $\sigma = 0$ axis at B, F, G and I to facilitate the constructions. The slope of the lines DA, EH and DJ are the same and equal to the target impedance Z_{T}, whilst that of OJ is Z_{W}. In each case an expression can be derived for σ_{spall} from consideration of the length of the particle velocity segment CH, which is just twice the value $\sigma_{\text{spall}}/Z_{\text{T}}$. In one case this increment may be expressed in terms of the difference between states 2 and 6, BI, and in the second in terms of that between 6 and 7, BF. Considering the former first:

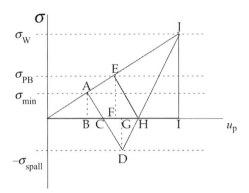

Figure 2.31 Detail of the P–u states of Figure 2.29.

$CH = BI - BC - HI$, which is to say

$$2\frac{\sigma_{spall}}{Z_T} = \frac{\sigma_W - \sigma_{min}}{Z_W} - \frac{\sigma_{min}}{Z_T} - \frac{\sigma_W}{Z_T}$$

$$\Rightarrow \sigma_{spall} = \frac{1}{2}\left[\left(\frac{Z_T}{Z_W} - 1\right)\sigma_W - \left(\frac{Z_T}{Z_W} + 1\right)\sigma_{min}\right]. \qquad (2.57)$$

Alternatively, one may express the equation in terms of the height of the reloading signal thus:

$CH = BG + GH - BC$, which is to say

$$2\frac{\sigma_{spall}}{Z_T} = \frac{\sigma_{PB} - \sigma_{min}}{Z_W} + \frac{\sigma_{PB}}{Z_T} - \frac{\sigma_{min}}{Z_T} \Rightarrow \sigma_{spall} = \left(\frac{Z_T}{Z_W} - 1\right)(\sigma_{PB} - \sigma_{min}). \qquad (2.58)$$

The form of Eq. (2.58) is particularly useful since the height of the pullback signal is scaled merely by the elastic transmission coefficient from the target into the window. The derivation above for the elastic case presented formulae for the spall strength which may be extended to hydrodynamic loading by replacement of the elastic with dynamic impedances.

In other experiments, a VISAR or other free surface system may be used to generate free-surface velocity measurements for a spall experiment. Note *free*-surface velocity (i.e. no window). Here the pressure jump is recovered from the measurement using the following:

$$\sigma_{spall} = \frac{1}{2}Z_T \Delta u = \frac{1}{2}\rho_0 c_0 \Delta u, \qquad (2.59)$$

where the tensile strain is given by $\varepsilon = \frac{1}{2c_0}\frac{\Delta u}{\Delta t}$. Here Δu is the jump in the particle velocity from the minimum to the peak of the reload pulse. Again a value can be deduced from the drop in the signal from the particle velocity at the Hugoniot state to that at the start of the pull-back signal and again related through suitable impedances. Here Δu will be the drop in velocity at the start of unload.

The last statements confirm that all of the above formulae contain impedances that the reader will have realised vary as one transits from elastic to plastic behaviour. Given that these are really indicators of values, it is advisable to use elastic impedance (elastic wave speeds) when the stress is modest and plastic impedance (bulk wave speeds) when in the strong shock regime. The various published data on spall measurements variously adopt elastic and plastic impedances (which are close in this region). However, the measurement itself is indirect and so claims for absolute values, particularly in the incipient spall regime, must be treated with caution.

Note that these formulae are the simplest that can be derived for use in these experiments. Romenchenko and Stepanov (1980) pointed out that failure at a release surface would give rise to a reload signal that was attenuating in its travel to the window surface. Thus more complex formulae are required if accuracy is important. A fuller treatment that gives all the corrections necessary to better interpret these signals is given in other texts; the book chapter by Grady and Kipp (1993) is particularly good. However, it is important to remember the initial caveats about the measurements made. This is a remote, integrated signal propagating from a plane deep within the target and is only an indication of the complex processes within. Microstructural development is a complication that must be accounted for in new models, as will be discussed in Chapter 5. Nonetheless the formulae presented serve to give a good indication of a quantity that a fully physically based and detailed failure model for a material must predict, and this should also indicate to the reader that new experiments and techniques must be developed further to fully understand the detailed processes that occur.

2.12   Burning and detonation

The advance of shock wave research in the latter part of the nineteenth century proceeded with developments and discussion with such key workers as Riemann and Rayleigh. However, it was Rankine (1870) and Hugoniot (1889) who placed a consistent framework around shocks in materials and it was this work on which others based a consistent theory of detonation, including Mallard and Le Chetalier (1881) and Berthelot and Vieille (1882). It was Schuster, stating that a 'detonation wave is a reactive shock', who laid the basis for the basis of the theory due to Chapman (1899) and Jouguet (1906, 1917; CJ theory). It was not until the Second World War that Zel'dovich (1940), von Neumann (1942) and Döring (1943) produced the building blocks that today complete the connection fully through to the unreacted state (ZND theory).

When materials experience any kind of dynamic loading, the temperature rises quickly since heat conduction is a diffusion process which will always be slower than the wave propagation that loads the material. This means that the process will be adiabatic and the release adiabatic and isentropic. In the case of shock loading, work done will be distributed between mechanical ($P\Delta V$) and thermal ($T\Delta S$) energies within the solid. If the thermal energy can induce a significant temperature rise, and that reaction proceeds exothermically, then the material may liberate sufficient energy to subsequently drive a shock. Heat may be created either at local hot points or within the bulk to start burning, and these grow and coalesce as a flame front travelling through the material. Burning kinetics can be described using the burning rate law of Vieille, which states that

$$r = \beta P^{\alpha}, \tag{2.60}$$

where β = burning rate coefficient; α = pressure index (typically 0.9 for gun propellants) and r = burning rate (typically a few mm s^{-1} in a direction perpendicular to the burning surface). Clearly the burning rate depends critically upon pressure. Thus once it starts in a confined energetic material the pressure and thus the rate of reaction will build and pressure will increase unless waves from boundaries release pressure before its completion. Thermal explosion occurs once energy liberation exceeds losses and a burning front then accelerates until it eventually forms a reactive shock; there is a series of processes by which this may happen discussed later in this book. The idealised state looked at here is one in which the material behaves in a manner where the flow is one-dimensional and there are no effects of boundaries. Observation has shown that detonation waves are steady and thus Chapman and Jouguet constructed a theory that required only the products' EoS and instantaneous energy release. This was later followed by ZND theory, which properly accounted for details of combustion in the reaction zone to allow greater insight into processes occurring. The procedure of conserving quantities across the front equates conditions in the manner adopted earlier for inert shocks.

To reflect the arguments of the sections above (particularly Eqs. (2.22)–(2.25)), and apply the same reasoning to reactive materials, requires the introduction of a means to track the degree of chemical reaction that has occurred. One simple way of doing this is to define a variable, λ, to track the reaction's course. This quantity is a scalar that takes the value 0 when the material is unreacted, progressing through to 1 when the reaction has completed and gaseous products have resulted. A further notation change replaces U_s with D to describe not merely a shock, but a reaction-driven detonation wave in the equations tracked above. Otherwise the three conservation relations and the completion of the set with an equation of state represent closure of the necessary conditions to describe detonation. Chapman and Jouguet considered only the fully reacted explosive state with $\lambda = 1$.

Assuming the same geometry as for the inert case but this time allowing the fluid to react (Figure 2.32), the derivation of the equations for conservation of mass and momentum (as in Eqs. (2.22) and (2.23)) can proceed in an identical manner for a reactive shock travelling at D as they did for an inert one travelling at U_s. The change in specific internal energy, however, must now refer to not only the work of compression, but also the energy liberated during reaction and so, to reflect this, the energy difference must equate that in the initial material with $\lambda = 0$ to one in a final state with $\lambda = 1$. This latter state is generally referred to as that of the products, considering the process as being the normal chemical progression from reactants to products. The two initial conditions represent conservation of mass

$$\rho_0 D = \rho_1(D - u_{\mathrm{p}}), \tag{2.61}$$

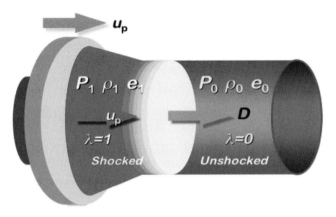

Figure 2.32 Propagation of a detonation wave down the cylinder of Figure 2.6.

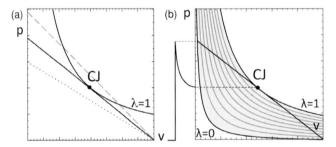

Figure 2.33 P–V plane representation of detonation. (a) CJ theory. The only state considered is the fully reacted state and the CJ detonation state is the only unique solution. (b) The ZND theory in which reaction proceeds from $\lambda = 0$ to 1 and details in the reaction zone are explicitly considered.

and conservation of momentum

$$P = \rho_0 D u_p, \tag{2.62}$$

which solved together yield a further representation with physical meaning. The equations can be combined to eliminate u_p to give

$$\rho_0^2 D^2 = \frac{p}{v_0 - v}. \tag{2.63}$$

Plotting this in pressure–volume space (Figure 2.33) leads to a line of slope $\rho_0^2 D^2$ which can be trivially reduced to the detonation velocity D. Thus again the slope of this, the Raleigh line, mirrors the velocity of the wave that results.

Conservation of energy takes the same form as for inerts, yet now the initial and final state differ in that chemical reaction has proceeded from one to the other. That is expressed in terms of the variable λ, the degree of reaction defined above. Thus the initial energy is at a state $\lambda = 0$ and the final specific energy, e_1, at $\lambda = 1$. The Hugoniot for the detonation products may then be expressed as

$$e_1^{\lambda=1} - e_0^{\lambda=0} = \frac{1}{2} p(v - v_0). \tag{2.64}$$

Completing the set with an equation of state as before allows solution for all relevant variables,

$$e = e(p, v, \lambda). \tag{2.65}$$

Thus the descriptions of a reacting material mirror those of an inert one with the Rayleigh line, and the plots of Hugoniots best observed in the p–v and p–u_p planes. Whereas it is generally best to track states in the inert case in the p–u_p plane, it is normally more useful to track reactive ones in the p–v.

In a Chapman–Jouguet (CJ) detonation, waves propagate at the detonation velocity, D, and instantaneous reactions occurring in the explosive are assumed to drive the front. Thus only the product equation of state is important and the Rayleigh line is a tangent to that curve at only one point (the CJ point). This defines the Chapman–Jouguet pressure, P_{CJ}, at the front and the detonation velocity, D. Simple manipulation shows that the flow is also sonic at the CJ point. Equation (2.62) can be simplified and an approximation allows estimation of the CJ pressure, P_{CJ}, and of the density at the CJ state, ρ_{CJ}, accurate to about 7% for ideal explosives thus

$$P_{CJ} = \frac{\rho_0 D^2}{4} \quad \text{and} \quad \rho_{CJ} = \frac{4}{3}\rho_0. \tag{2.66}$$

These constructions are best seen using the pressure–volume Hugoniots of Figure 2.33. To the left in Figure 2.33(a) are the key constructions describing CJ theory. The lower right point of the graph is the initial state of the unreacted explosive. Reaction in the HE on completion has a Hugoniot different and shifted to higher density than the unreacted one. A shock state must lie on this curve and a tangent to it has a single solution: the CJ state. This pressure and density define the mechanical conditions for the fully reacted explosive, whilst the velocity of the detonation wave taking it there is given by the slope of the Rayleigh line that connects from the initial state to that point.

One last detail is the single value for detonation velocity: why is there only a single permissible Rayleigh line? Two other potential solutions are illustrated. The light dotted line beneath does not intersect the product Hugoniot and so cannot represent a detonation: reaction does not complete. The heavier line above intersects at two points above and below the CJ point. These are sometimes called strong and weak detonations. Neither solution is stable and this relates to the release processes that would operate at them. Recall that the head of a release wave has a velocity which corresponds to a tangent to the Hugoniot. Thus the flow at the CJ point is sonic. At a point above the CJ point the release wave speed is faster than the detonation front (the slope of the Rayleigh line), whilst at the lower point it is slower. In the former case the detonation front would be overtaken by the release and in the latter the reaction zone would be spreading out making the wave unsteady. Both of these possibilities are unphysical leaving only one unique solution.

A modern hydrocode can solve for flow and use this theory for detonation needing to know only the detonation speed and the products equation of state. A thermochemical code (such as Cheetah from Lawrence Livermore National Laboratory) can provide these based on the components of the unreacted explosive, their density and chemical formulae. Since chemical potentials are known for the components of the explosive products, it is possible to minimise the Gibbs free energy of the mixture to get the products and the energy released, which defines D via the CJ condition. Then a simple criterion can be used in a code to describe the detonation. If the detonation point is specified then it spreads simply each timestep as the time multiplied by D. This scheme is called *programmed burn*.

Thus CJ theory contains the bare bones necessary to describe the major features of detonation and allows it to be simply incorporated into hydrocodes for prediction of performance. However, as it is a top-level approximation there are some facts about explosive performance that require modifications. For instance, as detonation passes down real cylinders with free surfaces, it is found that D increases with cylinder diameter until it eventually asymptotes at some maximum value. CJ theory works well for these large diameters since burning is complete by the CJ plane and there are no release effects from free surfaces. However, when burning is not complete by the CJ plane for thin cylinders or slow reaction rate then detonation is at lower pressure and velocity and the detonation is termed *non-ideal*. In these cases CJ theory has no explanation for the observations since the reason lies in the behaviour in the reaction zone where a more detailed treatment is necessary.

2.12.1   ZND theory

The CJ theory defines boundaries for behaviour in the explosive. For instance, the CJ plane is that which is sonic within the flow and this boundary condition remains fixed in the expansion behind this plane. The wave travels forward at a constant speed given by D and this value is steady. However, the assumption of an infinitely thin reaction zone now needs amplification so that details within it can be resolved. ZND theory explains the salient features that occur in this zone. The energy release from the detonation drives a shock into unreacted explosive and this shock ignites the explosive. This in turn releases energy and drives the reaction forward. To keep these processes coupled requires the velocity of detonation to be the CJ speed defined above.

This can be seen in the Hugoniots of Figure 2.33(b). The Rayleigh line and the fully reacted ($\lambda = 1$) product Hugoniot can be seen as before. But other levels of reaction are now shown, showing the transit of the explosive from its unreacted state. The unreacted state ($\lambda = 0$) is shown passing through the initial state of the material in the bottom right of the figure. It transits through a series of partially reacted states to the product Hugoniot. This process takes time as it is governed by the kinetics of the mechanisms acting within the reaction zone. An additional feature becomes apparent in this figure since the Rayleigh line must initially take the material to an unreacted state on the $\lambda = 0$ Hugoniot. This inert shock state is at higher pressure than the CJ one achieved after full reaction. Thus the detonation front transits in the manner described in the sketch to the right of the figure from a high point at the start of reaction to the sonic plane at the rear. This transient pressure peak is known as the Von Neumann spike, and the pressure itself relaxes over time as reaction proceeds to that at the CJ plane, which also defines a reaction zone length. This time of the order of 10 ns in a secondary explosive like HMX that shows ideal behaviour which corresponds to a zone width of the order of 100 μm. In a lower performance material like TATB the reaction time increases to around 500 ns and a reaction zone of several millimetres results. The ZND theory allows features of detonation to be described and provides some greater physical insight into new materials and how they might respond.

However, the theory is still not complete. Detonation fails along cylinders of small enough radius and further more detailed structure is observed within the fronts. There is evidence particularly from gaseous detonations of structure that corresponds to waves propagating laterally within the reaction zone. In pure materials this gives rise to pulsing effects along the front. Clearly the edges of the zone at an inert material or a free surface are regions where energy losses take place and this gives rise to, for instance, diameter effects in the explosive. Finally, reaction may not be complete by the sonic plane. Most explosives are in some respect non-ideal yet these principles serve as the building blocks on which to base understanding of energetics for the future. Further details of the effects observed in real explosives are discussed later.

2.12.2 Taylor wave

Finally, one must consider what happens behind a CJ detonation. The flow here is, as has been said above, sub-sonic so that the boundary conditions around the expanding gases have influence upon the flow that occurs. In the simple 1D case the CJ plane moves away from the piston and a release expands from the detonation point, which grows in spatial magnitude as the waves diverge. The flow is thus self-similar and a fan in an x–t diagram, as seen before.

Burn to violent reaction and acceleration to detonation is a fascinating area of research. The details of processes in the reaction zone and the chemistry that occurs there are an active research area that will be picked up again in Chapter 8.

2.13 Numerical modelling techniques

To map the consequences of dynamic loading in the physical world requires a simulation platform with two elements. The first is a scheme to simplify and describe structures and the forces that they experience and the second is a material description that determines the response that is tracked. These two components require a meshing scheme to describe geometry and models to represent the hydrodynamic and strength behaviours of the material as strain is applied. In all cases the descriptions derived assume that physical laws and operating deformation mechanisms describe a steady state in the volume considered. At the microscale, atomistic modelling calculates the forces experienced by individual atoms and then evaluates their response to them. The result of loading a body containing a population is then mapped in some manner to understand macroscopic behaviour. At the mesoscale, techniques are used to integrate atomic units to simulate flow. For instance, individual dislocations are tracked as a series of connected line segments moving in an elastic medium in the dislocation dynamics technique. Again the information gained is integrated into a volume element behaviour which is mapped up to the macroscale through continuum constants for comparison with observations. At the higher length scales, hydrodynamic wave codes and finally structural simulations ascribe continuum assumptions to materials behaviour. Wave codes consider steady stress, strain and temperature states which change across defined wave fronts and introduce material behaviour changes across different loading conditions using the continuum parameter strain rate as a barometer in the flow. Each of these classes of code, their applicability and some of their limitations will be described below. The details of the operation of such formulations are left to the reader; there is no attempt to cover the many numerical methods and material descriptions that have been derived or are in use in present codes. However, some notes on the numerical platforms will indicate where the formulations can be relied upon and how they relate to the simple theoretical framework presented earlier.

2.13.1   *Ab initio* models: kinetic Monte-Carlo and molecular dynamics

Ab initio as the name suggests uses only quantum mechanics as a root to derive behaviour for individual atoms interacting with one another. The forces on such atoms are then calculated using representations of the instantaneous electronic configurations that are instigated by the atomic configuration. Density functional theory (DFT) is a method used to approximate a spatially dependent electron density. It may then be used to describe interatomic interactions although it is still under development for some bonding types (for instance Van der Waals forces). Typically MD calculations use the embedded atom method (EAM). Here the energy between atoms is described by assuming the electrons to form a background charged 'jellium' with positive ions embedded within it.

Two flavours used at present use different means to assess the new state of the matter under load. In the case of the kinetic Monte Carlo method (KMC), random numbers are used to generate a series of possible states which can be assessed against criteria which fix whether the new configuration is adopted. Molecular dynamics (MD) assumes a potential field for an atomic configuration and applies Newton's laws in order to determine forces based on conservation of energy. In a further more accurate but computationally more expensive variant, quantum molecular dynamics (QMD) may be used to update positions of the nuclei. MD thus represents a robust numerical framework for solving equations of motion for a population of interacting particles in order to shed light upon physical processes occurring at the microscale. The operation of an MD code allows the determination of the forces on particles in the simulation from the potential field developed, as velocities and positions are updated on subsequent cycles of the calculation. Thus the relative motion of atoms, which changes their kinetic energy, subsequently changes the temperature of the ensemble of atoms. However, although ion motion contributes to temperature rise, electronic contributions are neglected so that temperature is not accurately recovered from simulations and errors accumulate over time.

A second issue with the technique is that defects cannot be adequately treated. This is tied to the need to better understand the physics operating at the atomic scale since treatments at present are inadequate and indeed measurements of quantities such as defect density are crude. However, the technique can be used to return macroscopic quantities to investigate phase boundaries, determine elastic constants and generate phonon spectra and these may be compared with measurement. At present the technique demonstrates qualitative representations of microscale phenomena in dynamic processes. Movies of lattice deformation can illustrate dislocation generation and interaction as well as atomic rearrangements in phase transformation. Such a window at present only exists through such simulations although new bright light sources are opening the way to techniques where real-time diffraction measurements will shortly deliver equivalent data. Limitations leave the technique primarily quantitative and operating only at the microscale since the defect boundary to the mesoscale cannot be adequately treated. Further, the potentials themselves are limited and applicable only to a subset of materials. However, this technique still remains the only means of visualising processes within atomic arrays and as such remains the basis for greater understanding of processes seen at the macroscale.

2.13.2   Dislocation dynamics

MD simulation allows qualitative understanding to be gleaned on the nucleation of dislocations and the onset of slip at defects within grains. However, plastic deformation is the integrated effect of large numbers of dislocations and to simulate realistic populations with accurate force fields and accurate defect distributions is beyond the present capabilities of MD. Understanding has been enhanced by many years of work on dislocation motion and interaction with corresponding theoretical frameworks developed and already in place. This theoretical framework is enhanced by experimental tools such as transmission electron microscopy (TEM), which has revealed details of the plastic deformation of crystalline materials. The dislocation-dynamics (DD) technique was created as a numerical tool to understand crystal plasticity. Individual dislocations are modelled as segments which can move in an elastic medium according to dislocation interaction forces driven by the external loading. Segments respond and interact by discrete movement according to a mobility function characteristic of the dislocation type and the material considered. Interactions are controlled by rules that are prescribed at the beginning and represent an adjustable parameter set in the simulation. The end result, however, may be compared with the macroscale response and simulations can produce stress–strain curves for materials as well as showing dislocation microstructure developing during loads. Of course TEM provides a picture of behaviour that shows qualitative connection between simulation and observed behaviours. However, as with the other microscale methods described above, the paucity of direct experimental information at this scale, coupled with the difficulties in mapping to continuum behaviour, severely restrict these techniques and at present they illustrate physical processes rather than acting as engineering tools.

2.13.3   Crystal plasticity

Crystal plasticity modelling describes the deformation of single and polycrystals assuming crystallographic slip as the main deformation mechanism. For a given initial texture and suite of materials parameters, the code computes the stress–strain response and evolution of the orientation of single crystals or of an aggregate of crystals under forces applied at defined boundaries. As load is increased the resolved shear stress on many of the available slip systems will rise until a critical stress is reached on the primary slip system. This is determined from the Schmidt factor (which is a multiplier for a resolved component of the stress field), which will have the greatest value on the primary system. At each time increment the activated systems are identified and slip (and rotation) are calculated for each crystal. Thus incremental plastic deformation of the aggregate is tracked as the time steps apply deformation to individual anisotropic components of the microstructure. Of course such calculations have complications. Hardening and softening rules must be input to adjust the response of particular systems to match results. However, these mesoscale studies of single crystal deformation and dislocations inform the further development of continuum field theories of deformation. It provides a vehicle for developing a comprehensive theory of plasticity that incorporates existing knowledge of the physics of deformation processes into the tools of continuum mechanics and represents a component in the development of advanced, physically based models for engineering applications at the macroscale.

2.13.4   Lagrangian and Eulerian hydrocodes

The hydrocode has become the most important workhorse in the analysis of structural mechanics problems over the past 50 years. It became possible with automated calculation and finite-difference methods to numerically determine solutions to differential equations. To solve the equations of motion and boundary conditions, the finite element method (FEM) allowed iterative computation to find solutions to the suite of operating partial differential equations (PDEs). The first crude steps down the road to the machines of today used devices that delivered performance in the kiloflop regime. At present the fastest machine is Sequoia (LLNL), which operates at 16 petaflops, but exaflop machines are already in the pipeline. In early calculations material descriptions had to be simple and the first models neglected strength, modelling materials at high pressure as dense fluids. Numerical platforms gave great flexibility since material deformations are non-linear as pressure increases and analytical solutions are only tractable if geometries are simple and boundary conditions known. Thus the hydrocode was born, a computer programme for the study of fast, intense loading on materials and structures. As the computer explosion unfolded over the years following their introduction, it became necessary to input more realistic material descriptions which would have wider applicability, and these require experiments to determine the operating mechanisms to be successful.

As has been alluded to previously, the definition of materials must be based on two basic classes of model: equation of state descriptions (EoS) and strength models (constitutive equations). The run-time of the simulation and the scale of the components within it determine the range over which these descriptions are required to remain accurate in stress, strain and time. The following chapters will show the form of response of materials and the regimes of behaviour where specific descriptions must be fixed for accurate response. Regrettably, however, it is all too common to use arbitrary fitted constants to match a particular experimental state, so losing general applicability. The form of the models and their particular parameters are not relevant to the present discussions but some will be mentioned later as the text develops.

The divisions in the form of the numerical platform come when the response takes sufficiently long and acts over components of such a scale that waves have equilibrated the stress state within the structure in a time step of the simulation. At this point codes need not track wave fronts and can proceed without analysis of mesoscale detail using Newtonian mechanics to define the state. There is thus a division between structural response and wave-propagation codes that separates the commercial platforms on the market today. The assumptions in operation lead to a series of differences found between each class of code. They employ different timescales since they operate on different assumptions about the continuum stress state. In one case the development of the state is tracked by wave propagation. But equally it is possible to fix the structural response by considering structural modes. Typically behaviour is elastic and strains of 0.5–2 are found in structural mechanics simulations. Thus the mesh used is almost always Lagrangian, deforming with the material under load.

In wave propagation codes strains of hundreds of per cent may be generated. Further, the events of interest typically occur over ns to μs with the pressure range many times the yield stress for the material. Thus a new suite of more specialised material models are required that are different from those appropriate for structural response. Further, the greater strains mean that deformation of a Lagrangian grid occurs rapidly and frequent remapping must be undergone to keep the solution convergent as time progresses. Such remapping is time-consuming and adds numerical errors; frequently a simpler solution is to use a Eulerian code where the material elements are tracked as they move through a fixed mesh. Here errors come in mapping material interfaces to a new position within the grid.

Consideration of lower length scale response means that deformation has regions of localisation and this means that details of the microstructure need to be adequately described in the assumed models. For instance, in rapid tensile loading, the failure criteria adopted are frequently based on instantaneous maxima or minima of field variables. This results in non-physical anomalies in simulations with some features of the flow correctly reproduced and others not represented. Such is the fate of a continuum formulation representing localised failure and whilst gross features are correctly represented in such platforms, details can never be. It is thus necessary to add specific subscale models which correct for the unphysical parts of the formulation in such schemes and since these represent behaviours important in specific regimes, it is they that limit generality in the application of the platform. When using hydrocodes for design it is important not to expect too much detail outside the range where continuum behaviour operates.

2.13.5   Meshless methods

In problems where it is necessary to deal with large deformations of the mesh there will always be problems with approaches that fix cell geometries. These may occur in computations where there is propagation of interfaces between phases or perhaps in simulations of failure processes where cracks, for instance, might follow arbitrary and complex paths. Such problems are not well suited to conventional computational methods since their underlying structure relies on a mesh that cannot adequately treat discontinuities which do not coincide with the original mesh lines. In order to avoid reapplying a new mesh for each step of the calculation, it is in some circumstances better to adopt a meshless method which represents the structure with a random array of nodes tracked as they deform according to the conservation laws. Thus it becomes possible to solve large classes of problems which are awkward with mesh-based methods. The most successful and best developed meshless approach is the smooth particle hydrodynamics (SPH) method, which was first used successfully to model astrophysical problems such as exploding stars and dust clouds but has now been expanded to address a range of continuum mechanics problems.

2.13.6   Codes and their regimes of operation

The brief overview above has described a wide variety of numerical techniques which describe material behaviour in geometries of interest with varying degrees of success. These range from *ab initio* potentials incorporated into atom-based MD calculations at the microscale to simulations with continuum models that can address processes in the cosmos at astrophysical scales. Continuum-based codes operating at the mesoscale and beyond require a series of inputs to define equation of state and constitutive models for the materials involved, as well as special features of the simulation that must be included, such as interfacial effects to describe friction or more rigorous failure models to depict spallation. All of these sub-models are describing localisation in a manner conducive to incorporation in a hydrocode that is based on concepts from solid mechanics. Of course localisation makes the stress state unsteady within cells and these anomalies must be fixed by defining the state within, or even deleting the cell. Equally features of the models are expressed in terms of solid mechanics concepts such as strain rate which are defined only when the stress state is steady. Thus wave and structural codes work in regimes of steady flow and, where that is not the case, sub-models with the missing physics must be input to satisfy this requirement.

In all cases the models used in present codes for describing materials use analyses to define one-dimensional constants. This puts an onus on experimentalists to design suites of tests which can populate such descriptions with similar one-dimensional continuum loadings. These tests will be visited through the next chapters along with the range of behaviours found in different classes of matter. In a very few special cases, and in the simplest geometries with well-defined boundary conditions, it is possible to find analytic solutions to test code predictions against. However, such closed-form solutions are generally one-dimensional and do not take into account lateral inertia.

The key to a global understanding of material response is in the transition from one regime of pressure or of loading time (or scale) to the next. This suggests that the space is not continuous but has zones of behaviour with boundaries where transit requires reevaluation of materials properties and these transition thresholds are those where new defect populations appear. Solid mechanics is derived to operate for ambient loadings and has been tested at the macroscale for centuries. Here the macroscale is divided into the component and the structural scales. A structure is a composite of components held together by joints that are either introduced by man or by nature at the ambient scale. The component scale is that of one of those elements and the division between them is the defect that exists at the interface. Of course this book will consider response and loading at scales beneath this range. The mesoscale is defined by the microstructure and the force laws found within a component which will determine the continuum properties observed in the laboratory. The constituents of the microstructure (grains within a metal for instance) contain stacked atoms with a particular packing and this constitutes the microscale. The grain boundaries and other particles at the mesoscale constitute a defect population at the grain scale in the metal. With the packing of the microscale grains will be another defect population ranging from vacancies to line defects once dislocations are nucleated. Each of these features is represented in Figure 2.34 and the different modelling schemes typically used to describe them (and discussed in the previous section) are marked at some point in the phase space of the figure. The coordinate of the plot is displayed as a time that defines the kinetics of the operating processes or the time step of interest for the calculation. It may be converted to a distance by multiplying by a wave speed to determine the length swept in the relevant time. The regimes of influence of density functional theory (DFT), kinetic Monte-Carlo (KMC), molecular dynamics (MD), crystal plasticity (CP), wave codes, structural codes and plasma fluid dynamics codes (CFD) are indicated. The anchors show the regimes where material parameters for the relevant formulation are conventionally derived. That to the left at atomic dimensions refers to the states used for derivation from quantum theory of the potentials to use in MD codes. That to the right refers to the derivation of constitutive modes for strength which are fit to static and intermediate-rate experiments conducted at a stress just beyond the yield surface and with a response affected by defects at that scale. It is clear that the parts of stress space where the derivations are made are very different to some of the regimes in which the codes are often applied.

The yield surface is represented in this space by a projection as a dotted line at lower pressures. It reaches lower values as the strength reduces at larger length scales. The darker triangles denote defect regions of influence that compress on loading in the inelastic regime. No process is instantaneous and there is an unsteady phase to response before steady flow occurs and this unsteady phase involves instability and localisation so that the stress state beneath the defining curve is not constant across a volume element. Solid lines for each class of material represent the division between the regimes. At the longest structural scales the interactions are no longer electromagnetic and gravity determines the response.

The regions of extension of solid mechanics occupy the regime up to the *finis extremis* beyond which core electron states become perturbed and the nature of strength changes.

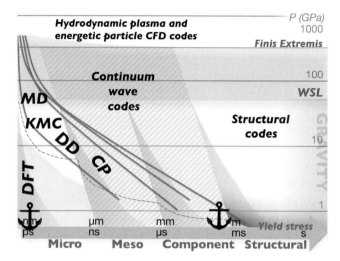

Figure 2.34 Numerical methods and their regimes of operation across a schematic region of space spanning pressure and impulse/process timescale. The yield surface limits the dotted line representing the elastic limit near the base. Darker triangles denote defect regions of influence. The lines of each class of material represent the division between unsteady and steady flow regimes. The regimes of influence of density functional theory (DFT), kinetic Monte-Carlo (KMC), molecular dynamics (MD), crystal plasticity (CP), wave codes, structural codes and plasma fluid dynamics codes (CFD) are shown. The anchors show the length and pressure coordinates where material properties for the material models are conventionally derived. That to the left (at the atomic dimension) refers to the states for derivation from quantum theory of the potentials used in codes. That to the right refers to the derivation of constitutive modes for strength that is fit to static and intermediate-rate experiments to a stress just beyond the yield surface and dominated by defects at that scale. WSL is the Weak Shock Limit.

The Weak Shock Limit (WSL) is the pressure above which all defects have been activated. At this point the shock becomes overdriven and behind the front a steady stress state exists which models can reproduce more easily. Now imagine looking from a point at some state within the macroscale and observing processes with a magnifier at the microscale which define the equation of state for the material. If the material is a low-density foam then compression will result in collapse of porosity at the mesoscale and these processes will dominate response. However, consider solids close to full density but with a population of defects typical of engineering materials. Under compression the defect sites collapse and cracks or dislocations travel out to define the hydrodynamic flow state and allow inelastic deformation to begin. However, the response to compression is determined primarily by the electronic repulsion at the atomic level since the volume fraction of these defects is negligible. Thus materials descriptions for pure substances at close to full density can describe the equation of state with some surety. Strength, on the other hand, depends upon the volume of material sampled in its measurement. The defect population is key and thus measurement and fitting of this quantity at the component scale will underestimate the strength of material at the microscale since the defect population is very different. Further, the crystal at the lower scale will be anisotropic, whereas at the component scale is likely not to be so, meaning that the two states will be very different. Mapping across scales where stress states are constant and defect populations are close to fixed is possible and in pure materials can be made to work well although another series of rules is being applied. However, crossing defect scale boundaries is a complex matter since new effects and mechanisms come into play and mapping must be done with care from one scale to next. No single numerical technique can be scaled across these boundaries since the physics changes with new modes of deformation introduced at each scale. Further, no single experimental technique can be mapped from the atomic scale to the macroscale without missing vital components of material response for the same reasons. Finally, the physics binding materials to one another changes at the *finis extremis*, which is at a few megabars in most materials since the energy density applied in the pulse becomes of the order of that of a valence electron. It is not possible to apply concepts derived above to regions at higher stresses than this since solid mechanics cannot be applied there. Thus the key to understanding lies in the physics of these inter-scale boundaries, and that is where future development of models and platforms should take place.

2.14   Final comments

This chapter presents a basic toolset of models and supporting concepts to describe behaviour that will illuminate the use of devices and the operation of mechanisms in the classes of material encountered beyond this point. The models derived are those that apply to materials in electronic states within the realms defined in the remit of the phase space of Figure 1.2 and populated with numerical platforms in Figure 2.34. In this respect the field for the most part attempts to extend the formality of solid mechanics derived under ambient conditions to high pressures and temperatures that result from rapid loading under compression and this defines akrology. However, variation in these effects with changes in bonding, for instance, are in general not explicitly addressed, except in assuming elastic constants can be extended through solid mechanics with suitable equilibrium thermodynamics, and so as states become more extreme these formalisms will become increasingly unphysical. Such approaches have some success, particularly in the strong shock regime, where their application has produced intuitive and predictive models. Thus application of codes using these concepts works well for the regime that, for example, includes detonation pressures for engineering design. Of course there are many assumptions that are incorrect in moving to the next level of detail. The processes occurring in the shock are assumed to be instantaneous, the material is regarded generally as isotropic and the yield surface, well represented in compression, is poorly represented in the tensile quadrant. Thus continuum theories at modest pressures can only have success in compression where waves equilibrate the stress state behind the shock. In tension, however, where localisation determines behaviour, models are not developed to the same level since localisation occurs. In particular, strength is poorly understood and applied with difficulty to these dynamic loads since it is a function of the volume element sampled by the wave in the application of interest. All of these comments make it clear that one must take great care in mapping behaviour at the smallest dimensions to those at the continuum or conversely applying rules formulated under ambient conditions on the laboratory scale back to behaviour at the microscale. In using any formalism, and especially in these extreme states, it is always wise to ensure that all assumptions are checked and the background to the theory is understood before extrapolating the results obtained.

2.15   Selected reading

Asay, J. R. and Shahinpoor, M. (1993) *High Pressure Shock Compression of Solids*. New York: Springer Verlag. (Particularly note Chapter 8 which covers dynamic tensile failure.)

Cooper, P. W. (1997) *Explosives Engineering*. New York: Wiley.

Davis, W. C. (1987) The detonation of explosives, *Sci. Am.*, 256(5): 106.

Davison, L. and Graham, R. A. (1979) Shock compression of solids, *Phys. Rep.*, 55: 255–379.

Fickett, W. and Davis, W. C. (2000) *Detonation: Theory and Experiment*. Mineola, NY: Courier Dover Publications. Reprint, originally published 1979.

Mader, C. L. (2008) *Numerical Modeling of Explosives and Propellants*. London: CRC Press, reprint.

Malvern, L. E. (1969) *Introduction to the Mechanics of a Continuous Medium*. Englewood Cliffs, NJ: Prentice-Hall.

Zeldovitch, Y. B. and Raizer, Y. P. (2002) *Physics of Shock Waves and High-Temperature Hydro-dynamic Phenomena*. Mineola, NY: Dover, reprint.

3 Platforms to excite a response

3.1 The scientific method

The present chapter gives an overview of experimental platforms showing how they may be used to populate models for materials behaviour. The condensed phase defines the pressure and temperature range of interest, which may be approximately fixed at less than 1 TPa and below 10 000 K. Indeed pressure has one of the largest ranges of all physical parameters in the universe (the pressure in a neutron star is c. 10^{33} Pa), so that most of the materials in nature are under conditions very different from those on Earth. The goal of shock experiments is to track response and mechanisms across the realms of stress and volume that are experienced by condensed-phase matter across the universe. At the highest pressures and temperatures, materials move from the solid to the liquid and then to plasma states as new correlations and bonding are formed. These high-density states have been termed warm dense matter (WDM) and lie beyond the *finis extremis* – outside the regime of extreme behaviour considered here. A summary of the phase space occupied by matter in these regions is shown in Figure 3.1.

The goal of experimental work is to provide adequate knowledge of the response of matter over the operating regimes of the relevant plasticity mechanisms. By this means, analytical descriptions can be constructed to try and capture the fundamental relationships between the independent variables – stress and stress state, strain and strain rate, and temperature – that determine the constitutive, damage and failure behaviour of materials. A shock impulse provides a pump to drive materials deformation and control of that impulse also allows a window into the operative mechanisms that lead to plasticity and damage evolution. This includes determining dynamic strength as a function of pressure as well as determining equation of state over the range of interest for particular applications.

It is also necessary to distinguish between types of experiment necessary to understand response to compression in a material and those necessary to describe failure processes within it under tension. Dynamic inelastic flow mechanisms are most easily studied with experimental techniques that develop homogeneous stress and strain fields by application of well-controlled loading. In such experiments the stress and deformation fields remain nearly homogeneous as the deformation evolves and a small number of measurements (typically single-point or area-integrated) are sufficient to obtain information about the deformation and stress fields of interest. However, once a failure process begins in the material, such as results from dynamic tensile or torsional loading, the

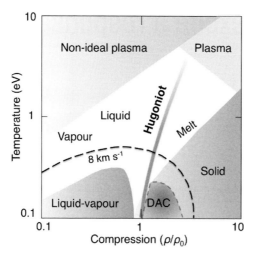

Figure 3.1 Phase space for dynamic loading of materials. Source: after Neil Holmes, LLNL.

deformation field rapidly becomes localised and the experimental techniques used to extract information using single point or area-integrated measurements become inadequate to track the processes. Thus, understanding localised failure processes requires a different suite of experimental techniques to those required to track homogeneous compression.

Advancing mechanistic knowledge of dynamic processes proceeds in the same manner as in all other physical science. The method of inquiry relies on gathering measurable evidence through experiment and then subsequent hypotheses are constructed which may then be suitably tested to determine the range of applicability of derived mathematical laws. Such principles relate back to the methods of the ancient Greeks most elegantly summarised by Aristotle in the third century BC. A key figure in its modern form was Ibn al-Haytham (Alhazen), born in Basra in 965, who directed its key purpose toward revealing absolute truth. The method contains four key components for the aspirant. The first involves gathering the assembled observations of other workers, and appreciating the totality of their message against the backdrop of the experience and knowledge accumulated from training and experience. Such a process leads to the second stage: construction of hypotheses that may be placed into analytical mathematical descriptions of the subject. These hypotheses lead to logical deductions which form predictions, the third phase of the method, which extend beyond existing observation and can be tested with further experiment. These trials constitute the final stage of the method, which tests the hypothesis and confirms, refutes or extends predictions and mathematical descriptions of the subject.

The method applied in the area of extreme dynamic loading is often influenced by the difficulty and expense of creating in controlled form the conditions necessary to simulate extreme events: whether the method chosen applies mechanical impulses of TPa for minutes to simulate a bolide impacting with the Earth, or tries to replicate the pressures and temperatures existing at the centre of Sun to overcome the barriers to fusion in deuterium and tritium. In such cases, the difficulty of fielding the experimental stage of the process shifts the focus of the method to the construction and testing of models. The application of those models then becomes the pivotal part of the effort made to understand the observations and there is a risk that exercising simulations becomes the primary means of investigating the physical processes that occur in materials and structures under dynamic loads. There are thus two classes of experiment that must be distinguished for the modern era. The first are designed to derive mathematical descriptions of materials to be inserted into mathematical test beds for simulation, and the second are those used to validate those models for use on specific design problems. The scientific method has evolved in this subject to focus effort around the powerful numerical platforms that dominate engineering for design.

3.2 Derivation and validation experiments and verification and validation of material models

The modern need for idealised experiments is then to serve one of two important purposes. The first is as a means of *deriving* mathematical descriptions of material behaviour; the second is to *validate* that the descriptions assembled contain correct formulations that truly represent all the loading modes that a particular application may supply. These are termed *derivation* and *validation* experiments. The exact approaches taken depend upon the stimulus that drives the problem of interest. In some cases an application exists for which one may wish to pick an optimised material. Here, the practical conditions of the loading direct a series of experimental tests which can only later involve material descriptions. In other theatres, one may know the particular material of interest (it may have been selected for reasons dictated by electrical properties, for instance) so that the challenge is in producing a physically based mathematical description to describe its yield behaviour. In constructing a model one must first analytically describe the operating deformation mechanisms which can then be used to construct an equation which connects stress, strain, temperature, and frequently strain rate (so that some means of representing time enters) for steady deformation at the continuum. Such a model will apply to a class of materials and as such needs a suite of experiments to populate it. These will define the array of constants peculiar to an example of such a class that will allow the equation to be used within a numerical scheme such as a continuum hydrocode. The present formulation of semi-empirical engineering models requires relatively few derivation experiments necessary to populate such a model. They define yield points at particular strain rates and temperatures so that constants may be fixed. The quasi-static tests on load machines are well known and will not detain us here. However, more specialised, and less well-documented, experiments conducted under dynamic conditions will also be described below since it is clear from the discussion of modelling at the end of the last chapter that current models are frequently inadequate in regimes where defects define length and timescales where different physics applies.

In a world in which the virtual testing environment has become increasingly important, new procedures must be adopted to ensure a method is in place to ensure the veracity of their results. Once a model has been derived and populated it must ideally undergo a series of procedures designed to allow it into service. One means of doing this has been to adopt methods from the fields of quality control and assurance to allow procedure to qualify a simulation for use. Verification is the control process that is used to evaluate whether or not a model, or system or class of models, complies with the specifications set out at the onset of the development. Validation is a quality assurance procedure that provides a high degree of confidence that the platform accomplishes its intended requirements and satisfies acceptance of its fitness for purpose with end users and stakeholders.

Brannon *et al*. (2007) have given a more erudite summary of the processes:

Although scholarly definitions are available, the distinction between verification and validation is frequently explained as follows: verification ensures that we are solving the equations right, whereas validation ensures that we are solving the right equations. Verification is a purely mathematical and comprehensive demonstration that the equations are well posed and the numerical implementation will yield an accurate solution (preferably relative to simplified analytical solutions, because a converged result for this class of models cannot safely be presumed to actually solve the governing equations). Validation, which should always come after verification, assesses the physical merits of the equations by confirming that they adequately reproduce all available data using a single material parameter set. Of course, what constitutes 'adequate' validation is rather subjective, in that the answer depends on the class of problems to be solved as well as on what information is sought (e.g. averages or distributions). A linear elastic model, for example, might be adequate for routine service conditions, but inadequate under abnormal conditions, in which failure might occur. Therefore, any assertion that a constitutive model is 'well-validated' must include a clear description of the model's domain of applicability. Finally, unless there are compelling arguments to the contrary, a verification and validation process must demonstrate that the governing equations are compatible with basic physical principles, such as thermodynamics and frame indifference, even in domains in which data are unavailable.

Thus while the concepts and the implementation of verification and validation procedures to computer models remain to be addressed routinely, the field must continue to appreciate the need to adequately experimentally test and compare predictions in order to determine accuracy given the reliance on materials modelling. The derivation and validation experiments that result are separate classes of test that define themselves purely on their ability to suggest operating mechanisms or populate parameters in existing models, or alternatively to exhaustively demonstrate that a mathematical platform is suitable for design use across a range of thermodynamic conditions. These will be described in detail in the next chapter.

3.3   Compressive stress

Loading materials places them under a range of stress and strain states according to the impulse applied and the particular operating mechanisms excited. To understand these at the lower length scales at which they operate requires application of an idealised (simple continuum) loading state where boundary conditions in behaviour at the laboratory scale may be fixed and details of response within a material under loading may be studied. A series of such techniques have been developed to study material response in compression starting with the massive presses of the industrial revolution when systematic investigation of material behaviour had its genesis. The other loading stress state frequently employed is testing in tension. Obviously in the case of compression the material shortens and strain becomes smaller whilst in contrast in tension the material extends and the strain increases. However, the strain state localises in tension and failure occurs in a different manner from that under compression, particularly under dynamic loads, which places restrictions on the information that one may gather from experiments.

For defect-free materials this behaviour is reflected at the atomic level: the atoms are moved apart when the target experiences tension and together when under compression. They move under forces that result from the interatomic potentials that exist between them, since in either case excursion from a minimum energy state results in forces throughout the lattice that oppose the applied motion. In reality, however, materials at the continuum scale will always contain some population of defects and these dominate mesoscale behaviour consistent with the macrosopic applied strain. In both cases the application of load will result in some mesoscale localisation (adiabatic shear in compression; ductile failure in tension), which will drive deformation and motion of the free surface (barrelling in compression; necking in tension) controlled by the boundary conditions imposed by loading devices (friction in a loading frame; end clamping fixtures in a tensile machine). These two manifestations of defect-controlled mechanics operating under different strain fields result in the macroscopic failure or buckling that the engineering of structures must avoid.

These limits define the compressive strength of a material to be that value of uniaxial compressive stress reached when the material fails completely. The compressive strength is traditionally obtained experimentally by means of a compression test and in simple, quasi-static engineering measurements it is achieved by employing a load frame of some nature which can generally apply tension or compression by straining in a particular direction. Further, confining the free surface of a loaded specimen, such as placing the sample in a pressurised bath of fluid, clearly increases the stress required before plastic deformation occurs. In what follows, the effects of compressive loading at variable speeds and confining pressure will be tracked from slow loading to generally higher amplitude impulses over a much shorter timescale where the loading is dynamic.

Application of pressure across a body defines a stress acting on an area of surface which gives rise to displacements or volume changes. Mechanical loading defined by pressure (compression in what follows) and its conjugate variable volume results in changes to both structure and density. The work done by the force defines energy deposition into the body, which results in changes in the other thermodynamic variables of temperature and entropy. The means of characterising these changes and relating them to the microstructural and atomic rearrangements occurring requires the development of new sensing techniques to track changes in real time as load is applied.

The application of rapid mechanical loading applies to a vast range of natural and engineering situations affecting structures as well as their constituent materials. There are a great variety of dynamic natural engineering loads, such as those encountered by structures placed in winds, or subject to vibration on land. With extra thermal energy in the atmosphere, wind speeds become more extreme and structures placed within them must be designed to encounter and withstand greater loadings. Atmospheric turbulence has an effect not only on aircraft flying within it but also the dynamic response of tall structures placed in surface winds. Further, earthquakes will load both structures and their foundations by subjecting them to rapid displacements and consequent dynamic torsional and tensile stresses. This may range from soil responding by partial liquefaction beneath a building, to physically transmitting ground movements into the foundations of built structures. To counter such threats, the response spectrum encountered in relevant areas and regimes may be analysed using Fourier methods to define the consequent need for vibration isolation. Here, better understanding of the loading envelope allows the design of seismic-resistant structures for future building in such areas. This volume is concerned with much higher amplitude loadings and in this regime the loads make it necessary to design with blast mitigation to protect vulnerable structures.

In the following discussion, the development of interest in problems at the engineering scale, developed through probes into the dynamic response of materials, will be tracked from the conceptually simpler static loads through to the ultrafast dynamic ones found today.

3.4   Experimental platforms

In order to investigate and characterise the operating mechanisms for a target material under a particular load, the deformation must be tracked with the application of a defined impulse which can be used to excite the array of inelastic mechanisms that classes of material exhibit. This pulse is delivered by a particular experimental platform and this chapter will review the suite available today (Figure 3.2). Lasers are capable of delivering the highest energy density but their loading pulse is short-lived. At the other extreme are load-frames which can quasi-statically deform a material, although in this case they only deliver stresses around those required to plastically yield the material. The principal determinant of deformation in these devices is strain. In all cases except chemical reaction the deformation mechanisms excited by the applied impulse are only those than can run to completion whilst the load is present.

The evolution of the interatomic bonding is a function of both the kinetics of the operating mechanisms and the amplitude of the driving impulse which deforms the material under load. In the earliest moments of loading the material is in an assemblage of non-equilibrium electronic and mechanical states which transit rapidly to compress the solid. Above a critical pressure, equilibration of the states reaches a regime in which localised ionisation becomes more prevalent with increased pressure until an homogeneous plasma state is reached. Beyond this high-pressure ionisation threshold, the realm of our concerns transits to that of plasma physics, where terminology like strength and anisotropy become undefined and the operating physics is dominated by homogenisation of the states achieved. This transit occurs at several hundred GPa in condensed matter, as discussed in Chapter 1. This text is concerned with what occurs

Table 3.1 Characteristic length and timescales for observing typical crystalline deformation mechanisms in a single-phase metal after applying step compression loading at $t = 0$

Mechanism	Representative timescale	
Defect activation	100 ps	Localisation
Phase transformation	1 ns	
Twinning	1 ns	
Slip	10 ns	Flow
Dislocations locking	100 ns	Interaction
Pinning at boundaries	100 ns	
Void growth	1 μs	Localisation
Adiabatic shear band growth	10 μs	
Slip, hinging	100 μs	Flow
Buckling, creep	1 s	
Collapsed component	10 s	Interaction

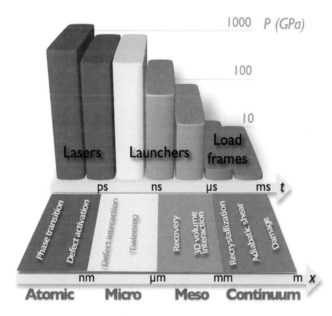

Figure 3.2 The range of available platforms to apply load to a material and representative metal deformation mechanisms.

in the stress space before this transit pressure, and in imaging and diagnosing the operating physical mechanisms to allow a fully physically based modelling framework to be constructed and verified. In order to do this it is necessary to view mechanisms that must be accessed by the platforms considered.

Table 3.1 shows a suite of typical mechanisms for a deforming crystalline solid with corresponding order of magnitude estimates for the timescales required for such processes and length scales over which they operate. The times themselves clearly vary markedly between materials of different structure and density and with the applied compression in individual impulses, so that these values are for ranking purposes only. This is since the wave speeds that drive the faster mechanisms are defined by the atomic mass of the atoms that are displaced. It is possible to place the mechanisms listed into groups that act at different times and scales and show that these are driven by the defect distributions at each scale that exist in the material and the structure constructed from it. For the purposes of the following discussion this will focus on the response of a single-phase metal, and these groupings will be labelled localisation, flow and interaction (LFI).

Localisation mechanisms operate at the atomic scale around point or line defects within the material or by shearing the unit cell to changes phase. These processes have a relaxation time of the order of a nanosecond within the metal. If the threshold for them to be activated is reached, these will be the first operating mechanisms observed within the solid during loading. Defects will nucleate dislocations and establish inelastic processes that allow slip and accommodate strain. Once defects are activated slip occurs until dislocations interact either locking or annihilating and ending a phase of plastic flow. In some metals at later times at this stage macroscopic hardening will occur where the dislocation density is high and the resistance to slip is low. These processes equilibrate strain states within crystals and define a state for a crystalline element between grain boundaries that inputs to a crystal plasticity code, for example. These crystalline elements interact once again to equilibrate a continuum stress state for the assemblage of grains. Within a component there are slower operating mechanisms accompanying processes driven by surface creation or deformation, but again these processes operate to achieve stress-state equilibrium within the solid. Slip occurs after strain localisation occurs, which increases plastic work in regions deforming the material as well heating it. This establishes slip boundaries within the solid that allow flow to occur in the structure and the stress state has equilibrated and plastic hinge and slip zones are defined. Again flow can occur until the structure collapses and is locked against some further boundary.

In each regime, the temporal estimates for typical single-phase metals reflect the speed at which the mechanisms operate. At each scale there is a delay before steady flow can commence where some type of localisation occurs. In all cases the connection to spatial scales comes through wave speeds in the materials (which are of the order of thousands of m s^{-1} in these metals).

Each impulse has some activation stress in order for it to operate and some characteristic relaxation time determined by the operating physics. In order for the process under investigation to be complete in a particular experiment, impulses of greater time than that of the mechanism will be required to observe the corresponding deformation since it is only for this time that shear stress exists to drive flow. The platforms used to investigate each mechanism and to construct continuum material models from observations in an experiment need to have suitable capability to deliver the correct impulses (amplitude and pulse duration) to observe the required response. The grouping of deformation mechanisms made on the basis of their regimes of operation and

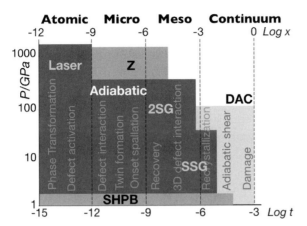

Figure 3.3 Operating metal deformation and damage mechanisms and their characteristic length and timescales. Timescales along the abscissa map to length scales along the top through the wave speed in the material (which is of the order of 1000 m s^{-1} in metals). Thus a pulse of 1 μs sweeps and loads a spatial dimension of 1 mm. Source: reprinted with permission from Bourne, N. K., Gray III, G. T. and Millett, J. C. F. (2009) On the shock response of cubic metals, *J. Appl. Phys.*, **106**(9): 091301. Copyright 2009 American Institute of Physics.

described above can be related to the impulse applied by individual platforms as well. Some isolate the localisation phase where the defects start atomic processes, others include the inelastic flow phase where slippage of planes can occur to accommodate the applied strain. There is then a further instability phase where waves propagate to equilibrate the stress state across the structure localising deformation to geometrical zones. Finally, there is a further flow phase where the structure with its component parts collapses around continuum features such as shear bands or hinges. The three phases of localisation, flow and interaction occur at the micro-, meso- and component scales as time progresses and each must be investigated to construct materials models for each process.

3.5   Tools for discovery science

One major goal of experiments is to derive materials models and platforms to investigate such behaviour must cover as large a pressure amplitude as possible, spanning a range of thermomechanical, electrical and magnetic states whilst controlling the period of their application from constant load to femtosecond pulse times. The devices must be capable of providing a pump with suitable diagnostics to probe the resulting response. At present the suite of platforms capable of generating shock response in materials of interest consists of a suite of developed methodologies that test behaviour over a pressure range of *c.* 1 TPa using impulses ranging from quasi-static high-pressure cells to nanosecond pulsed-laser sources and a temperature range of many thousands of degrees. A schematic of the range of devices and their operational regimes is illustrated in Figure 3.3. To populate the relevant material descriptions for all components requires models that cover a three-dimensional pressure–temperature–time space with validated certainty. At present it is only possible to load for nanosecond pulse lengths over a small target size to the highest pressures. This limits the mechanisms and microstructures that can be probed with such tests. As the compression amplitude decreases, it is easier to apply long pulses to microstructures representative of the continuum until, under ambient conditions, the diamond anvil cell provides high loading pressures but only over a sample dimension of the order of 1 μm at these high values.

There are four principal groups of loading techniques employed to recover material response over this regime, three of which are dynamic and including: laser-induced plasma loading, Z pinch devices, compressed gas and powder-driven launchers and energetic drives, and diamond anvil cells (DACs). The laser methods introduce short pulses (less than a ns) yet the largest can apply the highest pressures (TPa) and heat adiabatically to high temperature if a shock is launched. They cannot load target dimensions for long enough to deduce more than early time hydrodynamic response. Z pinch devices can provide pulses of hundreds of nanoseconds duration and can thus load targets for long enough to probe hundreds of microns under compression. They have the capability to allow stress to increase over a similar time period limiting temperature rise and leading (at worst) to a quasi-isentropic loading path. They may also launch plates that can shock targets.

Gas- and powder-driven launchers are capable of loading samples of a size and for a duration that are limited by the dimensions of the device. They can shock-load materials to hundreds of GPa for tens of microseconds over sample volumes of cm^3. They are precision devices providing well-defined loading impulse over times relevant to all the operating plasticity mechanisms that enables kinetics to be deduced from the response to the loading. However, they cannot provide pulses long enough to excite localisation mechanisms within the material.

Diamond anvil cells (DACs) have been used since the time of Bridgman to compress materials hydrostatically to hundreds of GPa. Clearly they can access impressive pressure ranges within a material but the size of the cell limits the loading to principally single crystal targets that at the highest pressures are only of micron dimensions. Although the DAC accesses equilibrium states, dynamic loading generally proceeds via metastable ones at different thresholds in state variables. New challenges are pushing the cells and the pressure-sensing methods to ever-higher pressures, yet beyond 100 GPa there is a limitation in assuring the stress state in the target and to sense pressure with great accuracy. This represents work in progress for the technique but the advent of ever-brighter light sources in synchrotrons and free electron lasers (FELs) has opened new possibilities for interrogating structure. At present synchrotrons are used to interrogate static experiments, but in the next few years a new beam-line at the APS in Chicago will be dedicated for the first time to dynamic experiments as well.

A series of pressure-induced deformation mechanisms, each with different respective kinetics, is triggered by shock acceleration of the impact face of the material. Figure 3.3 shows the pressure–time plane with shading for the temperature (the dark grey adiabatic; the lighter isothermal). The various operating regimes for the devices are dimmed so that a series of representative times for some important plasticity mechanisms in shocked

Table 3.2 Commonly used mechanical test beds. The table gives generic loading conditions for each device.

	P/GPa	Loading time/μs	Pulse rise/μs	Target/mm	Impulse/GPa μs
Large press	0.01	QS	–	50	∞
DAC	300	QS	–	1	∞
SHPB	0.1	100	10	5	10
SSL	150	10	<0	100	1500
TSL	300	3	<0	50	900
Z Sandia	500	0.1	<0	10	50
Large lasers	1000	0.001	<0	1	1

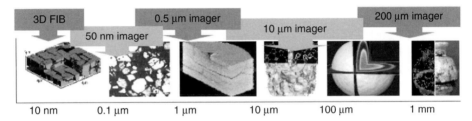

Figure 3.4 Imaging techniques and resolutions at different length scales.

metals might be superposed at values representing the order of magnitude for their kinetics. A short pulse length will not access the full extent of the distribution of defects or the representative microstructure within the material so that it will appear to display a higher strength than the continuum. On the other hand high-pressure states cannot be accessed for any length of time. Thus all techniques must be employed to understand material response.

New diagnostics and techniques are key to understanding condensed matter under load and they must be developed for greater spatial and temporal resolution. The technology in imaging includes a range of faster and more linear imaging tools under development with the continued advances made in solid-state sensors. Already non-invasive imaging using 3D X-ray tomography has allowed progression of techniques to observe real microstructures down to the micron scale and it may be expected to extend this to dynamic imaging over the next ten years (Figure 3.4). However, the ability to sense at the key length scales to fill the understanding at the micro- and mesoscales is vital to link advances in atomistic modelling to the developed continuum models in everyday use.

There are several platforms used to investigate mechanical response in the laboratory environment. They range from compression presses found in most engineering laboratories to the most intense laser platforms constructed to investigate nuclear physics including fusion. Table 3.2 lists a number of these and ranks them according to several parameters to characterise their response. Quasi-static loading devices include diamond anvil cells capable of achieving high pressure in small (tens of microns) targets, as well as Hopkinson bars which extend loading to the limit at which measurement can be made

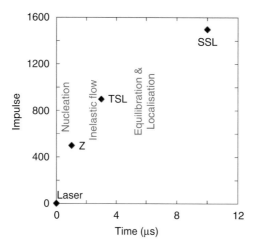

Figure 3.5 The impulse delivered by platforms that excite deformation mechanisms in targets, including lasers, Z pinch devices, two-stage launchers (TSL) and single-stage launchers (SSL).

of continuum properties (since the target is no longer in stress-state equilibrium after this platform). This group yield data concerning the continuum loading of the material under investigation since in all case stress states have equilibrated before data is extracted from the deforming target. In this respect they differ in nature from dynamic experiments in the wave regime where equilibration is occurring during the loading. The other devices in the table are operating when equilibration processes are underway. In the case of single- and two-stage launchers (SSL and TSL) a step impulse is applied to the impact plane. Z pinch loading applies a more gradually rising stress impulse to a target. In this mode different mechanisms are accessed at different stress thresholds that proceed with different kinetics, energy sinks and work done on the target. However, the magnetic interaction may also be used to accelerate plates to impact applying shock loading to the target. In this mode it can apply *c*. 100 ns pulse loading and this load time is plotted as the datum Z in Figure 3.5. Finally, laser pulses apply ns loading but to the highest amplitudes. This overview of platform impulse delivery shows the relative abilities of different platforms to excite deformation mechanisms within a material. It is both amplitude and loading time that are required from platforms needed to understand mechanisms and their timescales. The plastic state achieved in materials with high defect concentration or within liquids or gases allows determination of the equation of state parameters using these platforms.

To design for an application where structures are fielded in extreme environments requires a series of steps to be followed. Firstly the mechanical insult itself needs to be defined in terms of the amplitude and duration of the loading experienced. This defines an impulse experienced by the structure during the event. Each material has a potential suite of operating deformation mechanisms in response to that impulse which are triggered by the pulse as it passes. Knowledge of these responses for a range of materials fielded in a device allows the design of a structure optimised for an application. This includes suitable material types for particular component geometries as well as a range of jointing strategies between those components within the structure. At the level of the device it is the defects at each length scale within it that determine the strength of the whole. The principal defects activated by the insult relate to the jointing between components and then with the material during the load. Both of these features must be understood in idealised tests before design can occur successfully. Thus the construction of mathematical representations of the device response must include adequate material characterisation tests over the impulse range to be experienced including the correct stress amplitudes but also the operating kinetics in the deforming material.

Any test programme that wishes to cover a particular regime of mechanical behaviour must view a target material as a body that must be excited with pulses that isolate each kinetic effect in turn to properly represent it within a mathematical description. Pulse lengths of different duration at similar thresholds can then be used to map out the material response and construct valid models over the stress range applicable. In this view, each mechanism can be investigated to determine its threshold for operation and its relaxation time for completion. Clearly if there are several mechanisms operating simultaneously, then each occurs in parallel with the other and dissipates energy at a rate determined by the nature of the processes. In this picture an idealised pulse applied to the material is filtered by its properties so that the resulting output after some travel through it can be deconvolved to deduce the material mechanisms operating during the process. The measurement of states with suitable sampling rate at varying locations within it is thus a necessary part of the process which must be matched to the nature and kinetics of the mechanisms operating. An ideal form for a mechanical impulse in such an investigative process is a shock pulse with defined amplitude and duration since this ensures activation of all processes with threshold beneath its maximum stress level at the same time. Thus an input step with a defined duration is a necessary impulse for experiments designed to probe material mechanisms.

In order to determine the utility of experimental platforms in matching material models to deformation, it is necessary to find some means to characterise their usefulness for the derivation of models to design structures for future use. To do so, a dimensionless constant the Freya, F, has been introduced which offers a means of determining the totality of the deformation mechanism (see Chapter 1). It is defined here to represent the extent to which a mechanism is driven to completion by a stress or temperature excursion during the time for which the impulse is active in the following manner:

$$F = \frac{t_{\text{relax}}}{t_{\text{impulse}}}, \tag{3.1}$$

where t_{relax} refers to the characteristic relaxation time for the step in the rate-limiting process and t_{impulse} is the length of the impulse applied to the structure. Thresholds in stress, pressure or temperature need to be exceeded in order that the mechanism is activated and further when the critical process (since there will in general be a suite of them) is complete long before the impulse releases back to an ambient state, the stable final form will be an equilibrium state of the material under loading. If this is the case F will be small and the state will be defined. Conversely if the impulse is short relative

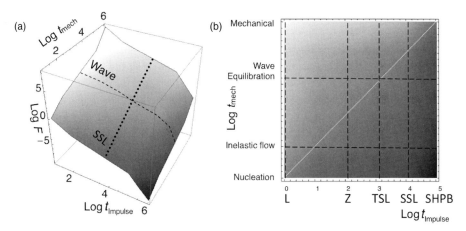

Figure 3.6 (a) F as a function of relaxation times for mechanisms and impulse durations delivered by various platforms. The impulse and mechanism times are measured in ns. (b) Platforms include lasers (L), Z pinch devices (Z), two-stage launchers (TSL), single-stage launchers (SSL) and split Hopkinson bars (SHPB).

to the completion time of a mechanism assumed to act (or the shortest available at that stress level) then F will be large and the state observed will be transient and similar to the initial state of the material at the start of the process.

Such a criterion can be used, for instance, to rank experiments to derive the quantities necessary to construct an equation of state for a material since it measures the degree to which loading has reached an equilibrium state within it. Thus it represents a litmus test of the ability of an experiment to excite a response to investigate a process on the one hand and provides a means of tailoring an impulse to optimise a material's properties for responding to dynamic deformation on the other.

Figure 3.6 shows the form of F for a series of impulse and relaxation times. There is a region in which it exceeds one and another where it is less than this value which are represented as greyscale regions in the contour map to the left of the figure. Individual characteristic times are used to indicate regions occupied by devices along the abscissa whereas individual mechanistic groups are indicated schematically up the ordinate. Since experiments aim to occupy a space in which they are complete, and F should be less than one, the experiments should sit below and to the right of the diagonal indicated. An example would be wave equilibration and localisation processes which can be studied using launchers and the Hopkinson bar to construct descriptions of the operating mechanisms.

The completion of processes defines a timescale which matches a length scale swept by a pulse. These are connected by some propagation velocity which depends upon the process considered but is limited in the fastest mechanisms driven at the highest amplitudes by a wave speed in the material. This region defines a minimum volume element for the completion of the mechanism under investigation. It also defines a voxel over which measurement must be achieved within the loading time of the experiment. All measurements of state evolution are integrated over some temporal and spatial domains and account of these defines necessary equipment operating over the correct stress regime sampled at the relevant scale.

From the continuum (engineering) perspective, the final geometry for a material or of a structure of several materials must be predicted by understanding constitutive behaviour as a function of time that can then be applied to a representative volume element within it. With this information a computer code can evolve the full stress state to a final form when the behaviour's dimensionality collapses to classical solid mechanics.

If the range of mechanisms available for deformation in that material can be assembled, then a representation of the development of the stress tensor with time can be constructed to describe the response of the material to load. Some means of switching on the mechanisms at appropriate times in the response must be incorporated in order to filter the suite operating at any instant during the deformation. In this respect, F may be used as a coefficient to activate a term describing a material mechanism within the evolving stress state. The development of the total stress field for a material, σ, can be described by a sum over the i developing stress states for each operating mechanism with their ordering controlled by F for each mechanism over all positions X_j in the following manner:

$$\sigma(P, V, T, t) = \sum_{ij} \exp[-F(t)]\sigma_i(X_j, P, V, T, t). \tag{3.2}$$

In this way a material's response is the sum of the mechanical effects due to a temporally stacked suite of mechanisms operating within it. The natural division of the total stress into a hydrostatic and a deviatoric component,

$$\sigma(t) = P(t) + \frac{4}{3}\tau(t), \tag{3.3}$$

also separates the hydrodynamic and shear components. Typical states are described by adding a constitutive model to an equation of state which is a step towards the description in (3.2). In the formulations used at present the hydrostatic component does not vary with time, and since such descriptions are generally applied for times short enough that these variations are not significant, the strength term does not either. However, as time extends to longer and longer values, experience argues that the strength term fades towards zero in the limit.

Of course, nature has no boundaries in time and scale; the operating behaviours map the potentials at the atomic level through to that at the planetary scale and beyond. To some extent the boundaries between the regimes discussed here refer only to conceptual considerations in the construct with which nature has been scientifically described over the last 500 years. However, there are scales at which there is a rapid change in the provenance of effects from neighbouring materials and it is here that these boundaries exist.

It is common in engineering to test small-scale structures under a range of defined loads as a means of quickly and cheaply asymptoting to a full-scale structure – this process is called scaling. It is finally worth noting that the processes of scaling can be assessed quantitatively using this methodology. Reducing the geometrical size of a test also scales the loading impulse. Since material deformation is a nested suite of operating mechanical mechanisms, scaling will only be valid where particular mechanisms relevant at one scale are reproduced at a different one. If this is not the case, the kinetics and amplitudes of the deformation will not match those in the alternate case and the behaviours will not be analogous. In particular, damage mechanisms will be seen to occupy a defined length and timescale which relates to defect activation and growth but which can be turned off by the incorrect choice of impulsive loading.

3.6   Static high-pressure devices

A basic requirement for using the formulations of solid mechanics is to derive and populate materials constants and this requires standardised tests to be done. One of the most basic of these is derivation of the engineering stress–strain curve for a material and to generate such characteristic behaviour requires standardised loading machines. Static hydraulic presses have been used in heavy industry to generate high pressures for centuries. The first static experimental studies on geological materials under high temperature and pressure were probably carried out by Sir James Hall (1761–1832). He studied melting and crystallisation in the furnaces used in the ceramic and glass industry, to reproduce the textures and mineral assemblages of natural basalts. In his studies on the recrystallisation of limestone, Hall sealed the sample together with some water into gun barrels and heated the sealed vessels in a furnace to reach pressures close to 0.1 GPa and temperatures above 600 °C. Under these conditions limestone was converted to marble and Hall therefore simulated metamorphic processes for the first time.

3.6.1   Load frames

A typical load frame consists of strong supports surrounding a hydraulically driven upper grip and a fixed lower one between which the sample is compressed. In order to record stress in the target, a well-calibrated force transducer (or other means of measuring the load) is required, as well as an accurate record of the displacement history to deduce strain. This may be monitored from the motion of a cross-head controlled to move up or down at (generally) a constant speed. Some machines can program the crosshead speed or conduct a range of programmed loading paths, including cyclical testing, application of constant force, testing at constant deformation, etc., using electromechanical, servo-hydraulic, linear or resonance drives. To ensure a well-posed initial boundary condition, tests must be made under controlled target environments (temperature, humidity, pressure, etc.) in chambers around the test piece. Finally, it is vital to ensure reproducible test fixtures including holding jaws for the specimen, and to keep to well-defined sample fabrication methods for precision mounting for a test. All of these features provide a stable, rigidly fixed and well-aligned structure for the experiment, and during the test the machine's construction prevents deformation and vibration within permissible limits under the action of applied forces regardless of their magnitude and direction. Modern machines can provide static loading as well as some degree of dynamic capability using a feed servo motor with an appropriate transmission system to allow speeds corresponding to target strain rates of up to $c.$ 10 s^{-1} in systems designed for samples of low strength and size.

The selection of the machine and test is determined by the force, speed and travel required to probe features of the material response. This then defines whether data are required in compression, tension, peel, tear, shear or flexural tests in bending. Loading large targets at high rates requires more specialised platforms, and in all cases the complexity of the devices increases as loading is applied more rapidly. Similarly, limits on the magnitude of the pressure that may be applied are soon reached with hydraulic testing machines once the yield strength of structural metals is exceeded. Further advance with high-pressure machines requires presses with hard anvils and systems where greater forces can be generated using leverage to create higher magnitude forces over small areas.

3.6.2 The diamond anvil cell

Higher pressures are most easily achieved by compressing a sample volume, loaded to a hydrostatic state, by applying uniaxial stress to two hard anvils cut to form two cones. This allows stress to be concentrated onto smaller, flat contacting surfaces. In the devices used at present, these surfaces are typically 300 μm in extent. Pressure is raised in the target by compressing one of the anvils against a second, subjecting the target placed between them to high loads. By this means, the isothermal response at a particular temperature may be tracked over a range of applied pressures. Such research was pioneered by Percy W. Bridgman (1882–1961) of Harvard University in the first half of the last century. In early experiments he used tungsten carbide anvils, which allowed pressures of only a few GPa to be achieved. In fact the majority of Bridgman's work was carried out at rather modest pressures in the range of up to a few GPa (using primarily hydraulic systems) and over moderate temperatures, reaching pressures in excess of 10 GPa on only a few occasions. He measured a wide variety of properties at high pressure, discovering a range of new phenomena including polymorphism in alkali metals. One notable discovery was of a metallic form of phosphorus at high pressure (Bridgman, 1914). This is probably the first example of the transition of a non-metallic polymorph (yellow phosphorus) to a metallic form under pressure. Further, it also demonstrates a principle often used in later years to guide research that the high-pressure polymorphs of a given element (or its compounds) resemble the low-pressure polymorphs of the heavier elements in the same column of the periodic table. Bridgman was awarded the Nobel Prize for physics in 1946.

The late 1950s saw the introduction of single crystal diamond anvils (the diamond anvil cell, DAC) invented by groups at the University of Chicago and at the National Bureau of Standards. In modern cells the diamond faces are separated by a gasket which confines a medium in order to transmit a uniform hydrostatic pressure to the sample. As the anvils are brought together, the fluid deforms hydrostatically in the confining

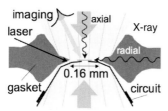

Figure 3.7 The diamond anvil cell.

field and a small crystal of a fiducial material or of ruby (which displays a pressure-dependent fluorescence spectrum) is used as a pressure calibration. This loads a small target of material under test (typically ten microns in dimension), which is then imaged using transmitted or reflected radiation to deduce its microstructure using X-ray diffraction, or anvils with sensors incorporated within them are used to study its electrical properties. Pressures in excess of 100 GPa were first achieved using these cells in the mid 1970s.

Static high-pressure cells are an important tool in the fields of materials science or geophysics for understanding the structure of materials and transformations of phase within them. High-pressure powder diffraction experiments constitute the major group of work using these devices. There are two sources of radiation applied to samples loaded in these cells: X-ray and neutron diffraction. Neutron and X-ray diffraction measurements are generally made down the direction of the loading axis, though in recent radial experiments, X-ray measurements have been performed perpendicular to it. This requires differences in the detailed design of devices for the various classes of experiment.

A cell typically contains a range of components (Figure 3.7). A mechanical mount holds two opposing diamond anvils and supplies the force required to push them together in compression. A gasket between them confines a working fluid which transmits pressure to a sample compressed between the culets. Gasket materials include steel, Inconel, copper alloy and rhenium, as well as pure beryllium, which is transparent to X-rays (when experiments require radial diffraction). The gasket confines a pressure-transmitting media (typically argon, xenon, hydrogen, helium, paraffin oil or a mixture of methanol and ethanol) with a pressure standard introduced. These standards include ruby for fluorescence or a structurally simple metal such as copper or platinum with known behaviour as a fiducial whose X-ray signature can be compared against. A desirable pressure medium should ensure that the stress applied to the sample is homogeneous, leaving it free of any differential stress or induced shear strain over the entire pressure range of the experiment. The whole assembly is compact and may be hand held, which makes it easily included within facilities such as synchrotron end stations where such experiments are typically fielded.

Typically the beam is probed axially with X-rays or visualised optically through the anvils. At the present time, the principal diagnostic in such investigations is imaging of the compressed sample using X-rays from a bright light source. The suite of accessible third- and fourth-generation light sources has allowed the development of a range of

X-ray diffraction and fluorescence techniques. Further, illumination with lasers allows optical absorption and luminescence to understand the pressure variation of band gaps within materials under load. Indeed, spectroscopy of absorbed or scattered radiation on a target allows a range of probes of electronic changes informing a wide range of chemical transits in the material based upon standard spectroscopic techniques. It is also found that altering the pervading electromagnetic or microwave environment around the sample excites a range of observable effects as pressure is increased. This may require electrical connection to the sample to investigate response and so recently a range of designer anvils has been fabricated, depositing sensors or conducting paths onto the anvil surfaces.

The present range of devices supply data from sample sizes too small to directly instrument, so that probing deformation is primarily by diffraction which tracks strain at the atomic scale. This informs deformation in single crystals of material and this scale defines understanding of a suite of operating deformation mechanisms at the microscale. The use of X-ray diffraction (XRD) has allowed the field to deduce structure through analysis of Debye–Scherrer diffraction patterns, at pressures up to 300 GPa. Interaction of X-rays with a sample creates secondary diffracted cones of X-rays related to interplanar spacings in the crystalline sample. Diffraction is observed when the Bragg equation is satisfied:

$$n\lambda = 2d \sin\theta, \tag{3.4}$$

where n is an integer, λ is the wavelength of the X-rays, d is the interplanar spacing generating the diffraction and θ is the diffraction angle.

The diffraction pattern consists of Laue spots for single crystals or diffracted cones for powdered samples recorded by a solid-state detector or on film. The spacings of the spots and the corresponding d spacings are used to determine the lattice parameters and the coordinates of the crystallographic unit cell as a function of pressure. This is measured using the R1 and R2 bands of ruby, which fluoresce (wavelengths of 694.24 and 692.92 mm at zero pressure) when excited with blue or green laser light. These bands show a strong wavelength shift as a function of pressure, calibrated using XRD measurements on metals whose equations of state are well known from shock-compression measurements. Measurements may be averaged since ruby grains as fine as 1 μm can be finely dispersed throughout the target. Additionally, optical measurements can directly measure dimensions of the sample as a function of pressure or of changes in the volume of the sample chamber. Knowing the pressure and volume by this means, it is possible for pure single crystal samples to determine isothermal equation of state curves. Further, adiabatic moduli and their derivatives may be measured using Brillouin spectroscopy on the pressurised targets (Figure 3.8).

In this standard form, the technique allows the determination of the pressure and volume to deduce equation of state for the material which, with associated diffraction, additionally shows the crystal structure of the material. Recently, however, radial X-ray diffraction has been used to investigate an intentionally non-hydrostatic stress state in the sample. Under these conditions the elastic lattice strain is measured as a function of the angle from the loading axis. The stress state in a polycrystalline sample under

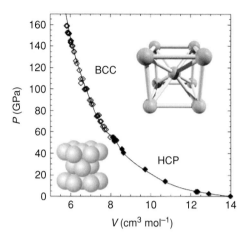

Figure 3.8 Magnesium in the HCP and in the BCC phase at higher pressure. Source: after Macleod *et al.* (2012).

uniaxial compression in the diamond anvil cell is described by a maximum stress along the cell loading axis, σ_x, and a minimum stress in the radial direction, σ_y.

The shear stress, 2τ, is related to these components for isotropic media in the manner discussed earlier

$$2\tau = \sigma_x - \sigma_y. \tag{3.5}$$

During the loading the sample is elastically and then plastically compressed. Grain size may also decrease with compression. So that incident and diffracted signals may be recorded, synchrotron X-rays pass through a transparent gasket such as beryllium and the *d* spacing and intensity of individual diffraction lines are determined as a function of the angle from the loading axis. The differential stress is deduced from the measured lattice parameters and used to determine the differential stress defined above. For most materials this increases with compression and eventually reaches a plateau at higher pressures. This behaviour is discussed in relation to dynamic measurements later in the chapter on metals. The technique has attracted recent attention and with development has great potential for future studies.

Although pressure and volume may be applied and sensed on loaded targets, it is necessary to also adjust their ambient temperature by heating or cooling to subsequently track the isothermal response of the volume as pressure is increased. It has been found possible to heat cells electrically to a few hundred degrees kelvin. Similarly, encapsulating targets in cryogenic liquid gases has enabled cooling to millikelvins. In recent years, laser heating has been used to reach temperatures of 7000 K. However, accurate measurement of temperature is subject to the difficulties typical of a diagnostic method that involves sensing grey body radiation, and this will be discussed later this book. Since one of the principal physical mechanisms under investigation is melting in the metal under pressure, this is also difficult to determine within the cell since the volume of the sample is so small and the radiation generally applied to it is so intense. Methods have been developed which observe speckle patterns from reflected coherent radiation, indicating surface optical diffraction effects which change in the target as the radiation is applied. Interpretation of these observations is controversial and at present the technique is in development. Also these measurements are confined to surfaces and it may be that the bulk has not yet reached melting during the loading. These difficulties result since heat conduction is a diffusion phenomenon and laser pulses are short resulting in non-uniform transient effects, analysed in this case using a technique appropriate to steady states in the material. Further, melting is not a phenomenon that occurs at a single temperature or position since no target is homogeneous. That having been said, this research offers a controlled means of accessing the locus of a change of phase across a pressure range and as such has important application for understanding several transitions in particularly metallic behaviours which have importance in a range of applications.

As shown earlier, to inform deformation at the macroscale it is necessary to acquire information from much larger volumes to address the polycrystalline behaviour of materials. It becomes more difficult to apply high pressures to larger volumes but nevertheless progress has been made using a range of devices that are presently available. At the same time as diamond anvils became common for small targets, the first versions of multi-anvil presses appeared. Over the latter part of the twentieth century there have been many improvements and modifications to these presses. Piston-cylinder and multi-anvil presses are used primarily for the synthesis and study of relatively large samples, whereas diamond cells were invaluable because transparency of the anvils allowed visual in-situ observation along with X-ray diffraction and spectroscopic measurements. Diamond cells have also reached the highest pressures of the two methods with pressures in excess of 100 GPa achievable in DACs.

The goal of larger targets at higher pressures has been achieved by using new capabilities to grow diamonds synthetically to large sizes. An intermediate device analogue to the DAC which sits in size between the smaller cells and multi-anvil presses is the Paris–Edinburgh cell.

The ability to use a spallation neutron source to illuminate targets led to a collaboration between Paris and Edinburgh Universities at the ISIS facility at the Rutherford-Appleton Laboratory to create a cell capable of compressing samples of c. 100 mm^3. The 'Paris–Edinburgh' (PE) cell is a compact larger-volume press with a 250 ton capacity and a mass of c. 50 kg (Figure 3.9). It was originally developed for time-of-flight neutron scattering experiments but has been adapted for a wide range of *in situ* measurements, such as neutron and X-ray diffraction, Extended X-ray Absorption Fine Structure (EXAFS), Compton scattering, inelastic neutron and X-ray scattering, and ultrasonic studies. While it was originally designed for powder diffraction on a spallation source, it has now also been used for inelastic neutron scattering on reactor sources. Pressures of up to c. 10 GPa may be regularly reached with tungsten carbide anvils but the advent of sintered diamond anvils has allowed the pressure range to be extended to 25 GPa. It is also possible to simultaneously heat the press to achieve temperatures up to 2200 K on several mm^3 sample volumes and then recover the macroscopic target after loading for further study.

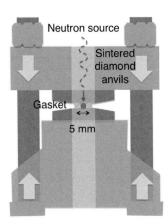

Figure 3.9 The Paris–Edinburgh cell. This large-volume cell is capable of loading targets of dimension c. 5 mm to pressures up to 25 GPa.

The cell is a hydraulic ram loading against a steel platen containing a screw-in breech. The breech and ram have anvils mounted on them between which the sample and gasket are compressed. The platen and the ram are linked in the original design by four large bolts, but this limits the angular aperture so that further adaptations have abandoned them to allow a panoramic view to the target. Further adaptation of the stage has allowed high pressures to be reached along with incorporation of thermocouples to measure temperature and instrumentation on the anvils for other measurements. Developments in the cell coupled with increased access to neutron sources will allow new techniques for powder neutron diffraction at elevated pressures to become a standard tool on continuous neutron facilities.

Other presses have been developed to uniaxially compress large volumes. These include piston-cylinder devices, Bridgman anvils and multi-anvil presses. The piston-cylinder operates in the manner its name suggests: accurate monitoring of the piston displacement allows volume measurement as a function of pressure, which at room temperature is limited to c. 5 GPa. Alternative apparatuses include Bridgman anvils and the Drickamer press. In these devices tungsten carbide anvils are forced together to uniaxially compress a sample contained within a retaining gasket ring. X-ray transparent rings (such as B or Be) my be used to allow diffraction measurements if required. Both devices allow greater pressures with a Bridgman anvil press capable of c. 15 GPa and a Drickamer press c. 35 GPa at room temperature.

Multi-anvil devices compress samples by application of force down several directions simultaneously rather than just uniaxially as with the DAC (for example, see Figure 3.10). A set of steel components transmit force through six or eight generally tungsten carbide pistons to the sample volume in a quasi-isostatic manner. This contrasts with opposed anvil-type devices, since in these the pressure is generated by the compression and confinement of a second, solid medium between the working surfaces of a pair of opposed-anvil dies. Multi-anvil devices are more difficult to use, since the anvils must be precisely aligned and synchronised in order to uniformly compress the

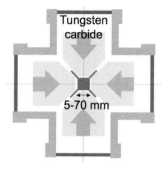

Figure 3.10 Multi-anvil press.

sample. However, in those meeting all of these requirements, the pressures reached in large samples can be much higher. There are a range of multi-anvil presses, but the tetrahedral and the cubic anvil devices are the ones most commonly used for equation of state measurements. These geometrically confine a sample within a cylindrical cavity inside the cube or tetrahedron and apply compression with four or six hydraulic rams. The oil pressure in the ram is calibrated against known phase transitions over suitable range such as in materials such as bismuth (I–II at 2.5 GPa and the V–VI transition at 7.7 GPa). Pressures reach *c.* 10 GPa in the cubic press with multi-anvil designs reaching 25 GPa. However, X-ray diffraction in such presses is difficult because of the external constraining equipment and this leads to a limited range of diffraction angles available to extract lattice parameters.

3.6.3　State of the art in static loading now and issues for the future

The previous sections have summarised a vibrant field for the study of extreme static loading which is developing all the time. By these means pressures in excess of 300 GPa have been achieved in the cell, although calibration of the pressure scale above 100 GPa and the stress state in the sample at such high compressions are sometimes in doubt. The measurement and application of a uniform temperature, so that it may be controlled for greater flexibility, has been attempted, but at present is a work in progress to attain a stable state. Further, raising the sample size and introducing dynamic deformation represents a new frontier for loading in cells.

High-pressure experiments wish to load as large a target to as high a pressure as possible, providing complete access to the loaded target whilst doing so. The size of the press, particularly its weight, is also a component feature in considering particular devices for the field. It has been found that a useful figure of merit (K) for 'compactness' of a press is the product

$$K = \frac{V P^3}{M},\qquad(3.6)$$

where V is the sample volume, P is the maximum pressure and M is the mass of the press. Looking over a range of static pressure devices of all types (piston-cylinder, diamond

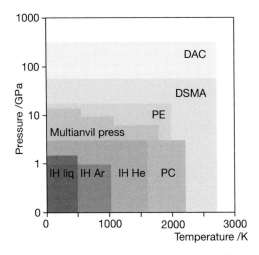

Figure 3.11 Pressure and temperature ranges of static high-pressure presses: diamond anvil cell (DAC); double stage multi-anvil (DSMA); Paris–Edinburgh cell (PE); end loaded PC (piston cylinder); internally heated vessel with liquid (IH liq); internally heated vessel with argon (IH Ar); and internally heated vessel with helium (IH He).

and multi-anvil cells) shows that K is typically 10^2 GPa^3 mm^3 kg^{-1}, and more or less independent of the technology involved (Figure 3.11). Thus the push to overcome these relations requires a change in the technology or materials strengths used in the cell. These limits are discussed below along with the present incremental approaches being taken to advance the use and performance of the cells. At present the major step improvement in utility of these platforms will come from the huge impact of fourth-generation light sources on the measurements that can be made rather than in their construction or use.

The small sample size in the diamond anvil cell requires a narrow, high-intensity X-ray beam which also has high energy in order to penetrate the anvils themselves through the limited opening available. These requirements are generally fulfilled at present with white synchrotron radiation and such bright sources allow a range of X-ray techniques to be adopted to monitor structural change in the targets with pressure. The intense hard X-ray beam from a third-generation synchrotron source allows the use of small sample volumes, and therefore gives access to a higher pressure domain, reduction of acquisition times and the possibility of using other complementary techniques. These included combining X-ray absorption spectroscopy (EXAFS) with X-ray diffraction to investigate a material under pressure. The challenge for loading devices is the provision of as wide an aperture as possible to allow a range of techniques to be applied over as many diffraction angles as is feasible along with provision of as broad an X-ray spectrum to the loaded sample. This latter requirement has meant that a range of anvil and fixture materials made of light elements has been adopted and further developments are taking place. Of these, anvil development is at the core of further improving both pressure achieved and sample volume raised to pressure. Diamond anvils provide the hardest and most durable systems reaching the highest pressures. However, clearly the sample size

Table 3.3 Hydrostatic limit in the behaviour
of pressure media (after Angel *et al.* 2007)

Medium	Hydrostatic limit (GPa)
4:1 Methanol-ethanol	9.8
Anhydrous 2-propanol	4.2
Nitrogen	3.0
Argon	1.9
Glycerol	1.4
Silicone oil	0.9

achievable is limited by available crystals. Thus, new techniques are becoming available that allow larger anvils to be made. Synthetic diamond has been available in various forms produced by a range of high-pressure and temperature syntheses including detonation and by chemical vapour deposition and advances in production methods continue. By these means sintered diamonds have been used in cells to extend achievable pressure range but also to achieve higher temperatures.

While large anvils can be made and high pressures thus achieved, a suitable fluid around targets that is good over the full pressure range is still to be found. An ideal hydrostatic medium will not support shear stresses over the entire range of the experiment. However, at some point this condition is violated for real materials. A non-hydrostatic medium can modify several of the measurements made in the experiment. Strain inhomogeneities in the sample broaden diffraction peaks whilst altering the relative evolution of cell parameters in deforming crystals. Such effects on pressure markers will result in consequent measurement errors for pressure increasing as compression proceeds and thus affecting results from equation of state and elasticity measurements. Finally, inhomogeneous stress states may promote or suppress phase transitions in some cases assisting amorphisation of crystalline samples. Thus, the stress state of a non-hydrostatic pressure medium cannot be easily quantified, and the resulting behaviour of the sample crystal cannot therefore be easily related to the experimental conditions. Indications from specific investigations indicate that the practical maximum limits to the hydrostatic behaviour of common pressure media at room temperature are of the order of 10 GPa and less (Table 3.3). Only hydrogen and helium can effectively provide such confinement at the highest pressures experienced by the cell.

Radial diffraction has addressed this by embracing the more complex shear state that develops under pressure increase to gain strength information from the sample. A second technique that has been used to apply shear directly is the rotational diamond anvil cell (RDAC). This allows one anvil to rotate about the compression axis relative to the opposite one under an axial load. This generates additional shear stress and strain on the sample, generating further plastic deformation. Under such pressure-shear loading processes such as phase transformation can be studied by application of shear strain to the target. This is distinct from the dynamic pressure-shear technique where the processes are kinetically controlled.

The DAC offers a further possibility that has been explored in part unintentionally by various workers. A sample may be compressed to a stable initial state of combined uniform high pressure and temperature using well-characterised apparatus, but may then be pulsed by a dynamic impulse of pressure or temperature to study departures from its elevated initial state. One means of applying that impulse may be using a high-powered laser either to heat or to mechanically load the compressed sample. Pulsed laser heating of targets in the DAC has been discussed previously and the interplay of the kinetics of melt coupled with the duration of the pulse has given a material state that is difficult to precisely interpret. Although the technique was conceived to overcome the difficulty of uniform heating, this new departure offers a means of controlling and investigating the dynamic melt process at pressure and should be embraced as a means of investigating rapid processes.

In a further evolution, Livermore researchers have developed the dynamic diamond anvil cell (dDAC), which compresses samples between two diamond anvils already compressed to an initial state by applying a further secondary dynamic axial load from an actuator. Using these dDACs, the compression rate is controlled by adjusting the rate at which the anvils expand or contract. The cell incorporates an electromechanical piezoelectric actuator to vary the pressure on the sample at a known rate, achieving compression rates of up to 500 GPa per second, or 5 GPa in 10 milliseconds, to study responses that include phase transformation kinetics under varying load.

Having run through a range of improvements in the technology for generating static high pressures, it remains only to say that these techniques have now advanced to the point where they are being incorporated in fourth-generation light sources. The 10^4-fold increase in beam intensity offers greatly increased precision in the measurements traditionally made, but also the possibility of using a new suite of techniques to inform high-pressure physics for the future.

3.7   Platforms for loading at intermediate strain rates

3.7.1   The development of faster loading techniques

To develop strength models for metals requires measurement of appropriate mechanical properties. Targets must be loaded in compression, tension or torsion over a range of loading rates and temperatures appropriate to the application of interest. However, to derive models describing purely material response requires idealised modes of loading which limit the range of techniques available. Such procedures are classified as derivation experiments. In the case of impact loading, testing from quasi-static to shock loading is necessary. A range of mechanical testing frames are available that achieve nominally constant loading rates for limited plastic strains at a constant strain rate. The standard screw-driven or servo-hydraulic testing machines achieve strain rates of up to five per second. Specially designed testing machines, typically equipped with high-capacity servo-hydraulic valves and high-speed control and data acquisition instrumentation, can achieve strain rates as high as 200 per second during compression loading. To achieve faster loading speeds, projectiles are impacted onto targets to induce stress-wave propagation in the sample materials. These results may be interpreted as higher strain rate tests, but to compare them with static machines on the basis of only strain rate is not possible since loading machines produce states where the target is in stress-state equilibrium, whilst in a wave front the material is not. Also, testing machines can test to large strain whereas impact tests only extend to small. Thus data acquired from small strain tests are frequently used to calibrate models that are then applied to large strain simulations beyond the limits of the calibration.

At intermediate strain rate, the principal dynamic loading technique is the split-Hopkinson pressure bar (SHPB), which can achieve the highest uniform uniaxial compressive stress loading of a specimen at a nominally constant strain rate of about 10^3 per second. The technique was first suggested by Bertram Hopkinson in 1914 as a way to measure stress pulse propagation in a metal bar connected to a target when a wave guide may be used to explore the response of the material (Figure 3.12) (Hopkinson, 1914). R. M. Davies later developed the technique further and the use of this device has been extended over the years to include a whole class of bar techniques. In compression the direct impact Hopkinson bar (DIHB) simply fires the incident rod onto the target and uses the information recorded in a strain sensor on the transmitter bar to recover stress–strain information from the target. Thus one wave is recorded and this leads to a derivation of the target deformation using a one-wave analysis.

H. Kolsky refined Hopkinson's technique by using two bars in series (now known as the split-Hopkinson bar) to deduce stress and strain in a deformed sample by monitoring the incident, reflected and transmitted components of the stress pulse; this is by analogy dubbed a three-wave analysis. Later modifications have allowed for tensile, compression and torsion testing. It is possible to reach nominal strain rates of up to 2×10^4 per second and true strains of 0.3 in a single test using the SHPB. The bar is at the limits of materials' testing where one can assume a constant stress state in the loading of a target. As discussed previously all measurement platforms have three phases of operation: a localisation phase where loading begins and plasticity is established; a flow phase with the material in stress-state equilibrium; and an interaction phase where deformation is halted. In the first and last phases of operation the continuum strain rate to classify the loading is not defined. In the SHPB the confined specimen takes some time to achieve a uniaxial stress state: several microseconds in a pulse length of maybe 100 ms for fully dense metals but much longer for low-wave-speed materials such as highly viscoelastic rubbers or soft high-density metals like lead. In some cases the target may not have achieved a uniaxial stress state by the time that the loading pulse has released. Thus the results must be interpreted with caution and signals collected during the early parts of the loading should be discarded until stress-state equilibrium can be shown to be established.

3.7.2   The Hopkinson or Kolsky bar

Figure 3.13 shows the mode of operation of a typical split compression bar system. The system is designed to use two elastic waveguides to deliver and transmit a pulse from

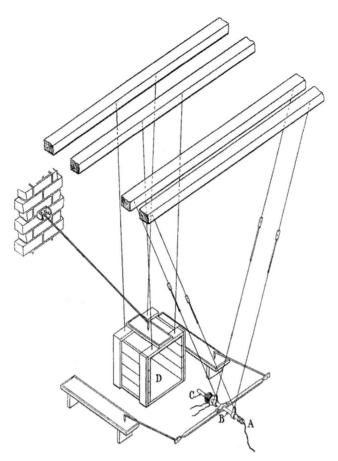

Figure 3.12 Apparatus developed by Bertram Hopkinson at the Department of Engineering, University of Cambridge, for the measurement of pressure produced by the detonation of gun cotton. Used to measure the shapes of pulses using momentum traps. Strain gauges were developed in the 1930s and only applied in the 1950s when amplifiers developed sufficiently. Source: after Bertram Hopkinson (1914), A method of measuring the pressure produced in the detonation of high explosives or by the impact of bullets, *Proc. R. Soc. A*, **89**: 411–413.

a lower impedance, deforming the target material. A bar is launched at a low velocity onto a second transmitter rod, the impact introducing a pulse which travels down it (indicated beneath the sketch of the apparatus in the figure). The pulse length is defined by the double transit time of a wave in the impactor rod and this defines the incident loading impulse seen by the target. The incident pulse arrives through a sample, which is compressed between the incident and transmitter bars. It is partially transmitted and partially reflected by the lower impedance interface between target and bar. It is important to note that as the target material in this technique is lower impedance it therefore yields before the bars do. The target is thinner than its diameter (generally a thickness to diameter ratio of 2:1 is adopted). This is to allow the pulse to bounce within the target and the stress to ring up in it and so attempt to ensure that the stress

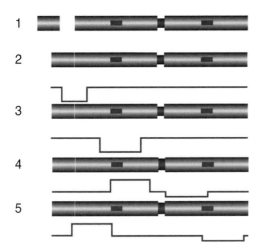

Figure 3.13 Operation of a split Hopkinson bar: the impactor comes in from left in frame 1 and impacts in frame 2. The pulse is formed by fronts travelling back into the impact rod and forward into the transmitter bar. The wave is split again by the lower impedance sample placed between the rods and shown. Typical bar dimensions give a time period of c. 20 μs between frames.

state is steady. Until a wave has traversed three longitudal passes through the sample it has not reached a uniaxial stress state which must be achieved if the simple experimental analysis is to be employed. In low-sound-speed materials (such as polymers), the signals recorded at the bars are not those of steady-state deformation unless the pulse length is sufficiently long. For this reason some researchers make use of Fourier transform techniques to deconvolve multiple bounces of the wave in the bar to recover large strain data from the deformation history. The forward pulse continues into the transmitter bar where it is sensed by a strain gauge mounted on the second. The incident pulse reflects and returns to a sensor on the incident bar to record a reflected pulse from the deforming target. The design of the bars and the impactor rod must ensure firstly that the material has reached steady-state conditions before the loading pulse releases, and secondly that that pulses at the incident strain gauges are not superposed so that complete information is recovered.

The following derivation using the notation introduced in Figure 3.14 follows the wave transmission and loading of the target of thickness. By definition the strain rate $\dot{\varepsilon}$ in the sample is given by

$$\dot{\varepsilon} = \frac{(\dot{u}_1 - \dot{u}_2)}{\lambda}, \tag{3.7}$$

where u_1 and u_2 are the displacements at the end of the bar.

Substituting for the incident, reflected and transmitted strains in those positions gives

$$\dot{\varepsilon} = \frac{c_R}{\lambda} \left(-\varepsilon_i + \varepsilon_r + \varepsilon_t \right), \tag{3.8}$$

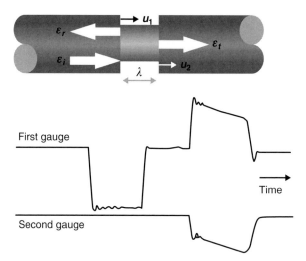

First gauge

Time

Second gauge

Figure 3.14 Wave propagation within the split Hopkinson pressure bar. Wave transmission occurs at the incident bar and the response of the target is picked up by the second. Sensors on the two bars capture the pulse history in the two wave-guides.

where c_R is the wave speed in the rod. Having derived these relations, it is possible to recover the stress and strain history in the target, and thus the dynamic response and also the material strain rate for the test conducted, as follows:

$$\dot{\varepsilon} = \frac{2c_R\varepsilon_r}{\lambda}, \qquad (3.9)$$

$$\sigma(t) = \frac{c_R Z_b \varepsilon_t}{A_s}, \qquad (3.10)$$

where the volume of the target remains constant ($A_0\varepsilon_0 = A_1\varepsilon_1$) throughout the test.

In real systems there are a series of attendant practical difficulties that limit the application of the technique. Firstly, the striker and input bars are almost always misaligned. This can introduce some off-axis loading modes unless carefully controlled and thus extends the initial unstable phase before the stress state settles. Secondly, the bar diameter is greater than that of the specimen so that it may expand under uniaxial stress as it is softer than the target. In reality this results in greater dispersion of the signal. A further key assumption in the analysis of the technique is that the gauges see what the specimen does. This means that the waveguides must be well characterised since their material response will also convolve the signal in some way. This is especially true when they are made of viscoplastic materials such as polymers. These effects and the travel of the pulse down a cylindrical section result in Pochhammer–Chree oscillations (who wrote an analysis of the phenomenon) transposed on the signal due to release effects from the rod. Here the boundary conditions (zero traction on the surface of the finite diameter rod) yield the Pochhammer frequency equation. Thus the wave front will smooth out, and high-frequency oscillations appear on the propagating pulse as the wave front travels along the bar. A finite diameter Hopkinson bar is thus a dispersive system and the pulse rises earlier than specimen strains, which effects the integration of the signal which must subsequently occur to recover the strain history. These uncertainties and lack of stress-state equilibrium mean that the signal, particularly at early times, represents the response of the measurement system rather than that of the material. Thus one cannot use the technique as an accurate measure for values of the Young's modulus, E, although the flow stress is measured accurately.

Comment was made earlier concerning the L/D ratio for Hopkinson bar targets. If the diameter of the specimen gets very large and very much greater than the thickness then friction may become critical and determine the expansion of the specimen. If present, this effect has the result of implying that strain rate hardening is occurring in the target. Thus it is necessary to employ lubricants on the sample under test suitable for the temperatures of the experiment, such as MoS_2, colloidal graphite, oil or petroleum jelly to ensure free movement of faces on the target.

The device itself has an upper bound to the strain rate achievable that is set by the dimensions of the target. The smallest sample sizes are determined by the number of inhomogeneous phases required to let the target be an isotropic representative of the continuum. This is typically 20 grains for a metal, which gives a minimum target size of $c.$ 1 mm. Radial inertia sets the upper bound on strain rate to $c.$ 10^5 s^{-1} in this case.

The Hopkinson bar test and derivation of the equations for its operation have a series of assumptions embedded within the experimental design and analysis which mean that it cannot be used as a black box to generate stress–strain data. There are a series of assumptions necessary to achieve a successful test and these are summarised below.

3.7.3   Assumptions of a valid SHPB test

Before using Eqs. (3.7)–(3.10) to infer the average stress–strain behaviour of the specimen material under high-strain-rate loading, the validity of the experiment and its assumptions need to be checked against the experience gained over the last century using the technique. Using a waveguide to load a specimen means that measurement stations are at some distance from the deformation in a SHPB test and thus deducing behaviour is contingent on assumptions about deformation and loading. For a valid test it is necessary to ensure the following:

(i) That the bars are good waveguides and the stress waves passing down them propagate in 1D. This implies that the bars should be homogeneous and isotropic in properties, which requires tight control of the material pedigree, and secondly that they are geometrically true with uniform cross-section and good axial alignment. Obviously the amplitude of the signal transmitted needs to be within the elastic limit of the bar material so that linear signals are maintained. Further, the signal must be steady (satisfied if the bar is $L/D > 20$) and ideally must not disperse with travel. However, clearly Pochhammer–Chree oscillations result since this is always the case, so the pulse must be long enough that this loading is fixed within the range for experimental errors.

(ii) The bar and specimen must be accurately flat and all interfaces with incident and transmitted bars must be planar throughout the loading period. This implies that the specimen should be softer than and its diameter less than that of the bar.

(iii) It is of course vital to ensure that the specimen is in stress-state equilibrium for the duration that valid data can be collected. This will inevitably not be the case in the first moments whilst the pulse within the sample is ringing up the stress state, but this can be checked by comparing 1-wave and 2-wave analyses for the experiment. These terms refer to whether a single or both reflected and transmitted signals are used in the analysis (see Gray III, 2000a) In all cases a thin specimen is necessary to minimise such effects and in general L/D should be < 1 whenever possible. In targets with low sound speed the time taken to equilibrate the stress state may approach the loading pulse duration, and therefore use of this method for such materials should be seriously considered.

(iv) If the specimen is compressible (which may be the case for soft or non-linear materials) then assumptions within the simple analysis are not met and special analysis techniques must be used.

(v) Finally, friction or inertia effects in the specimen and the loading must be minimised using suitable lubricants at the interface, providing the layer introduced is sufficiently thin that the properties at the bar–specimen interface are not changed.

Any materials test is always more valuable if the sample loaded can be recovered and examined after the experiment. In the case of the Hopkinson bar this requires some momentum trapping to ensure a single pass of the wave. Adaptations have been described for loading a target with one clean pulse in tension or compression and these give access to a deformed specimen that may be examined microstructurally or reloaded after the experiment.

There are other experimental devices and target geometries that have been adopted which recover similar information. More recently, testing in tension and torsion has been developed using waveguides in a similar manner. Tension testing in a split Hopkinson pressure bar (SHPB) is more complex due to a variety of loading methods and specimen attachments to the incident and transmission bar. The first tension device was designed and tested by Harding *et al.* in 1960; it involved a hollow weight bar, containing a threaded specimen, connected to a yoke. A tensile wave was induced by impacting the weight bar with a ram and having the initial compression wave reflect as a tensile pulse from the free end of the target and load the specimen on its return. There are many equivalent designs in operation, but the techniques all suffer from difficulties in trying to mount the specimen to the bars without forcing some restriction in maintaining the pulse form. In addition, tension is a more difficult mode to test than compression in all the examples considered since failure is by localisation and stress concentrations always ensue.

In the torsional variant, first developed by Duffy *et al.* (1971), the sample is held as a thin cylinder which receives a pulse launched down an incident bar when a bolt is rapidly removed from a torqued-up bar. Thin-walled cylindrical specimens are tested under dynamic shear and the torsion wave propagation in a circular rod is dispersion free. Torsional waves of large amplitude are generated in an elastic bar and waves with rise times of the order of 25 μs and maximum angular velocities of the order of 10^3 rad s^{-1} have been achieved. Tubular specimens have been loaded at shear-strain rates up to 10^4 s^{-1} using this method. The technique is limited by the need to have a thin-walled cylinder, which constrains the target size and thus the ability to represent bulk properties. It has thus been applied principally to the deformation of pure metals. Finally, in recent times a shear bar has been used. Here the target is a more complex top hat specimen placed into a compression SHPB where the transmitter bar is replaced by a transmitter tube. A hat-shaped specimen is then used with a traditional SHPB for shear testing.

In summary, the Hopkinson bar is a device in which the yield stress of a softer material may be determined deforming against a higher impedance elastic waveguide. It loads materials to intermediate strain rate by application of an impulse to a target by impacting it directly or via a transmitting bar and sensing the deformation remotely down a second waveguide. These two configurations are known as the split or direct impact configurations and produce data collected from a single transmitted wave or via two measurement stations, which is then reduced to stress–strain behaviour using simple assumptions concerning mass and momentum conservation. The Hopkinson bar can be designed to load materials in compression, tension and torsion, over a temperature range of –200 to 500 °C, to measure the yield stress under uniaxial stress conditions to an upper strain rate of 10^5 s^{-1}. The principal difficulties with the technique include radial inertia, friction (and thus lubrication) of the sample, dispersion of the pulse and an inability to accurately measure the modulus of elasticity.

Other tests have been developed to measure continuum (stress equilibrated) material properties under tensile loading. These include explosively or magnetically expanding rings, or firing a wedge into two targets to force them apart. In all cases, the attempt to measure high strain rate continuum tensile properties has proved difficult since some strain localisation inevitably occurs, rendering the stress state over the specimen in-homogeneous in the later stages. This often leads to confusions in interpreting the data from such tests and in comparing them with other wave-generated results where different localisation mechanisms operate.

3.8 Platforms for shock and quasi-isentropic loading

Although elastic waves displace material that subsequently returns to its original density after the front has passed, the shock permanently and irreversibly deforms it into a higher density state. Save the Z pinch loading discussed later, standard intermediate-rate techniques are primarily motivated by the need to determine the yield stress and load to amplitudes where inelastic deformation has just commenced. However, the shock and associated loading paths track the state of a material beyond yield to determine behaviour at compressions many times those found at ambient. The shock process occurs so quickly

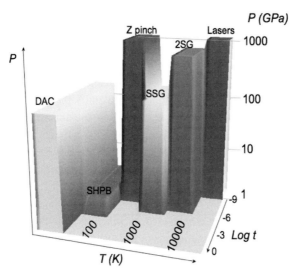

Figure 3.15 Loading equipment and delivered impulse on commonly used platforms. *P* is the maximum pressure achieved in the loading, *t* the pulse duration of the technique applied and *T* the temperature achieved in the pulse. SSG and 2SG are single- and two-stage guns respectively; SHPB is the split Hopkinson pressure bar and DAC the diamond anvil cell. The shade of grey indicates the temperature achieved with darker shades for loading at the highest amplitudes. Source: reproduced from Bourne, N. K., Millett, J. C. F. and Gray III, G. T. (2009) On the shock compression of polycrystalline metals, *J. Mat. Sci.*, **44**(13): 3319–3343. With kind permission from Springer Science and Business Media.

that it is not possible for heat transfer to occur in the time taken for the front to pass. Thus, it is adiabatic and accompanied by a step increase in entropy. At present the suite of experiments capable of generating shock loading of materials of interest consists of a suite of developed methodologies that test response over a pressure range of *c*. 1 TPa using impulses ranging from quasi-static high-pressure cells to nanosecond pulsed-laser sources. A schematic of the range of devices and their operational regimes is shown in Figure 3.15. The loading represented is mode I and reflects the final temperature achieved in the test as a shade of grey, with light colours representing the isothermal state (ambient temperature) and darker grey the adiabatic (shock temperature) achieved using the relevant platform. To populate the relevant material descriptions for all components of a device requires models that cover this entire three-dimensional space with validated certainty. At present it is only possible to load for nanosecond pulse lengths over a small target size to the highest pressures. This limits the mechanisms and microstructures that can be probed with such tests. As the compression amplitude decreases, it is easier to apply long pulses to microstructures representative of the continuum until, under ambient conditions, the diamond anvil cell provides high loading pressures but only over a sample dimension of the order of 1 μm at these levels.

There are five principal groups of loading technique employed to recover material response over this high-pressure regime, four of which are dynamic and include laser-induced plasma loading, Z pinch devices, compressed gas and powder-driven launchers, and energetic drives, with static diamond anvil cells (DACs) covered in the previous section. The laser methods introduce short pulses (typically less than a nanosecond) yet the largest can apply the highest pressures (TPa) and heat adiabatically to high temperature if a shock is launched. They ablate a plasma which loads the surface of a target and pressurises a face. The loading pulse is spatially and temporally difficult to control and this limits the impulse applied. The optical challenges of high-power lasers fix the focused spots to be small which means it is difficult to apply spatially uniform illumination across a large target area. They are consequently limited at present to probing subgrain dimensions laterally and microns longitudinally and this constrains the mechanisms that can be excited and probed. Thus they may be principally used to investigate the early time, shocked state of a material probing subgrain length scales. Recent developments have utilised reflected X-rays at the plasma–solid interface to recover the unit cell conformation under high pressures. Nevertheless, new facilities have been built over the last decade and if matched diagnostics and analysis techniques can follow, the possibility exists to study early time deformation in crystals to extremes of pressure at the theoretical limits of strength. Z pinch devices can provide pulses of hundreds of nanoseconds duration and can thus load targets of hundreds of microns dimension. They also have the capability to allow stress to increase over a similar time period, limiting the temperature rise and leading to a quasi-isentropic loading path with a shockless drive. They also have the ability to launch plates that can shock targets rather then employing direct magnetic interactions to apply the loading more slowly. The technology is becoming increasingly mature and the technique is being actively worked on in the laboratories possessing such machines.

Gas- and powder-driven launchers are capable of loading samples of a size and for a duration that is fixed by the dimensions of the device. The stresses they can achieve are limited by the velocity achieved and thus the propulsion method adopted, but typically they can shock-load materials to hundreds of GPa for tens of microseconds over sample volumes of cm^3. They are precision devices providing well-defined impulse loading over times relevant to all the operating plasticity mechanisms, which enables kinetics to be deduced from the response to the loading. Since they are a mature technique the analysis is equally pedigreed. The innovations in this technology are principally in the development, use and linkage to analysis of new diagnostics.

Diamond anvil cells (DACs) have been used to compress materials hydrostatically to hundreds of GPa. Clearly they can access impressive pressure ranges within a material but the size of the cell limits the loading to single crystal targets that at the highest pressures are only of micron dimensions. Also the timescales over which they load are much shorter than those in the dynamic cases discussed above, meaning that pathways may operate in diamond anvil cells with different kinetics than are observed under shock loading. Thus, whilst the DAC accesses equilibrium states, the dynamic case frequently jumps via metastable ones at different thresholds in state variables. New challenges involve pushing the cells and the pressure-sensing methods to ever-higher pressures. Beyond 100 GPa there is a limitation in assuring the stress state in the target is hydrostatic and to sense pressure with great accuracy. This represents work in progress for the technique but the advent of ever-brighter light sources has opened new possibilities for interrogating structure for the field.

The moment of impact or the application of a pressure pulse onto the surface defines the onset of impulsive loading in a solid. At that time the first plane of atoms on that surface is accelerated inward and the interatomic potentials provide an elastic response in the material to the insult. It is only after a characteristic time that the microstructure can respond in some manner to this step load by deformation by slip triggered from defects and it is the range of these processes and their kinetics that occupies what follows.

It is useful to classify the loading regimes described above into four groupings based on the length scales considered. Engineering design at laboratory scales requires experiment and modelling applicable to the continuum. On the other hand, to focus down to microstructural length scales requires one to probe a scale where inhomogeneities can be understood and their role in continuum mechanisms assessed and then quantitatively addressed. This microstructural length scale is dubbed the mesoscale. At a length scale beneath this again is the microscale, where the microstructural features addressed are recognised as groupings of atoms to study processes such as slip and twinning. Finally, the atomic scale accesses the physics that results from interactions occurring within the unit cell. By this means one can say that laser techniques can access the atomic and microscale, that launchers (Z pinch, gun or HE) are necessary to access the micro- and mesoscale, and Hopkinson bars or load frames are required to test the component-continuum scale.

Comparing devices to host investigation of the spectrum of deformation and damage evolution processes of interest is a complex task when considered in isolation from the applications to which they are relevant. What is more useful is to view all available experimental techniques from the standpoint of a community with a perceived need. At the present time, a primary requirement is to identify the mechanisms necessary to populate the modelling and simulation environments required to support design activities. By these standards, each can be judged as to its success in achieving a description of material performance over the operational stress range picked for condensed matter physics above. From this perspective, lasers and DACs load small volumes of material and probe the unit cell or nanometric substructure, whereas launchers and Z pinch machines investigate continuum dimensions by virtue of the larger target volumes and therefore microstructural scales interrogated. Although techniques that load a material using shockwave loading do so adiabatically, shockless techniques have the capability to take compression to lower temperatures so achieving higher pressure states more easily in the laboratory. On the other hand, laser platforms can load at the highest pressures, yet the loading times are sufficiently small that the transit from an elastic to a plastic state may be observed. Only launchers or HE lenses can load for the tens of microseconds required for localisation to occur to form shear bands, which is the channel for flow at the structural scale. Thus for a general need to populate materials descriptions over a wider operating range and for describing the bulk response of materials, all of these devices and both *in-situ* and post-mortem techniques are required to resolve the temporally and spatially non-equilibrium response of materials in extreme environments.

In the shock regime (up to the *finis extremis* at a few hundred GPa) where one is studying primarily the condensed state, loading has been applied in a configuration where the material is subjected to a state of uniaxial strain at the continuum. In this regime a planar shock wave is introduced into a target by the impact of a flat plate from a gas-driven launcher with precise alignment of impactor and target prepared to high tolerance. This technique is often dubbed plate impact and this geometry will be described below. Applications accessing stress levels induced by direct contact from a detonating high explosive, through blast and down to other lower amplitude impulsive loadings also occupy this regime. The response of metals here is dominated by thermally activated dislocation motion as a primary plasticity mechanism in the weak shock regime and this will be followed and assessed in subsequent chapters. In brittle solids the inelastic response is through fracture and the failed material is then compressed to high-density states behind the shock. Polymers and plastics deform through regimes where firstly the Van der Waals forces between chains are overcome and then by reforming of a diamond-like high-density phase through atomic rearrangement. Finally, energetic materials reach the transition beyond which chemical reaction is triggered within the shock front itself and high-explosive response is then defined by the kinetics of reaction at pressure within the material. To span this range there are three principal classes of device capable of achieving the high-pressure states required in materials of interest: launchers, Z pinch devices and lasers. Each of these classes will be described in turn below.

3.8.1 Specimen preparation and alignment

Before beginning to discuss the range of experiments and launchers available for dynamic experiments, it is worth quickly noting one of the primary needs for all plane wave experiments. That is that the faces of the flyer and specimen plates must be flat to high tolerances for the experiment to be effective. Typically in plate impact on launchers, it is necessary for the impact face to be flat across the surface to *c.* 5 μm. One means of doing this is to ensure that the flyer plate and specimen are lapped flat (for example, by first using a 15 μm abrasive, followed by a 6 μm diamond grit or equivalent). One means of confirming the flatness after working them is to place them onto a transparent optical flat and observe the surface through this whilst illuminated with a monochromatic light source. The interference fringes (Newton's rings) formed between the polished surface and the optical flat can be used to map the surface planarity. The procedure is continued until only two fringes are visible over the whole surface to ensure that adequately flat targets are produced. The rear surface of the target plate must also be polished using abrasives to obtain a reflective finish suitable for diffuse or specular reflections for interferometric purposes, according to the technique used. In experiments where a reflection is expected for interaction with other waves within the target, then it is necessary to ensure the surfaces are also parallel to a similar tolerance. For a pressure–shear experiment impact surfaces are roughened by lapping with 15 μm diamond paste. This is to ensure sufficient surface roughness to transfer the shear loading by dry friction. Finally, it should be noted that rear surfaces observed with velocimetry have slightly roughened surfaces to ensure diffuse reflection. This is necessary if the visibility of fringe systems is to remain acceptable in dynamic experiments.

3.8.2   Launchers

In order to introduce a square compression pulse into a material in the regime below the *finis extremis*, the loading of choice for the past 50 years has used gas-launched projectiles to impact onto a target. High-explosive systems are also used but neither put a square pulse into a target nor easily allow control over a large pressure range. Clearly the arbitrary impact of a plate onto a structure will put components into complex loading states, but here planarity is an issue. Thus for investigation of materials and their deformation mechanisms, an idealised, uniaxial strain geometry is preferred. This requires high tolerances in the construction of the parts of the gun system, since a projectile must accelerate a flat (to *c.* 10 μm) plate to impact a target of similar flatness. The accuracy of the alignment is fixed in construction to be better than 0.5 mrad at all velocities. Whereas such precision is difficult to attain experimentally, it does introduce a planar shock wave giving a mechanical state that is (in most cases) easily analysed; the whole is termed the plate impact experiment. Plate velocity and shock amplitude are accurately controlled, and the induced wave profile is measured at some distance into the sample by one of several techniques. There is a range of sensors employed that includes embedded devices measuring particle velocity, stress, and strain at a Lagrangian point as well as interferometric techniques measuring particle velocity or temperature at a surface. The centre of the sample material is uniaxially strained and then at a later time relieved in the direction of the shock if the impactor is thin enough, since this point is inertially confined. If release does not come in from a thin flyer then waves penetrate from the periphery of the sample and relieve the material by strains in the plane of the shock but at this point the uniaxial strain state is lost.

An embedded sensor designed to measure the stress at a position must use some coupling between its mechanical and electrical properties in order to obtain a signal. The most commonly used are piezoresistivity and piezoelectricity. The former requires the use of a power supply along with the gauge to provide a constant current so that a voltage may be recorded. Piezoresistance describes the change of the electrical resistivity of a material with applied external stresses. Bridgman used manganin (an alloy of 84% copper, 12% manganese, and 4% nickel that fits the requirements outlined above) as a standard pressure transducer up to 10 GPa in his static high-pressure cells. Manganin is the material of choice for high-stress ranges (up to 150 GPa). Other materials have also been studied as possible candidates, and of these carbon and ytterbium have proved the most reliable for use in the lower stress range (up to 2 and 4 GPa respectively) where manganin is less sensitive. Use of gauges in other configurations allows the determination of further parameters within the sample. A transverse gauge can be used in conjunction with a longitudinal one to measure directly both principal stresses in a sample allowing direct determination of the shear strength of the material. Additionally, both components of strain may be measured directly by embedding gauges. In the strong shock regime such sensors do not operate and surface interferometric techniques dominate.

The principal advances in dynamic instrumentation over recent times have come in developing methods to track the particle velocity state of a target using embedded gauges in the weak shock regime or by monitoring moving surfaces. In the latter case, non-invasive interferometric techniques may be used to record the velocity history of the external surfaces of shocked specimens. Mathematical descriptions of physical and chemical states developing in an energetic material have been greatly aided by measurements of wave evolution through a target. Such experiments have used an embedded sensor moving in a magnetic field. The particle velocity gauge is uniquely able to survive for the times of interest within heterogeneous materials. Clearly this class of experiment involves firing inert flyer plates to provide the pressure inputs to the material. Traditionally measurements made were time histories, using various sensors fixed within the target, the motion of boundaries, or emitted light, with the latter being the most common. These studies all generally supported particular explosive models, but details of the build-up process remained unknown. On the other hand, the use of multiple, embedded particle velocity and impulse gauges allows one to look at the evolving reaction as a function of space and time.

In summary, the plate impact geometry allows for detailed information to be obtained directly about the processes occurring after impact, including the chemical kinetics (in the case of HE) as the material responds. The data it produces allow the direct determination of parameters that can be used in the construction of a mathematical description of the material to define its dynamic response. Such a loading pulse may then be used to test these models using further experiments including other test geometries and these will be covered in more detail in Chapter 4.

There are two principal classes of launcher: single- and two-stage gas guns. It is the gas flow from the breech or propellant that determines the velocity attained. The maximum practicable velocity for a projectile is the speed of sound. In constant-pressure gas breeches the volume of gas in the chamber becomes the limiting feature. Using gun propellants gives a large pressure pulse but the gas pressure drops during the travel down the barrel as it occupies such a small volume. Thus, in practical situations, conventional gunpowder artillery is impracticable above 2 km s^{-1} and difficulties in propellant loading and variability in performance make powder launchers relatively rare for laboratory work at these velocities. However, single-stage gas guns achieve around 1.5 km s^{-1} maximum speed but are more accurate in velocity. In the lower velocity range gas guns are preferred whilst in the higher velocity regime propellant still gives optimal performance. Transverse flow gas guns (the electrothermal ramjet, the ram accelerator *et al.*), where the flow of propellant is transverse to the projectile movement, can accelerate large projectiles to a much higher velocity than conventional gas guns. They, however, have other problems in performance and operation.

Electromagnetic guns (coil-gun and rail-gun) are heavy and expensive due to the high cost of the electric power supply and switches. In such guns, a conductive projectile is fired into the rail-gun and slides between two parallel conductive rails and closes the electric circuit. The large current flowing in the circuit generates a magnetic field and Lorentz force, accelerating the projectile. In all such systems constructed, rail erosion has been severe and the demand for electric power and the cost of switching

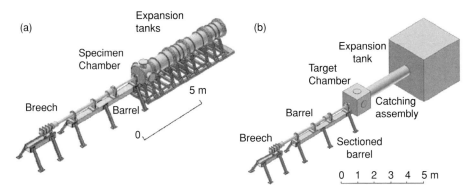

Figure 3.16 Single-stage launchers with alternative expansion tank designs. Source: reprinted with permission from Bourne, N. K. (2004) Gas gun for dynamic loading of explosives, *Rev. Sci. Instrum.*, **75**(1): 1–6. Copyright 2004 American Institute of Physics.

circuitry high. Physical contact between the projectile and the rails ablates both and produces a plasma which short-circuits the rails and limits the maximum velocity to about 6 km s^{-1}.

3.8.2.1 Single-stage launchers

The design of any launcher must account for the velocity and the mass that is to be delivered for impact on a target. The starting point for determining these quantities, however, is the pressure or stress and the loading time (the incident impulse) that the experiment requires. The history of ballistics has used black powder as a propellant and the engineering of a metal tube to encase the powder and projectile. The burning of the propellant induces a fast pressure pulse which accelerates the projectile down the barrel with a short impulse applied during launch. Over the history of development of the chemistry and ignition of propellants, it became possible to make the burning process sufficiently controlled that projectile velocity was reproducible to a precision that allowed use of artillery in more demanding applications. However, there is still some variability that results from the complex interplay of the chemistry and packing of the propellant bed within the gun breech. For laboratory use, the powder gun has been largely replaced by the compressed gas gun (Figure 3.16). The latter provides a precise, high pressure to the rear of a projectile by rapid opening of a constricting valve. These restrictions are the principal variable in the operation of the gun and the velocity achieved since they choke the flow of gas by some means or other from the high-pressure reservoir to the accelerating projectile. In what follows, the means of operation and analysis of such a gun will be described to illustrate the limits that exist for plate impact as a means to probe material response.

Each launcher, as will be seen below, has a range of velocities that can be obtained simply and accurately. However, this range has lower and upper practical bounds and one gun cannot cover the entire desired spread of speeds for a large range of projectile masses. Further, special systems must often be constructed to overcome the often more difficult need that is to operate a device with precision at lower velocities. An upper

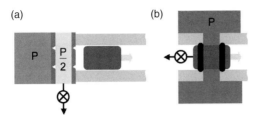

Figure 3.17 Firing designs for gas-driven launchers: (a) bursting disc and (b) wrap-around breech (b). In (a) a high pressure P is kept from the projectile by two discs which burst at a pressure of just $> P/2$. The small central volume is kept at pressure $aP/2$. A valve empties this central region and fires a projectile with the pulse P. In (b) the projectile seals the high-pressure cylinder with two O rings. It is held in place with a rear vacuum pulling the projectile back. When this restraint is removed, the projectile is sucked forward, clearing the sealing rings and releasing high-pressure gas to the rear of the projectile.

bound for the speed is fixed by the propellant gas and the projectile mass. Thus for a typical 5 m barrel length (typical laboratory system) a velocity range of greater than $c.$ 400 m s^{-1} and less then 1400 m s^{-1} is typical for accurate operation for projectile masses of hundreds of grammes. Of course, the projectiles to be launched will determine the mass and velocity achievable with the gun. To induce the maximum stress the user must launch high-impedance materials at the target and that often results in heavy, metal flyers. It is thus often useful to use the fact that ceramics typically have high impedance and low density to reduce mass. Also, projectiles are going to be lightweight and if embedded sensors such as particle velocity sensors require a magnetic field for their operation, data would not be perturbed by a metal in the field.

Once a projectile is loaded into the breech, the gun is fired by applying gas pressure to its rear, so accelerating it down the barrel. The greatest velocities are achieved by applying the pressure pulse instantaneously when a firing impulse is received and ensuring that the force is uniform across the rear area of the sabot. For safety reasons the means of firing the gun should only be operational when a firing pulse is received and for practical reasons the firing mechanism needs to be reproducible. The theories of the gun described below all assume that the firing process is instantaneous and there are various methods used to ensure that this happens in a controllable manner, of which two common ones are described below.

In the first instance (Figure 3.17(a)) the breech may be separated from the barrel using two bursting discs, where the first confines the gas in the breech and the second seals the projectile in the barrel. By this means accurate velocity control may be achieved. A second method uses a wrap-around geometry (Figure 3.17(b)) with no consumable parts. The sabot seals the reservoir and when released opens the rear space behind the projectile to the full pressure contained. However, it is less easy to control velocity using this technique. For lower speeds, the bursting discs open reproducibly when there is a high breech pressure and thus projectile mass. To attain higher speeds, the lack of a need to support any load before firing means that lightweight sabots may be used.

Clearly the breech could potentially be pressurised using any desired gas. However, a requirement for its use on modern, safety-conscious sites is that the highest sound-speed inert gas (which is helium) be adopted. When they expand, gases from the breech reservoir are collected in expansion tanks constructed around the end of the barrel. These tanks, and the barrel ahead of the projectile, are pumped down to approximately 1 mbar before shooting. This is necessary since the room should not experience overpressure due to the firing of the gun and the sample should not be disturbed by the impact of a column of air expelled ahead of the projectile. Finally, the gun fires almost silently since no air shock travels ahead of the projectile. Additionally, there is no adiabatically compressed cushion of gas heating the impact face of a target. A good vacuum is thus necessary to avoid spurious events resulting from ignition temperatures achieved during adiabatic gas compression. The bore of the gun fixes the loading time within the target: the useful time under uniaxial strain before release waves penetrate from the periphery (around 4 μs after impact for a Cu target in a 50 mm bore gun).

The bore of the gun also fixes the cross-sectional area of the sabot and thus its mass. The maximum speed is fixed by the sound speed of the gas propelling the projectile and it is this around 1500 m s^{-1} that also determines the size of the gas reservoir and its working pressure. The target is mounted directly onto either a machined end surface in a suitable location on a disposable end section or onto an adjustable plate which is aligned to the gun axis with alignment threads.

A simple screening of design parameters may be achieved using an analytical interior ballistics model. The interior ballistics of a single-stage gun may be simply described as the expansion of gas into the barrel transferring kinetic energy to the projectile. Equating the volume of gas expanded at pressure P into a barrel of cross-section A and length L to the kinetic energy gained by a mass m in its acceleration, one can solve for the velocity v thus

$$v = \sqrt{\frac{2PAL}{m}}. \qquad (3.11)$$

This approach is too simplistic since it predicts a monotonically increasing velocity for barrels of increasing length that results from the assumption that the pressure accelerating the projectile remains constant throughout its travel. While this is not the case in reality, the reservoir can be made sufficiently large to approximately satisfy this condition until the projectile exits the barrel. Of equal importance to the acceleration of the projectile is the shedding and reflection of release waves from its rear. As it accelerates down the barrel, the sabot leaves an area of lower pressure behind it that propagates as a release back toward the breech as high-pressure gas from the reservoir rushes forward to balance the pressure. Such a wave, on reaching the breech, sees a change in cross-sectional area in the throat separating the barrel from the gas reservoir and is thus partially transmitted and reflected at the interface. The wave returning from the breech changes phase and arrives back at the base of the projectile as a compression pulse that will thus act to further accelerate it. The sound speed in the compressed gas is elevated to several times its ambient value and many wave reflections may occur during the passage of a projectile down the barrel.

The operational dimensions of the working space available fix various parameters for the gun system while leaving others to be optimised. This may be done using a simple interior ballistics model. The size of the breech is fixed to ensure that the volume of compressed gas is sufficient that there is little drop in pressure behind the projectile as it travels down the barrel. On the other hand, the volume of the expansion tanks is fixed by the space available. Typically, the volume of the breech is c. 20 litres for the gun outlined previously, which ensures that the pressure drop does not result in appreciable velocity differences to those predicted analytically. Substituting all these factors into the model, one obtains a relation for the velocity in terms of pressure, the dimensions of the gun and the thermodynamic properties of the propellant gas. The maximum working pressure of the breech is typically fixed to be c. 700 bar to achieve the high velocities required. One means to solve the equation for the position of the projectile analytically is to adopt an approximation valid for the case where the ratio of the mass of the gas in the breech to the mass of the projectile is small. In this approach, the effect of the many wave arrivals at the base of the projectile could be described as the asymptote to an oscillatory function and thus it is possible to derive the dependence of the projectile velocity v on the pressure P, in the form

$$V = \frac{2c_0\sqrt{a_0}}{\gamma - 1}\left(1 - \left(1 - \frac{\gamma P A L}{c_0^2 G}\right)^{1-\gamma}\right), \text{ where } G/m < 1, \qquad (3.12)$$

where P is the pressure in the breech, c_0 is the sound speed in the reservoir at that pressure, a_0 is a function fitted from the numerical integration of Eq. (3.13), G is the mass of the gas in the chamber, A is the cross-sectional area of the barrel, L is its length and γ is the ratio of specific heats of the gas.

The function a_0 results from the original analytic solution and depends upon G/m and γ. The function, though small in magnitude, is at the very heart of the dependence of the velocity upon pressure:

$$\frac{G}{M} = \frac{2\gamma}{\gamma - 1}a_0(1 - a_0)^{\frac{2\gamma}{\gamma-1}\int_0^1 (1-a_0\mu^2)^{\frac{1}{1-\gamma}}}\,d\mu. \qquad (3.13)$$

One needs to combine several factors including the polynomial for the parameter in Eq. (3.12) to determine the dependence of velocity on pressure. Further, one needs to substitute an Abel equation of state for the gas sufficient to account for the high gas pressures. Finally, one needs to correct for the area mismatch at the breech (chambrage) of the gun. In order to account for this it is necessary to incorporate a correction for the additional driving force provided by the reflected wavelets. This may be done by scaling the driving pressure and the initial sound speed c_0, derived using the assumption that the flow into the barrel is always sonic:

$$p = p_0\left(1 + \left(\sqrt{\frac{\gamma + 1}{2}} - 1\right)\left(1 - \frac{A_1}{A_0}\right)\right)^{\frac{2\gamma}{\gamma-1}},$$

$$c = c_0\left(1 + \left(\sqrt{\frac{\gamma + 1}{2}} - 1\right)\left(1 - \frac{A_1}{A_0}\right)\right), \qquad (3.14)$$

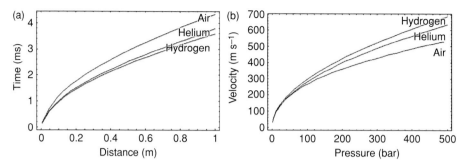

Figure 3.18 (a) Calculated paths for a projectile as a function of time down a short barrel. (b) Velocity achieved as function of breech pressure. Source: reprinted with permission from Bourne, N. K. (2004) Gas gun for dynamic loading of explosives, *Rev. Sci. Instrum.*, **75**(1): 1–6. Copyright 2004 American Institute of Physics.

where γ is the ratio of specific heats, A_1 is the area of the barrel and A_0 is the area of the reservoir.

It is now possible to analyse the transit of the projectile down a barrel with some accuracy. Figure 3.18(a) shows calculated paths for a point on the rear of the projectile as a function of time down the barrel. The histories for different gases are shown, with the lowest of the curves that for hydrogen, the centre one for helium and the top one for air. The curves are derived by numerically integrating the velocity–time behaviour derived in the model. The motion of the projectile is qualitatively similar for each gas. Its acceleration is slow, rising to within a few per cent of its maximum speed over the first part of its journey. Figure 3.18(b) shows the expected velocities as a function of pressure for all three gases. As expected, hydrogen is the most efficient propellant gas (since it has the highest sound speed).

Clearly, using a given launcher involves using a range of projectiles with different masses since even if the sabot remains the same weight the range of flyer plate dimensions and materials mounted to it will ensure that the total projectile mass is not constant. Also one can adjust launch tube dimensions and driver parameters to optimise performance. It will be seen from Eq. (3.11) that the velocity scales as the inverse of the root of the mass. Figure 3.19(a) shows a plot for a 75 mm bore gun of velocity as a function of breech pressure for a range of projectile masses and this shows the dependence of velocity on mass. The tightest spread of calculated velocities is given for the ratio P/m, where P is the pressure and m the mass. This relationship is shown in Figure 3.19(b) and experimental velocity data are plotted along with it. It has been found for comparison that with a 6 mm barrel and 50 mm bore, the maximum velocity achieved is *c.* 1500 m s^{-1} with a 50 g projectile and 300 bar of helium.

The performance curves for air and helium have experimental data for a range of shots plotted with them. The individual points represent the average of several firings conducted at a range of velocities and with different masses, so that the curves represent performance over projectiles that can include many different impedance flyer plate materials according to the experiment conducted, including alumina, copper, aluminium, tungsten alloy, glasses and PMMA in thicknesses varying from 1 to 20 mm. Helium has

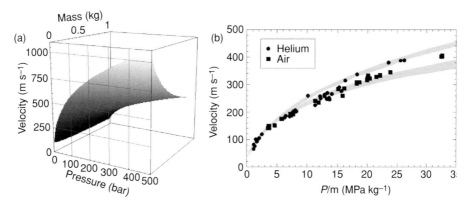

Figure 3.19 (a) Performance curves for various masses of projectile fired down a 1 m long barrel at a range of pressures. (b) Performance curves for projectiles fired with air and helium. The lower is for air and the upper is for helium. Source: reprinted with permission from Bourne, N. K. (2004) Gas gun for dynamic loading of explosives, *Rev. Sci. Instrum.*, **75**(1): 1–6. Copyright 2004 American Institute of Physics.

the better performance as expected and the theory gives good results. It is emphasised that the velocities calculated are those for heavy projectiles where G/m, the ratio of gas to projectile mass, is small as is the case for large mass projectiles on the large guns found in major laboratories for plate impact.

When the mass of the projectile is much less than that of the gas, an alternative formulation is necessary since the analytic solution breaks down. The approximation of Eq. (3.13) no longer holds and the velocity v is given by

$$v = \frac{2c_0\phi}{\sqrt{\gamma - 1}} \left(1 - \left(1 - \frac{\gamma PAL}{c_0^2 G} \right)^{1-\gamma} \right), \qquad (3.15)$$

where

$$\phi = 0.2423 + 0.3462\frac{G}{M} - 0.0989 \left(\frac{G}{M} \right)^2 + 0.0154 \left(\frac{G}{M} \right)^3 - 0.00094 \left(\frac{G}{M} \right)^4.$$

This representation holds when G/M takes higher values (0.4–6 when $\gamma = 1.4$). In the modern era such approximate solutions are rarely used with the onset of computer flow calculations for the interior ballistics. Nevertheless, they provide good tools to assess the effect of varying parameters to design launchers and show useful trends in gun performance. This approach nevertheless ignores a raft of effects that slow the velocity in real systems such as friction, gas flow effects and finite valve opening time to vent gas systems. As illustrated above there are many ways to achieve performance and the discussion here should encourage users to think carefully on these parameters to achieve the required results in their designs.

3.8.2.2 Two-stage hypervelocity launchers

Despite the possibility of new design, physical limits will always cap single-stage launcher performance. In order to address problems of relevance over a greater range of impact loading scenarios it is necessary to achieve velocities in a higher domain, for instance c. 10 km s^{-1}, which is the realm of application for impacts upon space vehicles. In the case of gas-driven launchers, pressure and the sound speed of the gas used to propel the projectile limit the peak velocity attainable by a single-stage device. For those in which standard nitrocellulose gun propellants are used, it is not possible to launch projectiles at speeds above 3 km s^{-1}. Even with very large breeches, strengthened gun walls and low-mass projectiles, the performance of a conventional gun is limited by the amount of energy expended in accelerating the combustion products or driver gases. That is, once the projectile accelerates to velocities that are greater than the sound speed of the driver gases, the efficiency of the acceleration process falls off dramatically and most of the chemical energy from combustion is expended in accelerating the gases. For those heavy gases produced during the combustion process of most propellants, this energy can be considerable.

It was soon recognised that much higher projectile velocities could be achieved by replacing the heavy driver gas produced by combustion by a high-temperature gas with low molecular weight and high sound speed such as hydrogen or helium. This realisation led to the development of two-stage light-gas hypervelocity launchers. The two-stage device rapidly compresses a gas reservoir to attain transient pressures many times those achieved by pumping a cylinder up to a static value. The small compressed volume bursts a disc and the high-pressure gas released then launches the sabot with a high-pressure impulse that accelerates it to velocities in excess of the sound speed in the uncompressed gas. These conditions of high pressure and rapid flow mean that the gas states are very different from ambient fluid mechanics and such a flow of gas is thus classed as hypersonic. The word *hyper* means *more than* in Greek and convention defines hypersonic behavior to a class of flows in which the flight Mach number is greater than five. However, this classification is not precise and there are several observed features of the flow at flight speeds greater than the speed of sound which affect gun performance. The compressed gases will include oblique shock waves and indeed highly curved shocks with entropy layers. Further, the pressures and temperatures reached mean that there will be a number of equilibrium and non-equilibrium real-gas effects including vibrational excitation, dissociation, ionisation and radiation. Any of these physical features will make hypersonic flows distinct from simple systems and thus account of these effects must be taken in designing for these cases.

Before the 1950s the basic characteristics of hypersonic flows were not well understood. However, the rocketry and space exploration programmes of the USA and Soviet Union stimulated intensive research in understanding high-speed fluid dynamics. These programmes involved vehicles re-entering Earth's atmosphere from space travelling at velocities of 8–11 km s^{-1} and thus subject to high temperatures nearing 11 000 K. In order to meet the demands of space vehicle and rocket design a number of centres around the world commissioned launchers capable of achieving these velocities.

The first two-stage light-gas hypervelocity launcher was developed at the New Mexico Institute of Mining and Technology in the late 1940s. The gas used was hydrogen, which allowed the gun to launch 4.5 g spherical projectiles with velocities nearing 3.7 km s^{-1}. By the 1950s and 1960s there were several light-gas guns in use in

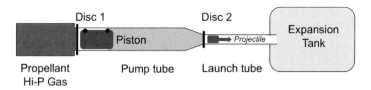

Figure 3.20 Schematic of a typical two-stage light-gas hypervelocity launcher.

aeroballistic ranges around the world and velocities in excess of 11 km s^{-1} were obtained. Although many other high-speed launcher designs were considered, the two-stage light-gas gun provided good performance and has found universal acceptance in hypervelocity ranges, particularly since it is possible to achieve high velocities with relatively low accelerations.

A typical simple two-stage light-gas hypervelocity launcher is shown in Figure 3.20. The basic configuration consists of a pump tube and a launch tube. As the gun is fired, a piston, usually made of a high-strength plastic such as polyethylene, is rapidly accelerated to high velocity by gases produced in the breech either by burning a granular solid propellant or by a conventional high-pressure gas. This piston then generates a series of compression and shock waves that heat and pressurise a light gas, such as helium or hydrogen, contained in the pump tube. At a predetermined pressure, a diaphragm located at the junction of the pump and launch tubes ruptures. Subsequently, the compressed light gas expands and propels a projectile along the launch tube and into flight at hypersonic velocities.

Although the piston accelerated by the hot propellant gases is decelerated as its kinetic energy is transferred to the compressing gas, in many cases it has a significant residual velocity when it reaches the end of the pump tube. In order to decelerate it, it is allowed to intrude into a conical area reduction section. This extrusion process rapidly decelerates the piston to rest. In many launchers, the extraction of a piston from this tapered bore section after a firing is a difficult task requiring compressed gas systems to free it.

By the 1970s, the space research programmes of the USA and Soviet Union had dramatically scaled down, leading to reduced interest in hypersonic aerodynamics. Many of the facilities, including the aeroballistic ranges, were mothballed or scrapped. Although some light-gas guns continued to be used in studies of hypervelocity impact, terminal ballistics and material penetration, most of the existing two-stage light-gas hypervelocity launchers became inactive and the development of new ones with higher performance has largely ceased.

3.8.2.3 The operation of a two-stage gun

Figure 3.21 shows the basic operation of a gas breech, two-stage launcher. The positions of the pump tube, tapered section (the grey rectangle) and launch tube are shown on the figure. The pressure P in the pump and launch tubes of the gun are calculated at various positions down the device and integrated over the delivery of a projectile at the end of the launch. Two simulations are shown with different initial pressures of helium in the

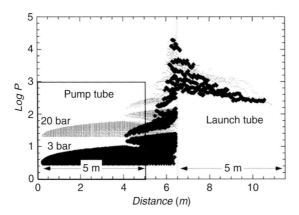

Figure 3.21 Pressure down the length of a two-stage gas gun during two launches with identical conditions in breech. Black is 3 bar initial helium pressure in pump tube and grey is 20 bar. The projectile mass is 300 g.

gun's pump tube. As the gas breech pressure accelerates the piston, it drives a shock into the gas in the pump tube and the pressure is increased as it travels down the barrel. The pressure achieved is almost the same (10^4 bar) in both cases calculated here even though the initial pressures are different, and this shows the interplay between the drive of the piston and the resistive force from the gas ahead of it. It will be seen that the pressure increases in steps as shocks propagate into the stationary gas, reaching its final values at minimum volume in the tapered section. At this point the second bursting disc opens and the pressure drops as the projectile is launched down the launch tube.

There are five key physical processes that must be accurately represented in order to understand two-stage light-gas gun performance. These are the effects of pressurised gas or the burning of the solid composite propellant and subsequent motion of the hot propellant gases in the combustion chamber; the complex unsteady compressible flow of the pre-compressed light gas (usually helium or hydrogen) in the pump and launch tubes; the unsteady compressible flow of the low-pressure air ahead of the projectile in the evacuated section of the launch tube; the motions of the cylindrical piston and projectile/sabot package in the constant area sections of the pump and launch tubes; and the extrusion of the deformable piston as it impacts, deforms and finally stops in the tapered section between the pump and launch tubes. The sequence of events and their relation with one another must be understand to predict the launch and more detail can be found by considering a one-dimensional model showing a more detailed simulation of the operation of such a gun. The following discussion shows these details of behaviour for two classes of launcher: one with a gas-driven piston and the second with a propellant drive.

The pressures at various positions within the pump and launch tubes calculated as a function of the time for transit of the piston and projectile down the barrels are shown in Figure 3.22(a). Initially the pump tube pressure is contained with a bursting disc, and this does not open until the pressure exceeds a fixed value. The pressure, P, is shown

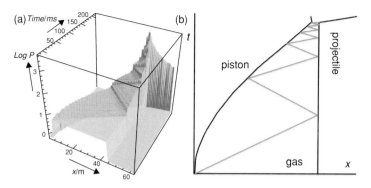

Figure 3.22 The operation of a 10 m pump tube initially pressurised to 10 bar with helium, connected to a 5 m long launch section in vacuum. The projectile is of mass 100 g and the piston is of mass 10 kg. The piston is fired with 700 bar of helium.

vertically and its logarithm is plotted as the pressures become very large; however, this only happens for small times. As time progresses the pressure in the pump tube increases to c. 0.5 GPa, at which point it ruptures the bursting disc. At this time, some 200 ms after the gun initially fires, the final sabot is launched to much greater velocity than the piston by the compressed helium trapped in the tapered section.

It is instructive to view the motion of the piston and projectile as purely a function of time. The x–t diagram is shown in Figure 3.22(b). The piston with inert gas pressure driving it accelerates to a constant speed over the first half of its flight down the barrel; the dark line trajectory shows its front face. As it does so it drives a shock wave into the trapped helium, increasing the latter's pressure in steps as can be seen from the wave fronts displayed. The wave is reflected at a tapered end section which acts to introduce two-dimensional flow into the gas. At a later time, the disc ruptures after the piston has intruded. The projectile exits at a velocity of c. 5 km s^{-1} and its rear surface position is plotted. It will be noticed that the piston front surface is in fact moving back on itself after rupture. It is even possible to find an initial pressure for the pump tube gas that will compress as the piston travels forward and reach a peak where the piston rebounds back down the pump tube, so obviating the need for its removal. This, however, generally reduces the launch efficiency and leads to a lower final projectile velocity.

A different mode of operation may be seen if a propellant charge is used to drive the piston (Figure 3.23(a)). Here the drive is pulsed rather than the constant pressure supplied from the gas breech. The impulse rapidly accelerates the piston away and compresses the contained helium, as seen previously, but the pressure is initially highest at the moments of burn in the propellant. However, the impulse ceases as the compression continues and the piston slows towards the latter parts of its motion. This gives a different mode of operation, with much stronger shock waves produced in the compressed gas.

The rapid acceleration at the start of the process sends a strong shock into the gas confined in the pump tube. Notice that there is a pressure pulse at the initial moment of piston motion which corresponds to several hundred bars – an order of magnitude

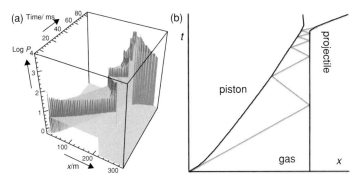

Figure 3.23 The operation of a 10 m pump tube pressurised to 10 bar with helium, connected to a 5 m long launch section. The projectile is of mass 100 g and the piston is of mass 10 kg. The piston is fired with a quantity of propellant which accelerates the piston rapidly at the start of the process.

increase in pressure. This contrasts dramatically with the more gradual compression seen when accelerated by a constant-pressure gas breech. Clearly the piston is loaded with large shock pressure jumps in its wave interactions during motion down the pump tube and this means the piston must be mechanically strong enough to withstand these excursions. After a much shorter time than in the previous simulation, the disc bursts and the projectile is fired off at a similar velocity to that seen in the case of an inert breech.

Looking at the x–t diagram for the piston motion, gas shocks behind the projectile travel may be clearly seen. The plot of Figure 3.23(b) shows in particular how the piston motion differs in form from that of Figure 3.22(b). There the trajectory of the front face of the piston is concave, whilst in the case of propellant the path is convex. This clearly reflects the different driving impulses supplied in the first case by a breech at a constant pressure of several hundred bars, whilst in the propellant case a pulse accelerates the piston down the pump tube in the initial stages. The optimisation of all these features – the intrusion of the piston, the bursting of the confining disc and the launch of the package – is a complex process and dependent on many variables in the design of a shot. Further, the packages launched must be capable of withstanding accelerations of millions of g. However, when all these factors are taken into account it is possible to launch hundreds of grammes at over 10 km s^{-1}.

The development of hypervelocity launchers was very rapid in the years between the war and the early 1970s. At the present time their use is limited, however, as designs and modes of use have not moved on in the past 40 years. It is possible to further increase velocity in several ways, particularly using the new array of strong and light materials that have become available in recent times. Thus the multistage launcher is a successful means of accurately delivering projectiles at high velocity and advances are to be expected in future years.

3.8.2.4   Pressure-shear impact

Pressure-shear plate impact is a means of generating dynamic shear stress–shear strain data under high-rate conditions with precisely controlled loading. Typical shear strain

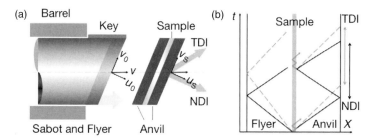

Figure 3.24 The pressure–shear impact experiment: (a) experimental configuration for the test; (b) Lagrangian X–t diagram for the loading with elastic waves in compression (solid line) and shear (dotted line). The windows for loading with normal and transverse interferometry (NDI and TDI) are indicated.

rates of the order of 10^5 s^{-1} are obtained for specimens thinned to thicknesses of c. 200 μm, and rates of up to 10^7 s^{-1} can be obtained for very thin specimens (with thicknesses of 2.5 μm) prepared by vapour deposition. The limitation on the technique is that the microstructure of the material must be such that the grains are small enough to make the measurement representative of bulk properties since the targets are so very thin.

In typical experiments, an inclined flyer plate is launched at a slender target which, on impact, becomes confined between itself and a rear anvil whose impedance is lower than that of its surroundings. The stress wave bounces within the target and the stress is said to ring up between the two plates around it, and since they remain elastic for the duration of the loading the signals transmitted through the rear anvil can be monitored and deconvolved to deduce the states developing (Figure 3.24(a)). The target is constrained inertially in pressure-shear and uniaxial flow is preserved. Impact launches compression and shear waves into the specimen, which travel at different speeds and arrive at the rear of the anvil at different times (Figure 3.24(b)). Application of a grid to the rear of the anvil allows lateral displacement to be monitored with laser illumination, allowing normal and transverse displacement measurement by standard interferometers (normal and transverse displacement interferometry, NDI and TDI).

The shear strain rate on the target is recovered from the lateral motion at the anvil face via

$$\dot{\gamma} = \frac{v_0 - v_s}{h}, \tag{3.16}$$

and the shear stress τ is given by

$$\tau = \frac{1}{2}\rho c_2 v_s, \tag{3.17}$$

where v_0 is the transverse component of projectile velocity, v_s is the free-surface velocity and ρc_2 is the elastic shear impedance in the flyer and anvil.

Thus measurement of the transverse free-surface velocity v_s allows the shear strain rate and the shear stress in the specimen to be determined once sufficient wave reflections have occurred for a nominally homogeneous states of stress to be established. The shear strain can be established by integration of Eq. (3.16), which enables the shear stress and shear strain and thus the dynamic stress–strain curve to be obtained. As with Hopkinson bar loading paths, the initial, steeply rising part of these curves corresponds to times before homogeneous states of stress are established, and should be disregarded.

After equilibration, the normal compressive stress, σ, is approximately equal to the hydrostatic pressure, P, and thus

$$P \approx \sigma = \frac{1}{2}\rho c_1 u_0, \tag{3.18}$$

where u_0 is the normal component of projectile velocity and ρc_1 is the acoustic impedance for elastic longitudinal waves in the flyer and anvil. Thus the pressure at which the loading has occurred is also defined. In other versions of such experiments, the loading is applied to non-metals and in this case particle velocity gauges mounted in the target can be used to directly sense in-material normal and transverse particle velocity histories within the target.

Of course the technique has several limitations that result from the integrated loading applied. It cannot be used to derive a model directly in this configuration but rather tests the predictions of an already formulated constitutive relation. In this respect it is more a validation experiment in an idealised geometry. Nevertheless, the geometry controls the shear applied in a controlled manner to focus on the continuum material response.

3.8.3   Explosive generation of plane waves

The need to load materials with idealised loading pulses allows a range of devices to be considered. The previous section has described a series of those used to drive flying plates to induce pressure by impact. However, chemical reaction may also produce pressured gases sufficient to drive controlled impulses into targets. The precision of such wave forms provides an easily engineered resource for understanding material response and the next sections enumerate the means of doing so using high-explosive lenses.

The detonation of a high explosive leads to a propagating shock with a characteristic profile that defines a characteristic mechanical response within a target (Figure 3.25). It consists of a series of zones determined by the chemical and mechanical conditions found in the explosive molecule on the one hand, and in the confinement of the charge on the other. The unreacted Hugoniot for the explosive defines the Von Neumann spike, which is the highest pressure reached instantaneously in detonation, whilst the reaction and flow then decrease the pressure through the reaction zone to the Chapman–Jouguet (CJ) point. At this point, releases in the flow determine the steady decrease of the pressure away from the detonation wave front since the flow of the hot product gases is determined by the mechanical confinement of the explosive. This form of pulse is typical for all ideal high explosives and the decrease in pressure due to the expansion of the products is given the generic name, the Taylor wave. The parameters that determine its form were derived and discussed more fully in the previous chapter; however, briefly,

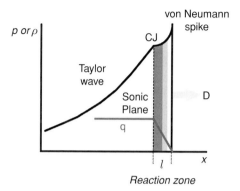

Figure 3.25 The detonation front in an ideal explosive (from ZND theory). The pressure at the Von Neumann spike decays to that at the CJ plane. Expansion of the hot products forms a Taylor wave with the final pressure determined by the boundary conditions.

the spike point is determined by the intersection of the Rayleigh line and the unreacted Hugoniot whilst the product isentrope is used to match the CJ point to the mechanical boundary materials.

This form of loading is generally more prevalent in the engineering arena than for long-duration impulses, such as those induced by the devices described above to derive materials parameters. To understand a material's response, a wave must be produced over a large volume of material onto which an explosive loads to induce the plane strain conditions described above. The simplest device capable of producing such a pulse is the plane wave lens discussed below.

A shock impulse suitable for experiments capable of derivation of operating physical mechanisms requires a planar wave front to be delivered to the target to give one-dimensional strain loading at the continuum. An explosive is an obvious means of doing so but conventionally devices are initiated by lighting at a single point, generally inducing a spherical wave front that propagates out from a point into the energetic material. An explosive plane-wave lens enables a spherical waveform to be converted to a planar shock that may be delivered to an inert or another reactive target.

Current explosive lenses, although successful, have problems; they tend to be expensive, require rigid tolerances and often prohibitive machining costs result in great expense in their production and use. Further, complex explosive formulations often make the uniform fabrication of lenses difficult. Finally, the pressure states particular at large diameter may be different even if the shock arrival is simultaneous. The first frame (Figure 3.26) shows the arrangement of low- and high-velocity explosives in a typical plane-wave lens. Most operate by transforming the spherical wave from a single detonator to a plane wave using a central cone of explosive with a slow detonation velocity (S), bounded by an outer sheath with a faster detonation velocity (F). The fast detonation velocity of the outer explosive expands on a spherical front, driving a flat wave in the central explosive that is moving at its detonation velocity. The two explosives are chosen such that the detonation velocity of the fast explosive, D_F, is related to that of the slow

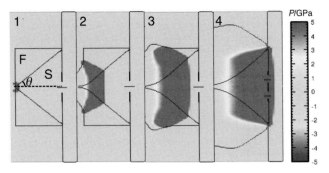

Figure 3.26 The operation of a plane-wave lens. Fringes of pressure at intervals of 0.8 μs. An explosive sheath with a fast detonation velocity surrounds a cone with a slow detonation velocity of half angle θ. The device is initiated at the apex of the cone.

explosive, D_S, through the half angle of the cone θ, where

$$\theta = \arccos\left(\frac{D_S}{D_F}\right). \tag{3.19}$$

Since the detonation in the central cone is driven by that in the sheath, the detonation velocity of the lens system is not diminished by attenuation from waves coming from the rear. The faster external explosive usually overdrives the inner one and this again should result in a relatively constant velocity. However, as well as difficulties with the manufacture of the device, there are other problems. In particular, the pressure across the flat shock plane is not constant as can be seen from the simulation. An inhomogeneous pressure distribution in the outer regions can be accounted for but is not ideal. Finally, the peak pressure in the pulse may be attenuated by allowing it to propagate through plates of some inert material of accurately known properties. Since it is possible to calibrate the performance of the device, it is then possible to produce tables of pressure vs. gap thickness that can accompany it to a test for experimental design purposes.

In summary, an explosive pulse can be produced with a controlled planarity and defined peak pressure using an engineered explosive device. It suffers from issues in reproducibility of the HE, inhomogeneous pressure profiles, expense in manufacture, and, from the point of view of model derivation, that the pulse is a decaying shock not a square impulse. Nonetheless, the plane-wave lens or generator (PWG) has been one of the most important workhorses for shock research since the Second World War and will continue to be so into the foreseeable future.

3.8.3.1 Regimes accessed by gun- and HE-based research

Gun facilities for generating a range of pulses in loaded materials in the range from hundreds of ns up to 10 μm and from the yield stress up to hundreds of GPa are important capabilities required to cover the range of behaviours observed in condensed matter deforming under extreme loading (Figure 3.27). The pulses applied are long enough that inelastic flow mechanisms have operated and the volumes of material loaded are

Figure 3.27 Gun facility constructed by the author. Single- and two-stage launchers firing projectiles from 10 to 100 mm diameter (50, 75 and 100 mm plate impact projectiles shown to the left).

sufficient that defect densities representative of condensed matter at the continuum scale have been sampled in order to gain information that may be applied to design or extrapolated to behaviour at greater scales. High explosives loading metals are important in the engineering design of a range of devices. The understanding of the effect of Taylor waves on targets is vital to explain the details of the response of these devices.

To conduct experiments over a range of pressures for advanced engineering applications requires a suite of single- and two-stage launchers capable of handling inert targets, high explosives and toxic materials, since plane-wave lens are limited by the explosives' CJ pressure. The highest pressures will be achieved by launching high-impedance plates supplying long pulses to the target, which will mean operation below the maximum velocity of the gun limited by the projectile mass. For a typical device, it should be possible to launch high-impedance plates at velocities of 4500 m s^{-1} at high- and low-impedance materials. Figure 3.28 shows the phase space accessed by such impacts in stress particle velocity space. Tungsten projectiles are assumed to impact upon the typical range of materials found in most applications going from the lowest impedance metals like beryllium to the highest impedance targets such as tungsten. A Hugoniot for stainless steel is included for reference. Single-stage launchers (SSLs) occupy a part

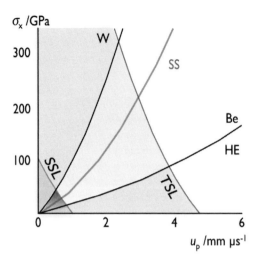

Figure 3.28 Phase space in stress particle velocity space accessed by impacts of SSL and TSL on two metals of greatly varying density.

of the phase space here (in darker grey) but one that includes the strength information for most materials of interest. For reference, a typical CJ pressure for an explosive is to be found at the upper end of this range. Two-stage launchers (TSLs) occupy a higher stress band up to 300 GPa in tungsten and around 100 in beryllium or a high explosive. These curves represent the shocked states; in addition there are a range of release states at lower pressures and temperatures as well as those achieved through shockless loading with ramp waves.

The great majority of engineering applications experience pulses in these regimes and apply loading onto targets at the laboratory scale which these techniques access. Further, they approach the limits at which strength is defined so that they closely map the regimes of extreme loading discussed in earlier chapters. However, more extreme states can be accessed using other platforms and these will be described below.

3.8.4   Z pinch quasi-isentropic loading

There are various means to introduce shocks into solids and over the past century much high-pressure equation of state data has been obtained from shock compression that maps the principal Hugoniot of materials, but these only represent the states produced by the passage of a steady, single shock wave. Other thermodynamic regimes either below the principal Hugoniot, or cooler states at a given compression, have not been extensively explored since techniques to do so have been difficult to engineer. Consequently, a significant portion of a material's complete EoS surface is inadequately probed and for the most part is deduced by mapping the Hugoniot to the new state using a mapping equation. There is thus a need for accurate measurements of states which are not found on the Hugoniot curve and this has driven the development of several experimental approaches to produce well-controlled platforms that deliver continuous ramp loads to condensed matter that take different thermodynamic paths in the loading of the material. These states once measured are dubbed off-Hugoniot states.

Ramp loading of materials generally produces thermodynamic states close to an isentrope since irreversible effects produced by viscoplastic and plastic work are usually small. Techniques to produced these loadings are often referred to as isentropic compression experiments (ICE); however, it is more accurate to refer to the thermodynamic states achieved by ramp loading as quasi-isentropic or more simply shockless loading.

Early efforts to create quasi-isentropic loading used the impact of plates onto buffer materials such as fused silica which compressed and slowed higher-pressure parts of the pulse. This technique results in shockless loading in target samples monitored at the rear with suitable sensors. However, accurate control of the loading rate is difficult using such methods. A development of this technique made use of graded density impactors called pillows that produce a small shock upon direct impact with a sample, followed by a gradual increase in loading stress. Recently, this technique has been further advanced through computer-designed graded-density impactors that improve control of the input stress history. The primary perceived complication with such techniques is the formation of an initial shock that precedes the ramp and this must be accounted for in analysis of the results.

Recently, lasers have been used to deposit energy onto a surface and produce a stagnating plasma on the surface of a planar specimen at the microscale. This can be tailored to load a target to hundreds of GPa pressure without inducing a shock. The technique has been demonstrated to produce smooth compression but is limited to short loading times of c. 10 ns at the microscale. It augments the suite of high-power laser techniques available that provides impulses at the highest pressures for compressing solids at the present time.

The most versatile of recent platforms able to load targets at the mesoscale is the magnetically driven isentropic loading technique using the Z accelerator developed at Sandia National Laboratory. This facility has matured significantly over the last decade and recent upgrades have given it unrivalled capabilities for studying mesoscale targets at extreme pressures. The Z accelerator produces smoothly increasing compression of materials to pressures of several hundred GPa over times of up to c. 300 ns and has been used in studies of a range of materials to high pressure and density (Figure 3.29).

3.8.4.1   The phenomenon of Z pinch

When an electrically conducting filament flows in a magnetic field the conductor experiences a compression force acting upon it. Such constraints occur naturally in situations where the conductor is a plasma and in nature such electrical discharges occur in various situations such as lightning bolts, the aurora, current sheets and solar flares. Z pinch is an application of the Lorentz force in which a current-carrying conductor (with the current flowing axially down the z-axis) in a magnetic field experiences a force. Such a force is responsible for two parallel wires carrying current in the same direction pulling towards each other as it flows. If current flows in a plasma, it acts to compress it axially and confine it into a constrained path.

Figure 3.29 Low-impedance water insulated transmission lines breaking down at Sandia's Z machine, Sandia National labs. Source: copyright Sandia National Laboratories (photographer Randy Montoya).

The effect was actively pursued as a technique to confine reactants as an inertial confinement for fusion. To this end, research was begun at Sandia in the 1980s to employ a different approach to using the effect. Rather than using an external magnet to generate the induction field, a powerful electrical discharge (several tens of millions of amperes for less than 100 nanoseconds) was discharged into an array of thin, parallel tungsten wires. These were vaporised and formed conductive plasma and the current flow caused it to pinch. Further, the imploding plasma produced a high temperature and an X-ray pulse which could create a shock wave in a target structure.

In order to allow shockless, quasi-isentropic loading at lower strain rates on large samples, magnetically driven isentropic loading has been induced in target materials using Z pinch accelerators. Targets of interest are mounted directly onto a flat anode plate of either aluminium or copper and a short between the parallel anode and cathode plates allows a current pulse (typically 20 MA, with 100–300 ns rise-time) to flow from one plate to the other generating a planar, dynamic magnetic field varying in time between the conductors (Figure 3.30). The resulting magnetic pressure launches a high-pressure ramped compression wave with strain rate of c. 10^6 s^{-1} into the anode conductor and hence into the planar target. In a second configuration (Figure 3.30(b)) the pressure is used to launch a plate at the target. Such plates are of the order of 10 mm in diameter and hundreds of microns in thickness. By this means extremely high velocities can be achieved which then impact onto specimens accessing high shock pressures. In the latest

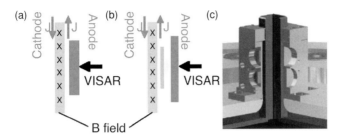

Figure 3.30 Z pinch loading of targets. Pulse of current J through rectangular coaxial, shorted electrodes induces magnetic field B. (a) Magnetic force $J \wedge B$ transferred to electrode material; (b) force launches flyer to impact on sample; (c) engineered target mounting: four panels in a square geometry holding two targets in each.

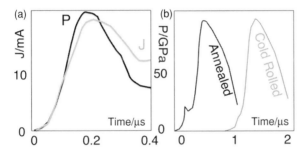

Figure 3.31 (a) Typical current and pressure history on the front of the target in the Z machine. (b) Ramp loading of a tantalum target (after Asay *et al.*, 2009) showing development of the IHEL in an annealed material. Source: Asay, J.R., Ao, T., Vogler, T.J., Davis, J.-P. and Gray III, G. T. (2009) Yield strength of tantalum for shockless compression to 18 GPa, *J. Appl. Phys.*, **106**: 073515. Copyright 2009 American Institute of Physics.

upgrade to the Sandia machine (ZR), peak currents of *c.* 25 MA allow peak isentropic compression pressures of *c.* 600 GPa and flyer velocities in excess of 45 km s^{-1}.

The current J and magnetic field B interact to induce a force on the target which exerts a pressure P to load it in compression, where

$$P = \frac{\mu_0}{2} B^2 = \frac{\mu_0}{2} J^2. \tag{3.20}$$

Thus the loading rate is determined by the current history in the electrodes, which itself is controlled by the discharge rate from the capacitors in the multiple Marx banks placed around the target. The form of the current pulse and the force on the target are shown in Figure 3.31.

To reconstruct the loading path in isentropic compressed targets, the Lagrangian velocity histories are analysed. The conservation relations for isentropic loading of a solid or liquid from an initial state are derived from the equations

$$\frac{\partial \sigma}{\partial h} = -\rho_0 \frac{\partial u_\mathrm{p}}{\partial t}, \tag{3.21}$$

$$\frac{\partial V}{\partial t} = -V_0 \frac{\partial u_\mathrm{p}}{\partial h}, \tag{3.22}$$

$$\frac{\partial E}{\partial t} = -\sigma \frac{\partial V}{\partial t}, \tag{3.23}$$

where ρ_0 is the initial density, V_0 and V are the initial and final specific volumes $(1/\rho)$, respectively, E is the specific internal energy, u_p is the *in-situ* material or particle velocity, $\sigma = \sigma(V, S)$ is the longitudinal stress, which is a function of specific volume and entropy, and the derivatives are taken with respect to Lagrangian position h and time t. If one assumes that the material response is strain rate independent and dissipative contributions to entropy change are neglected, the flow becomes self-similar and subsequent analysis in a Lagrangian frame becomes straightforward. Flow in an inviscid fluid, where longitudinal stress is equal to hydrostatic pressure, is assumed to be isentropic, which simplifies the governing equations thus

$$\mathrm{d}\sigma = \rho_0 c \,\mathrm{d}u_\mathrm{p}, \tag{3.24}$$

$$\mathrm{d}V = \frac{V_0}{c} \mathrm{d}u_\mathrm{p}, \tag{3.25}$$

$$\mathrm{d}E = -\sigma \,\mathrm{d}V. \tag{3.26}$$

This assumption means that strictly the flow is only isentropic for an inviscid fluid so that in real materials it is only quasi-isentropic.

As the ramp induced on the impact face of the target propagates through it, the wave front will steepen since for most materials the sound speed increases with increasing pressure. The front eventually evolves into a shock front with the jumps in state variables that are normally associated with such waves. In real materials, non-linear response will result in shocks forming in particular pressure regimes; this will lead to the appearance of transition thresholds such as the isentropic elastic limit (IEL) within the material (Figure 3.31(b)). In order to extract equation of state measurements from the compression isentrope, two target thicknesses are monitored and if the analysis and assumptions are correct, there is then no need to know the actual input pressure history on the impact face of the sample.

The development of microstructure follows different routes in quasi-isentropic loading to that observed after shock since the processes are initiated at different levels and propagate with different kinetics to those at high temperatures excited in a single pulse (Figure 3.31(b)). Thus features observed and threshold measurements made are complementary to those observed in plate impact experiments if the boundary conditions for the input pulse are well known. The technique is more recent than shock loading and devices much less common. Thus the new behaviours observed are less appreciated than is the case for shock. However, this mode of loading represents a new phase of development for this field and indicates a new arena for experiments to launch novel loading paths across the equation of state surface for a material.

3.8.5   Laser shock loading

Platforms capable of introducing large impulses have been described in preceding sections. In one case high-pressure gas drives plates to impact, in another detonation waves driven by chemical reaction transmit to compress target materials. In a third magnetic interactions compress and drive stationary targets. In a final driven system, intense coherent radiation interacts with a target region which vaporises under the applied pulse and forms an inertially confined plasma. These high pressures are possible since lasers are uniquely able to focus large amounts of energy into microscopic length scales over short times. Plasmas at high temperature and pressure are driven by the rapid heating rates to expand against the target material and drive the surface inward. The only other means of generating such extreme conditions is to use magnetic pinch facilities to discharge large currents but none yet have reached the extreme energy densities available from laser drives.

Vaporising part of the target itself to drive its deformation represents the most recent and the extreme form of idealised loading applied in the laboratory to solids. The technique additionally offers a means to not only generate high temperatures and pressures but also accelerate matter to high velocities offering the potential to drive collapse under inertial confinement. Facilities capable of fielding such loading pulses include NIF at LLNL in the USA, Orion at AWE and Vulcan at RAL in the UK, and LMJ at CEA Bordeaux, France and these lasers will deliver several MJ in each pulse. Relevant compressions to those found in neutron stars may be studied with short pulse lasers opening study to the superextreme conditions found in astrophysics and beyond the *finis extremis* for condensed matter. These states exist for a nanosecond or less, probing the limits of equilibrium thermodynamics and beyond the limit where temperature is steady and the conservation equations can be applied with certainty.

3.8.5.1   **Laser shock experimental techniques**
The devices themselves have developed over the past decades to extraordinary power and complexity. A high-quality seeding beam is amplified in power and intensity from neodymium doped, defect-free glass. By this means, the beam is prevented from depositing energy at homogeneities or cracks or by stray diffraction so damaging the laser components. This allows the beam intensity to grow to high levels, extracting a large fraction of the stored energy. Using intense radiation in this manner deposits radiant energy without chemical or moving media intervening (explosives or accelerating masses) and this allows the targets themselves to sit in clean and well-controlled environments before, during and after the loading is applied. Since the impulse applied is so small, recovery of targets is easier and they are thus free from later loading by stray momentum from impacts into components or from product gases as is the case with other techniques. This is because the momentum imparted by radiative heating is orders of magnitude

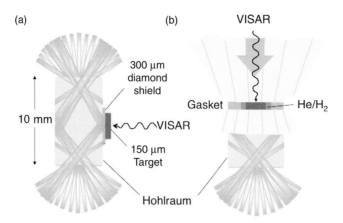

Figure 3.32 Laser techniques for the generation of high-pressure pulses in targets: (a) shock and (b) half of a Hohlraum mounted on a diamond anvil cell used for precompressing and then shocking.

lower than for the other platforms. Targets are thus typically of millimetre size and constructed with high precision so that control of the plasma can keep shock planarity good in the target. State measurements are typically deduced from temporally and spatially resolved free surface measurements observing reflectivity or using line-interferometric measurements.

Immediate laser illumination of a region on the surface of a target is known as direct drive. Typically the pressure induced by direct drive can be varied from a few GPa to around 1 TPa, with a typical shock duration of nanoseconds. Thus the drive stimulates and samples the shortest relaxation time physical mechanisms operating in the target. Laser-launched flyer plates can generate shocks with pressures of up to several hundred GPa, and this offers pulses similar to those that can be generated by Z pinch where the plates are accelerated using magnetic impulses. A second class of experiment that lasers drive creates the plasma remotely to load the target. It illuminates a hohlraum, which is a cavity whose walls are made of a high atomic mass metal and are in equilibrium with the radiant energy applied within. Laser beams are fed into the cavity through small holes to irradiate the inner surface and create an inertially confined plasma within. The loading on adjacent targets produced by this technique is known as indirect drive. In either case, the high-pressure plasmas and vapours produced during irradiation induce compression pulses within condensed matter, which may be either shock or more slowly rising pressure waves with ramps typically rising over c. 10 ns. Hohlraum and shock wave techniques are illustrated in Figure 3.32.

The diagnostics fielded on such lasers are adapted for high recording speeds from standard devices and techniques used for longer pulse loadings, typically sampling at c. 100 ps. The motion of the free rear surface of a loaded target is recorded using point or line VISAR, which has to be sampled at extreme rates to record an adequate signal. These are recorded by photomultipliers and streak cameras which will be reviewed in the next chapter. They are used to record equation of state information whilst X-ray diffraction

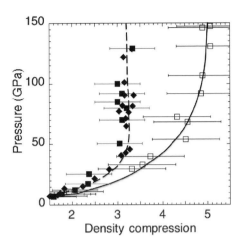

Figure 3.33 Equation of state of deuterium to 150 GPa derived using different techniques including laser (solid) and *Z* pinch (dashed) loading. Sources: after Knudson, M. D. *et al.* (2001), Da Silva, L. B. *et al.* (1997) and Collins, G.W. *et al.* (1998).

is used to determine phase, lattice compression and other related measurements in such experiments.

3.8.5.2 Laser shock measurements of equation of state and strength

The two drive configurations, direct and indirect, are coupled with other techniques in order to make measurements of EoS and constitutive parameters. It is well known that applying an acceleration to a surface between two materials of different density produces a Rayleigh–Taylor instability at the interface. This is also the case when the junction is driven by reflection of a wave at an already rippled surface on the back face of a target. These will grow at a rate determined in part by the strength of the material. This growth can be recorded by dynamic streak or framing cameras and compared with a simulation using an equation of state and a constitutive strength model in which model constants can be adjusted to fit to the experiment. By this means a typical strength parameter is fitted for the metal.

Further, many astrophysics problems require data on equation of state for fluids at extreme pressure. Models of the interior structure of Jupiter for instance are sensitive to the details of the high-pressure EoS of hydrogen. The Hugoniot of deuterium has been determined by a series of techniques including launchers, Z pinch and laser experiments, and a summary of all these techniques and the equation of state up to 150 GPa is shown in Figure 3.33. Clearly the laser experiment (to the right in the figure) showed greater compressibility, not evident with the other experiments and the disagreement between this technique and the others quoted has yet to be explained.

Tensile loading pulses in laser experiments are the optimum means of approaching the theoretical strengths of solids since the pulses applied are localised and intense. The shock pulse reflects from the rear surface of targets and the localised wave forms a tensile region close to it. The pulse applied is extremely short in such experiments, beyond the

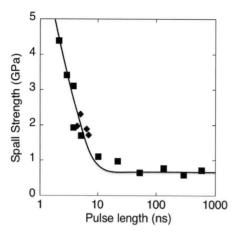

Figure 3.34 Spall strength as a function of pulse length in aluminium. Source: after Bourne, N. K. (2011) Materials' physics in extremes: akrology, *Metall. Mater. Trans. A.*, **42:** 2975–2984. With kind permission from Springer Science and Business Media.

thresholds required to fail at inhomogeneities but not long enough to grow voids in a metal as is the case with other platforms. Thus the triggered failure is a localised region and bond scission is triggered at interactions with the smallest scale defects. In such experiments continuum concepts such as strain rate become redundant and it is more useful to rather think of pulse durations and the regions of interaction. However, the laser drives the shortest of pulses and can probe local failure in a manner no other platform can compare with.

In all cases the observed laser-driven spall strengths show much higher values than those encountered with other driving systems. At pulse lengths below 10 ns, the spall strength climbs rapidly towards the theoretical strength measured in pure aluminium (Figure 3.34).

3.9   Final comments

This chapter has given an overview of platforms available to load solids to extremes both statically and dynamically, from the elastic limit to the threshold where the nature of the bonding state changes. There is an equally diverse range of diagnostics to probe such experiments and the next chapter will cover a selection of these to tie the loading paths introduced by the platforms discussed here, with the state-sensors fielded in targets during load.

To construct physically based models for material response requires platforms capable of accessing the range of states necessary to populate the materials models used today. At present these are mainly quasi-static machines but also include devices that may apply a faster loading, albeit one that keeps the stress state in equilibrium across the whole sample volume as the amplitude increases. The limit for reaching a uniform developing stress state across an entire volume is reached in the Hopkinson bar where the crossover to wave-dominated loading occurs. The term strain rate on either side of this boundary has different meanings and must be used with care in these different regimes since, whilst the pressure loading hydrodynamically links to states at the atomic scale, strength is defined over the volume element sampled. The density of defects changes the routes for slip or fracture and gives different values for measurements taken. Thus new devices for transient loading must explore not only amplitude and pulse length but also the sampled volume in order to fully map behaviour and identify all the operating mechanisms. The ability to apply uniaxial stress and strain states for response to single impulses is but beginning in designing devices to populate the next generation of constitutive models.

3.10   Selected reading

ASM Metals Handbook, Vol. 8: Mechanical Testing and Evaluation (2000). Materials Park, OH: ASM International.

Cooper, P. W. (1997) *Explosives Engineering*. New York: Wiley.

Duffy, T. S. (2005) Synchrotron facilities and the study of the Earth's deep interior, *Rep. Prog. Phys.*, 68: 1811–1859. (Static high pressure.)

Remington, B. A., Bazan, G., Bringa, E. M., *et al.* (2006) Material dynamics under extreme conditions of pressure and strain rate, *Mater. Sci. Tech.*, 22: 474–488.

Seigel, A. E. (1965) *The Theory of High Speed Guns*, AGARDograph 91, May.

4   Tools to monitor response

4.1   What do you need to measure?

The platforms described in the previous chapter access a range of states via a number of thermodynamic loading paths taken by a material as it deforms. Some load to the elastic limit, some up to the *finis extremis* where electronic bonding changes its nature, and some beyond that. What follows will concern loading from the elastic limit to the point at which ambient descriptions of strength cease to apply. A few of the loading paths necessary to define an equation of state for a material are shown in the schematic of Figure 4.1. There are a range of outputs which may be sensed to give insight into the response of materials under load. Experiments should aim to map their states beyond the yield point statically and dynamically. In the first case they induce an ideal stress state to define operating mechanisms represented in suitable models, which are later tested against other loading down more complex paths. Thus shock experiments map out Hugoniot curves but can also yield information that allows one to deduce compression isotherms and isentropes. Isotherms are generally measured using static compression experiments at some fixed temperature in the diamond anvil cell (DAC). To briefly recap, the isentrope generally lies between the isotherm and Hugoniot curves and is in fact tangent to the Hugoniot at the common starting state. Although shock experiments generally yield only a final $P–V$ state on the Hugoniot, an ideal isentropic compression experiment (ICE) yields a continuous locus of points along a different loading path. Although not precisely following the isentrope, it is certainly possible to load more slowly and avoid the adiabatic conditions of shock, and so this is better dubbed *shockless* loading. To record this data demands sensors capable of acquiring pressure, density and temperature as a function of time, which requires sub-nanosecond data collection under the fastest loadings. To measure deviatoric quantities entails measures of the stress state in the target which is itself directional. Thus a series of accurate, time-resolved sensors has been developed to make such measurements in these experiments. Another means of recording the data is to use a quantitative imaging technique (such as X-rays) to deduce state parameters from the flow. Imaging itself allows the visualisation of geometries changing under load whist offering non-invasive measurements of flow parameters.

    The ideal experiment will map the full state at every point in the material volume loaded as a function of time. An array of sensors will record all aspects of the flow at a point of zero volume with arbitrary time resolution and output unaffected by

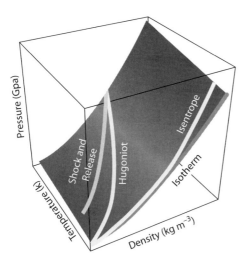

Figure 4.1 Dynamic loading paths in $PT\rho$ space.

variation in any other flow constant. These variables are principally the pressure, volume, particle velocity, strain, temperature or a stress component. There are also electrical and magnetic states of interest in some cases but these will not be described in detail in what follows. Finally, there is the need to look at not only mechanical but chemical pathways excited during loading. Although there has been mention of the techniques available for tracking displacement, the electronic structure of the bonds requires fast spectroscopic techniques to understand the development of reaction in materials. This chapter will focus on the mechanical effects, but some of the results, particularly those of fast spectroscopy in reactive materials, will be returned to later when energetic materials are considered.

This paradigm of defining the evolution of the full state tensor for a material at every position, non-invasively and at an arbitrary rate, can never be realised, but new sensors and imaging capabilities are being developed to move closer to this goal. Any dynamic research programme must aim to sense material state, image deformation and recover the post-event microstructure. In the following pages, this work will scan the range of techniques available and point to some areas where development for the future should guide the subject.

4.2 Derivation and validation experiments

As mentioned in the last chapter, experiments perform two important purposes. The first is as a means of *deriving* mathematical descriptions of material behaviour; the second, to *validate* that the descriptions assembled contain correct formulations that truly represent all the loading modes that a particular application may supply. This defines the two classes of experiment that are used for the construction of an operating code design environment – *derivation* and *validation* experiments.

Material models are constructed by conducting experiments across a range of different loading paths and producing a mathematical description with constants determined from them. The methods used to analyse displacement (or strain) data from diamond anvil cells and split Hopkinson pressure bars have been discussed previously. However, it is the shock that excites all available mechanisms at a defined time and allows their kinetics and behaviour to be best mapped, and so experiments which introduce them are described below. In the stress range to *c.* 100 GPa the plate impact experiment is the workhorse for deriving material behaviours and a suite of diagnostics is arrayed to measure properties. At the highest stress levels and shortest times in laser and Z pinch loading, the brevity of the pulse and small size of the targets means that measurements are confined to velocimetry or streak photography. Measurements are dominated by two classes of experiment in this field. These are remote interrogation of external surfaces interferometrically with the addition at lower stress levels of embedded sensors that intrude at some station in the flow and track a Lagrangian state. It should be pointed out from the outset that embedded sensors are physically mounted within a target and will perturb the stress state within the flow. If their dimensions are small and their mounting is such that stress equilibrates quickly, then they will track the state around their location. However, embedding requires a cut to be made in a sample into which the sensor is inserted, and it is a prerequisite of any experiment to ensure that this must not affect the state that they monitor. All such gauge mountings localise deformation in tension and fail the target, and thus yield little useful information in this loading mode. But in the small-strain, compressive loadings considered here they can be shown to operate with precision and have been so calibrated so that one may rely upon the state parameters generated. When loading extends into the strong shock region, material is compressed so that volume defects are compressed and the flow is homogeneous after the rise of the plastic front. However, below the Weak Shock Limit (WSL) an experiment that can match to the relaxation time of the mechanisms operating in the material under load is necessary so that F remains < 1. In the weak shock regime the lattice is at close to full density for metals and brittle materials but the mechanisms are not complete within the shock front. The inelastic mechanisms of choice can be probed with the plate impact experiment in this region.

Figure 4.2(a) shows a schematic configuration for the plate impact experiment. A flyer plate of accurately known dimension is launched down a gun at a velocity accurately measured using shorting pins, laser beam gates or with a channel of velocimetry. Impact time and planarity are tracked by sensors embedded into a target which will contain either further embedded sensors or be probed by a laser non-invasively to measure the velocity history of the rear surface or in some cases the motion of an interface with a transparent window. The plate and the target are aligned to less than a milliradian of tilt and the sensor records are monitored until the volume of the target that they sample is released from the outer surfaces and the continuum one-dimensional strain state down the impact axis is lost.

The impact induces a region of flow and high pressure in which uniaxial strain (at least at the continuum) applies (Figure 4.2(b)). This state is maintained until the flyer plate is

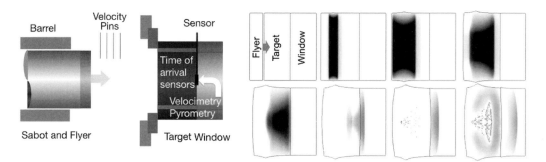

Figure 4.2 Standard configuration for the plate impact experiment and simulation of pressure state (dark for high values) as function of time. Frames are a microsecond apart, and run left to right, top to bottom.

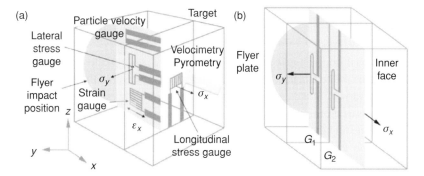

Figure 4.3 (a) Configuration of sensors in a plate impact experiment. Impact is on the rear surface of the target. The shock sweeps forward and meets embedded Lagrangian sensors measuring components of stress, strain or tracking the particle velocity. Free surfaces or those with mounted windows are interrogated by velocimetry non-invasively. (b) Several gauges may be mounted on inner planes to track the wave evolution with time. Lateral and longitudinal stress gauges are on perpendicular planes.

decelerated by longitudinal release waves from the rear of the plate or by lateral waves from the periphery of the target. After a time these re-establish the ambient state, but if longitudinal releases have interacted, the target will fail in the plane over which they interact and ductile or brittle failure will initiate at defects in that region. In most cases the window material is of lower impedance than the target and these wave interactions will occur in the same way as if the target rear surface were free. However, clearly the amplitudes of the tensile pulse will be less in this case. Thus, the key to observing one-dimensional states to construct models for the material is to confine sensors to the central axis of the impacted targets and limit their sampling area to reduce the effects of lateral release.

Figure 4.3 shows typical instrumented targets that include a wide selection of the available sensors that may be used to sample shocked flow. The simplest target preparation is to interrogate surfaces with velocimetry; this only requires flat, parallel surfaces with sensors to indicate impact planarity (or multiple velocimetry channels). If the target has to be sectioned so that gauges must be introduced, then the task is somewhat more involved. Since the flow is globally one-dimensional, strain sensors in the plane of the shock will not respond to any lateral motion. This simplifies the choice of sensors required. On the other hand, the stress state (for a material with strength) is biaxial in perpendicular planes so that typically the target must be cut to accommodate measurements in two directions. The final requirement is that the embedded package must be thin enough to equilibrate the stress as quickly as possible to the external target. Equilibration will happen after about three wave transits through the layer and this provides a limit on the sampling technique that can be achieved using a particular technique.

One commonly used validation experiment, containing a range of compressive and tensile deformation mechanisms, was suggested by G. I. Taylor. A right cylinder impacts upon an ideally perfectly rigid target, deforms and can be simply recovered. Taylor, in the initial use of the technique, made measurements of its shortening in length to directly calculate the dynamic yield strength. Today the technique is used as a validation of constitutive models of metal yield since the deforming cylinder can be observed using high-speed imaging and the idealised geometry is simple to apply in continuum code platforms, which makes the results obtained sensitive to the models employed. In a second variant, identical right cylinders impact one other and this symmetry ensures the impact is as if on a perfectly rigid target in a frame containing the centre of mass moving at half the impact velocity.

On the impact face the state in the very first moments of the process is uniaxial strain. As the wave passes down the cylinder it induces a multiaxial one which changes with distance until it reaches that of uniaxial stress. Clearly, it takes the waves time to sweep the cylinder and penetrate from the boundaries, which limits the duration of the compression pulse and evolves an axisymmetric stress state. It is thus useful as a validation test since it encompasses several mechanisms operating within the loading time; for instance at high enough amplitudes, release fronts fail material at the central axis and open the cylinder near the impact face as the flow expands radially outwards. The highest rate is induced by the progress of a shock front through a material, yet at later times the rod reaches large strains on the impact face. The capability to sense these developing states across the target is continually being developed. At present the principal diagnostic tools are high-speed (optical) imaging and recovery of the deformed target for microstructural investigation (Figure 4.4) but there is suggestion that it may be improved to better test the constitutive models with improved diagnostics. Finally, sensors applied to the anvils may supply Lagrangian state information or reverse impacts onto diagnosed stationary rods can track states developing in the cylinder. Work in metals undergoing phase transformation will be particularly relevant to this effort.

The examples discussed divide the needs for diagnosis in experiment into two principal groups, that is imaging of deformation and sensing of state. In what follows these two groups will be elaborated upon with examples of various systems employed to accomplish these tasks.

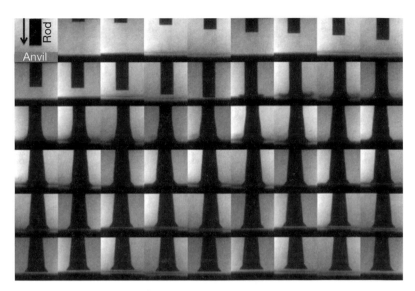

Figure 4.4 Taylor impact on a copper cylinder travelling from top to bottom; high-speed shadowgraphy frames run top to bottom, left to right.

4.3   Imaging and sensing

The common ideal in sensing material's state is to non-invasively determine the relevant variables at any point in space and as a function of the loading time of interest. Equally an ideal imaging system must penetrate structures made of a material to three-dimensionally retrieve its changing geometry at arbitrary resolution and over any part of the electromagnetic spectrum. Both of these wishes are mediated by the available technologies and the data acquisition and storage restrictions placed by the engineering of today. However, although compromises are always made in gathering the data, considering the nature of the impulse, we can at least optimise the measurement made.

A sensor measures a physical state at a particular point in a material. It samples a volume element within and records the value of a quantity as a function of time with a particular bandwidth. The length scale of the sampled element determines the provenance of the variable tracked. Measurements made on waves propagating over tens of millimetres sample continuum states in single-phase materials with defects at the micron scale. States are steady in compression as stress equilibration has occurred and the volume element sampled by the sensor and of course the sensor itself should be in state equilibrium before its output can be relied upon. Similarly loading at the microscale requires diffraction measurements of displacement at the atomic scale from which strain can be inferred. However, this cannot be easily read across to the laboratory scale save in pure, defect-free elements. If the states have equilibrated so that local thermal equilibrium is established and the wave is steady, measurements at lower length scales will inevitably be part of a statistical distribution of compression states with

Table 4.1 A selection of fast operating Lagrangian sensors

Technique	Operating characteristic	Quantity measured	Comments
Manganin	Piezoresistive	Stress	Up to 150 GPa
Carbon	Piezoresistive	Stress	Up to 2 GPa
Ytterbium	Piezoresistive	Stress	Up to 4 GPa
Strain gauge	Piezoresistive	Strain	Constantan
Quartz	Piezoelectric	Stress	Up to 5 GPa
PVDF	Piezoelectric	Stress	Up to 2 GPa
Lithium niobate	Piezoelectric	Stress	Limited range
PZT	Piezoelectric	Stress	Limited range
VISAR	Differential velocimetry	Particle velocity	The original non-invasive technique
Photon Doppler velocimetry (PDV)	Heterodyne velocimetry	Particle velocity	Known as HetV in UK
Particle velocity gauges	Induced EMF in magnetic field	Particle velocity	Insulating materials
Thermistor	Change of resistance	Temperature	
Ellipsometry	Interferometric	Temperature	Difficulties in interpretation
Pyrometry	Planck radiation	Temperature	Difficulties with emissivity

one-dimensional conditions only satisfied at the continuum which makes interpretation of measurements at local volume elements more difficult.

Embedded sensors track continuum scales and are typically millimetres in size averaging states over corresponding-size volume elements. Examples are shown in Table 4.1 where a range of sensors and the variables they sample is given. The mechanical states of pressure and volume are the easiest to measure whilst the most difficult, since equilibrium over large areas takes time, is temperature. However, over recent years pyrometry has advanced to measure temperature from emitted radiation and its use is reviewed below. With any embedded sensor, the device tracks a local state that will move with the flow and thus these measurements are Lagrangian. Similarly Doppler-shifted surface velocity histories track a moving interface. It is imaging technologies where the camera or the source are fixed that render Eulerian information from the phenomena observed.

For loading using the shortest pulses which excite deformation at atomic scales, the measurement tools available have been confined to observation of the diffraction patterns that develop during loading. With the advent of brighter light sources, new diffractive image techniques have become possible and this has opened up new possibilities for probing deformation in these regimes.

The strength behaviour of materials has only recently assumed the importance that it deserves, but is now at the point where measurements can be understood and compared across pressures and from shock to quasi-static loading. There are several major techniques that can be used to measure the quantity and each delivers a different result with different errors, including the following:

(i) Measure the Hugoniot and then calculate the offset to the hydrostat at a particular stress level, where the hydrostat is measured in DAC experiments or calculated from an analytic form such as

$$\sigma_x - \sigma_y = 2\tau = Y; P = \frac{3}{2}(\sigma_x + 2\sigma_y) \quad \text{so} \quad Y = \frac{3}{2}(\sigma_x - P). \quad (4.1)$$

This method does not account for shock heating and relies on the accuracy of the hydrostat measured or calculated and on the equation of state assumed.

(ii) In the previous chapter a gun was introduced which allows the inclined impact of flyers onto inclined targets, and this has been used to generate longitudinal and transverse waves in the target. Monitoring the in- and out-of-plane wave arrivals allows measurement of the bulk and shear wave components, which in turn can be used to deduce strength. This pressure–shear technique is limited to less than c. 20 GPa by the complexity of the experiment and diagnostics. Firing an isotropic crystal at a target is another means of generating shear waves, but in this case the impactor fixes the ratios of the longitudinal and shear amplitudes.

(iii) Recent discussion of the use of laser methods to generate bulk shock compression data has proposed monitoring the growth of Rayleigh–Taylor instabilities as a means to measure strength. This assumes a model for perturbation growth that includes a term of some form to describe strength which applies a slowing of their development. Such observations are really a validation test of existing models and so strength is not directly measured without assumptions.

(iv) The development of brighter light sources has also allowed some progress in the use of electron diffraction to directly measure deformation in single crystals. These techniques are under development but, in principle, offer a means of delivering strains in the unit cell from which some bound on strength may be derived.

(v) The DAC technique has developed over the years, assuming a purely hydrostatic pressure load on the targets in the cell. Realisation of the difficulties in ensuring truly hydrostatic loading has inspired new research, allowing the anvils to contact the specimen and generate shear loading of the target. Two methods have been used to generate strength data from such measurements. The first involves measurement of pressure, P, as a function of the radius of the anvil and to relate the shear stress to this radial variation of P. The second rotates the cell in an incident X-ray beam and probes the lattice at multiple angles to recover lattice deformations. These methods may be used to deduce material strength as a function of pressure. The technique shows great promise towards determining the high-pressure strength of materials under quasi-static conditions.

(vi) Asay and co-workers proposed a *self-consistent method* for estimating dynamic yield strength of materials in the shocked state (Asay and Lipkin, 1978). The method uses two experiments where a target is first shocked and reloaded, and then shocked to the same pressure but simply unloaded. The resulting behaviour is used to estimate the yield stress in the precompressed state, assuming a generalised yield and hardening behaviour and without requiring any knowledge of the hydrostat for the material. The advantage of the method is independence of knowledge of the hydrostat. The method implicitly assumes that the yield surface is fixed for the duration of the loading. If that is not the case, then it can be regarded as recovering a value for the strength at a time corresponding to the end of the first pulse.

(vii) As seen earlier it is possible to embed piezoelectric or piezoresistive foil stress sensors into the flow in such a manner that the lateral component of the stress field might be measured as well as the longitudinal stress history. This technique will be discussed here since it offers the only means of recording a history of the strength through the shock front. Since the gauge must sit in an insulated package within the material that might short the sensor above conduction thresholds, this has limited their use to *c.* 25 GPa at the present time. The package must also be swept before stress equilibration has allowed recording to begin which limits the response time. The technique nevertheless represents the only one that does not require any knowledge of the material under investigation in order to deliver results. In the weak shock regime it and the self-consistent method provide the principal means of measuring deviatoric behaviour.

Other than the lateral gauge, all of these techniques require assumptions that represent an integrated set of constraints upon the value of the yield stress recovered. In particular, since they are not time-resolved measurements, the yield surface is assumed to reach a steady value instantaneously so that pulse length is not regarded as an issue, and this, as shown below, will constrain the result of the experiment to determining strength at a time corresponding to that at the end of the pulse applied, which may or may not be the final yield strength or may average the value delivered.

Imaging technologies vary across the wavelengths that may be generated to probe deformation. These include a range of sources across the electromagnetic spectrum including optical, X-ray and particle beam imaging. The former is principally used to examine the deformation of structures, but with some adaptation can be used to measure front positions and recover quantitative pressure fields in transparent media. X-rays and particle beams are principally used to recover density information and their ability to probe denser materials and larger structures is always an issue. The modern generation of FELs allows X-ray pulses to be generated to image structures with coherent beams and these offer the most exciting prospects for the future. Clearly the ability to pulse or continuously illuminate dynamic events is determined by the power needs for the source and the available energy density across the field of view. However, the present time has seen great breakthroughs which will be followed into application over the next decade.

The twin goals of real-time tomography and the determination of a dynamic state tensor for a range of points in a deforming solid are realisable with present technologies. The next few years will prove the promise of the techniques and the twin needs of imaging and sensing will drive the development of the next generation of diagnostics.

4.4   Sensors

Over the past century a range of sensors has been developed to follow the state of a location within a material that has been dynamically loaded. It would be impossible to list all of them here, but an overview of the most commonly used ones, grouped according to the parameter that they measure, will be presented. One aim of an impact experiment is to characterise the variables that describe the state of the impactor or target, in other words to define the pressure, volume and temperature as a function of time. However, these parameters are not generally measured directly. Instead, related quantities such as stress or particle velocity can be inferred and calculations used to recover state variables. In some cases when the loading is simple it is sufficient just to measure the velocity of waves passing through the target and then to employ analysis to deduce the state variables.

4.4.1   Time-of-arrival sensors

The earliest (and the simplest) measurements of the state of matter under high-rate loading conditions were performed with time-of-arrival sensors which determined shock velocity assuming waves in the solids were steady. Particle velocities could be inferred from impactor velocities for instance. Various conducting pins, short-circuited by the arriving metallic projectile, were used to record arrival times at accurately known positions using storage oscilloscopes. Alternatively, a thin conductor could be severed to break a circuit. These techniques still find application for the measurement of velocity, where pairs of fine pins may be sequentially shorted or thin foil screens broken. Further, such signals may be used to interrogate alignment of gun systems. More recently, piezoelectric pins, producing a sharp voltage spike on impact, have been used with some success.

Although such mechanical methods are robust in many circumstances, it must be remembered that any measurements in an impact experiment are made in a very hostile mechanical and electrical environment, and it is sometimes necessary to use optical methods and a high-speed imaging or detection system. One such technique is the flash-gap method in which a small volume of gas (argon is frequently used as it emits strongly in the visible spectrum) ionises and luminesces when the shock wave reaches it. A streak camera viewing the target records the flash as the wave reaches each gap and a velocity can be inferred.

4.4.2   Stress sensors

The sensor designed to measure the stress at a position in a target must possess some coupling between its mechanical and electrical properties in order that an emitted charge or resistance change may be converted into a voltage signal than can be recorded. The most commonly applied of such properties are piezoresistivity and piezoelectricity. Additionally, in order to select a material for constructing a stress or pressure gauge, the following criteria are ideally met. Firstly, a candidate should have some property that has high sensitivity to pressure so that a large signal may be measured. Secondly, the material should show low sensitivity to the variation in other variables in the flow field, particularly temperature. Thirdly, the material's properties should remain stable so that a unique calibration may be established. This also implies that it shows low sensitivity to

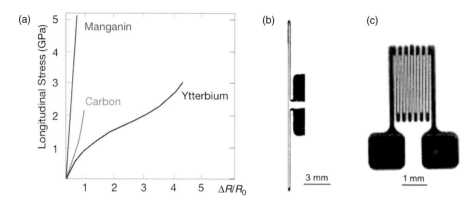

Figure 4.5 (a) Manganin, ytterbium and carbon resistance change as function of pressure. (b) MicroMeasurements manganin gauges J2M-SS-580SF-025. (c). MicroMeasurements C-951213-C.

its exact composition (if an alloy) and to its manufacture. Finally, a monotonic (or very nearly so) response to pressure is necessary, implying that it has no phase transitions or other thresholds in behaviour in the working range.

4.4.2.1   Piezoresistive transducers

A transducer should approximate a point sensor, responding accurately and reproducibly to an applied stress or strain whilst being insensitive to variations in other properties. Piezoresistance describes the change of the electrical resistivity of a material with applied external stresses. It was discovered by Lord Kelvin in 1856 and thoroughly investigated by Bridgman throughout the first half of the twentieth century. Bridgman used manganin (an alloy of 84% copper, 12% manganese and 4% nickel), which fits the requirements outlined above as a pressure standard up to 10 GPa in his static high-pressure cells. The relative change in resistivity may be expressed as a piezoresistive coefficient multiplied by the stress. Since a crystal generally has anisotropic piezoresistive properties, and the stress is generally expressed in terms of its components in various defined directions, the full formulation is necessarily of tensor form.

Sulphur was the first piezoresistive material to be used as a dynamic stress sensor in a shock experiment, but manganin soon replaced it as the material of choice for high stress ranges (up to 150 GPa statically). In shock experiments the range of pressures in which it may be used is limited by the electrical response of insulating materials which short the sensors made from it. Other materials have also been studied as possible candidates, including carbon and ytterbium (carbon resistors are cheaper but offer much more limited accuracy), and these have been used in the lower stress range (up to 2 and 4 GPa respectively) where manganin is less sensitive. One caveat in using these latter materials is that the change of resistance with temperature, found negligible for manganin, is not so in their case. Their performance curves, showing resistance change as a function of applied stress, are shown in Figure 4.5(a). Typically such metals can be drawn as thin wires or sputtered onto substrates (typically fibre epoxy) to act as thin sensors. Two examples of different geometries are shown in Figure 4.5(b) and (c).

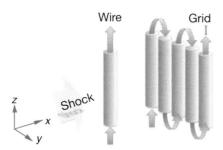

Figure 4.6 Shock sweeps a piezoresistive wire and grid elements.

The resistance of the section of metal fabricated into a gauge element changes with pressure in proportion to changes in the cross-sectional area and length of some piece of gauge material. The active elements of a gauge are made to be long and thin. This is obviously so for the case of a wire (or a hoop such as Figure 4.5(b)) but a grid gauge (Figure 4.5(c)) consists of relatively thin elements aligned parallel to one another with thick curved end pieces (and thus negligible series resistance) to join them. The means of mounting of such gauges means that changes in the length of an element are rendered negligible compared with those in the cross-sectional area in an element. The stress–strain curve may thus be expressed in terms of piezoresistive constants relating resistance change in the metal element directly to the stress field surrounding it. If the metal is fabricated into a gauge element mounted as part of Wheatstone bridge, a suitable constant current passed through it can be used to generate a voltage signal that may be recorded using a high-speed data storage system and a connection will then be made directly between a resistance history and the pressure pulse observed by the sensor.

Of course any sensor is an intrusion or defect at some scale within the target material it samples. The dimension of the sensor thus defines a mesoscale structure within the target. If this scale is allowed to equilibrate in stress over time by the action of waves from the boundaries it will attain the global stress state found around it as all such volume elements do in real materials. Thus on the scale of the gauge element, the local stress and strain states will be different from the continuum in the first moments a wave arrives. Consider two geometrical forms for the sensors: a single wire or a grid of elements closely aligned to one another (Figure 4.6).

Locally a wire experiences a hydrostatic stress and strain state around the element. However, in the case of a grid, other elements close by distort the local field and the integration of the strain across components on the larger (continuum) scale cancels these effects from element to element. Thus the strain state around a wire and a grid gauge are different although the same pressure is applied to both. The strength of the gauge material will be significant at this scale for its response. Consider the two swept by a shock down the x direction in the figure. The continuum uniaxial strain and biaxial stress systems will define stress and strain tensors for conditions within a grid gauge as follows

$$\begin{pmatrix} \varepsilon_x & & \\ & 0 & \\ & & 0 \end{pmatrix} \quad \text{and} \quad \begin{pmatrix} \sigma_x & & \\ & \sigma_x - Y & \\ & & \sigma_x - Y \end{pmatrix}, \tag{4.2}$$

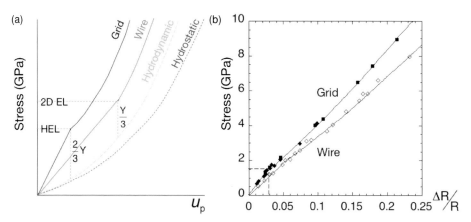

Figure 4.7 (a) Shock loci for loaded grid and wire sensors (Y is yield stress of sensor material). HEL (1D conditions) is shown along with elastic limit under biaxial state (2D EL). (b) $\Delta R/R$ vs. σ_x for grid and wire manganin transducers.

and for a wire

$$\begin{pmatrix} \varepsilon_x & & \\ & \varepsilon_x & \\ & & 0 \end{pmatrix} \quad \text{and} \quad \begin{pmatrix} \sigma_x & & \\ & \sigma_x & \\ & & \sigma_x - Y \end{pmatrix}, \tag{4.3}$$

where σ_i and ε_i refer to the local stress and strain states, respectively. The grid and wire geometries will cause the metal element to follow different shock states in one- and two-dimensional compression (Figure 4.7(a)). This means that the pressures in the wire and grid gauges and the dynamic confined elastic limits will be different when compared with a common stress within them, since

$$P_{\text{grid}} = \sigma_x - \frac{2}{3}Y \quad \text{and} \quad P_{\text{wire}} = \sigma_x - \frac{Y}{3}. \tag{4.4}$$

Further, since annealing will reduce the value of Y, the measured resistance change should be higher at a given stress than that of an as-received gauge. Further, the orientation of a grid gauge will affect the output recorded from it, particularly if it is placed in a plane perpendicular to that of shock front. To define orientation it is conventional to talk in the following terms. A grid gauge will be in the longitudinal orientation when it is emplaced perpendicular to the flow direction in the target. A gauge will be in the lateral configuration when a wire geometry is perpendicular to the flow direction, or when a foil is placed parallel to this direction with the electric current perpendicular. This leads to several differences between gauge types, such as different calibration curves and hysteresis not being observed with wire sensors in a lateral geometry.

The resistance changes of wires or grids will be close to the same when the pressure or stress becomes large compared with the yield stress of the gauge. However, it is important to consider the latter if the gauges are to be used in low-stress regimes and to understand the observed calibration curves. For the rest of this section the commonly used sensor alloy manganin will be discussed and for this material the yield stress is

0.75 GPa. However, all metals display some plasticity mechanisms under shock and dislocation generation and storage in this alloy causes shock hardening to occur. Thus the yield stress will generally be a function of the stress level to which it is shocked.

Figure 4.7(b) shows the calibration curves for applied shock stress for manganin grid and wire gauges. The stress is plotted against the relative resistance change of the gauge. It can be noted that the grid gauge curve lies above that for wire gauges since even though manganin responds in the same manner to pressure, its response to stress differs. A value of the Hugoniot elastic limit for manganin is indicated by the dotted lines and appears as a break in slope between the linear elastic and the concave plastic response. This occurs at around 1.5 GPa for these manganin foil gauges (which corresponds to a yield stress of 0.75 GPa, assuming a Poisson's ratio of 0.33).

The performance of a material in an engineering application, particularly its resistance to penetration during impact, is reflected in the strength it displays when loaded with different impulses. It is a pressing present concern to better understand the wealth of data that has been collected on all materials in order to map their dynamic strengths under one-dimensional, weak shock conditions. Such behaviour has been studied in several ways using a range of techniques. These include using the measured stress and shock velocity vs. particle velocity data to construct Hugoniot and hydrodynamic curves (corrected for temperature) to determine their offset and then calculate inferred strength, or employing the pressure-shear technique to directly map the deviatoric behaviour. Another common method is to load the material and then allow release from states induced by single and double shocks (Asay technique) to determine by integration the stress offset of the Hugoniot and isentrope. Finally, it is possible to monitor directly the longitudinal (σ_x) and lateral (σ_y) stresses, and (in isotropic materials) assign the difference between the two to the compressive strengths that exist behind the shock front at high pressures.

The direct measurement of lateral stresses with piezoresistive gauges is now a mature technique that has resulted in a better understanding of strength development within materials under load. The response of the gauges to various loading conditions in both longitudinal and lateral configurations has now been analysed and understood. Mounting the sensor in the direction of the flow defines a stress tensor around the elements as follows

$$\begin{pmatrix} \sigma_y + Y & & \\ & \sigma_y & \\ & & \sigma_y \end{pmatrix}, \tag{4.5}$$

where σ_y is the lateral stress in the matrix, which is also the stress transmitted to the gauge through its thickness. This stress state results in a hydrostatic pressure component in the gauge, which is equal to

$$P = \sigma_y + \frac{1}{3}Y_g. \tag{4.6}$$

With this simple approach the reduction of the lateral gauge data consists of two steps. First, one needs to use the calibration curve of the gauge to deduce the volumetric strain for a particular experiment and then use the compression curve to find the pressure in the gauge from which the lateral stress in the matrix (σ_y) can be obtained. Again, the geometry of the gauge has an effect in the low-stress region, which must be accounted for, but these calibrations have been conducted and published.

4.4.2.2   Practicalities of gauge use

Several different geometries have been adopted for piezoresistive transducers and these may be grouped according to whether the gauge has two or four terminals (examples of these types can be seen in Figure 4.5). The two terminal configuration is used for 50 Ω gauges in the lower stress range, 0–20 GPa, whilst much smaller, four terminal, 50 mΩ gauges (often dubbed π gauges because of their shape), are used for higher (20–100 GPa) stresses. One of the principal difficulties of working with these sensors, and one that limits their response time, is the need to insulate them from their environment. It is readily seen that this is necessary for metallic targets, but it should be noted that many insulators become conducting when shocked including the mounting polymers (insulating sheets and epoxy bonding).

Piezoresistive gauges are used either as in-material sensors, embedded to record shock wave histories directly, or as gauges on a target back-surface (as illustrated in Figure 4.3) with a window material bonded to them. In the embedded configuration the gauge is sandwiched between two insulating sheets and is glued with a low-viscosity epoxy under applied pressure between two tiles of the target material. The polymer sheets and epoxy form a layer of the order of 100 μm thickness of low-impedance material around the gauge and this sits within the surrounding matrix. Any impedance mismatch between gauge package and matrix will compromise the response of the gauge to the stress in the target since the whole low-impedance package must equilibrate with the external field before the gauge records the matrix stress. In order to improve the response time, the sensor may be mounted in a rear tile of poly-methyl-methacrylate (PMMA) bonded in place of one of the target plates. The gauge now sits in a medium to which it is almost perfectly matched and response time is now determined only by a 5 μm layer mismatched to the surrounding material and its electrical properties (inductance being the most significant for foil gauges). This mounting is frequently called the back-surface configuration. The penalty accrued for this move is that the measurement of the stress is no longer direct as the gauge senses a transmitted wave (of which a portion will have been reflected back into the target as a release). The measured stresses must thus be scaled by the dynamic transmission coefficient between target and backing. The result of these measures is that the stress state equilibrates in the sensor over a time of 10 or 20 ns in a back-surface gauge and takes 50–100 ns to do so in an embedded configuration. Skilled and careful mounting is essential to minimise equilibration times.

Clearly changes in resistance in piezoresistive gauges can only be detected if a constant current is passed through them and the voltage is monitored. The small changes of resistance that the materials undergo require that a large current (tens or hundreds of amps) is passed through them to produce measurable voltages. On the other hand, their thin elements mean that the resultant Joule heating would vaporise the gauge if the power dissipation were not limited by pulsing – providing it is live at the time at which the measurement should be taken, of course. The simplest and most robust power supply

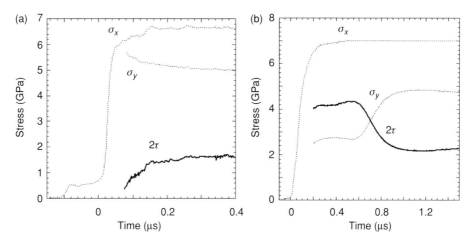

Figure 4.8 (a) Nickel and (b) glass showing piezoresistive sensors mounted 4 mm from an impact face showing development of the stress field behind a shock front. Source: reproduced from Bourne, N. K., Millett, J. C. F. and Gray III, G. T. (2009) On the shock compression of polycrystalline metals, *J. Mat. Sci.*, **44**(13): 3319–3343. With kind permission from Springer Science and Business Media.

places the gauge in one arm of an initially balanced Wheatstone bridge, which becomes mismatched as the gauge resistance changes due to the stresses across it. Clearly the current must be triggered, which requires some synchronisation of the pulse, yet this is easily done and such sensors represent one of the few ways to obtain embedded data to fully explore the stress or strain state of a material.

Piezoresistive gauges are thus vital components in the armoury of an experimentalist aiming to understand the complex physics occurring in materials under load. To accurately mount and power such sensors, and to understand their response and limitations, is a complex technical task that must be understood before interpreting the data. Yet the benefits of doing do are manifold to all who try to understand response, since these techniques are the only means of observing the development of a continuum state *within* a macroscopic volume. Other techniques such as interferometry are surface measurements that sense components of the compression fields that exist within a material at the time it is loaded. Further, deducing state information in this manner risks using steady wave linkages between state variables that may not apply for the loading impulse applied. Nevertheless, calibrations are now accurate to less than a few per cent on these sensors and with gauges well mounted within a flow a means exists to sample continuum lateral and longitudinal stress and strain states within a shocked flow. The limits are reached on temporal resolution and breakdown of insulation, which limits the stress range over which the sensor responds. Two examples are shown in Figure 4.8 where shocked nickel and glass illustrate the development of longitudinal and lateral stress states (along with the calculated shear stress) with time as the shock passes the sensor. The time for initial equilibration of the lateral sensor in these experiments is *c*. 180 ns and the signal received over this time has been removed from the figure.

4.4.3   Strain sensors

Changes in crystal structure under shock have been tracked with time-resolved X-ray diffraction. These experiments showed that crystalline order does exist behind shock fronts to extreme states and additionally allows accurate measurement of atomic position from which strain can be deduced in a bulk single crystal. However, this technique requires specialised equipment and bright light sources (synchrotrons, Free Electron Lasers, etc.) and so the few studies of this type that have been done are limited to the strong shock regime and laser platforms. A simpler and direct technique using constantan gauges to measure strain in the weak shock regime has been developed for use at the continuum scale. The longitudinal strain gauge is embedded in a specimen in such a way that its length (current direction) is parallel to the shock direction (as shown in Figure 4.3). The gauge is thin enough to only cause minimal disturbance to the specimen, and since it is bonded to the faces of the surrounding target it is shortened as the matrix compresses behind a shock. Since constantan is an alloy insensitive to either pressure or temperature, the only resistance changes are due to the shortening of the gauge. Constantan strain gauges are commonly used in uniaxial stress experiments, where their resistance changes express the measured strain through the relation

$$\frac{\Delta R}{R_0} = K\varepsilon, \tag{4.7}$$

where K is a gauge factor that for commercial gauges ranges between 1.9 and 2.1. Constantan gauges may also be used in uniaxial strain loading and the same analysis can be used in shock. However, calibration has shown that the gauge factor in this case should be taken to be 2.0 and the strain ε is the longitudinal strain, ε_x, in the specimen.

The technique has been demonstrated on metals such as aluminium and copper, and also in glasses. It has even been combined with a thermistor to yield simultaneous temperature and volume measurements, allowing a determination of the Grüneisen gamma. An example of a strain gauge trace in a glass target at different stresses levels is shown in Figure 4.9. The differences between the strain histories in these two experiments are clearly seen. In both cases the signal drops as the gauge elements are compressed, which is to be expected since their cross-sectional area increases. The shock takes some finite time to sweep the gauge, which explains the ramped appearance. Although the slopes of the elastic waves are very similar, the plateau at the top is different, and the residual strain in the high-stress shot is clearly evident. Also evident is the complete release to zero residual strain of the low-stress record, which is in accord with the elastic nature of this experiment. The residual strain at the top of the full release in the high-stress shot is $c.\ 0.025$.

The elastic shock strains the glass to 0.065 at which point there is a marked change in slope and the strain increases more slowly to a maximum strain of 0.075. This change in slope is interpreted as the dynamic yield point (the Hugoniot elastic limit; HEL). There is then an elastic release to 0.035 at which point the slope of curve changes to a slower plastic release leaving a residual strain of 0.025. Notice that the difference between

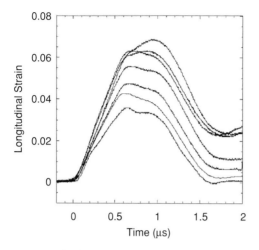

Figure 4.9 Strain records from a gauge in soda-lime glass shocked in a series of experiments between 2.9 and 7.3 GPa. The impactor was a 3 mm aluminium flyer in each case.

the elastic strain on loading to the HEL (0.06) and the elastic strain unloaded (0.04) is exactly the residual strain. This is a direct measure of the compaction of glass at this stress level and a demonstration of the elastic-inelastic behaviour in this glass.

An embedded sensor, if properly mounted, can gain vital tracking information on the evolution of the continuum state at different locations within a loaded material. The mechanisms tracked are often only operating within inertially confined target volumes so that observations at interfaces only record echoes of the behaviours occurring within. It is vital to couple all measurement types in order to observe the full suite of mechanisms at work within a loaded target material.

4.4.4 Piezoelectric transducers

The Curies in France first discovered the piezoelectric effect, in which charge is generated by the application of stress to a material, in 1868 in their studies of tourmaline. Indeed it was embedded tourmaline crystals that were first used in shock wave measurements in iron. However, the development of synthetic, flaw-free quartz for the electronics industry standardised the piezoelectric sensor and lead to development of the Sandia quartz gauge in the early 1960s. This has only recently been surpassed by piezoelectric polymers, which have a wider range of use and are more easily mounted. Although widely available, the low conversion efficiency of quartz led to the development of various ceramic ferroelectrics. Examples of these include lithium niobate, lithium tantalate and lead zirconate/lead titanate (PZT) compositions. The latter polycrystalline materials cannot be produced as accurately as the quartz gauges, which led to the latter's dominance as a gauge material in the 1970s. Most recently Bauer has developed techniques to reproducibly enhance the properties of a semicrystalline film of a polymer, polyvinylidenedifluoride (PVDF). This material is based on the monomer CH_2CF_2, but copolymers of PVDF with C_2F_2H and C_2F_2H are also under development.

Thick quartz gauges were developed in the 1970s at Sandia. In order to keep the field uniform within the gauge, their design incorporated electrodes placed in concentric cylinders (guard rings by analogy with capacitors). As a wave crosses a piezoelectric crystal, such as quartz, the strain induces an electric field in the material due to the direct piezoelectric effect. Through a non-linear coupling, this field leads indirectly to further mechanical stresses, which then induce a complex electromechanical state within the gauge that can only be analysed approximately. Fortunately, the deviations from simple analysis are small and the stresses can be measured to within a few per cent without undue complexity. A piezoelectric gauge can be employed in two basic geometries for which the electrical and stress analyses are different. In the first, the gauge is made thick with respect to the length of the pulse to be measured and the shock sweeps through the crystal; in the second, it is thin and the stress wave reverberates within it until equilibrium with the target is achieved. Any analysis of the thick gauge response must thus incorporate shock propagation effects within the material, and since this can be complex, practical operation is limited to the elastic range (0–4 GPa for quartz, for instance).

Piezoelectric gauges may be operated in one of two electrical configurations – charge mode and current mode. In charge mode the gauge sees a hardware charge integrator and the output is sent directly to a scope to measure a voltage proportional to the stress. Alternatively, a hardware integrator consisting of a resistor and a capacitor may be used to record the stress directly. In current mode, a resistor is placed across the gauge and the voltage across it is monitored directly. In this case the stress derivative is recorded and a time integration must be carried out to recover the stress history. Here one must ensure that data are recorded at adequate frequencies since any aliasing will clip the signal recorded and recover the wrong signal.

The PVDF polymer is a thermoplastic consisting of a mixture of crystalline and amorphous phases. Electrically it is classed as ferroelectric but it can be regarded as a piezoelectric material for the purposes of analysis. The material was first reported in the late 1960s but it has taken more than 30 years of development to produce reproducible sensors since it is only after an involved process of mechanical and electrical conditioning that a useful gauge can be made. The PVDF film is stretched to induce crystallinity (the polar β-phase). Biaxial stretching is found to be the most effective (producing 50% crystalline material) but processing must result in uniform properties (thickness, etc.) across the film. A stretched film contains randomly oriented polar crystallites which are aligned by applying an external electric field, alternated cyclically until a reproducible remnant polarisation is produced. Material produced in this manner can range from 2 to 25 μm in thickness and may be used as an embedded gauge. Such sensors are produced in larger quantities by mechanical conditioning of the film, depositing nickel or gold leads on to either side of a film with legs separated from one another but with two parts crossing, and then poling through the deposited legs. The advantage of using this method is that only the small area directly between the electrodes is active, making the gauges spatially accurate. The sensors require no power supply, have a fast response, and are now repeatable in manufacture, so that accuracies are comparable with other gauges. They have a low impedance and match well to materials such as other polymers or explosives.

4.4.5   Particle velocity sensors

Measurement of the velocity field in a flowing material has always been one of the principal goals in dynamic experiments. Early researchers used high-speed streak photography to monitor the movement of external surfaces and began to develop electrical methods, for example using a moving metal interface as one plate of a capacitor and recording the variation of the voltage, to measure the displacement. Soon afterwards, the particle velocity gauge was developed using the interaction of a moving conductor in a magnetic field. There was early work in Russia in the 1960s with further improvements in the USA through the 1970s and 1980s. Development in the UK and France occurred in the 1990s and there is now widespread use worldwide, particularly in gauging high explosives.

From the very early stages, interferometry represented an alternative way of accurately measuring the small displacements involved although the speed at which fringes moved represented a challenge in detection. It was only when a robust, suitably sensitive instrument, allowing diffuse surfaces to be monitored was developed that non-invasive particle velocity measurement became routine and velocimetry now occupies the paramount role in shock wave instrumentation although measuring the velocity of interfaces rather than within the bulk of the flow.

4.4.6   Electromagnetic gauge

For insulating specimens, the electromagnetic particle velocity gauge is conceptually the most simple and straightforward of techniques to apply. A fine wire (or foil) is embedded in the specimen in such a way that a fine element moves in a defined direction within an accurately defined, perpendicular magnetic field. If more than one element is included, Lagrangian velocity histories at multiple positions may be recovered giving a multiple position picture of the evolution of the flow within the target. Since there is a requirement for only a conductor, there is only calibration of the applied field to be done prior to a shot and no need for any power supplies to energise anything other than the Helmholtz coils used to apply the field. Even this is not necessary if permanent magnetics are used. Difficulties with the technique come in the accurate assembly of the complex targets used and the sensitivity of the technique to electrically active materials under load since each sensor acts as an efficient aerial. Nevertheless, for explosives and polymers it has provided an unrivalled method for gaining full target information on wave evolution and material response at multiple stations for each experiment.

A constant magnetic field is oriented parallel to the shock and perpendicular to the wire. As the shock reaches the wire, it carries it along at the particle velocity within its following flow and moving it relative to the magnetic field. An electromotive force

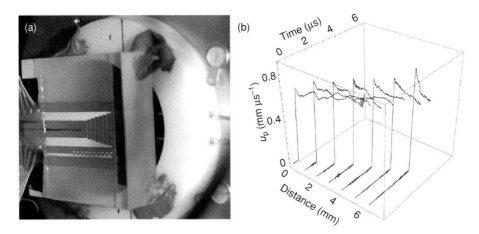

Figure 4.10 (a) Mounted PV gauge package on $30°$ plane cut in a glass target. (b), Velocity histories from and experiment in a PBX.

(EMF) is generated across the wire which, by Faraday's law, is given by

$$E = \boldsymbol{L} \cdot \left(\boldsymbol{B} \wedge \boldsymbol{u_p} \right) = BLu_p, \tag{4.8}$$

when velocity, field and gauge elements are orthogonal, where E is the induced EMF, B is the magnetic field strength and u_p is the velocity of the wire (vector element L) and thus the flow. Typically, the length of the active element is 10 mm, which requires a field of around 1 T to induce 1 mV per m s^{-1}. The accuracy of the measurement is governed by the magnitude of the magnetic field and the quality of the alignment of the wire with it. The gauge can be used over a very wide velocity (and thus stress) range and within inert or reactive insulators. Several wires can, of course, be simultaneously embedded, and indeed recent developments have seen the fabrication of etched hoops onto a glass fibre backing to create a gauge package with multiple gauge stations along it. An example built by the author is shown in Figure 4.10(a), embedded into a glass target. The outputs can be coupled with Lagrangian analysis to yield a complete stress–strain characteristic for the specimen. A particle velocity (PV) gauge package is typically mounted at an angle to the impact axis so that individual elements are not shielded by flow around an upstream element.

The gauge package illustrated consists of seven thin strips, 25 μm thick (the gauge elements), orientated orthogonally to the magnetic field, with leads normal to it so that they do not experience an imposed voltage during shock loading. As can be seen at the bottom there is an eighth gauge element, consisting of a castellated loop running the length of the package. This is a shock tracker, so oriented that the wave propagates along its length. The output from such an element consists in alternating positive and negative potentials according to the orientation of each individual bar element, which are separated by 1 mm intervals. The result is a castellated voltage–time trace, the periodicity of which gives a direct measure of the shock velocity. Figure 4.10(b) shows an example of a series of velocity histories from a reactive material where the transit of the wave from the front to the rear of the target causes reaction to grow at the shock front. In other experiments, particle velocity gauges have been used in the pressure-shear experiment to monitor both the longitudinal and transverse particle velocities in insulating targets.

To summarise, embedded PV gauges have found increasing use advancing particularly study of the shock to detonation transition in condensed phase and liquid explosives. It is a continuum measurement that gives unrivalled information and detail throughout the material under study. The technique is robust and easy to understand and interpret, yet the difficulty of manufacture of targets and the restriction of study to insulating materials has limited its use. However, this author has used the technique in liquids, polymers, brittle materials and explosives and the information gained is unrivalled in the applications for which it can be fielded.

4.4.7   Velocimetry

The measurement of velocity is an important requirement for processes in which waves equilibrate states and accelerate surfaces. The data acquired can be used to validate wave codes or combined with steady wave conservation relations, to deduce pressure or density states within materials. However, it has only been with the measurement of velocity histories that it has been possible to deduce mechanistic detail from analysis of structure within waveforms.

The principal advances in dynamic instrumentation have come in the development of non-invasive interferometric techniques for the monitoring of the velocity of the external surfaces of impacted targets. Incident light on a moving surface is Doppler shifted but the magnitude of the resulting frequency is too high to be detectable routinely. However, if the velocities of interest are much less than the speed of light, interference of the two signals will produce a beat frequency that may be recorded with modern equipment; the term heterodyning is used to describe such a process. Two devices have resulted from this requirement that are used extensively through the subject. The first is the Velocity Interferometry System for Any Reflector (VISAR), which splits the Doppler shifted reflected beam and interferes one part of that with a component delayed in time. The second is the Photon Doppler Velocimeter (PDV; known as the Heterodyne Velocimeter or HetV in the UK), which interferes an incident beam directly with the Doppler shifted reflection from a moving surface. PDV is thus a displacement interferometer since the signal changes as the position of the reflector does. In the case of VISAR, light is interfered with a second packet that left the target at a different time, giving rise to fringes of lower frequency which depend on the relative displacement at times t and $t + \tau$. This implies it is a differential displacement interferometer since it produces fringes that depend on the relative positions of the moving target at two closely spaced times. Thus VISAR is a velocity interferometer responding as surface speed changes and the beat frequency of the interfered beam is lower than is the case in PDV.

The lower frequency in the interfered signal allowed use of 1960s' recording equipment, which saw VISAR developed ahead of PDV. However, the latter has become feasible since fast, affordable digitisers have become available with advances in fibre

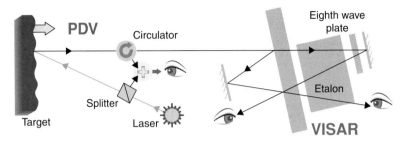

Figure 4.11 Schematic of the PDV and VISAR velocimetry systems.

optic data transfer. Other interferometric techniques have also been developed in parallel. In the 1960s a system using a Fabry–Pérot elaton to create an interference pattern of concentric circles and a streak camera to record their diameters was established. Displacements in the plane of the shock, necessary for detecting lateral motion in elastic anvils for the pressure-shear experiment, were also measured. This technique uses a diffraction grating etched onto the surface of the anvil within which the sample rings up, and off which laser light interferes, to deduce the transverse displacements (the transverse displacement interferometer; TDI). However, it is VISAR and PDV that dominate this class of devices and it is these that will be described below.

A schematic of the two techniques is shown in Figure 4.11. A laser illuminates a target which is moving into the beam causing a diffuse reflection, Doppler shifted in frequency, to travel back from the surface. In the case of PDV the Doppler shifted reflection is mixed with the incident beam and the two interfere and produce a signal intensity I at a detector thus

$$I = I_{\mathrm{in}} + I_{\mathrm{refl}} + 2\sqrt{I_{\mathrm{in}}I_{\mathrm{refl}}}\cos(\phi_{\mathrm{in}} - \phi_{\mathrm{refl}}), \qquad (4.9)$$

where the beat frequency is

$$f_{\mathrm{b}}(t) - \frac{2v(t)}{\lambda},$$

and I_{in} and ϕ_{in}, and I_{refl} and ϕ_{refl} are the incident and reflected intensities and phases, respectively. The first two intensity terms act as an offset to the signal, whilst the square root term is a convenient amplification factor to boost it. The velocity information, $v(t)$, is contained in the beat frequency, f_{b}, and analysis of the signal to extract this frequency information is the most challenging component of the use of PDV. The construction of the system is very easy since the advent of modern telecommunication equipment. Fibre beam splitters and couplers allow the incident laser light to be mixed with the Doppler shifted reflection. One other key component is the circulator. This transfers an optical signal from one port to an adjacent one with little leakage of light to other ports apart from the intended destination. It separates the Doppler shifted light from the incident laser beam, which allows a single fibre to both carry light to the target but also collect the reflection. Once the data are collected off the recorder, the analysis routines extract frequency information from the signal to yield velocity. There are several techniques which have been employed but Fourier analysis (using particularly the short-time Fourier transform) is robust and has been written into several automated computer packages that can be run quickly on desktop machines.

VISAR is a Michelson interferometer (to be accurate a wide angle one) with a delay etalon in one leg to superpose light with a time-delayed transformation of itself. Incident Doppler shifted light is partially transmitted and reflected with one leg travelling through an etalon to delay it in time before recombination. In Figure 4.11, the beam also passes twice through a wave plate so that two outputs from the beam splitter are of opposite phase. Recording in quadrature allows the signal to indicate the direction of movement of the surface whilst also allowing subtraction of noise from stray incoherent light that may result from loading or reaction. The recorded fringe patterns can be quickly analysed automatically to deliver velocity–time histories.

Contrasting the two devices highlights the ease of manufacture and operation of the PDV device, on the one hand, against the more difficult analysis methods necessary to recover velocity histories, on the other. VISAR is a more complex optical device with a long pedigree and easy data analysis. However, there are a series of known difficulties with VISAR that PDV can overcome. Amongst these are resolution in the signal and an inability to handle multiple velocities in the input signal. The new developments made with PDV are quickly overcoming some of its shortcomings as more complex systems are being made. These include quadrature PDV, which combines signals at fixed phase offset and frequency mixing, allowing PDV to recover the direction of motion from the signal.

Both of these velocimeters most often determine the response of the surface at a single point sampled by an incident beam. However, an extension allows the evaluation of data down a line or across a plane. Line VISAR has been used as the tool of choice to observe heterogeneity of flows at free surfaces and windows illustrating the statistical nature of the defects and states found across the wave front. The development of PDV (HetV) allows a similar capability to be fielded with these techniques and this offers many advantages given the relatively cheaper components for the system. The limitation at present is in recording the high-frequency signals, as noted previously, but new techniques will be developed as new sampling technologies come on line.

Although there are occasions where the surface whose velocity or displacement of interest is simply moving in air (or vacuum) towards the detector, such as a projectile travelling down a gun barrel, in others a transparent window of particular impedance is added so that the interface is better matched to the target. In this case the movement of an interface between target and window will have been tracked. However, the incident and reflected light from the target interface will now have to pass through the compressed window material and the extra optical path must be added to that introduced by varying position or velocity in an interferometer. The variation in refractive index, n, can be related to the change in density, ρ, thus

$$\frac{\mathrm{d}\rho}{\rho} = \frac{\mathrm{d}n}{n-1},\qquad(4.10)$$

and a range of windows have been calibrated to allow interpretation of the signal emanating from the interface with the target. Window materials commonly used include PMMA, fused silica, z-cut sapphire and lithium fluoride, and the relevant necessary shifts to recorded data are tabulated for users.

4.4.8   Measuring temperature

In a system to which equilibrium thermodynamics is applied is it important to measure all the intensive variables describing properties of the material and to define its equation of state across as wide a pressure range as possible. Although pressure and volume (or related variables) have been discussed above, the measurement of temperature remains the critical unknown in the fast loading of materials. It is intimately linked to the structural changes observed in a material deforming inelastically and to the entropy created as a result through the statistical thermodynamics expression

$$S = \frac{\partial}{\partial T} (k_B T \ln Z), \tag{4.11}$$

where k_B is the Boltzmann constant and Z is a partition function that describes the equilibrium statistical distribution of the energy microstates within the system. Classical thermodynamics and assumption of a Mie–Grüneisen equation of state allows one to calculate temperature for a shock state by integrating (4.11) along the Hugoniot

$$dT = -T \left(\frac{\gamma}{V}\right) dV + \frac{1}{2c_v}[(V_0 - V)dP + (P - P_0)dV]. \tag{4.12}$$

However, such a calculation assumes that the Grüneisen parameter γ and the heat capacity c_v (and their dependence on the thermodynamic state) are known accurately – which is not the case. Thus better measurement is the only way forward to allow the accurate formulation of equations of state for extreme states.

Yet it should be appreciated that these relationships are those of equilibrium thermodynamics and the volume under consideration may not be so if the loading is dynamic. Local regions will be after certain times in local thermal equilibrium (LTE) over defined volume elements within the flow. Considerations of larger volumes than this will consist of assemblages of non-equilibrium states defined by the impulses delivered to a loaded target and this must be borne in mind in interpreting measurements of temperature at lower length scales where the mismatch of relaxation times for thermal and mechanical equilibria must be considered. Measurements and (analytical treatments) of mechanical and thermal localisation are thus of equal importance to equilibration; hot spots (local thermal ignition of reactive media), nucleation at defects in inelastic flow and adiabatic shear banding in dynamic compression are all failure mechanisms of importance that are poorly considered at present. Finally, as pulses become short, thermal equilibrium becomes unattainable within the loading time and the relaxation of states occurs after the pulse is over. However, on longer timescales and with larger volume elements, temperature controls the kinetics of physical and chemical processes, connecting mechanical to electronic state changes within matter, and defines the kinetics within many processes. In particular, it governs bonding states and thus martensitic or bulk phase changes such as melting or freezing as well as further larger strain processes, which makes it vital to observe within matter under load.

Two approaches to track the thermal state have been taken by experimentalists. The first is to non-invasively observe the spectrum of the radiation emitted, or the vibrational states accessed by shocked material. The second is to attempt to embed thermistors or thermocouples which can be used to record output potentials as their temperature equilibrates with their surroundings. The difficulty of measuring temperature compared to pressure or volume is that the mechanical wave processes are around a thousand times faster than thermal diffusion ones. Measurement on macroscopic scales requires sensors sampling significant volume elements defined by the sensor area and the pulse applied to equilibrate with the flow. Thus the physical thickness of the sensor foil must equilibrate with the temperature field applied in order to record a steady reading and this process takes hundreds of nanoseconds or microseconds for the thinnest gauges. On the other hand, the simplicity of interpreting the data is much greater with sensors than optical emissions. The use of optical effects by sampling excited states in transparent media allows temperature states to be sampled using Raman techniques such as the observed Stokes–anti-Stokes lines in spectra. Recently, neutron generation from spallation sources has been used to measure temperature using Neutron Resonance Spectroscopy (NRS), although recent work is developing corrections to the technique. All these approaches are of interest in specific circumstances, yet in the general case targets are opaque and if they are metals, have a small skin depth from which grey body radiation is emitted. The range of methods that are used in this case are collectively called pyrometry. The technique then involves recording the radiance, I, of the surface at several wavelengths and applying Planck's law to determine its temperature thus

$$I = \frac{2hc^2}{\lambda^5} \frac{\varepsilon}{e^{(hc/\lambda kT)-1}},$$

(4.13)

where h is Planck's constant, ε is the emissivity, c is the velocity of light and λ is the wavelength of the radiation. Since metals are both opaque and have a short optical depth, the observed spectral radiance is actually that of a layer whose depth for infrared light in a typical metal is of the order of a few nanometers. It is thus vital to ensure that the surface state of the target is well controlled since measurements at a releasing surface are at best difficult. The key property to determine is the emissivity for the interface. Methods exist to determine static values for the coefficient, but it is difficult to determine dynamic or pressure-dependent values for its behaviour although ellipsometry, or equivalently polarimetry, does allow determination of the dynamic emissivity by measuring the real and imaginary parts of the index of refraction. In practice in dynamic experiments it is generally assumed that loaded surfaces are grey bodies (the emissivity for each wavelength is the same), which is, however, unlikely to be true.

Nevertheless, proponents of the technique believe that careful preparation and control of the target surfaces allows measurement of surface temperature with an accuracy of 5–10%. Measurements have been demonstrated on materials that span a range of temperatures between less than 400 K up to more than 2000 K. Free-surface measurements are problematic because the release at the interface puts the material into an undefined state. Further, it is potentially an imperfect surface within a small skin depth and potential clouds of ejecta may be present in the beam of the emitted signal. Thus windows are preferred yet this requires adequate determination of their optical properties and a need to ensure that fractoemission or other light emission does not occur from the target under load. Practically this means that there will be a composite interlayer where the temperatures probed by the pyrometer interact. Metal skin layers will interact with adhesive layers in a region where thermal bodies at changing temperature and with varying thermal conductivity coexist. Thus it may be that a metal layer, sputtered onto a window and thermally equilibrated with the body under load, may be the fiducial necessary for interrogation under load. However, it is vital to understand the interaction of these multiple interfaces if the uncertainties of pyrometry can be tied down. Work is going forward to better define these difficulties to make pyrometry a standard technique for dynamic experiments, but cross-comparison with other techniques is necessary to give assurance on the measurements being made.

4.4.9   Dynamic X-ray diffraction

X-rays have a wavelength between 0.01 and 10 nm and thus couple well with atomic spacings in materials. This spawned the discipline of X-ray crystallography, which has developed over the past 100 years. Diffraction patterns map Fourier transforms of the atomic arrangement in crystals to the laboratory scale. The Bragg diffraction condition for incident radiation at an angle θ and wavelength λ for both diffraction by electrons and neutrons relates displacements in the reciprocal plane to atomic spacings in the lattice d thus

$$n\lambda = 2d \sin\theta. \tag{4.14}$$

Recording the spacing of different order peaks in the pattern allows the full unit cell geometry to be mapped. Diffraction is thus used widely in the static high-pressure field to determine structure at high pressure. The information that can be recovered has developed particularly with the advent of bright, synchrotron and now FEL X-ray sources found in national facilities throughout the world. However, such techniques have only recently been used for dynamic experiments.

   X-ray diffraction is the natural means to make atomic scale measurements of the dynamic deformation of crystals. Transmitted and reflected beams can be used to gain information on longitudinal and lateral strains in the lattice as a result of dynamic deformation from the Laue diffraction pattern that results. Two configurations have been used in experiments: an impact face measurement of the state on a surface in the moments that a shock is launched; and a downstream measurement as a rear surface is probed as a shock breaks out across it. In the latter case, an impact drives a compression wave into a target and a pulsed X-ray source to the rear records a diffraction pattern on a camera triggered when the shock reaches the penetration region to the rear of the target (Figure 4.12(a)). There is a finite X-ray probe depth depending upon both the sample

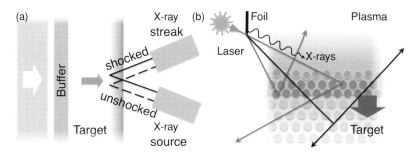

Figure 4.12 X-ray diffraction: (a) steady wave configuration; (b) impact face configuration.

material and the wavelength of the radiation, and this will be of the order of several microns for shocked metals. Break out of a steady wave at the rear of a target will produce a rich spectrum of shocked states that can be probed by the beam.

A laser source may both drive plasma to load a lattice as it drives a shock into a crystal but also produce an X-ray pulse by timing a beam to vaporise a metal foil close by. This means that the X-ray pulse and the shock need to be synchronised to nanosecond accuracy in order to record both static and shocked diffraction patterns. The point source projects a wide range of angles which diffract when the Bragg condition is met. A Laue pattern is recorded from planes parallel to the surface, giving axial response, and normal to the surface (for thin samples), giving the deformed shocked response. By this means in- and out-of-plane strains can be recorded directly by fast imaging in the first moments that the shock compresses the lattice (Figure 4.12(b)).

These diffraction patterns show agreement with molecular dynamics simulations, calculating the expected results and fit to the observed image plates from experiment. The work has been applied to both single and polycrystalline targets and across the martensitic phase transition in iron and has confirmed the structural transformation route suspected from continuum state measurements. These atomic scale measurements will couple to those at greater length scales to track the response of crystals to dynamic deformation at high pressure.

Other long-impulse experiments driven by a flyer plate have been conducted in which diffraction has been observed on the rear surface of a target from elements put into states driven by a steady wave within a solid. However, the pulse has travelled some distance and a distribution of these states exists at the length scales probed by diffraction. The laser impulse, however, forms a plasma which loads a surface for a small time, and diffraction through this loading plasma shows the effects on the face in the first moments for which the compression pulse is established at the surface. States have not developed into a distribution, the results are clean and short timescale processes such as phase transformation may be probed. Thus coupling the techniques for unsteady and steady flow can illuminate the atomic-scale processes operating in shock compression of a lattice and brighter light sources offer the possibility of employing a new suite of techniques for imaging deformation at the microscale.

4.4.10   Spectroscopy

X-ray illumination also allows other techniques to be employed to gain information from the lattice. One of the X-ray Absorption Spectroscopy (XAS) techniques includes Extended X-ray Absorption Fine Structure (EXAFS). This probes lattice short-range order in both polycrystalline and single crystal samples and potentially allows inference of phase, compression and temperature within the loaded samples. Further, when lasers are pulsed onto targets, it is possible to undergo a range of dynamic measurements that have spectroscopic techniques as a basis. These include methods that give insight into chemical reaction pathways such as vibrational spectra and related information, and also the techniques that are used to measure shock temperatures including ellipsometry and Raman. In these investigations it is vital to probe optical properties including polarisation and emissivity (which relate to the complex refractive index).

4.5   Recovery

One may sense a state, deduce structural change or picture a transient, but a piece of material captured from a defined loading pulse is worth many histories. If one can locate a sample that has been recovered from a known loading excursion one can capture the changes that the microstructure has undergone during the loading process and deduce much from the far greater range of techniques available in the ambient state. Any research programme should ideally include sensing state, imaging deformation and recovering deformed microstructure.

Populating material models with physical descriptions of behaviour requires knowledge of operating mechanisms at the mesoscale. To realise this, targets must be recovered for microstructural examination and have a pedigreed loading history that accurately defines and isolates the impulse applied to the target. Clearly it is desirable to consider targets loaded globally in one-dimensional strain in order to generate new theory. However, in real impacts other processes operate which complicate the interpretation of the recovered microstructure. These include spallation and triaxial late-time inelastic deformation due to lateral release. Materials respond not only to high-amplitude pulses but also need significant time for the suite of deformation mechanisms to operate. Thus high stresses applied for nanoseconds produce local deformation on surfaces which can be relatively easily examined after load since significant momentum has not been imparted to the target. Longer pulse lengths load greater volumes of material and the momentum transferred accelerates targets to high velocity after loading, which makes catching them without damage difficult.

Shock recovery techniques have been developed over the past years using varying means of trapping momentum and different devices to introduce the pulse. In order to trap radial release waves coming in from the periphery of the target, it is necessary to place momentum trapping rings which contain the radial releases before they can enter (see Figure 4.13). The general requirement for them to trap radial unloading is that the target must have released longitudinally in one dimension (which proceeds via

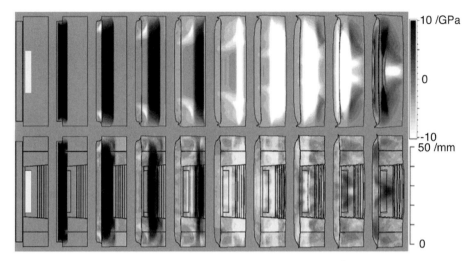

Figure 4.13 Impact of copper impactor onto a copper target at 500 m s^{-1}. Two simulations are shown. The topmost shows the flyer plate impacting a solid block of copper. The lower shows the same copper impactor impacting a recovery assembly. The grey levels indicate the pressure in the assembly as a function of position. The lighter rectangles in the first frame before impact indicate the position of the recovered target within the sample. Source: after Bourne and Gray III (2005a).

longitudinal release propagating from the rear) before radial waves arrive laterally from the periphery. Two rings optimise the trapping of the lateral momentum around the central target region. The central inner momentum trap is in addition tapered, since this, as well as the several nested guard rings was found to be necessary to reproducibly keep residual strains less than 2%. The complex target arrangement to realise this is shown in Figure 4.13. The comparison between a solid and a sectioned target shows that trapping components of momentum allows a sample of material that has undergone a precise loading history to be recovered for further experiment. This has been demonstrated on ductile but also brittle solids where any lateral strain causes failure.

After loading and release are complete the targets must be gently decelerated and the induced shock temperature rise rapidly dissipated before the microstructural defects generated and stored in the sample during the shock-loading excursion are significantly altered by thermal relaxation and static recovery. Deceleration and thermal quenching is generally achieved by plunging the centre of the target into a liquid bath immediately after the shock pulse has released.

The soft recovery of materials is proving of increasing importance to the understanding of mechanisms operating in materials during shock loading. Up to the present, this technique has been successfully demonstrated for stress loads up to 10 GPa and pulse lengths of microseconds for metal targets of this size. Pushing stresses upward to view processes acting over microseconds is clearly a necessary development for future work in the next years. In addition, increasing pulse lengths to observe longer timescale mechanisms is necessary to understand the range of processes at work in deforming solids.

4.6   Imaging

Images are formed after illumination with some specific wavelength of radiation or by scattered nucleons and will be recorded on some medium that will respond to the incident beam physically or electronically so that the picture may be stored. Once captured, the image needs to be permanent, both so it may be reproduced easily and shared with others but also so that the behaviour it displays may be analysed by others and stored so that further information may be gleaned at later time as analysis tools become more sophisticated.

Photography, which was first developed for operation in the visible region, is now extending into the infrared and to other wavelengths as new imaging sensors become available. To observe the internal details of structures or to quantitatively obtain density information, X-ray imaging is a mainstay of modern effort, and although the analysis of radiographs can be precise and informative, the greater energy of the radiation makes it less controllable than visible light, which can be imaged with optical techniques to quantitatively measure refractive index change and thus visualise fronts and density variations at least in transparent objects. Projecting an image onto a plane produces a two-dimensional representation of an object, yet with the advent of digital technology and the ability to process multiple images from many viewpoints, it is possible to construct three-dimensional representations of objects of interest. The development of such tomographs is moving ahead at speed, yet the means of using and analysing the plethora of information recorded is progressing more slowly. Finally, there is a need to image processes at different length scales in materials. Photography and optics have developed to handle imaging at the laboratory scale with light and (using phosphor screens) higher energy illumination. Once the wavelength of the radiation becomes commensurate with atomic dimensions, diffraction within the lattice occurs. More difficulty has come in imaging at the microstructural (meso-) scale and this frontier represents the next boundary for modern technology in this field.

Snapshots of spatial information provide a great deal of information if the resolution and dynamic field of the image can be optimised for the application of interest. However, for dynamic events or processes occurring under load it is frequently optimal to have sequences of images at defined times. On the length scales of human function there are important groups of processes and scales that are connected intimately by the velocity at which information can travel in each case, e.g. human action at the lab scale, wave motion at the lab scale, wave motion at the mesoscale and electromagnetic waves (Table 4.2). To observe human activity requires observation of processes that occur at speeds of the order of 1 m s^{-1}. To observe phenomena governed by the propagation of mechanical waves in substances, such as shock fronts or surface waves from earthquakes, requires the ability to image events on the laboratory scale at the order of 1000 m s^{-1}. Finally, electromagnetic waves are travelling much faster again at $c. 10^8 \text{ m s}^{-1}$. These will be observed at different rates according to the length scales of interest for each phenomenon; humans will operate on a nominal 10 m scale, waves will be studied over 10 mm or at the mesoscale at 10 μm. If a sequence of nominally ten frames is required to capture each process the following rate of image capture will be necessary in each

Table 4.2 Required rate of image capture, assuming 10 frames per sequence

Process	Length scale of interest (m)	Speed of event (m s^{-1})	Required image rate (s^{-1})
Human activity	1	1	10
Wave at lab scale	10^{-2}	1000	10^6
Wave at mesoscale	10^{-5}	1000	10^9
Electromagnetic wave	10^{-3}	10^8	10^{12}

case. It will be seen that microsecond imaging gives the ability to study mechanisms that operate at or less than the wave speed in materials and these are available at present commercially. However, the next hurdle will require order of magnitude faster imaging at giga- and teraframes per second (Table 4.2).

In the following sections, high-speed optical, X-ray and nucleon imaging will be described and explained, emphasising the common techniques present across technologies.

4.6.1   High-speed photography

The development of high-speed imaging closely followed the course of the development of photography itself. Images had been captured transiently by pin hole imaging since the earliest times, but it was the *camera obscura* (Latin for 'dark room') which formally allowed projection onto a two-dimensional plane where the image might be used. Joseph Niépce in 1826 was the first person to successfully fix an image using a pewter plate coated in bitumen, but it was Louis Daguerre in France, John Herschel (astronomer and originator of the term *photo-graphy*) and Henry Fox Talbot in the UK, who became the driving forces behind the early development of the technique in the mid 1800s. Edweard Muybridge (born Edward Muggeridge) was the first to take a moving sequence using cameras with shutters set to a speed of one five-hundredth of a second and then released by threads broken in one case by a horse (or in other work by clockwork) in 1877. This first sequence proved Leland Stanford (founder of the university and at that time governor of California) correct in his hypothesis that a horse travelled by 'unsupported transit' with its hooves off the ground for a period in the gallop (Figure 4.14(a)). Muybridge later went on to develop the zoopraxiscope, a precursor to the projector, which displayed a series of images sequentially.

In 1888, George Eastman invented photographic film and the Kodak cameras that housed it. Kodak developed a 16 mm cine camera in the early 1930s which eventually reached 5000 frames per second. The development of this wet film technology continued up to the 1980s with the development of rotating prism cameras to preserve shuttering and accurate imaging for the devices. Harold Edgerton popularised high-speed imaging through his work freezing motion on a single image with multiple exposures. His strobe light (operating at *c*. 120 Hz) was used to create images that were as much fine art as high science, with his work appearing in *Life* magazine and *National Geographic* and on display in art galleries throughout the world. His work for the US government developed

(a)   (b)

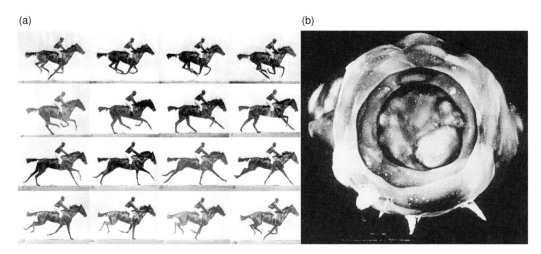

Figure 4.14 (a) Time-lapse photographs of a man riding a galloping horse, by Edweard Muybridge, 1872–1885. (b) One of a sequence of photographs taken within milliseconds of initiation of a nuclear explosion at the Nevada Proving Grounds, *c*. 1952, photographed by by Harold E. Edgerton; 1 ns exposure, taken from seven miles away with a lens 10 feet long.

cameras with rapid shuttering for events of high luminosity, including the nuclear tests of the 1950s (Figure 4.14(b)).

Image tube systems appeared in the 1950s and allowed conversion of an image into an electron beam which could be shuttered and deflected. This enabled images to be rastered across phosphor screens to produce sequences of tens of frames in some cases. The phosphor decayed over milliseconds, long enough to be optically or fibre-optically coupled to film. Although these tubes were used to produce framing sequences they could also be configured with one or two sets of deflector plates. The converted electron beams could be swept across the phosphor screen at high speeds limited only by the driving electronics; these electronic streak cameras could reach rates of the order of 10 picoseconds per mm.

In 1969 Smith and Boyle invented the CCD (charge-coupled device) which allowed electronic capture of colour images. By the mid 1980s the first handheld electronic cameras were being marketed and by the mid 1990s intensified high-speed CCD cameras were becoming available for dynamic imaging. The introduction of CMOS sensors in the 1990s allowed a further step forward since they were cheaper to build than CCD and easier to integrate with memory and processing on chips. At the present time, video cameras with 1280×800 sensors are able to deliver framing rates in excess of 15 000 frames per second, replacing film in professional media and film arenas. However, in the modern digital world the limits of high-speed imaging still rest with the quality of the optics necessary.

When one decides to obtain any stream of high-speed data on a dynamic event, but particularly a stream of images, one needs to know four key facts. Firstly, when the event will begin – preferably to have some warning so that equipment will be active, light can be fixed to the correct level, etc. Secondly, an idea of the duration of the phenomena under test, to allow the choice of the correct equipment capable of capturing several images at the right rate. Thirdly, some idea of the field of view necessary to capture all the action. Finally, the illumination required to correctly expose (or alternatively the filters necessary to limit the flash from) an event. Common experience suggests that shutter speeds of tenths of a second correctly illuminate film or digital photography in daylight. Anything shorter or darker requires artificial lighting of some kind. These requirements are simply met in a self-luminous sequence, such as an explosion. The bright flash of light means that the fast shutter speed and short exposure time necessary to capture the event can be met without recourse to extra lighting. With older film cameras, if the system is left running continuously at these fast rates the tiny exposure times for ambient light hitting the recording medium are so short that no darkening of the emulsion results. Thus leaving a camera running until the event is over and processing the resulting medium is all that is necessary to observe a sequence of the event. This method was used by the British in the 1950s to record their atomic weapon tests. To introduce some technical terms, if a camera requires no trigger to start recording then it is *continuous access*. The design of a device controls two features of its operation: the time for which light interacts with an image plate – '*the exposure time*' (ET); and the time between images before the next frame is triggered – the '*interframe time*' (IFT).

There are four components to a dynamic imaging experiment: the light source; the imaging system employed; the recording medium adopted; and finally the time sequence that marries the other components together. Dynamic light sources come in three types: high-power continuous flood lamps; strobing flash sources of particular luminosity and repetition rate; or a single flash of specific power, rise time and duration. The last of these requires a trigger pulse to allow it to respond, warm up to peak brightness and illuminate the period of interest for the event. The imaging system consists not only of the required optics to image the subject onto the required internal planes but also components to ensure fast shuttering of the image on the film planes. In the past high-speed cameras have used rotating mirrors or prisms to sweep images across a series of stops placed around a strip of unexposed film. In modern cameras this feature has been replaced with a series of mechanical stops allied to the fast shuttering electronics on the CCD or CMOS arrays themselves. Recording media too have changed over the years with phosphors matched to the wavelengths of incoming radiation and emulsions optimised for them. In the modern age the search for suitable matches with digital arrays is key to matching the excellent resolutions gained over the 100 year development of wet film. However, connecting together all of these features into an integrated experiment requires sensors that trigger flash systems before turning on recording equipment, or in suitable circumstances devices that initiate an event when all devices are operating. And of course it is necessary to understand enough of the event and its kinetics to have matched a sequence of frames of the correct exposure time to interframe delays to allow the process to unfold.

There are two basic types of image recorded: a digital sequence of frames or an analogue image of an axis in an event. Figure 4.15 shows a schematic series of three frames of a dark wave travelling downwards over three times, t_1, t_2 and t_3, spaced equally

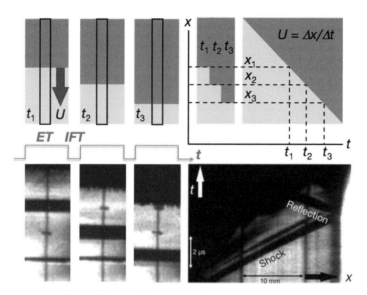

Figure 4.15 Streak photography. Upper schematic shows three frames to the top left, with pseudo and analogue streak images to the right. The square wave in the centre describes the switching of the camera pulses with an exposure time (ET) adding to an interframe time (IFT) to give the difference between the start times of each frame. Three frames and a streak from an experiment on borosilicate glass are shown underneath for comparison. Source: lower frames reprinted with permission from Bourne, N. K., Rosenberg, Z. and Field, J. E. (1995) High-speed photography of compressive failure waves in glasses, *J. Appl. Phys.*, **78**(6): 3736–3739. Copyright 1995 American Institute of Physics.

apart. Beneath the frames is a square wave representing a timeline for the sequence. The positive parts of the pulse show the frame being exposed; the shutter is closed in the zero sections and no light exposes the film. The times between the frames are such that ET + IFT must equal the frame time for each exposure. If a slit obscures the centre of each frame, a thin sliver of each of the three frames is recorded and these are figuratively transposed to the right-hand side where they are stacked next to each other in the figure. Now each slit is plotted on a horizontal axis at the specified times and clearly the distance coordinate is preserved for each. Thus an artificial sequence of steps is constructed that would be a continuous line if the image of the slit were infinitely thin and the frequency of sampling were increased, as if the image was merely being pulled or streaked along the time axis. The slope of the dark front preserved in the streak would now have a gradient U, the speed of the wave travelling down the frame.

Observe the real images beneath the schematics. Three frames from a sequence in which a front travels down a borosilicate glass block are shown in the figure. Behind the shock front travels a front preceding fracture from the impact face of the target. This wave drives cracks into highly compressed material and is known as a failure wave. It represents the head of the inelastic process by which glass fails under compression and has slow kinetics relative to metallic plasticity processes by virtue of the slower crack speed, higher bond strength and lower critical defect density. This will be referred to later in Chapter 6 on brittle failure.

Finally, in the bottom right, the streak image of the wave travelling down the glass block is shown. The shock wave is visible as a dark line by virtue of schlieren optics applied to the experiment (described below). Note that in the discussion above where the slit ran to the right, the time axis occupied the abscissa of the plot and the distance axis the ordinate. It is conventional to plot wave data of this sort in a different manner in this subject. Here time runs up the page and the abscissa is now the distance travelled. Nonetheless, the shock and failure front following it may be seen in the streak picture, and later discussions will clarify the mechanisms at work in the glass in this experiment. Notice that this method of presenting the data is not immediately obvious and requires some study before features within become familiar. Also note that this is a one-dimensional representation of a (nominally) one-dimensional experiment where interpretation is relatively simple. In the case where features develop down arbitrary directions, streak can be difficult to interpret and counter-intuitive without planning and experimental design. Finally, precise alignment of the streak axis down the travel direction of any wave is vital if quantitative information is to be gleaned, otherwise unphysically high wave speeds will be recorded.

4.6.2 Flow visualisation techniques

This final section will enumerate a range of techniques employed to quantify flow variables that allow the user to record quantitative density or displacement information from a sequence of frames. When flow occurs within transparent media it is possible to track any variation in refractive index, n, by using optical techniques. The phase difference between a beam that moves through a medium with a changed index and one that has travelled though one as yet undeformed is proportional to n. The technique that can be used to combine and analyse differences between these states is interferometry. The displacement of the beam in a medium shocked to higher refractive index is proportional to the second derivative of the index in the direction of the change. Finally, its angular deflection is proportional to first derivative of the index. Displacing a beam perpendicular to the direction of density variation in a travelling wave deflects light away from the wave front to a position to the rear of the disturbance. Thus illuminating a target with collimated light is sufficient to image density discontinuities within a transparent flow and this technique is known as *shadowgraph* (see Figure 4.16, top). Angular deflection at lower rate can be filtered by placing a knife edge in the conjugate optical plane and then reimaging the result to produce an image that shows an intensity that is a differential of the refractive index changes in the direction perpendicular to the edge (see Figure 4.16, bottom). This technique is known as *schlieren* (from German, meaning 'streak') and is a sensitive method used to image density changes in transparent media. Coloured filters and suitable optics allow more complete density fields to be reconstructed from flows. To the right of the wave diagrams for each technique are shown two images using each technique for the same experiment where a bullet travels in air. The dark and light form of the bow shock can be seen with the collimated light in the case of shadowgraph whereas much more detail can be seen in the flow for schlieren imaging.

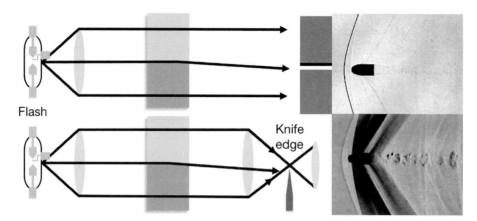

Figure 4.16 Shadowgraph and schlieren. A localised (point) source is used to create collimated light which passes through a target with a density discontinuity. Deflected light is imaged onto a plate (shadowgraph) or optically filtered on a knife edge (schlieren).

In both cases the set up of the lighting is key to success since both require a localised source (generally a discharge tube) in order that such effects are observed. Other materials (e.g. transparent polymers) show photoelasticity when stressed and polarised light shows colour contours on images that have been loaded.

For opaque materials, techniques based around Digital Image Correlation and Tracking (DIC/DDIT) have been developed to measure surface displacements. This mathematical tool analyses image registration for accurate two- and three-dimensional measurements of changes within images. This is primarily used to measure deformation, displacement and thus strain. There are flavours of various techniques used to make these motions visible. A grid and then deformation analysed using Moiré fringes is one simple method but only gives one-dimensional information. More useful is imaging of speckle patterns created by lighting rough surfaces with coherent illumination (laser speckle). This has developed into imaging individual positions by seeding the flow with tracer particles (particle image velocimetry). Imaging the flow and correlating the motion of the seeding particles from frame to frame allows the calculation of the displacement and thus the speed and direction (velocity field) of the flow being studied.

Each of these techniques fails as soon as tracers or added surface texture are unable to follow the flow or otherwise corrupt the properties under investigation. Non-invasive, fully three-dimensional measurement is always the paradigm to be achieved, but this as yet is confined to only a small number of special cases. Future developments will aid the development of high-speed imaging since its cost has come down markedly and resolution climbed rapidly over the past 20 years. However, there is still a need to combine these advances with improved correlation techniques to turn a qualitative tool into a quantitative workhorse.

Table 4.3 X-radiography applications

Domain	Medical imaging	Weld control	Flash radiography
Photon energy	50 keV	500 keV	4 MeV
Required dose at 1 m in air	1 rad	100 rad	500 rad
Exposure time	100 ms	3 h	50 ns
Dose rate at 1 m	10 rad s^{-1}	10^{-2} rad s^{-1}	10^{16} rad s^{-1}

4.6.3   X-ray illumination for dynamic experiments

In 1895, the German physicist Wilhelm Roentgen (1845–1923) discovered a new highly penetrating radiation which, lacking obvious explanation, he dubbed X-rays. The discovery sparked a rapid explosion in research, since following the publication of Roentgen's paper, more than a thousand books and articles were written on the subject with the number of publications rising to more than 10 000 before 1910. Within a year of the paper appearing the first medical experiments used X-ray illumination. The American physiologist, Walter Bradford Cannon used a fluorescent screen to follow the path of barium sulphate through an animal's digestive system. Modern advances have seen computed tomography (CT) largely developed by Hounsfield and Cormack in the 1960s who shared the 1979 Nobel Prize in Medicine. CT is also commonly used for the non-destructive testing of materials in research and industrial applications.

It is the interaction of electrons and photons with nuclei that determines radiation interactions with matter. Electrons do so by elastic or inelastic collisions, with consequent deflection and excitation of states within atoms. Nuclear electric fields may slow down an electron, however, which transfers its loss of kinetic energy to emission of a photon. This 'radiation during braking', *bremsstrahlung*, becomes more intense and shifts toward higher frequencies when the energy of the incident electrons is increased and the mass of the interacting atoms gets larger. Photons, on the other hand, interact elastically (Rayleigh–Thomson scattering), through absorption (the photoelectric effect) or inelastic collisions (Compton scattering). Alternatively they may create an electron–positron pair. Thus electrons generate photons, and photons generate electrons; the two particles play a pivotal, coupled role in radiography.

At low energies it is the photoelectric effect that dominates and stops the photons. At intermediate energies Compton scattering is dominant; at high energies it is pair production that is predominant. In practical terms this absorption results in a 500 keV X-ray flux undergoing attenuation by a factor of around 10 when passing through 35 mm of steel. Table 4.3 summarises source characteristics for some typical X-radiography applications.

X-rays are at the short-wavelength, high-energy end of the electromagnetic spectrum (0.01 to 10 nm); indeed only gamma rays carry more energy. Hard X-rays are of the highest energy, while those of lower energy are referred to as soft. The distinction between hard and soft X-rays is not well defined, however, and hard X-rays are typically those with energies greater than around 10 keV. The energy of the photons in a particular part of the spectrum affects the quality and information that the images recorded by the pulse can yield so that the spectral range must be matched to the applications discussed here.

In order to probe dynamic events, the pulse length of the illuminating radiation needs to be short enough to freeze motion in the object of interest. If the image to be acquired is at the millimetre length scale, waves in the target (travelling at kilometres per second) fix the timescales of interest to microsecond intervals. A rough estimate of the exposure required to freeze motion for such events is a tenth of this or 100 ns. In the case of observations of information at the atomic scales, the pulse needs to be of the order of 10 fs. Recent work on dynamic X-ray diffraction (DXRD) has used a technique in which a longer pulse of $c.$ 1 ns is used to backlight crystal targets. The penetration depth of the X-rays (which for these materials is typically several microns) ensures that recorded diffraction patterns record both uncompressed material lattice parameters, compressed plane information which indicates normal strain, and diffraction from compressed planes to give lateral strain.

X-radiography requires a source of high-energy photons that can be gated to image dynamic events. Although gamma photons, having energies greater than a few tens of keV, may be generated using radioactive elements, the dose rate is insufficient for the exposure times required for dynamic applications. Thus, various means of creating a brief, intense pulse of high-energy electrons that can produce a burst of bremsstrahlung as they decelerate in a metal target have been optimised over the years. At low energies (< 10 MeV), this is effected by way of a diode; at higher energies (> 10 MeV) generation involves a particle accelerator with a beam, produced in an electron gun, that is accelerated through successive cavities up to a final energy. X-ray sources are generally compared according to their spatial distribution (focal spot), angular distribution, intensity, spectrum and duration.

The highest powered systems use different means to achieve high intensity. In one generation a large capacitance is discharged through a diode to deliver a large current which is accelerated onto a high atomic number anode that liberates an X-ray pulse. These systems will produce an X-ray pulse of 1000 rad at 1 metre with a 2 mm spot size. In another system an injected current of electrons amplifies and accelerates it down a linear pulsed system. These linear induction accelerators accelerate electrons to produce intense, bremsstrahlung X-ray pulses for flash radiography.

The most exacting applications require X-ray sources capable of extremely brief exposure times and high photon energies in targets that will apply heavy attenuation to the pulse. The incident X or gamma ray photons pass through the object, being partially absorbed as they pass through. Clearly absorption varies according to the nature, density and thickness of the materials under test. The dose reaching the far side of a target includes photons that directly traverse it but also a scattered population, which are integrated into noise on the pulse that transmits. However, it is the direct absorption, characterised by a defined attenuation coefficient for each material passed through, that allows the analysis of recorded flux to deduce density information from the image and to optimise the system resolution for each shot. Suitable detectors must maximise spectral response and sensitivity and four classes are now typically used: screen–film combinations; photostimulable screens; CdTe (cadmium telluride) semiconductors; and scintillators coupled to cameras.

The outputs from these deliver raw images which can be analysed and image processed to recover sharp, accurate information from the experiments. Typically, Monte-Carlo codes are used to deconvolve the image which includes some statistical representation of the paths and interaction probabilities in the illumination of the object. Clearly any analysis must aim to account for the source size, the detector resolution and the blurring due to motion of the target during the exposure.

4.6.4   The next generation of light sources

Modern facilities have exploited various other new means to generate X-rays, such as synchrotrons and free-electron lasers. These devices use conventional accelerators to generate an electron beam and then long arrays of magnets to wiggle the electrons to generate bremsstrahlung. Lasers can also be used as X-ray sources and such advances in compact, high-power beams offer the opportunity to use such pulses for dynamic imaging. Clearly one of the other advantages of the X-ray source is not merely in the available power, but also the short exposures times which can be generated. This means that femtosecond pulses are achievable, allowing high resolution in both space and time to allow atomic scale investigations of ultrafast deformation.

MeV electrons and ions are observed from beam interactions both with gaseous and solid targets since electrons are driven in the electric field of an intense high-power laser to relativistic speeds. When such relativistic electrons collide with the ions of the solid, they rapidly accelerate, causing the emission of hard X-rays through bremsstrahlung as they decelerate in the material. However, if the beam interacts with a gas jet, collimated high-energy electrons are generated in the direction of the beam. Experiments conducted on existing petawatt lasers have shown that it is possible to achieve five times greater dose using gas rather than solid targets.

These sources offer great hope for future application in dynamic imaging, delivering picosecond pulses to the target but with a spot size of tens of microns, compared with present sources which are several millimetres. Further, targets can potentially be strobed from different viewpoints to enable three-dimensional radiographic sequences to be obtained.

The free electron laser (FEL) uses a relativistic electron beam projected through a magnetic field which can be geometrically and temporally designed to optimise output radiation (Figure 4.17). The magnetic environment which it encounters gives the ability to generate electromagnetic radiation from microwaves to X-ray via bremsstrahlung. Further, the source thus created has short pulse length (femtoseconds), full coherence, high brightness and high repetition rate. It may also be tuned as a source for appropriate wavelengths for different applications, which makes it versatile over a range of applications; soft X-rays (such as FLASH in Germany) are already used for biological imaging, but, for penetrating condensed matter discussed here, new hard X-ray sources (such as LCLS at Stanford) are coming on line. Such a device exceeds the brightness of

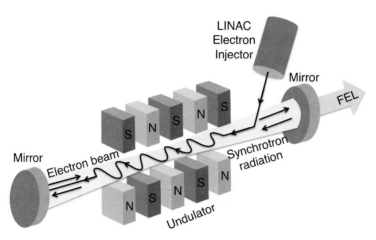

Figure 4.17 Principles of the free electron laser (FEL).

synchrotron sources by a wide margin, so that volumes of material may be probed with higher power pulses over smaller time intervals. Hard X-ray sources with short wavelength (*c.* 1 Å) mean that images have both atomic spatial resolution and high penetration into matter. By this means is it possible to investigate the suite of physical and chemical mechanisms that occur at the atomic scale in much greater detail than has been possible up to now. Time resolution in the femtosecond regime and high peak brightness allows both single- and multishot imaging and also the investigation of non-periodic structures by means of coherent scattering techniques.

The first free-electron machines built in the 1950s produced microwaves with the shorter wavelengths of visible light not achieved until the early 1980s. Meanwhile, wigglers (magnets designed to periodically laterally deflect the beam of charged particles) a few metres long have been used for decades to produce non-laser X-ray beams in synchrotrons. A particle accelerator directs electron bunches through a magnetic section (Figure 4.17). The orientation of the magnets alternates in this wiggler which causes the electrons in the magnetic field to slow and emit photons travelling down the same trajectory through the device. Interaction of photons and electrons down the radiation path slows them further increasing the photon flux. By tuning the speed of the incoming electrons, it is possible to synchronise the electron clouds with the photon bursts to produce a train of highly concentrated pulses of coherent radiation. The new free electron lasers, however, will have wigglers up to 150 m long and use highly concentrated electron clouds. Taken together, these features produce highly focused and coherent beams with X-ray pulses down to a femtosecond duration. Further, injection of electron bunches means that the output can be pulsed down the accelerator giving a strobing source at the target.

The interaction of an X-ray pulse with an incident photon flux of such proportions with target materials offers an exciting range of possibilities for condensed matter physics. The power of course is so high that the absorption typically destroys the microstructure being probed. However, assuming information can be extracted before this happens, such a source can complement imaging diagnostics for the understanding of atomic to macroscopic structure with both incoherent and coherent imaging. There is extraordinary potential for atomic-scale imaging techniques and different flavours of ultrafast spectroscopy to offer unique insights into the competing interactions between the system's constituents (and associated inhomogeneities). These include electron–phonon, electron–electron and spin–lattice interactions. Yet key aspects of the scattering and interference of coherent pulses give a range of issues for diffractive imaging that must be resolved, since the strong coupling and quantum effects that will occur in such an intense pulse have not, as this is written, been fully solved to allow complete analysis of the data acquired from such experiments.

The FEL shows great promise for dynamic imaging, combining the ability to tune wavelength with many orders of magnitude increase in brightness from a coherent light source. All of these known benefits sit alongside the unknown payoffs that run in parallel: the opportunity to access new physics that will inevitably accompany the fielding of fourth-generation imaging technologies within the national research communities. Over the next 10 years the means of producing coherent radiation by pumping electron plasma waves either with an electron beam or a laser will offer a further, more compact technology that will have wide application; the drive towards fifth-generation sources has already begun.

4.7   X-ray microtomography

X-ray microtomography (XRT) is a non-destructive material imaging technique that illuminates the sample with a cone of X-rays and allows the internal structure to be represented by reconstructing the spatial distribution of the local linear X-ray absorption coefficients of the materials/phases contained within. The 3D target volume is reconstructed via a filtered back-projection algorithm calculated on a computer and once created it can be interrogated to provide 2D cross-sectional slices. This reconstruction is typically done with a selection of Fourier techniques that provide a virtual, three-dimensional representation of the interior of the object from which two-dimensional cross-sectional slices can be viewed through the three orthogonal directions of the volume. There are two imaging techniques that are typically used: bright field, which employs in-line techniques; and dark field, where the detector is placed off-axis. These give better resolution in particular classes of material, e.g. phase contrast computed tomography in polymers.

In conventional radiography, the image plane is approximately normal to the X-ray beam, and the image represents the total X-ray attenuation through the object. Computed tomography (CT) creates a digital representation of a thin slice of the object parallel to the X-ray beam. This image is reconstructed from a series of two-dimensional radiographs taken at different orientations. The CT slice is stored as an array representing local X-ray attenuation values for each of the small volume elements (voxels) that make it up, and represented in a reconstructed image as a series of grey level values. Three-dimensional volumes are reconstructed via filtered back-projection algorithms, also providing cross-sectional 2D slices. A typical X-ray microtomograph images targets a dimension of centimetres with a resolution of $c.1$ μm. As time progresses laboratory equipment of ever-increasing power is becoming available, but the best images are still obtained using synchrotron X-ray sources which have ideal attributes. The means to do such imaging within a solid in real time will be the next stage of development for XRT: dXRT, which would allow the remote, 3D visualisation of microstructure to be observed non-invasively within the specimen as it develops.

4.8   Particle accelerated imaging

4.8.1   Neutron radiography

Neutron and proton radiography offer a series of alternative options for imaging of dense solids. The primary difference between X-ray photons and particle imaging is that whilst X-rays are scattered at electrons, neutrons and protons are also scattered by the nucleons in the atomic core. Although X-rays are electromagnetic waves or equivalently high-energy photons, and interact with electron clouds around atoms, neutrons are massive, slower moving and possess magnetic moment. In many senses neutrons offer the perfect imaging radiation in that they can penetrate more easily through high-impedance targets (80–90% of neutrons penetrate a sample) and deflect from magnetic regions as they pass. However, radiation sources are weaker and more complex than those used in producing X-rays. The interaction of each source with targets of different atomic number is reflected in the scattering cross sections for atoms to each radiation. Thus proton or neutron radiography is a technique of choice for non-destructive imaging where low atomic number materials (particularly fluids) are to be visualised. Further, there is greater penetration depth into most materials, considerable variations in contrast between chemical elements and isotopes, and weaker radiation damage than from other penetrating radiations.

Accurately controllable neutron sources are in two classes; radioactive sources are low flux and not accurately constrained so that the two most potent are highly controlled research fission reactors or spallation neutron sources. The latter devices are high-flux sources in which protons accelerated to high energy hit a large atomic number target. At the heart of such a source is a high-power accelerator repetitively producing intense pulses of protons injected into a synchrotron storage ring. Protons are released onto a high atomic number target, which may be solid (tantalum or tungsten) or liquid (mercury, lead and lead–bismuth alloy have all been used). In all cases neutrons are produced in the target by spallation which gives an intense neutron pulse, with modest heat production in the target.

Large neutron beamlines around the world achieve greater fluxes than other sources. Reactor-based sources now produce 10^{15} n cm^{-2} s^{-1}, and spallation ones generate greater than 10^{17} n cm^{-2} s^{-1}.

4.8.2   Proton radiography

Protons may be used in the same manner as neutrons to image materials states since both particles possess magnetic moments whilst protons additionally interact electrodynamically. pRAD in Los Alamos has used high-energy protons from the LAN-SCE accelerator facility as a probe in flash radiography for many years.

Imaging with nucleons offers advantages over X-rays given their interaction with matter by both strong and electromagnetic interactions. Since the strong interaction has a short range (10^{-15} m), protons interact with other protons and neutrons by collision and the collision probability is indicated by a material's cross section, which is dependent upon the number of protons and neutrons in the nucleus. Although a proton can interact elastically or inelastically with a nucleus via a strong interaction, it is more likely to do so electromagnetically with both the nucleus and atomic electrons. A proton generally ionises atoms which slow it at each interaction and this may be calibrated to yield density. However, nuclear interactions will result in multiple Coulomb scattering which may blur the image significantly. Protons travel tens to hundreds of millimetres through condensed matter before they undergo significant interaction with the object either through strong or electromagnetic forces. In comparison, X-rays have a maximum attenuation length of the order of 10 mm. This results in one in three protons as opposed to one in a million X-rays penetrating a 100 mm target. This results in the need for lower power source strengths to achieve similar intensity to an equivalent X-ray device.

A linear accelerator takes a proton beam and accelerates it to high power; such beams achieve 800 MeV in the LINAC at LANSCE, for instance. Protons are then fed into a storage ring for subsequent exit to suitable target sites or fed to other neutron production targets. In a proton radiographic source, the beam passes through a thin tantalum sheet, which spreads it to illuminate the entire test object. It then passes through a set of quadrupole electromagnets that give the protons the angular and positional correlation required for sharp imaging. The beam then passes through three proton lenses, each consisting of four quadrupole electromagnets, with a collimator at the midplane of each lens. The resulting radiograph can be collected electronically to yield image sequences comprising tens of frames. Examples of dynamic tensile failure in several metals are shown in Figure 4.18.

4.9   Future imaging platforms

The review above has shown the present means of delivering the high radiation doses of X-rays and particle beams to target materials and geometries. There are a series of other technologies which do not presently meet the same dosage characteristics but which may, over the next decade, come to rival present technology in their ability to image and diagnose dynamic experiments in the field. These include laser-irradiation and FEL to produce X-rays, and proton or neutron radiography. Nucleons offer vast reductions in attenuation through materials and these attenuation lengths are optimum for radiographing objects to extract precise material physical characteristics such as density. Further, the scattering processes mean that the image has a higher signal-to-noise ratio improving resolution. Finally, protons have the ability to discriminate between similar materials since X-rays are sensitive only to density.

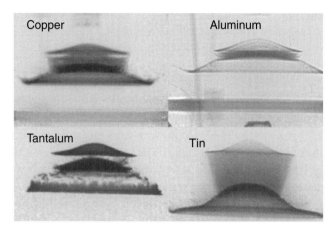

Figure 4.18 A comparison of shock-induced spallation in targets of different materials. Source: reprinted with permission from Holtkamp, D. B. *et al.* (2004) A survey of high explosive-induced damage and spall in selected metals using proton radiography, in *Proceedings of the Conference of the American Physical Society on Shock Compression of Condensed Matter*, Portland, OR, 20–25 July 2003. American Institute of Physics, 477–482. Copyright 2004, American Institute of Physics.

The discussion above has highlighted several of the features of optical, X-ray and proton light sources. Clearly a major advantage is the ability to discriminate and image materials of different densities. For the future, clearly the power of a source should be increased as far as practicable in order that denser and larger structures may be probed by the beam or smaller objects may be imaged at high resolution at fast speeds. It has been noted that X-ray attenuation lengths peak at 4 MeV whereas proton equivalents increase with energy. Thus future applications with higher power will require a switch of technology away from X-ray sources. Further, a proton beam may be split, giving the ability to image simultaneously down multiple axes in dynamic experiments.

Over the next few years, new pulse power devices will deliver an X-ray dose per radiograph of 1000 rad. Of course the total amount of radiation required to produce a proton radiograph is much less than that required with an X-ray source. At present, the equivalent figure for a proton source like pRAD is about 2 rad per radiograph. Thus more powerful accelerators will be required in order to push the performance of proton radiography to that of the new X-ray sources.

For diagnosing dynamic events it is key to develop capability for imaging sequences. In principle it is possible using a beam source to programme repetition rates to control exposure and interframe times. This will be done by synthesising a pulse from a continuous succession of microbunches each lasting about 100 ps and spaced 5 ns apart. Typically, a series of 8–20 microbunches can be used to expose each radiograph. At present a system such as DARHT may take four images, whereas pRad is capable of 20 but one may see advance in the control and in the length of these pulse sequences with new systems.

The high-power laser and particle beam sources have defined a new suite of potential technologies capable of imaging dynamic experiments. There is now a range of sources and detectors capable of meeting the challenges of imaging over all ranges of wavelength and coherence, but also capable of providing the suite of dynamic images that will make clear the mechanisms by which condensed matter responds in its interaction with a loading impulse. The range of materials to be investigated determines the provenance of the devices considered. For a range of projects the ability to probe with pulsed X-rays is a key requirement. However, other sources such as ion beams are also often needed. The radiation sources are found in two classes; incoherent sources that result from beam interactions with high Z target metals, and coherent ones from bremsstrahlung.

Neutron and proton imaging show great promise as a means of backlighting and offer new possibilities over X-ray in resolution and dynamic range. However, the brightest and most promising new X-ray technology is the free electron laser, which over the coming years will prove its versatility as a tool for long-term developments in dynamic condensed matter imaging.

4.10   Future experimental techniques

There are a range of key needs that define future challenges for research in the science of extreme states. Material characterisation at the mesoscale must advance to the next stage in producing validated operating models for deformation and failure. For instance, at present, models do not still fully describe physics in a complete enough manner to properly represent the operating mechanisms that control strength within solids. Further phase transformation under conditions of high temperature and pressure cannot be properly predicted unless more is understood of the thresholds and kinetics for melting, freezing and martensitic transformations within the classes of material encountered and how they depend upon stress state evolution in the sample material. Finally, energetic material deformation will lead to localised heat generation within a solid and will act as a precursor to chemical reaction pathways in the explosive. All of these physical mechanisms require an ability to make a range of experimental measurements across length scales and timescales to understand their origin and deformation paths. Constructing physically based models then links behaviour, particularly at the mesoscale at which microstucture is developing to the continuum where designs must operate.

There is thus a need for new focus to define the necessary suite of experiments to address length scales and timescales under static and dynamic loading paths, and to correctly assign the operating regimes of platforms that may be used. Local static high-pressure states can be achieved between diamond anvils but only on single crystal targets at micron scales at the highest pressures. The volume loaded currently only samples pure elements in compression, and must be increased to compress dimensions commensurate with alloys with larger scale microstructures if the full characterisation for the continuum scale of amorphous or defective materials, or even liquids, is to be realised. With launcher platforms the impact velocities achievable in the laboratory or the materials used in experiment act as limits on the stress levels reached. In particular future systems must provide the ability to push the limits of available platforms to probe the *finis extremis* for sample volumes that represent a continuum structure for seconds to track material response and push the limits of akrology to another level.

To diagnose such an array of experiments requires a step in capability in the real-time diagnostics available to fully characterise the 3D thermal and mechanical states within a deforming sample. Full field imaging of deforming surfaces to micron accuracy and microsecond rates is now routine. But in the future there is a requirement to track the stress–strain fields and the thermodynamic states at the meso- and then at the microscale within the material at the rate that the loading develops. In energetic materials the temperature field within the target generated under deformation induces chemical reactions probed by non-invasive spectroscopic measurements within the bulk in order to understand both the reaction pathways and electronic structure within the material and how these develop with the loading path taken during the experiment. With a full suite of tools at one's call it will be possible to measure in three dimensions the thermomechanical state with the ability to track the microstructure evolving in situ under load. Sensor advances will speed the process; for example new photonic crystals embedded into the flow respond to temperature excursions and offer the means to record a full three-dimensional thermal field for the loading path taken. Forging the connection between the progression of the state of the material and the development of the microstructure within will allow the physics of mechanisms to be completely determined to construct fully predictive physics-based models over the pressure and temperature range of interest in application.

One key major advance will occur in imaging, where the development of new light sources, including brighter X-ray pulses from the FEL and in the longer term from fifth-generation systems, will herald the means to fully resolve defects at nanometre scales. The availability of a coherent X-ray pulse will fuel adaptations in non-linear optics to push imaging resolution and increase the number and detail of their capture. Further, the advent of dynamic tomography will allow deformation to be mapped with fully 3D orientation imaging to resolve evolving defect structures within the deforming solid under load. Finally, there will be cross-fertilization from advances in theory and simulation where new means of extracting useful information from terabytes of tomography or simulation data will be developed along with the means to compare results of theory and experiment in a semi-automated manner.

The design requirements of the twenty-first century need new platforms and diagnostics capable of supplying the relevant impulses for design applications as well as longer, more intense loads for discovery science in order to access different loading pathways. By these means it will be possible to design and diagnose a new class of experiments that take the stress state beyond the one-dimensional planar tests of today, and towards controlled loading paths across stress space that excite particular deformation mechanisms within the target to isolate key physics and chemistry within processes. In the final reckoning, however, imaging and sensing the unfolding deformation states using the suite of tools available will still require the pool of trained scientists and engineers capable of grasping meaning from the data and using it to advance physical science to another level and it is always new minds working on the most challenging problems that must dominate attention for advance in the field.

4.11   Selected reading

Bell, J. F. (1973) The experimental foundations of solid mechanics, in *Encyclopedia of Physics*, Vol. VIa. Berlin: Springer Verlag.

Cooper, P. W. (1997) *Explosives Engineering*. New York: Wiley.

Meyers, M. A. (1994) *Dynamic Behavior of Materials*. New York: Wiley.

Zukas, J. (1990) *High Velocity Impact Dynamics*. New York: Wiley.

5 Metals

5.1 Introduction

In the next chapters, four groups of materials will be introduced and discussed. These will embrace metals, brittle solids, polymers and energetic materials. Some of these will be pure elements in various microstructures, others will be composites of several in different conformations. A lot of what follows has been described and studied by materials science and much terminology and commonplace understanding will be borrowed from there. Appendix A at the end of the book summarises some key concepts for those trained in other disciplines. At the most basic level, materials can be classified as metals or non-metals according to their ability to conduct electricity. The metals consist of cations in a delocalised electron cloud with structure determined by electrostatic bonds formed between the ions and the electron cloud. As pressure increases this bonding changes nature and above the *finis extremis* localisation of the electron density away from the nucleus occurs leading to new states.

Metals are the most common class of elements in the periodic table (Figure 5.1). Atomic stacking rules define a lattice of ions surrounded by a delocalised cloud of electrons, but from the point of view of the electronic states, one may equally consider them as materials where conduction and valence bands overlap. This definition opens the descriptor to metallic polymers and other organic metals and, considering the context within this book, one must consider the behaviour of materials that change their characteristics under high pressures and cause them to achieve metallic states (to conduct) at pressures below the *finis extremis*. A diagonal line drawn from aluminium (Al) to polonium (Po) separates the metals from the non-metals, and within that region the elements order themselves into subgroups defined by their electronic structures. These include alkali metals, alkaline earth metals, rare earths (lanthanides and actinides), transition metals, poor metals and metalloids; all of these have different mechanical characteristics that are ultimately a consequence of their electronic structure. The mechanical properties of metals differ from covalently bonded materials in that their strengths in both compression and in tension are high enough to be useful but low enough that they may be worked. This durability makes them both easy to form and strong enough in service to give them an unrivalled portfolio of properties that has made them the class of materials of choice in engineering. Their use as projectiles or targets where impact loading is an issue originated in the human need for tools to hunt or protect. With faster transportation and heavy industrialisation it is now a practical necessity to understand

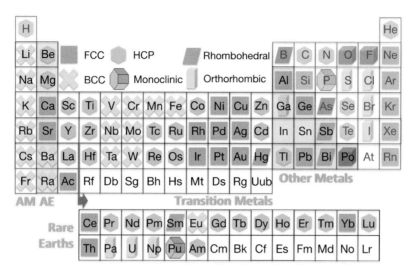

Figure 5.1 Periodic table of elements; particularly showing the metals at ambient pressures and temperatures. AM; Alkali Metals. AE; Alkali Earths. Polonium is simple cubic and indium and tin tetragonal.

metallic behaviour under extreme loadings to meet ever more stringent safety concerns and greater needs in performance. In the future, the versatility of metals offers the possibility of engineering new materials to withstand yet more extreme conditions. For instance, over the past few decades, the ability to form amorphous metals (bulk metallic glasses; BMGs) to engineering dimensions has been realised. These materials possess unique mechanical properties, including superior strength and hardness, good fracture toughness and improved wear resistance amongst others. Nevertheless, lack of significant room-temperature plasticity has limited the application of BMGs thus far, although these properties have many advantages in dynamic loading. To exploit such potential will require an understanding of the response of the microstructure to rapid loading and its effect upon continuum behaviour.

Across millennia humans have attempted microstructural control to tailor metallic properties to fit practical needs. Since the ancient Greeks coined the descriptor 'metal' (*métallon*, Greek for 'mine, quarry'), the structural transformation in steel has been used to harden swords by rapid cooling. With the ability to work metal came the realisation that the processing itself could be responsible for the strengthened article produced. Over time processing has developed from patient blacksmiths working sword blades for Samurai warriors to the industrial, steam-driven forges of Victorian Britain. At the root of many of these concerns was a military interest in understanding projectiles and armour systems subject to dynamic loads to oppose the offered threats. However, more recently other industries have adopted the same knowledge in their drive to maximise safety. Automotive companies, for instance, have optimised onboard protection through crashworthiness testing, whilst the aerospace industry designs against and investigates bird strike, foreign object damage and blade containment. A recent goal has been the development of space armours to protect satellites from hypervelocity impact. In all cases, a three-dimensional body hitting a curved target creates a complex, evolving stress and strain state which is fixed by the material, impact velocity and incident angle as it hits an arbitrary geometrical form. This means that it is nearly impossible to gain meaningful understanding of material behaviour from reproducing such impact events, and hence that it is necessary to load in simpler geometries where mechanisms can be deduced or mechanical constants may be extracted. The ultimate goal in such a venture, particularly in an age where complex simulations may be conducted on computers, is to derive valid, physically based, analytical laws that describe the deformation behaviour of materials. Ultimately it is to be hoped that numerical codes are approaching such a level of accuracy and proficiency that multiple simulations may be used to examine and iterate design requirements cheaply and quickly compared with laboratory testing programmes. However, populating materials descriptions found in such codes with suitable analytical models to truly describe a metal under all possible loading conditions requires a knowledge of operating physical mechanisms at the mesoscale that simply does not exist at the present time.

Research over the last two centuries has shown that metals show greater strengths as the rate of rise of the loading pulse is increased. This strengthening effect is measured in terms of the strain rate (the strain applied divided by the rise time of the loading pulse). The first investigations of material strength began in 1864 with the work of Tresca, who experimentally measured yield criteria for several metals. The first mathematical description of plasticity by Levy and Saint-Venant appeared in the late nineteenth century. These early studies mapped the behaviour of pure metals and their alloys and showed that their static properties were different to those obtained under impact conditions. By the end of the nineteenth century it was proven that metals showed higher flow stresses and increased strengths as loading rate increased. By the first part of the twentieth century the explanation for these observations was formed, developing the concepts of dislocation generation and motion. Orowan realised that the macroscopic strain rate could be related to the speed of the travelling dislocations which allowed plasticity under load to be related directly to slip at the microscale (a classic reference is Orowan, 1934). By the start of the Second World War, there was a need to determine the yield stress of metals to allow screening for their potential use as armours, and simple impact testing, such as that adopted by Taylor and coworkers, gave a means of quickly ranking materials under relevant conditions. He developed an analysis that allowed the determination of average dynamic yield strength by this means.

A further experimental advance came with the development by Kolsky after the Second World War of the original waveguide due to Hopkinson that was designed before the First World War (Hopkinson, 1914; Kolsky, 1953). It has been further adapted over the last 40 years to use analogous methodology to study materials loaded in tension and torsion. The Hopkinson bar has assumed a key role as the intermediate strain rate test of choice in (after stress equilibration) uniaxial stress to map the yield surface in compression. As seen in previous chapters, adaptations to the specimen form have allowed it to be used to measure the fracture toughness and shear properties of various metals. The latest work extends the analysis to allow larger strain data to be obtained for use on particularly soft materials where the slow wave speeds restrict the stress equilibration and the strain attained using traditional analyses. The device and its use was reviewed in more detail in Chapter 3.

It was seen in earlier chapters that at quasi-static strain rates ($< 10^1$ s^{-1}) samples can be loaded in uniaxial stress either in compression or tension using servo-hydraulic test machines or under conditions of plane strain for fracture toughness measurements. But as the strain rate increases (up to c. 10^4 s^{-1}), uniaxial stress conditions are only established in the final stages of the loading, using devices such as the split Hopkinson pressure bar (SHPB), since the effects of inertia dominate until releases from the free surfaces propagate into the target and waves equilibrate the stresses. When impact or energetic loading creates a stress which exceeds the elastic limit, the material behind a planar shock reaches a plastic state where a continuum uniaxial strain state exists. Yet that pulse must load for long enough that state is steady behind the front, and in the laboratory setting only launchers or explosives will probe the length scales of relevance to the continuum for the times necessary for the operating plasticity mechanisms to complete.

Experience has shown that the models used today to describe extreme materials behaviour capture much of the response but only to first order and with most success in compression. Thus to probe the mechanisms giving rise to plasticity and failure in metals across a range of microstructures a coupled sensing and imaging programme is necessary over available platforms. At the present time the two accepted techniques available to probe deformation mechanisms above the elastic limit in dynamic deformation are to use shock to load, unload and recover targets trapping lateral waves so that the target has seen a one-dimensional pulse, and to sense continuum behaviour at Lagrangian stations to build up a picture of the material's response. Shock recovery techniques have been developed using a range of means to trap momentum and different devices to introduce the shock-loading pulse. These have already been discussed in detail, but a summary is included here for completeness. The soft shock-recovery method is required to reproducibly yield recovered samples possessing low residual radial plastic strains. The inclusion of guard rings surrounding the sample serving as lateral momentum traps yields samples recovered after loading in which uniaxial loading dominates radial release processes. This can reproducibly deliver shock prestrained samples possessing residual strains $< 2\%$. Simultaneous deceleration and thermal quenching is necessary, achieved by plunging the shock-recovery assembly into a liquid bath immediately after the shock pulse has released. After removal from the catching tanks, the target can be sectioned and examined to give a window into the processes occurring behind the shock to understand the operating mechanism in metals subjected to impulsive loading. Such techniques are not used as much as they might be and in the future more regular use will better elucidate the operating mechanisms and allow physically based models to be constructed.

In what follows the stress and strength histories gathered from continuum sensors and the evidence deduced from these will be compared with the recovered response of face-centred cubic (FCC), body-centred cubic (BCC) and hexagonal close-packed (HCP) metals, emphasising the microstructural features that contribute to structural development and lead to the properties observed in the recovered targets. It should be said from the outset that each metal has individual character and metallic bond strengths, and the response of even elements in the same period to pressure are unique to each material as it deforms. Thus it is not merely the packing of the atoms that determines response, yet this has a controlling and overarching effect upon behaviour under loading, as will be shown below. Further, most metals in engineering are not single-phase but alloys so that interpretation of response in a production model must take into account multiple phases and microstructure at different scales.

The majority of applications in engineering lie in the regime where the pressure is less than *c*. 40 GPa (the CJ pressure of an explosive and below) but there is a pressure range with application in planetary science where loads are more extreme up to the megabar region and the *finis extremis*. Below this threshold electronic bonding controls behaviour and the observed plasticity is dominated by dislocation interactions akin to those observed in thermally activated dynamics in the lower pressure regime accessed in the laboratory. Up to this boundary the terminology and laws of equilibrium continuum behaviour may be extended with adaptation of concepts formulated for the ambient state to conditions at higher pressure and temperature. Strength develops, and phase changes between states occur and the formalism of the continuum again describes response. Above this boundary, matter exists as a dense fluid and terminology from the extreme states beneath and distinctions such as liquid and solid are not appropriate for the dense fluid that exists. This response will be discussed in more detail below and new research accessing multiple megabars will be described in later sections.

This work will present response to a step-shock and extend to longer durations in the loading. A timeline runs throughout the sections presented here. The response is described from the first moments at which material starts to move, rising in stress to a peak held whilst the microstructure rearranges. In this time period the shear stresses induced by the shock pulse begin to relax as slip occurs in the lattice, yet interactions of such moving planes harden the microstructure if their density is sufficiently high. A mechanical drive at some point unloads the material at which time shear stresses are relieved to be replaced, in some cases, by tensile forces that drive a region within the material to failure. Equally structural response, established after wave equilibration, sets the scene for localisation as well within some metals the formation of adiabatic shear bands in compression. These later time responses act on metal preconditioned by the loading of the initial shock, which has frozen in a microstructure determined by the loading history up to that time; the metal will thus not possess the same material properties as it did at the beginning.

5.2   Shock compression of FCC, BCC and HCP metals

Understanding of dynamic loading is based principally upon three classes of measurement platform that will be described in what follows – loading machines, intermediate rate devices and shock launchers. Quasi-static testing machines determine continuum yield stress and stress–strain behaviour at the laboratory scale. Such measurements, culminating with Hopkinson bar histories, span the range of continuum strain rates accessible and are used to define the yield surface for the material to which a constitutive model may be fit, defined for the regime where the stress state has equilibrated. Beyond that rate, wave loadings are limited to discrete volume elements that depend upon the amplitude of the drive as well as the duration of the pulse. Shock loading provides a probe to excite and differentiate between operating mechanisms in the response of materials to compression. As microstructure restricts the available slip systems or inter-atomic bonding becomes ever stronger as electron shells are overlapped, the evolution of the strength of a material is restricted under dynamic loading. In the weak shock regime, the passage of a shock front activates and potentially generates defects within the flow, and these are propagated and stored. Defects include a range of crystallographic features, including point defects at the atomic scale, dislocations within grains and deformation twins and/or stacking faults and phase-transformed regions that do not return to their original structure. The evolution of the stress state on a target causes effects to occur in distinct regions at different length scales determined by the impulse's form. The key part of plastic deformation occurs in the rise of the pulse but in the weak shock regime kinetics may be slow enough that a mechanism may continue to later times. In the end the impulse itself defines completion by removal of shear stress at the end of the pulse. Finally, another suite of processes occur during the unloading (the release) phase. The sections that follow aim to describe and classify the generation and storage of defects in the load, hold and release phases of the pulse, and for different microstructures. Finally, the plastic excursion introduced leaves a remanent, post-shock substructure which fixes the recovered material's macroscopic response to subsequent loading. These effects include increases in hardness and reload strength found for metals conditioned by shock.

There are three steps in this process that the following sections will follow. Firstly, it is necessary to distinguish between the responses of different metal microstructures. Secondly, one must apply real-time measurements to understand the development of the substructure and its affect upon the flow. Finally, one needs adequate analytical description of defect generation and storage so that mechanical models can be constructed, verified and applied.

To present an overview of metallic response is a daunting task given the number of metals, their different microstructural and electronic states and the variety of loading modes that need to be covered. Thus, an approach is taken here where response is followed through the duration of impulsive load, starting with shocks that compress the metal into an equilibrium state. Response will then be tracked as a function of pulse length applied to the metal. Each section will discuss response observed as time increases from the initial compression on the loading face of the metal and the presentation will follow behaviour from the elastic limit through the weak shock and then later the strong shock regimes, up to the point at which the material behaves as a condensed, close-packed fluid where concepts of strength become ill-posed. Further, rather then attempting a review of the disparate range of results obtained on elements and alloys, different crystal stacking systems for metals will be followed through a range of impulse loadings to highlight the broad-brush features of metallic response. The targets chosen from each structure are all of a known pedigree, and data is presented, generated on a range of experimental platforms to show features that make it representative of its crystal class. That is not to say that all materials of the same structure will behave in precisely the same manner, but the examples chosen will serve to illustrate the characteristic features, particularly at the mesoscale, that contribute to the observed continuum response. Thus the metals chosen for this task are FCC nickel, the ordered FCC intermetallic Ni_3Al, the BCC metal tantalum, two alloys based on the intermetallic phase TiAl and the important engineering titanium alloy Ti-6Al-4V (which is composed primarily of the HCP α-phase).

There are a series of key features that will be seen to influence the observed behaviours. The first, and most obvious, is the stacking of the atoms. Others include the defect distribution in the material, the key parameters that determine dislocation transport and velocity (such as the Peierls stress), factors affecting whether twinning is favoured over slip such as the stacking fault energy in FCC materials, and phase transition thresholds in the metal. This results in substructures which accommodate the applied strain but also affect subsequent reload strength and hardness. These include dislocation cells, twinning, stacking faults and defect loops. All of these provide the means for a metal to accommodate the compression applied by the shock choosing the most energetically favourable route and appropriate kinetics that shape the resulting recovered material. The first and most ubiquitous symbol of the shock physics discipline is the wave profile, determined either from the longitudinal stress recorded by an embedded sensor as a wave runs by or from the free-surface velocity of a loaded interface as it reflects from it. The Lagrangian shock history of flow in an elastic-plastic metal is the most well known of these profiles yet details of the constituents of that form contain key information about the processes at work within the microstructure that the next section will discuss.

The observed shapes of the longitudinal stress histories in the various metals in the weak shock regime are presented in Figure 5.2. These histories were recorded using back-surface piezoresistive gauges, i.e. the traces were recorded with a sensor backed by a PMMA block to the rear of the target, which allows the gauge to respond quickly with an optimal equilibration time (limiting the rise time recordable) of $c.$ 20 ns. The travel of the waves allows the elastic and plastic fronts to disperse so that an elastic precursor with a stress level at the elastic limit at that point precedes a plastic wave that rises to the Hugoniot stress. A longitudinal stress measured in the PMMA can be converted to that of the original pulse in the target using impedance matching at the interface. These impacts were at a stress at least five times the Hugoniot elastic limit (HEL) of the metal in question in order that the plasticity processes were driven such that the plastic wave had a steady rise time. In this regime, the plastic front rise time reflects the operating kinetics in the plastic deformation processes occurring in the metal.

Traditionally, such histories were used to infer the maximum stress achieved at different impact speeds that allow one to determine the principal Hugoniot and deduce the thermodynamic equation of state of the material. These data determine the steady hydrodynamic response of the material to shock since they define the possible states accessed. The advances in diagnostics now permit the acquisition of Lagrangian profiles that allow more features to be studied such as the rate of rise of the plastic part of the wave mentioned above, and secondly the times taken for it to release from the peak plateau value. The first histories presented here show the response in the weak shock

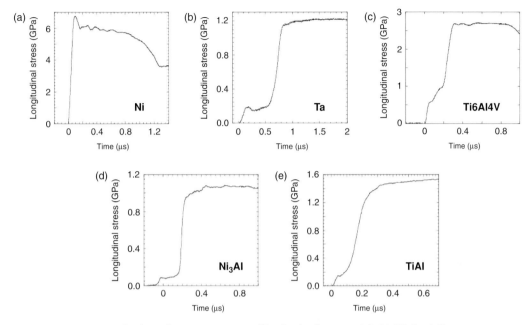

Figure 5.2 Back-surface stress wave profiles for the five materials highlighted. Source: reproduced from Bourne, N. K., Millett, J. C. F. and Gray III, G. T. (2009) On the shock compression of polycrystalline metals, *J. Mat. Sci.*, **44**(13): 3319–3343. With kind permission from Springer Science and Business Media.

regime (before the plastic wave speed exceeds that of the elastic wave) and capture features of response in the elastic and the plastic regimes and critically, the threshold for yield in the metal that develops as the wave travels though the target.

It is instructive to look at the FCC and BCC metals, nickel and tantalum, shown in Figure 5.2(a) and (b). These show that the rise of the pulse for nickel is an order of magnitude faster than that for tantalum. In fact, velocity interferometry shows that the pulse in nickel rises faster than the response time of the gauge (to which it is limited here) and this causes an overshoot and electrical ringing on the top of the trace that results from the rapid mismatching of the reactive bridge circuit. This contrasts with tantalum where the pulse rises over several hundred nanoseconds and then rolls over, reaching the peak stress more gradually. The plastic wave in Ti64 has a rise time intermediate between those for Ta and Ni, reflecting the different features of the slip kinetics in the FCC, HCP and BCC lattices. In contrast, the two intermetallics show similar features, with a fast rise but a slower roll over in the case of Ni_3Al and the slowest rise and longest time to peak stress in the case of TiAl. This will be returned to later in discussions of the global response of these metals.

5.2.1   The elastic precursor

The elastic limit, measured from the precursor travelling through the target, is the most obvious of indicators of the strength of the metal. At the moment of impact, the response

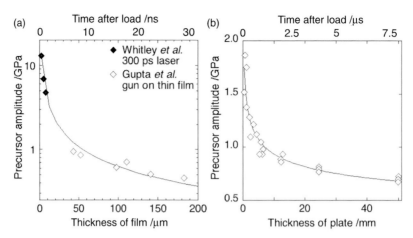

Figure 5.3 Elastic precursor decay: (a) in FCC pure aluminium, 300 ps laser pulse loading to *c*. 24 GPa (Whitley *et al.*, 2011), gun to 4 GPa (Gupta *et al.*, 2009); and (b) in BCC Armco iron (Taylor and Rice, 1963).

is entirely elastic with the motion driven before slip or twin boundaries can propagate from defect sites allowing dislocation motion to begin; thus the stress impacted is characteristic of the load on impact onto a perfect single crystal. The stress then begins to relax as shear stress is relieved by twinning and/or slip; indeed molecular dynamics calculations have shown that it takes around 50 ps for this relaxation to occur consistent with defect separations of the order of hundreds of nanometres. The Hugoniot elastic limit (HEL) is the yield stress of the material in one-dimensional strain. The elastic precursor wave, however, is not steady until twinning, phase transformation or slip processes are fully established and this may take different times to establish according to metal stacking. It thus travels forward into the target with its amplitude relaxing as it does so, since information from the plastic region relieves stresses at the front until the yield surface has stabilised. This process takes 50–100 ns in FCC metals with many slip systems and low Peierls stress, leading to rapid and easy slip. However, in BCC metals where the latter stress is higher and the dislocations travel more slowly, the decay is much slower taking hundreds of nanoseconds to microseconds. This is reflected in the distances over which elastic precursors are observed to reach a steady value.

Figure 5.3 shows elastic precursor decay in two metals: in (a) pure FCC aluminium and in (b) BCC Armco iron. The former experiments were conducted using 300 ps laser drivers and gun impact on thin foils, the latter deduced from experiments published in the 1960s using large diameter explosive plane wave lenses for the loading.

The results in Figure 5.3(a) show the amplitude of an aluminium precursor at 12 GPa, 2 μm from the face of a foil vaporised by a laser pulse. The later measurements at 5 and 8 μm show the rapid decline of this elastic pulse (note the scale is logarithmic). Gun-driven foils also indicate decay between 50 and 200 μm to a value of 0.4 GPa at the thickest foil. Gun work with microsecond pulses had previously indicated a steady HEL of c. 0.1 GPa for this material. In the case of BCC iron (Figure 5.3(b)), the decay occurs over a distance of 50 mm and the precursor amplitude drops from c. 2 GPa at 1 mm into the target to 0.7 GPa at 50 mm. The decline observed is far from linear, however, and by 10 mm the majority of the decay has occurred with the stress level at c. 0.8 GPa. This corresponds to a time of around two microseconds for the yield surface to begin stabilising and even by then there is further decay observed at a slower rate. A similar time for the FCC aluminium case might be c. 150 μm, which corresponds to a time of c. 20 ns for the precursor to settle. Thus the stabilisation of the stress state behind the shock, settling to an elastic value on the yield surface is of the order of 100 times faster in FCC metals than BCC ones. In both cases the initial stress depends on the magnitude of the driving impulse. The initial state is fixed by the driver, the final strength by the yield surface of the target metal.

The transformation of distance to time is a key relationship in wave phenomena. The sweep distance of a wave front corresponds to a deformation time and the elastic wave in the precursor region is continually lowering in strength as the shear stress is relieved by slip and twinning within the grains. In BCC metals the kinetics of the deformation mechanisms are slower than those in FCC since the Peierls stress is higher and this is reflected in the long precursor decay distances observed in these materials. That similar curves for FCC metals show precursor decay times of tens of nanoseconds as opposed to the microseconds reflects the lower Peierls stress that allows easier dislocation motion in the FCC case and the larger number of available slip systems activated by the shock. The phenomenon of precursor decay is found in all materials that deform under impact on a surface if the loading pulse rises more quickly than the deformation mechanisms can accommodate. It represents the equilibration of the stress state behind the shock as the plasticity processes operate to accommodate the applied strain. Each mechanism must complete to allow this to happen and if the driving impulse is too short no deformation can occur. The results presented indicate that deformation mechanisms are complete in FCC materials in tens of nanoseconds and stress-state equilibrium is achieved even at shock pressures of c. 20 GPa. The same is not true of BCC metals and pulse lengths less than a few microseconds will catch the material in a state before stress-state equilibrium has occurred at these stress levels. Of course, kinetics is driven faster at higher pressure and temperature states and this will be noted in the response curves of later sections.

5.2.2 Shear stress through the shock front

Methods exist for tracking the longitudinal stress in the material primarily developed to obtain an EoS for a metal either directly or indirectly using interferometry, and recently developments in gauge technology have allowed other stress components to be directly measured in the range up to c. 30 GPa, in order to construct constitutive models. However, there are other ways to deduce strength. The two most commonly used are the Asay self-consistent method and the direct measurement of the lateral stress. The former can be used at all stress levels but does not track evolution of the shear stress; the latter is limited to conditions in which the sensors can survive in the flow. Examples have been given in the previous chapter of techniques that monitor longitudinal and lateral

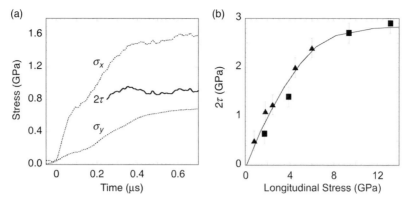

Figure 5.4 Shear strength of TiAl: (a) impact of aluminium alloy flyer onto TiAl at 194 m s^{-1}; (b) strength for two different variants showing the same trends. Source: reprinted with permission from Bourne, N. K., Gray III, G. T. and Millett, J. C. F. (2009) On the shock response of cubic metals, *J. Appl. Phys.*, **106**(9): 091301. Copyright 2009 American Institute of Physics.

stresses at a Lagrangian location in the axisymmetric stress state behind the shock. These can then be used to directly determine the shear strength. Once the stress state has equilibrated in the flow behind the front, the stress components track one another giving a constant strength at a particular pressure. An example is shown in Figure 5.4 where the longitudinal and lateral stresses and shear strength at different amplitudes are shown for an impact onto TiAl. The upper, dashed longitudinal stress history shows the elastic precursor preceding the plastic front. Since the uniaxial strain impact induces a biaxial stress state, the lateral component of stress acts normal to the impact direction. The shear strength of the material may be reconstructed by taking the difference between the longitudinal and lateral stresses at a Lagrangian station in the target. The histories presented correspond to one 2 mm from the impact plane in this experiment. The solid curve is generated from the difference in stress between the two pulses corresponding to twice the shear stress (2τ), determined from the difference between the longitudinal and the lateral stresses. It is known that this gauge has a finite response time and so the first 200 ns is not plotted, although fascinating physics is clearly occurring over that interval.

The longitudinal stress pulse rises first to an elastic stress of *c.* 0.8 GPa which corresponds to yield in one-dimensional strain, the Hugoniot elastic limit; the lateral stress history also shows an elastic signal which rises to *c.* 0.3 GPa, which is as expected in the elastic range since

$$\sigma_y = \left(\frac{\nu}{1 - \nu} \right) \sigma_x. \tag{5.1}$$

Substituting the Poisson's ratio for TiAl gives the correct numerical relation between the signals. There is a second rise to the peak of the pulse over around 100 ns during which dislocation generation, propagation and interaction occur, and this defines the plastic rise of the pulse. The shear strength is seen to be constant within the plastic region behind the shock in this material.

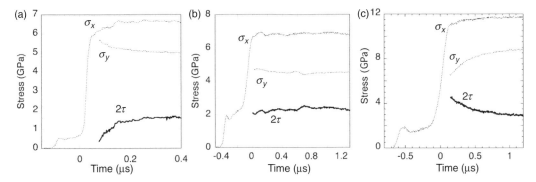

Figure 5.5 Longitudinal and lateral stresses and shear strength for impacts onto FCC, HCP and BCC metals. (a) Al6082 flyer impacting a Ni_3Al target at 570 m s^{-1}. (b) Ti64 flyer impacting a Ti64 target at 550 m s^{-1}. (c) Copper flyer onto tantalum at 498 m s^{-1}. Source: reproduced from Bourne, N. K., Millett, J. C. F. and Gray III, G. T. (2009) On the shock compression of polycrystalline metals, *J. Mat. Sci.*, **44**(13): 3319–3343. With kind permission from Springer Science and Business Media.

Repeating such experiments over a series of shock amplitudes allows the locus of the strength of the material behind the shock as a function of the increasing pressure state there to be constructed. Figure 5.4(b) shows the final shear strengths (after the stresses behind the shock have equilibrated) attained in two alloys, similar in elemental content but different in microstructure, as a function of the longitudinal stress recorded. Both materials harden as stress increases. In the general case all metals show such behaviour but then plateau out above a transition pressure as defect generation becomes limited to the shock rise. As pressure increases across a shock front two effects sum to deliver the material response. Higher compression results in faster wave speeds and faster kinetics in the plasticity processes operating within the metal, and the temperature of the region increases, pushing the material towards melt.

This combination of measurements of the stress histories at varying positions to quantify observed behaviour in individual experiments, combined with an overview of the strength at a series of applied stress levels, characterises the response of the metal in this regime and allows comparison of the effects of packing.

The strength history of Figure 5.4 is typical of many metals in which the stresses at a point swept by the shock have been monitored. Materials that have been extensively worked, metals that readily twin, alloys that have second phases within the grains, or indeed, metals in which the grains themselves are small, all show behaviour similar to that shown above. In a pure, single-phase metal, however, the situation is somewhat different. Three candidates will be shown typical of FCC, HCP and BCC response. To say typical is misleading for at the time of writing the number of metals investigated is small. Nevertheless, within the bounds of what has been found so far it appears that these show representative behaviours for the three stackings.

Figure 5.5 shows the response of nickel, Ti64 and tantalum under the impact loads given in the caption. All three are shocked to a peak stress of approximately five times the HEL or greater and at this level all three deform by slip with no twinning triggered.

This stress is low enough that an elastic wave travels faster than the plastic front so that this lies within the weak shock regime. Here the weak shock response of the three metals is different, with the FCC, the HCP and the BCC structures showing different developments in strength as the yield surface is defined behind the shock. For plasticity to occur requires slip and/or twinning. Twinning is fast and in general is observed above a threshold greater than that for slip. Further, this process can only accommodate limited strain by rearrangement of the lattice. In the regime shown here the plastic response is dominated by slip. Thus it is key differences in the slip kinetics that are reflected in the shear strength histories shown as solid lines in the figure. The FCC metal shows an increasing shear stress in the flow behind the shock. The predominantly HCP Ti64 also shows slight strengthening at early times, but then a constant response thereafter (like the TiAl of Figure 5.4). Conversely, BCC tantalum shows softening as seen in Figure 5.5(c).

For the FCC metal (Figure 5.5(a)), the lateral stress at the point at which the gauge has equilibrated is dropping whilst the longitudinal stress is increasing but more slowly. The shear stress thus rises over a period of c. 100 ns over a range of c. 1 GPa in this case, to a final value for 2τ of 1.5 GPa. The longitudinal stress for Ti64 shows a rapid rise to the HEL with a slower, rounded ascent in a plastic wave to the Hugoniot stress. Meanwhile, the lateral stress shows a slight decrease over the first 100 ns or so but then plateaus, producing a shear stress that quickly becomes constant. In contrast, the BCC metal Ta shows the shear stress behind the pulse decaying after 1 μs to around half its initial value. This reduction in strength occurs over a time interval which is an order of magnitude slower for the BCC material than was the case for the rise in strength occurring during the FCC one.

In the first moments, precursor decay indicates that metals show an unsteady stress state equalising behind the shock. In these measurements over microseconds BCC tantalum shows this relaxation as stress-state equilibrium is attained. In HCP materials slip is more difficult but fast twinning can accommodate small strains, quickly reaching a steady state. Finally, FCC nickel shows rapid precursor relaxation over tens of nanoseconds but on the microseconds considered here strengthening occurs as dislocations interact with one another as they propagate and entangle, thus strengthening the metal to further compression. Thus different mechanisms are illustrated operating here in the three metal packings: relaxation of the applied stress by slip and increase of strength by interaction and locking of dislocations. Although there are few basic processes at work it is the interplay of the pulse applied, together with the kinetics of the operating mechanism, that defines the response of the metal.

Thus strength measurements show characteristic behaviours for the different stackings in each of these metals, whereas Hugoniot measurements do not. The resulting interatomic energy barriers for each system affect both the ease of slip and therefore the numbers of accessible slip directions and the speed at which the dislocations can move through the lattice. Both FCC and BCC have many available slip systems but the number in the case of HCP lattice is reduced to three, of which only two are independent. Slip in FCC crystals occurs on the close-packed plane and there are

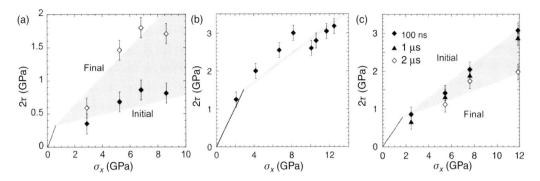

Figure 5.6 Shear strength of (a) Ni_3Al, (b) Ti64 and (c) Ta as function of maximum stress amplitude in the pulse. The shear stress develops over time and between an initial and a final state which are shown as filled and open data points respectively. Source: reprinted with permission from Bourne, N. K., Gray III, G. T. and Millett, J. C. F. (2009) On the shock response of cubic metals, *J. Appl. Phys.*, **106**(9): p. 091301. Copyright 2009 American Institute of Physics.

12 such slip planes available. BCC metals do not have truly close-packed planes, thus some energy input is required to activate slip. Finally, there are few active systems available in HCP metals. Thus the Peierls stress for BCC metals is higher than that for FCC ones so that many fewer are activated at any given stress level. Although the Peierls barrier in HCP metals is relatively low, the asymmetry of the lattice restricts dislocation activity. Thus over the first microsecond the BCC yield surface is still relaxing whilst the FCC metal has relaxed from the elastic to the plastic state and is then hardening as the many dislocations interact within the material. The HCP case is analogous to the BCC one where the number of dislocations interacting is limited by the reduced symmetry. Further variations in the hexagonal unit cell (the c/a ratio) have an effect upon the ease of slip and the prevalence of twinning. This behaviour is mirrored in other single-phase FCC, HCP and BCC materials and reflects the speed at which dislocations are generated and stored behind the shock within the two microstuctures. It is these defect activation and equilibration times which differentiate material classes and lead to differences in the observed dynamic response in continuum experiments. Defects control the strength of the metal and the kinetics with which the stress state can equilibrate within the material. Deformation processes may induce these during the loading, or even by vacancies generated during heating. Recently work by Kanel *et al.* (1991) has shown the strength increases close to the melting point of metals, where vacancies are created as the material begins to fail (this is analogous to cavitation in liquids before boiling). As the number of defects increases, slip becomes pinned so that near melt the HEL of the metal increases.

Having measured these changing shear strengths at different stresses, one can view overall hardening response over the range up to *c.* 10 GPa (Figure 5.6). The initial shear stresses are displayed as filled diamonds, the open symbol data are measured when the shear stress has steadied after stress-state equilibration. In the case of Ti64, where the strength is constant during the time of the shock pulse, only the final state (open) symbols are displayed. In the case of the FCC and BCC metals, the shaded region shows the strength increasing for the FCC and decreasing for the BCC metal over this time interval. To show that the yield surface does not develop at a constant rate, values taken for the strength at different times are plotted for tantalum. Again as expected all three classes of structure show hardening over the stress range up to 10 GPa, and this is seen over all the metals where measurements have been made by this or any of the other commonly used methods.

5.2.3   The pulse peak: the Hugoniot stress

Historically, Hugoniots were deduced from measured transit times and free-surface velocities in explosively driven targets so that other variables such as pressures were deduced from the conservation relations. Thus the compendia of data produced from the national laboratories give information that allows deduction of materials' equations of state (EoS) for the principal Hugoniot from the shock parameters derived. As seen previously, the Hugoniot is thus merely a locus on an EoS surface describing possible states attainable after a material has been shocked. The most often-used means of exploring regions of the EoS space away from the Hugoniot is to adopt a Gruneisen representation of the equation of state surface, accounting for temperature and taking the Hugoniot as a reference curve.

In previous chapters, the shock parameters and their manipulation were introduced to allow the combination of the linear U_s–u_p relation with the conservation of momentum to give the hydrodynamic curve for a material

$$P = \rho_0(c_0 + Su_p)u_p, \tag{5.2}$$

which is a quadratic relation between P and u_p, where the constants c_0 and S are the shock parameters described previously, which is valid in the strong shock regime.

For the metals discussed, the measured values of the shock parameters are presented in Appendix D. Using these values the hydrodynamic curves for the five metals can be superposed with states recovered from experiments similar to those discussed above and these data are plotted in Figure 5.7.

As expected the principal Hugoniots are seen to rank with their elastic impedances. The lowest of the five is that of TiAl whilst the steepest is that of tantalum. These are of course strongly influenced by the densities of each metal. The materials tested over the pressure range accessed in the experiments below show no evidence of martensitic phase transformations. Further, there are none visible at higher stresses for these metals in published EoS databases. The Hugoniot of the material reflects a material's ability to inertially confine the high pressure generated behind the shock front. Higher atomic mass and stronger interatomic attractions are reflected in steeper curves in this plane. The response reflects resistance to compression as atoms are brought closer to one another and whilst the packing of the metal is also controlled by this, it has no characteristic effect on the recorded trajectories.

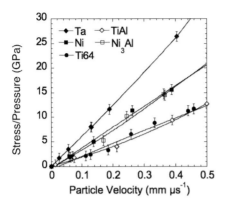

Figure 5.7 Hugoniots for the five metals discussed below. The solid curves are taken from the compendium of Marsh (1980). Source: reproduced from Bourne, N. K., Millett, J. C. F. and Gray III, G. T. (2009) On the shock compression of polycrystalline metals, *J. Mat. Sci.*, **44**(13): 3319–3343. With kind permission from Springer Science and Business Media.

5.2.4   Recovered microstructure

In order to understand the development of the microstructure metals must be loaded, unloaded and soft recovered for subsequent examination. The targets should be cooled immediately after shocking in order to preserve the microstructure in the state it achieved during the shock. These unloaded and recovered defect substructures are presented in Figure 5.8. After recovery, the samples can be prepared for further mechanical testing or for TEM and each shows clear differences highlighting their microstructure's response to the shock loading process. The recovery of shocked metals in the thermally activated region shows a limited number of plastic deformation modes. These relate to slip and twinning as outlined below, with defect interaction and aggregation mechanisms all operating to accommodate the applied shear behind the shock. The responses of the metals discussed previously shocked to *c.* 10 GPa with a pulse held for 1 μs are described below.

The TEM micrograph for the shock prestrained nickel is shown to the right of Figure 5.8(a). The observed features show uniform dislocation cells in common with other polycrystalline FCC materials which form on recovery from the shocked state. Much more work has been done on materials subjected to lower strain rate loading and there the deformation substructure is similar to that observed here in shock. Thus it is clear that the material responds with similar micromechanical processes. Cell formation is a temperature-driven phenomenon and thus is favoured in the high-pressure shock process itself. Clearly development of the cells is a time-dependent process, as shown by Murr in nickel, where he examined recovered specimens that had been subjected to different pulse lengths and showed that the cell structure developed until it achieved a stable form after 500 ns. This is precisely mirrored in the increase of the shear stress (Figure 5.5(a)), which reaches a stable value after the same time. Thus shear stress equilibrates with the development of a stable microstructure behind the front.

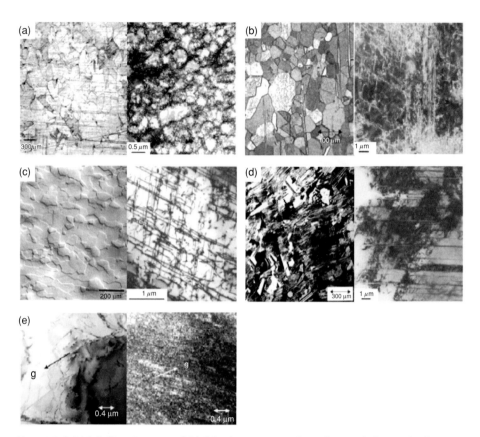

Figure 5.8 Initial (left) and recovered (right) microstructures from the metals for nominally 1 μs pulses: (a) nickel shocked to 10 GPa; (b) Ni$_3$Al shocked to 14 GPa; (c) tantalum shocked to 7 GPa; (d) TiAl shocked to 10 GPa; (e) Ti64 shocked to 10 GPa. Source: reproduced from Bourne, N. K., Millett, J. C. F. and Gray III, G. T. (2009) On the shock compression of polycrystalline metals, *J. Mat. Sci.*, **44**(13): 3319–3343. With kind permission from Springer Science and Business Media.

Figure 5.8(b) shows a substructure for Ni$_3$Al that contains a complex mixture of dislocation tangles, coarse planar slip bands, a few deformation twins and planar stacking faults. Deformation twins were also observed in samples shock loaded at stresses above a critical threshold. However, planar slip seen in the shocked Ni$_3$Al is much coarser than that observed in this material when deformed at lower strain rate. Micrographs of the recovered tantalum are shown in Figure 5.8(c). The deformed material shows sets of long, straight screw dislocations interspersed with dislocation tangles, small dislocation cusps on the long, straight screw dislocations, and small dislocation loops and dislocation debris. Figure 5.8(d) and (e) shows the grain structure before loading and the substructure after in shocked TiAl and Ti64. Features observed were deformation twins and planar slip. In all of these loadings the residual dislocation density increases with applied stress amplitude and the deformation consists of planar slip in the square

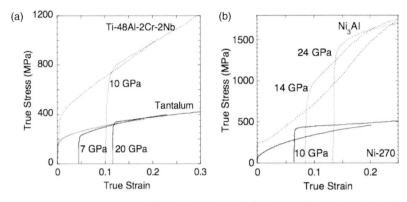

Figure 5.9 Reload stress–strain behaviour at 0.001 s^{-1} for annealed and pre-shocked material: (a) TiAl and Ta showing no hardening behaviour; (b) Ni and Ni$_3$Al showing hardening on reload from the shocked state. Source: reproduced from Bourne, N. K., Millett, J. C. F. and Gray III, G. T. (2009) On the shock compression of polycrystalline metals, *J. Mat. Sci.*, **44**(13): 3319–3343. With kind permission from Springer Science and Business Media.

pulsed specimens. In this case there were no twins when targets were loaded to 5 GPa, but in the 10 GPa pulsed samples about 3% of the grains twinned.

In summary, the five metals show characteristic deformation features. The FCC nickel shows developed cells whereas Ni$_3$Al has responded with tangles and stacking faults. In BCC tantalum, on the other hand, dislocations show long screws whereas in the intermetallic, TiAl, the structure shows long, straight screw dislocations but also in this case some twinning. Ti64 shows similar microstructural features to the TiAl with planar arrays of dislocations and some twinning.

The extensive work done using the pulses applied has resulted in very different microstructures before and after loading, and the key to the effect upon the continuum response of the materials is their reloading after shock. Tensile tests done at quasi-static rates on the targets recovered after shock are shown in Figure 5.9.

In order to understand the effect of the impulse on the materials, the annealed and recovered samples were retested at rate of 0.001 s^{-1}. Figure 5.9 shows the stress–strain behaviour recorded. If comparison of the effects of loading is to be useful, some account of the residual strain due to the process must be included to offset the reload curve and such curves are displaced along the strain axis by the total shock strain plus that from release (Eq. (2.3)). It is found that the reload curves for the shocked metals have a higher yield stress than annealed materials in all cases, but when the curves are displaced by the residual effective strain, different classes of behaviour are apparent. The figure is divided into two: Figure 5.9(a) shows TiAl and Ta, where the loading curves lie on one another, and Figure 5.9(b) shows Ni and Ni$_3$Al, where the curves all lie above the corresponding annealed ones at an equivalent strain level. FCC metals show the affect of shock strengthening whilst BCC metals do not; in engineering texts this is sometimes called path-dependent behaviour for FCC and path-independent behaviour for BCC materials.

Collecting this miscellany of data is necessary to understand response in a regime where intuition is of limited use. A complete picture emerges only from gathering together shock histories, microstructures before and after loading and then recovery and reloading of the recovered targets. The behaviours observed are markedly different between the FCC systems (with low Peierls stress and many accessible slip systems) and the BCC and HCP metals (with fewer activated or accessible slip systems due to, in one case, high Peierls stress, or in HCP metals to low symmetry). As slip becomes more difficult, twinning is favoured and *vice versa*. At least in the weak shock regime, increasing the pressure in the shocked region strengthens the metal in the plastic state and this effect appears to occur in quasi-static loading using the diamond anvil cell (DAC) as well as behind a shock, but only up to a defined compression.

5.2.5   Strength at higher pressures

Strength is a key parameter that tracks the response of a material from yield to megabar compressions and materials possess significant resistance to shear macroscopically at stresses of many times the HEL of the material. This has been neglected in the past due both to the technical difficulty of obtaining this data and its small magnitude relative to the pressures applied, and so the focus of research was into generating equation of state parameters for the existing analytical framework based around the conservation relations. In recent times it has become clear that strength is of key importance and could be around 3–10% of the shear modulus in many materials. It also tracks the interatomic bonding in the metal as pressure moves towards the *finis extremis*.

All of the available measurement techniques make assumptions that represent an integrated set of constraints upon the value of the yield strength recovered. Further, all of them have particular limitations that make them applicable only in particular ranges or specific measurements. Interpreting the data generated requires detailed knowledge of the defect generation and storage properties of the microstructure in the particular loading of the experiment, and thus in some classes of material care is necessary in interpreting the results. In particular, strength data produced for materials exhibiting path-dependent mechanisms (such as the Bauschinger effect or undergoing phase transformation) must necessarily be treated with caution if any technique that contains significant assumptions is used. All that being said, some assemblies of material data exist that allow comparison of the global behaviour of the strength of a solid as stress moves beyond yield until it reaches the *finis extremis*.

The response described above has shown data from the weak shock region in which the metals tested all show hardening behaviour. At higher stress amplitudes, there is a single shock front as the elastic wave is overtaken, which takes the material from the ambient state directly to the Hugoniot stress. In this strong shock regime plasticity mechanisms are driven to completion in the front and the strength reaches an ultimate value with all defects activated, generation and storage mechanisms exhausted and plastic flow driven at the maximum possible rate. Various methods exist to track behaviour into this regime, which will be illustrated below for two BCC materials. Figure 5.10 shows the strengths measured at 1 μs after loading commenced in the metals followed above as

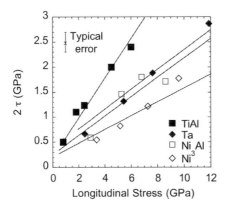

Figure 5.10 Strength achieved at 1 μs into the pulse for the metals illustrated as a function of pressure. The solid lines show the hardening behaviour for Ta, Ni and Ti64. Source: reproduced from Bourne, N. K., Millett, J. C. F. and Gray III, G. T. (2009) On the shock compression of polycrystalline metals, *J. Mat. Sci.*, **44**(13): 3319–3343. With kind permission from Springer Science and Business Media.

a function of the Hugoniot stress reached by shock. Although precursor decay may not be complete in BCC materials when the wave sweeps the location of the gauge, the strength histories (such as those shown in Figure 5.9) show that 1 μs represents a time by which shear stress equilibration (the components have developed) appears to be nearly complete in these materials for pulses of microsecond duration. Some of the metals are strengthened and some softened behind the shock, but comparison is made at a time where the strength is stable. Over these stresses the metals all show a small elastic rise and then strength increases monotonically with pressure in the plastic region for each. The pure metals are displaced from one another but appear to harden at much the same rate. The intermetallics on the other hand (particularly TiAl) work-harden with pressure at an increased rate which stems from their lower symmetry.

The metals chosen and their behaviour demonstrated above show the effect of stacking upon characteristic behaviours. While some show pronounced shock strengthening behind the shock front (like the FCC metals Cu and Ni), others show less rapid increase with applied pressure (the BCC Ta discussed here, but also W). The metals accommodate the applied strain by slip or twinning. Twinning occurs several orders of magnitude faster than slip mechanisms but can only accommodate limited strain by this means; this is minimal in the metals discussed above.

However, there are two properties that reflect metal behaviour in the weak shock regime: the stacking fault energy (SFE) in FCC metals and the Peierls stress that must be overcome for slip to occur. High values of the Peierls barrier (such as found in Ta and TiAl) reduce dislocation density and slow slip dramatically. In low Peierls stress materials such as Cu and Ni, dislocations nucleate easily, travel and interact quickly and with such interactions strengthen the material as the pulse progresses. Of course there are different responses amongst FCC metals and an indicator of the preferred flow mechanism is the SFE: a high value favours dislocation glide and leads to rapid strengthening and equilibration behind the shock front; when it is low, metals find it difficult to deform by slip and may twin. Cu and Ni have values for the SFE of 80 mJ m^{-2} for Cu and 90 mJ m^{-2} for Ni. Aluminium, on the other hand, has a higher value (180 mJ m^{-2}); silver has one much lower (20 mJ m^{-2}). Indeed, shock-recovered silver shows twins, twin-faults, stacking-faults and microbands with only dislocation tangles and loops. Thus SFE is an indicator of the deformation mode that may be expected for FCC metals in the weak shock regime. However, another indicator is needed for BCC stackings since stable faults are not known in a BCC lattice. In this respect, the Peierls stress is a better indicator. The Peierls stress for FCC Ni is 85 MPa whereas for Ta (252 MPa) and for W (940 MPa) the value is much higher, reflecting the difficulty of slip in a lattice that is not close-packed. Thus the decrease in shear strength behind the shock front in the case of tungsten, tantalum and their alloys reflects the lack of interaction of dislocations and hence stress is relieved as slip progresses. However, niobium shows no such relaxation behind the front since its Peierls stress is much lower (70 MPa) and dislocation generation is easier. Thus BCC metals have as wide a range of mechanical response to shock loading as do FCC metals and alloys in the weak shock regime. The key observation is that shock experiments give a means of not only observing micromechanical effects upon continuum behaviour but also of probing the kinetics of the operating mechanisms and filtering operation of each by changing the impulse applied.

The behaviour in the weak shock regime has been described above. However, in the strong shock regime defect generation may be regarded as saturated and an asymptotic strength is reached. In this case bulk heating occurs and this peak strength relaxes as the state approaches melt. Thus, at higher pressures, shock heating is ever greater but this is also coupled to the hydrostatic strength dropping as the material approaches melt. Further, the same behaviour has been shown to occur in a range of metals up to $c.$ 100 GPa loaded in a diamond cell where the conditions were isothermal and only single crystals were sampled. Figure 5.11 shows the values of strength plotted as a function of pressure for gold and a series of BCC metals. The shock data for Ta and W are shown alongside that taken using high-pressure DAC loading at ambient temperature for tungsten, tantalum and molybdenum, near neighbours on the periodic table. The strength increases with pressure for the range of materials investigated eventually asymptoting to a maximum value at higher stresses. The higher compression and hydrostatic pressure relax shear stresses faster and indicators such as sound speed in the high-pressure region increase. In shock loading there is a particular threshold beyond which plastic processes are so fast that a front is launched which exceeds the ambient elastic wave speed. This threshold is of the order of 20 GPa for copper but closer to 100 GPa for tungsten and this threshold is reached when compression has driven out defects, leading the material to exhibit a maximum stable strength. This is the limit of the weak shock regime and is called the Weak Shock Limit (WSL).

The FCC metal gold is a standard in DAC work; its strength is much lower than that of those materials considered earlier. Rhenium by contrast supports the highest differential

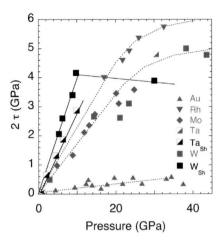

Figure 5.11 Strength as a function of pressure from the DAC work of Duffy and Weir (Weir *et al.*, 1998) and shock data for W and Ta from Figure 5.10. The filled points represent BCC metals. Those under shock conditions are black and those loaded in DAC grey. Au and Rh have triangular symbols and are included for completeness.

stresses in these experiments; it is has an HCP structure at ambient conditions and also has the largest shear modulus of any of those considered. Tungsten and molybdenum are both BCC and support differential stresses up to *c.* 25 GPa. That tantalum hardens in shock more rapidly than in the DAC may be an indication that the BCC yield surface requires greater time to relax than the microsecond over which these data were recorded. The high Peierls barrier not only reduces the slip available but also slows dislocation velocity by an order of magnitude compared with FCC materials. Certainly this is supported by the precursor decay data shown above for the iron.

To attempt to compare these data it should be remembered that the DAC is not in a uniaxial strain state as is the case under shock. This actually suggests that the curve should be below that of shock data since the loading under these conditions is inertially confined. In the elastic range the two can be related through the identity

$$\sigma_{\text{UStrain}} = \left| \frac{1 - \nu}{(1 + \nu)(1 - 2\nu)} \right| \sigma_{\text{UStress}}, \qquad (5.3)$$

where σ_{UStrain} is the stress under uniaxial strain shock loading and σ_{UStress} is the stress under quasi-static uniaxial stress loading (Eq. (2.5)). This relation concerns only the expected difference in the recorded levels due to inertial confinement. The DAC measurements of this type are recent and indications show that, whilst the stress state is axisymmetric, it is not precisely controlled and unlikely to be in uniaxial stress. Nevertheless, using values for ambient conditions in tungsten and tantalum gives the ratio of elastic stresses due to confinement alone in shock to be 1.27 times that in uniaxial stress loading for tungsten and 1.55 times that in uniaxial stress in tantalum. These represent the maximum deviation between the stress levels and are due to geometric considerations alone. However, it will be seen that for both these materials, the high-pressure DAC data

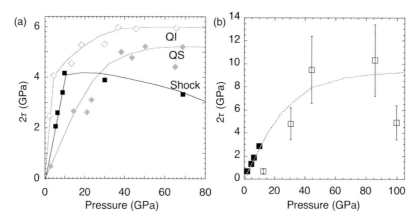

Figure 5.12 Quasi-static, quasi-isentropic and shock strengths for (a) tungsten and (b) tantalum. The black squares are shock data and the open points DAC.

show higher values than shock in the region above 20 GPa, underlining significant errors in measurements above this value.

Over the range of pressures for which there is (very limited) data it appears that isothermal loading gives higher strengths than shocked material due to the heating effects discussed above. In the weak shock regimes (< 10 GPa) the situation is somewhat different. Shock loading delivers higher strengths in materials at an equivalent pressure. The operating kinetics and the movement of the yield surface are key factors in what these values become since loading time is an issue that has not been addressed at present. Simply put, comparison is not possible quantitatively until $F < 1$, and strengthening mechanisms have completed.

Figure 5.12 shows data to c. 100 GPa for tungsten and tantalum showing the behaviour discussed above for shock, quasi-isentropic and static loading in the DAC. The tungsten data in Figure 5.12(a) gives the most complete picture of response across techniques. The impulsive loading gives higher strengths in the BCC W in the region below 20 GPa than is the case above that value. This suggests that the thermal effects are lowering potential barriers to slip and having a large effect on the observed strength.

In the dynamic loading of materials, strength at the continuum is the manifestation of resistance to dislocation transport at the microscopic scale. In the region below 20 GPa the behaviour appears to be dominated by thermally activated dislocation dynamics, whilst in the region above there is a regime of phonon drag. However, when all defects are generated in the shock front and overdrive occurs the strength asymptotes to a constant value as seen in the figure. The lower thermal activation in QI loading compared to shock is manifested in higher strengths over these loading levels.

In the case of tantalum similar observations may be made. Again the shock data show lower strengths than the DAC data even given that uniaxial strain loading should result in confinement and higher recorded values. It is useful to show the error bars on the data in the DAC. Given the novel nature of the technique and few workers active in this area these results should be regarded as an indication of the strength under these extreme conditions rather than an accurate determination of the precise value at this stage.

As a general observation, these data show that strength increases with compression at a rate greater than the shear modulus in many metals, implying significant strain hardening even under diamond cell loading conditions. At pressures of 20–100 GPa, this new work has shown that metals typically exhibit strengths of 1–3% of the shear modulus in this mode of loading. This discussion shows the importance of strength considerations since even in extreme amplitude dynamic loading, where previously it has been regarded sufficient to assume the loading was hydrodynamic, this assumption is manifestly not the case, which begs questions as to the operating mechanisms in these different modes and very different rates of application of loading. Comparison with tantalum under shock shows the complexities that kinetics brings to these discussions, making read across between loading speeds more complex. In the final chapter it will be shown that the end of the weak shock regime corresponds to the metal reaching its theoretical strength behind the front. The transition from weak to strong shock behaviour (the WSL) and the value of applied pressure at which the strength plateaus, is the point at which compression field overcomes the bond strength of the metal and at this point vacancies or larger volume defects have been overcome. In the regime above this stress, dynamic compression within a front obeys an interesting correlation discussed below.

5.3   Energy balance in shock and the fourth power law

Consider the energy dissipated by the shock as described earlier in this section (Figure 5.13). The lighter area beneath the isentrope for the metal accounts for stored energy due to the hydrostatic compression. The Rayleigh line takes the material from the ambient to the shocked state and the triangular region beneath it defines the total internal energy in the metal under compression. The difference between these two components defines the thermal part of the response behind the front.

The energy between the Rayleigh line and the Hugoniot is the dissipated energy in the shock front and does not correspond to the equilibrium conditions set up behind it. In this region the processes which define the shock and generate the entropy behind the front operate. These include deformation mechanisms such as dislocation motion and interaction, twinning and fracture as well as defect generation by the shock. These all contribute to the lattice kinetic energy and ultimately increased temperature behind the front. Deformation may also lead to defect generation which contributes to the entropy of the system. As will be seen this energy has two components shaded in the darker greys (α and β). In the weak shock regime the isentrope and the Hugoniot may be regarded as coincident and the thermal energy is represented by the upper region α.

From the moment that the lattice begins to deform to the point at which the stress state is steady behind the shock front, the thermal part of the deformation is defined by the shaded energy illustrated. It is found that the energy dissipated here multiplied by the time taken to establish the shock is an invariant of the wave front even when the wave is developing.

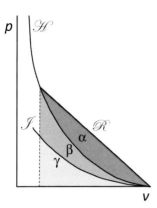

Figure 5.13 Energy balance in the shock process. The three shaded areas α, β, γ represent terms in the total energy behind the shock. $\mathcal{H}$ is the Hugoniot, $\mathcal{I}$ the isentrope and $\mathcal{R}$ the Rayleigh line. The sum of the grey-shaded areas accounts for the total internal energy behind the shock. The lighter coloured area beneath $\mathcal{I}$ accounts for the compression component of the energy whilst the darker two regions represent the thermal component. These correspond to the changes in internal energies ($P\Delta V$ and $T\Delta S$ respectively) during the shock process.

From this relation Grady (2010) has shown that the fourth power law found experimentally for the rise of steady shocks (Eq. (5.4)) in crystalline materials can be explained. This connects the strain rate $\dot{\varepsilon}$ to the maximum stress behind the shock thus

$$\dot{\varepsilon} \propto \sigma^4. \tag{5.4}$$

The stress jump is from the elastic limit to the peak of the shock and the time from the arrival of the elastic wave to the point at which the stress has equilibrated behind the front. It has been shown to apply to strong shocks, but can be shown to have some applicability even in the weak shock regime.

Interestingly there are several cases where this law is found *not* to apply. It appears that only certain materials in particular stress ranges show this behaviour. Granular and powdered materials, for instance, show only a linear dependence that appears to exist over a small stress range, whereas 1D composites show a square relation that appears to operate at modest stresses, although the fourth power applies again when the stress is fully distributed in 3D composites. The law is thus a property of loading above the WSL where defects have been removed from the solid. Whereas the stress applied in a pulse may be sufficient to do so in the low-density materials above, the experiment or measurement time need to be long enough to observe a steady state. Thus the fourth power law appears as discriminant of steady shock behaviour.

Thus constancy of the product of energy contributing to dissipation and a temporal duration of a structured shock wave is a fundamental property of a steady shock propagating in condensed matter. The behaviour has bounds from the elastic limit at the lower bound and the *finis extremis* at the upper portion where strength becomes ill-defined. As electronic bonding changes at this threshold the energy partition changes in the process as the volume state of the material is transformed. It has, however, been shown to work most successfully in the strong shock regime in fully dense materials. In the weak shock regime the division between thermal energy and defect generation may confuse the situation since the two are separated temporally with elastic and plastic waves separated. Also the threshold between weak and strong shock behaviour has at its root the saturation of defect states which limit the strength (as will be seen below). The relation suggests several insights that illuminate shock behaviour and offers results which can be used to assess impact situations. Clearly if it is not obeyed in the condensed phase the shock is not steady with implications for the states that exist behind the front. Further work must track this behaviour in other solids to further to understand the result and complete a fascinating puzzle that drives to the heart of the nature of a shock.

5.4 Amorphous metals

Amorphous metals offer high hardness and superelastic response combined with high free volume that impacts a diverse array of applications spanning from coatings to hydrogen storage. Absence of fundamental understanding of the spatially and temporally dispersed structure of this class of material currently precludes physically based structure-property models, which greatly inhibits the ability to understand and model their behaviour under extreme thermomechanical conditions. Increased understanding may also allow innovative processing using extreme fields; the use of intense laser pulses to introduce patterned amorphous structures for tailoring surface properties is but one example.

In addition to the mechanical effects discussed above, the electrical response of disordered materials is also expected to change with high applied pressure and temperature and under the constraints applied by transient loading. Electrical fields set up in semicrystalline materials under shock give a piezoelectric response that is utilised for simple sensing applications; future applications may involve geometry control that can be linked to more sophisticated electronic sensors. Further mechanistic information, obtained in fundamental studies of disordered materials, will set the stage for future exploitation of such interactions with many states including radiation.

An amorphous metal has a disordered atomic-scale structure. Such metallic glasses are tougher and less brittle than silica glasses and ceramics. Since the metal is non-crystalline dislocation motion cannot occur and there are no grain boundaries. Both of these features mean that the metal does not possess any of the defects of their crystalline analogues and thus show higher strengths and better compression properties. Metallic glasses tend to be alloys containing atoms of significantly different sizes. This again impedes crystallisation and allows the formation of amorphous microstructure at cooling rates slow enough that the metals can be fabricated in large quantities.

Under load metallic glasses show some plasticity in compression but quasi-brittle behaviour in tension and simple shear. This has limited their use in applications where safe operation is an issue for components, particularly if there is any possibility of transient tensile or shear loading. One of the principal deformation mechanisms at high strain rate is localisation into shear bands through which cracks may propagate after the loading has ended. This feature of the response of the materials is returned to below.

The descriptions above have given an indication of a least some of the important physical mechanisms occurring in a metal when a shock is launched on impact. An indication of four scales is given, which reflects the dimensions over which processes occur on the one hand, and the times typical of the processes occurring at these scales on the other. Each is associated with one or two techniques which probe materials over particular stress and temporal ranges and excite a defined suite of mechanisms determined by their loading times. For example, lasers can probe the atomic scale and the establishment of the first plastic states, launchers are required to probe growth of defects to induce failure, Hopkinson bars can demonstrate localisation if it occurs. No device can cover the necessary pressure, temperature and temporal loading space with sufficient breadth that others can be ignored. This section has focused on the mechanisms occurring in compression. Mention has been made of plastic deformation behind the shock and phase transformation is one means of accommodating the applied strain in crystalline metals. The next section will detail work on metal phase transformations. Additionally, rapid release and release wave interaction can lead to dynamic tensile loading, and this will also be reviewed below.

5.5   Phase transformations

The application of motion to a surface in a solid necessarily results in immediate reordering and the transfer of kinetic energy. This may lead to a range of first-order diffusionless changes to the solid including melting or vaporisation as well as structural rearrangements of the lattice. In particular, the martensitic transformation, a diffusionless first-order phase change, is important in shock problems and proceeds by nucleation and growth of a new phase. This displacive shuffle typically results in a 4% volume decrease; in contrast melting usually involves a volume increase. Of course all transitions involve a latent heat absorbed or released in the transition since the free energy of the two phases is different. Thus whilst the mechanics of the process are similar, martensitic transformation and twinning differ. Although twinning produces a new orientation of the same crystal structure, martensitic transformation produces a new orientation of a different one. Second-order transitions are also observed in shock and include magnetic and order–disorder transformations, but these will not be treated further here.

All of the pressure-induced transitions at thresholds between the elastic limit and the *finis extremis* take the microstructure through ever-denser forms as porosity and vacancies are eliminated from the microstructure with martensitic transformations limited to the weak shock regime. Above this threshold the structure forms a packing of a different electronic nature. As in previous sections, it will be not be possible to follow all of the investigations over the full range of metals that have been conducted over the last 50 years. Rather, examples of different microstructure will be used to show operating behaviours. The treatment will cover martensitic transformations, melting and freezing, which define the bounds of the condensed state below the *finis extremis* in some elements. There are many excellent texts on static phase transitions and the reader is referred to those for background. A basic physics framework has been sketched in Section 2.9,

Table 5.1 Martensitic transformation thresholds in selected materials.

Material	Threshold (GPa)	Initial structure
Antimony	10	Cubic
Bismuth	2.5	Rhombohedral
Carbon	23	HCP
Germanium	14	FCC
Iron	13	BCC
Phosphorous	3–10	HCP
Tin	40	BCT
Silicon	10–14	FCC
Titanium	9.4	HCP
Zirconium	23	HCP
AlN	23	HCP
BN	12	HCP
CsI	28	BCC
KBr	2.4	FCC
KCl	2.0	FCC
NaCl	23	FCC
Quartz, SiO_2	14.5	Trigonal
Steels	12–14	BCC/FCC

which describes the form of the Hugoniot for a material undergoing transition and sets the stage for what follows.

There have been a number of works detailing martensitic transformation under shock and a brief summary of thresholds is presented in Table 5.1. Some materials have multiple phases and the first transition stress or range over which transitions occur is given here.

It should be emphasised from the outset that the nature of the martensitic transition involves a shuffle of atoms under a deviatoric stress, and as such the distances they move are small and the effect of shear on the transformation is marked. However, workers in this field have assembled pressure thresholds for the transformation that have been deduced from experiments conducted to achieve one-dimensional strain loading at the continuum, and these are the values reproduced here where the shear applied is controlled. In what follows transformations of BCC, FCC and HCP are tracked and differences in their response are emphasised. The treatment is not intended to be exhaustive but to indicate that a rapid reordering of the structure to minimise its energy can accommodate strain quickly within the compressed lattice.

5.5.1   Transformation in BCC iron

The phase transformation in iron from the BCC α to the HCP ε phase is probably the best-known martensitic phase transformation to the shock community (Figure 5.14). Most of the experimental work has been done on the polycrystalline metal in which there will be a distribution of grain orientations and grain boundaries able to serve as

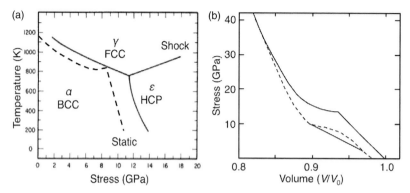

Figure 5.14 (a) *P–T* phase diagram for iron constructed from shock and DAC experiments. (b) *P–V* Hugoniot (solid) and a release isentrope (dotted) through a state at 40 GPa. A release path is shown (dotted line); a rarefaction shock (solid line) jumps the stress to obscure the reverse transition (see discussion on KCl transition). After Duvall, G. E. and Graham, R. A. (1977) *Phase transitions under shock wave loading*, *Rev. Mod. Phys.*, **49**: 523–579. Copyright 1977 by The American Physical Society

heterogeneous nucleation sites. The transformation leaves low strains transverse to the shock propagation direction in both phases and the lattice is compressed by up to 6% in the shock direction during the transition. The collapse of the lattice is consistent with shuffling of alternate planes in the BCC lattice to form the HCP phase. Of course iron is a metal of great interest since it represents the most stable, abundant element and exhibits magnetic as well as structural transitions. Work has shown that the magnetic threshold is lower that that for the structural transition and that it slightly precedes it, suggesting that the origin of the instability of the BCC phase in iron with increasing pressure may be due to the effects of magnetism.

The transition stress for the forward transition is at 12.8 GPa. However, there is hysteresis on unloading which means that the reverse transition occurs at 9.9 GPa (Figure 5.14(b)). Recovery of the target and sectioning shows that the transition is not completely reversible and there is transformation 'debris' remaining (trapped in the ε phase). Minshall (1955) assumed in the first experiments that he had accessed the α (BCC) to γ (FCC) transition, which had been observed in shock experiments but at elevated temperature. To access this requires the target to be preheated to a high initial temperature so that the shock-loading path then crosses the α–γ phase line.

Above 13 GPa there is also an α to ε pressure-induced phase transition in ferritic steels. Adding carbon to iron creates the most common and the cheapest range of steels with materials properties sufficiently enhanced over the pure metal that they can be used in varied engineering applications. Mild steel contains 0.16–0.29% carbon with the atoms fit into the interstitial crystal lattice sites of the body-centred cubic (BCC) lattice of iron atoms. Shock recovery and reloading has shown that increasing the peak pressure increases the volume fraction of second-phase material, which leads to enhanced shock hardening after traversing the transition shock stress. The material trapped in the new phase shows different shock properties and these are reflected in

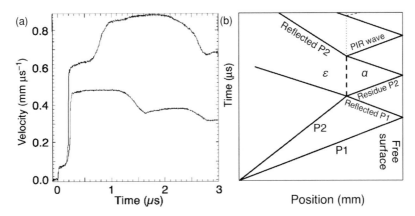

Figure 5.15 (a) Copper flyer hits mild steel target at 536 m s^{-1} and a tungsten flyer hits mild steel target at 709 m s^{-1}. (b) X–t diagram for an impact such as the upper of (a) showing the plastic waves P1 and P2 and the interaction with releases from the rear target surface.

different strength as well as different hydrostatic behaviour. On release from a peak stress the metal follows an isentrope back to ambient conditions and this will trigger a reverse transformation which typically shows hysteresis with respect to the forward path. Energy considerations require the trajectory followed by the release to keep to the left and beneath the isentrope just as the loading path must lie above and to the right of the Hugoniot. When the Hugoniot and isentrope possess such a cusp the stress follows a tangent to the isentrope on unloading until it can rejoin the curve at some lower stress and release in the conventional manner. This jump on release is called a *rarefaction shock*, and its peak and foot straddle the reverse transformation in the metal (Figure 5.14(b)).

Figure 5.15(a) shows free-surface velocity histories recorded in impact onto mild steel targets at two velocities that generate stresses of around 10 and 18 GPa respectively. In both cases the elastic wave rises to an HEL which is *c.* 1.4 GPa at 6 mm, nearly twice that for pure iron. The two pulses rise together, although the plastic front does so faster at the higher stress level since the shock speed in the untransformed material is higher. The lower amplitude pulse plateaus whilst the higher breaks at a level that shows the transition stress reached in the target corresponds to *c.* 13 GPa in this steel. It does not rise higher until the transition front arrives and in this geometry this has interacted with a wave from the free surfaces whose velocity is monitored with VISAR. The X–t diagram in Figure 5.15(b) shows a Lagrangian representation of the loading. The plastic wave (P1) reflects off the free surface and interacts and decelerates the phase transition front, P2. This partial reflection and transmission at the phase interface sends back a reloading pulse to the surface, which, after a further reflection, returns as a second reload to the interface. The VISAR picks up each of these waves as they arrive at the free surface. But although such histories have been obtained for many years, full description of the loading within the target is still not developed in numerical models since the mechanisms and particularly their kinetics are not fully described.

The effect of alloying on the transition stress has also been investigated in some detail and BCC iron alloys all show effect on the transition stress on the addition of alloying elements. Alloying with manganese or nickel decreases the transformation stress whilst with cobalt, vanadium or chromium, the value is increased. Although the addition of other elements alters the threshold in different ways, adjusting the concentration of additives in a continuous manner has a proportional effect on the transition. The magnitude of the change can be large since the transition stress can be halved when ten atomic per cent of solute has been added. These effects are to be expected considering the alterations to the bonding and to the Peierls stress that result in the lattice. They also rely upon the determination of the shear stress at the transition point that depends upon details of the evolving microstructure, particularly since nucleation is critically shear dependent.

Other properties change with pressure across structural transformations as may be expected. One such is the electrical resistivity of the metal as pressure increases. There is a smooth change in resistance with stress below 8 GPa and an increase at 17.5 GPa. However, it is the large increase in resistance associated with the 13 GPa transition that dominates response. Changes in shock-induced magnetisation are less sensitive to details of plastic deformation than changes in resistance. There is a decrease of magnetisation between 18 and 22 GPa and indeed iron is not ferromagnetic at all above 32 GPa. The key to these changes in mechanical, electrical and magnetic response is found in the generation of shock-induced defects which have a key effect in changing electromagnetic response in the bulk material.

Finally, polycrystalline materials have a second defect population at larger (grain boundary) scales compared with their single crystal counterparts. The threshold for a shock-induced transformation is thus lowered relative to perfect single crystals since nucleation can take place more easily at boundaries between crystals of different orientation of the BCC lattice.

5.5.2   Transformation in FCC metals and ionic solids

There are various metals of FCC structure under ambient conditions that undergo martensitic transformations under shock loading; two of present interest are cerium and lead. Ionic solids, particularly the alkali halides such as KBr, NaCl and KCl, also show martensitic transitions from FCC structures and potassium chloride will be mentioned here to illustrate this response. The most complete work has identified several features of the stress state, and the kinetics of the transformation. The motion of a shock front through a dielectric also produces changes in electrical conductivity associated with complex coupling between mechanical and electrical states.

The dynamic phase transformation of potassium chloride takes the material from the low-pressure B1 (NaCl) structure to the high-pressure B2 (CsCl) structure (Figure 5.16) and the threshold has been reported by several authors at stresses between 1.97 GPa and 2.34 GPa. At this transition there is a volume compression of 7.8–9.5%. Experiments on single crystal targets showed the rate of the transformation was significantly faster along the [111] direction than the [100] direction and that the transition was complete in a time of the order of 10 ns. The measurement of shock properties has traditionally relied

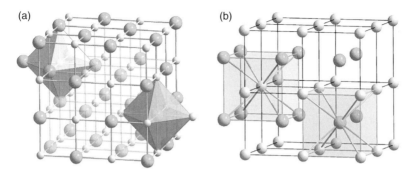

Figure 5.16 NaCl (B1) structure CsCl (B2) structure.

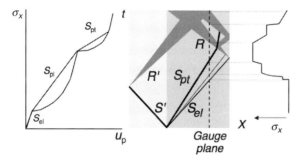

Figure 5.17 The Hugoniot to the left shows the Rayleigh lines for three waves (elastic, plastic and phase transition) which run at speeds S_{el}, S_{pl} and S_{pt} in the target. In the centre, a schematic Lagrangian X–t diagram shows the propagating waves. To the right is shown a stress history recorded at a station indicated by the dotted vertical line.

on either rear surface velocity measurements such as VISAR or direct measurement of stress using embedded stress gauges. These direct measurements allow one to sense the details of the waves arriving at a station within the target directly and without the complicating reflections of free-surface velocity measurements.

Figure 5.17 shows an experiment of this form. The measurement station monitors longitudinal stress and shows the effects observed within a target in the bulk. Figure 5.18(a) shows the stress signal sensed in experiment. The material shows a small HEL and then rises to the 2 GPa transition pressure where there is a delay before the arrival of the phase transformation wave. Clearly over this time there is an evolving substructure since the stress is not flat but rising. The transition itself reaches a new stress maximum where it stays for a few microseconds.

The rise of the transition wave is of the order of a few hundred nanoseconds and this feature is a common order of magnitude for such times observed in continuum, polycrystalline targets. Hayes (1974) measured a time an order of magnitude greater for single crystals of the same material and diffraction techniques have fixed kinetics to be faster than that on the impact surface of iron. These measurements show that the transition time for martensitic transformation is very fast and that the rise times are determined by the dispersion of the pulse through ever more inhomogeneous material

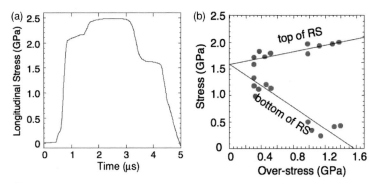

Figure 5.18 Stress gauge history for the phase transformation in pressed KCl: (a) stress history at a Lagrangian station; (b) variation of the amplitude of the rarefaction shock with stress level used to deduce the reverse transition stress in the limit.

as scale increases. This would be consistent with the observed wave arrival sequence, since the times measured during laser shock suggest that phase transformation should lead the plastic front. It must also be appreciated that shear is required in order for the martensitic shuffle to take place, and this requires defect generation and motion before transformation can occur. Thus, fast though the process may be, the defect motion must also precede transformation, which orders the processes in time.

The unloading proceeds in a regular manner with slow release until a plateau is reached. The reverse transition occurs after a hold at a threshold and is hidden by a rarefaction shock (described earlier for iron). At the base of this there is again a slow release back to the ambient state. This profile clearly indicates hysteresis in the process since the reverse transition is at a lower level than the forward stress reached in the loading phase. It is of interest to determine precisely where the reverse transition occurs since it is hidden by the rarefaction shock. The reader will remember that its peak and base span the reverse transition stress so that a series of experiments on a material, done at different shock amplitudes, allow the stresses at the top and at the base of the rarefaction shock to approach the reverse transition stress as the amplitude of the initial shock reduces. This indicates values for the forward transition stress of 2.0 GPa and for the reverse of 1.6 GPa for this polycrystalline target (Figure 5.18(b)). This large hysteresis shows the cumulative effects of defects generated in the loading on the transformed microstructure between the loading and unloading phases.

Phase transformation results in the rearrangement of the lattice to a new conformation which affects the generation and motion of further defects in the transformed and compressed phase. Thus the material will display a higher strength in the new phase and show different behaviours as pressure is increased than did the ambient, as-received microstructure reflecting the new structure adopted. Attempts have been made to investigate strength in material phases yet such a quantity is difficult to measure directly, as seen in earlier discussions. It has been possible to directly gauge the various stress components behind the phase transformation front in KCl, which to the author's knowledge is the only direct measure of strength to date across a transition front. Other

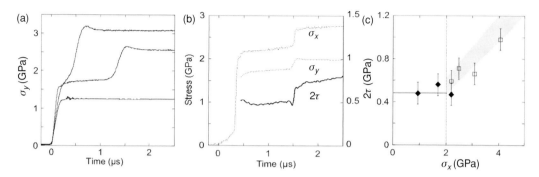

Figure 5.19 Shock loading of pressed KCl: (a) lateral stress at three levels, (b) stress components of flow at *c*. 3 GPa including the jump in strength at the phase transition, (c) shear strength above and below B1:B2 transition.

work has recovered targets that can be shown to have considerable hysteresis in the transition so that the reload strength of the transformed target can be determined and this has been done for the HCP series Ti, Zr and Hf discussed later.

Figure 5.19(a) shows lateral stress histories at three different stress levels in a target of pressed polycrystalline KCl at near to theoretical density. The lower history is in the region below the transition whilst the upper two rise to stresses exceeding the 2 GPa threshold stress for this transformation. The elastic wave is not resolved but the reader will notice the arrival of the phase transition front in both upper histories with an increasing transition front wave speed as the highest stress is achieved. Both histories show different behaviour in the transformed phase over the initial one. In particular, the lateral stress relaxes immediately behind the transition front indicating a strengthening of the material over a 500 ns period after it passes (Figure 5.19(b)).

Figure 5.19(c) shows the strength as a function of incident stress with the dotted vertical line at the transition stress. At stresses below the transition the strength is constant at around 0.5 GPa, yet it rises markedly with increased loading after transformation. The shaded region gives an indication of the magnitude by which the strength increases over the 500 ns after the transition front has passed. Thus whilst the thermodynamics and structure are interesting features of martensitic transitions, it is vital to appreciate that a new material with different properties and behaviours has been created. Further, while the material is strengthening during the hold time of microseconds in the high-pressure phase, the release process also develops the substructure since the target reverses the transition. In engineering materials, the reverse transition is rarely fully completed so that generally some fraction of the high-pressure phase remains within the target. This means that recovering and retesting a shocked sample shows a new series of behaviours which may harden the material but in some cases soften it, as will be seen below in the case of HCP alloys.

There are other features of the transition in KCl that may be measured. Dynamic X-ray diffraction has been used to better map the kinetics of the transformation and this information suggests that it proceeds via a non-equilibrium rhombohedral phase. In other work conductivity was measured as a function of shock pressure. At stresses up to *c.* 12 GPa the material remained an insulator, but above 14 GPa it conducts, having a resistivity of *c.* 7 Ω m.

These features show that electromagnetic and mechanical phase changes, whilst clearly intimately connected, do not necessarily occur at the same thresholds. Further work is necessary to understand the links between the band structure and the atomic packing in order to identify the necessary mechanistic links to construct analytical models to describe such response. A great challenge for the field is to understand electrical and mechanical response of materials to form a coupled description of response.

5.5.3   Transformation in HCP metals

Discussion of the effects of stacking on the ambient phases of metals has shown characteristic behaviours operating behind a shock. FCC metals show high concentrations of dislocations. This results in increased hardening in these materials with a higher reload strength after suitably long shocks compared with annealed metals. Pure BCC metals have higher Peierls stress and thus a lower density of dislocations on a reduced number of active slip systems. These are observed in planar arrays which can glide relatively unimpeded by defect interactions. Indeed cold working BCC tantalum *reduces* the yield strength by providing a greater density of slip systems available to accommodate slip. HCP metals in contrast have fewer systems available for slip. Thus pure BCC and HCP metals are more sensitive to strain rate and temperature than FCC materials and it is thus twinning that dominates plasticity in their response. It should thus be no surprise that the quasi-static and dynamic phase transformations in pure titanium and other group IVA alloys are known to be sensitive to strain rate, temperature, texture and chemistry.

Phase stability in the transition and the rare-earth metals has been shown to be controlled by the number of valence *d* electrons per atom. The pure group IVa transition metals (Ti, Zr, Hf) have all been shown to exhibit transitions from the HCP α crystal structure at room temperature and pressure to the ω phase which has a distorted HCP structure. The shift in packing in this transformation results in a restacking to a second HCP microstucture with the *c/a* ratio of the unit cell going from 1.59 to 0.62. At high temperature and ambient pressure they transform to the BCC structure before reaching the melting temperature. Although all shock transitions exhibit some hysteresis (as seen above for the α–ε transition in pure Fe and the B1–B2 in KCl), the α to ω transformation in Ti exhibits a particularly large one that results in the retention of the high-pressure ω phase on unloading back to atmospheric pressure. These behaviours are typical for this group of metals.

Figure 5.20 shows the phase diagram for another of the series, zirconium. In Figure 5.20(b) three shock histories are stacked indicating the appearance of the phase transformation wave at *c.* 7 GPa. The data obtained illustrate the behaviour in the HCP metals to be similar to that observed in the FCC and BCC materials seen earlier. In these materials, however, where slip is more limited, the purity of the metal has a key effect upon the transition. Positioning impurity atoms on sites in the planes on which deformation preferentially occurs alters the stress threshold at which the transition occurs as well as introducing hysteresis into the process. Eventually, with a sufficient concentration of

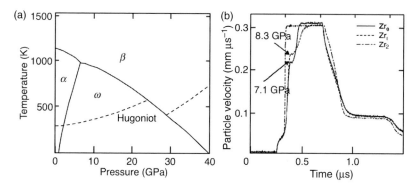

Figure 5.20 (a) Zr phase diagram. The dotted curve shows the locus of the Hugoniot. (b) Three plate impact experiments on zirconium with different oxygen impurities. Zr_0 has < 50 ppm, Zr_1 390 ppm and Zr_2 1200 ppm (after Cerretta *et al.*, 2005).

interstitials, the energy barrier for rearrangement may be increased sufficiently that the transition may be prevented entirely. Oxygen is known to suppress the α to ω phase transition in titanium through the interstitials geometrically constraining the lattice to shear. The histories in Figure 5.20(b) are for three materials Zr_0, Zr_1 and Zr_2 with different quantities of oxygen in the lattice. The first has less than 50 ppm, the second 390 ppm and the third 1200 ppm. Adding trace proportions of oxygen first elevates and then entirely prevents the transition in the three materials. The link between the atomic motions (strains) necessary for phase transformation and those needed to allow twinning may also be appreciated in these HCP metals. If interstitials reach critical levels then both may be prevented. Moreover, there is a correlation between lack of deformation twinning and ω phase formation in titanium alloys with interstitial oxygen hindering both shear processes. A similar effect is seen in iron alloys, which show twins suppressed with increasing carbon content. In summary interstitials effect the α–ω phase transition in HCP metals and the ease of twinning and slip in the loading phase of shock on the target. On unloading, the same interstitials frequently trap strain around them, leading to a decreased flow stress on subsequent reloading. This behaviour is known as the Bauschinger effect in loading behaviour. Such effects are a clear indicator of the complex microstructural behaviour that metals with limited modes of slip adopt as they respond to the shear component of the loading. This is reflected in anisotropy and sensitivity to metal pedigree in HCP response. Again the close relations between the shuffle rearrangements in martensitic transformation and twinning as a means of accommodating compressive strain are clear to see in these examples.

5.5.4   Freezing and melting

Other structural transitions occur during a loading process as pressure and temperature are changed. These include melting and freezing of the metal under pressure. Techniques that allow dynamic compression at a slower rate, so that temperature rise is limited (quasi-isentropic loading), show similar values to static DAC data. In such experiments, both melting and freezing may be sampled – in some cases with a single impulse. However, metals undergoing such transformations in shock are notoriously difficult to study with present techniques since the volume change is small, making observation difficult. Until recently, the relatively few studies of these transformations opened the question as to whether there was sufficient time in a shock pulse to allow these processes to even occur. It is now clear that in some systems at least it is possible to observe and verify such transformations, although there is need for more work and new techniques to better understand the operating processes. Using equilibrium thermodynamics to calculate thresholds or to extend static data is difficult and these problems largely arise from the nature of the processes themselves. Nevertheless, transition pressures allow temperatures to be estimated from simple hydrodynamic considerations and show reasonable agreement with pyrometry measurements made in experiment. These values for melting temperature and transition pressure can be calculated from the transition pressure and the Grüneisen equation, as seen earlier.

Melting under shock compression is not measured directly but rather inferred from other observations. The most successful route has been to determine the sound velocity in the shocked medium by measuring the release wave speeds and noting the state at which the velocity collapses to the bulk sound speed, indicating that strength has been lost. These experiments suggest mixed phase regions are formed at stresses below the threshold where fully melted material unequivocally exists indicating inhomogeneous response at these short timescales. Such behaviours, coupled with uncertainty in the interpretation of the data, mean that large differences typically exist between static and dynamic loading.

Observations suggest that melting and freezing occur by mechanisms which are driven by seeding at defect sites within the materials under load. This heterogeneity of behaviour thus requires some forced local nucleation to be included, although most theory assumes an homogeneous response in a material that is in thermodynamic equilibrium.

Two examples of melting under shock loading in pure metals are shown in Figure 5.21. It is interesting to view behaviour for group 14 of the periodic table (C, Si, Ge, Sn, Pb) which cross the boundary between metallic and insulating behaviour. Under ambient conditions diamond-structure solids occur for Si and Ge that transform into metals as the pressure is increased. Tin lies below these elements whilst lead at the bottom of the group is always metallic. Under laboratory pressures and temperatures tin has a metallic β-Sn structure with a body-centred tetragonal (BCT) lattice. However, at 40 GPa it appears to transform into a body-centred cubic (BCC) structure, and both BCT and BCC phases apparently coexist over a wide pressure range up to 52 GPa. The melt transition in lead (see Figure 5.21(a)) has been studied using pyrometry techniques. In order to be able to use such data, however, it was clear that dynamic target emissivity must be better defined in order to improve accuracy. At present these techniques are reaching maturity and have shown that the melting curve of lead is higher than that suggested by DAC static measurements at high pressure building on previous sound speed measurements such as those discussed earlier. Further, there is still uncertainty in the phase diagram for tin (see Figure 5.21(b)). The solid BCT and BCC phases may

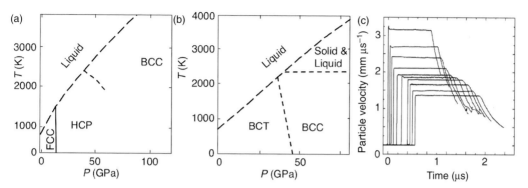

Figure 5.21 Phase diagrams showing (a) the melting curve for lead and (b) tin. (c) Release histories for tin; quasielastic release disappears above 2.5 mm μs^{-1} (after Hu *et al.*, 2008). Note uncertainties in the boundaries for these metals over the regimes shown. Source: reprinted with permission from Hu, J., Zhou, X., Dai, C., Tan, H. and Li, J. (2008) Shock-induced bct-bcc transition and melting of tin identified by sound velocity measurements, *J. Appl. Phys.*, **104**: 083520. Copyright 2008 American Institute of Physics.

exist together over a pressure range at low temperatures whereas the melt line itself is poorly defined. To tie this transition a series of shock experiments were conducted (see Figure 5.21(c)). The onset of melt is deduced by the disappearance of an elastic plastic-unload and the release proceeding at the bulk sound speed. The inhomogeneity of the sample volume element in the shocked regime cautions direct read over from static to dynamic thresholds for behaviour.

Although the shock measurements suffer from difficulties in defining the emissivity of the surfaces from which radiation is collected, the laser heating and the surface optical speckle effects interpreted as melt in static experiments give conflicting results between some experiments and sophisticated DFT calculations. The next challenge is again to better define the existing techniques but preferably develop new, less ambiguous ones to home in on the operating physics. There are two conflicting features of this transition which have contrary effects on the measurements made. Firstly, when one is considering a large sample volume, inhomogeneities will control the start and finish of the melting leading to a range of local temperatures recorded over a stress range in a rising pulse. Secondly, there are two timescales that operate in dynamic melting. The actual equilibration to a normal distribution of local atomic motions will be fast to define local thermodynamic equilibrium (LTE) and take of the order of hundreds of picoseconds. However, interfaces or second phases will require heating or cooling governed by thermal conduction and these times will be three orders of magnitude slower. Thus fast measurement requires careful consideration of when thermal equilibration is attained in the volume element of interest to the sensor. Nevertheless, this area of research offers great opportunities for fundamentally new interpretations of the processes involved.

Although melting has caused some debate as to the best measurements of the states achieved, the detailed study of freezing has provided difficulty in capturing more than extrapolations of numerical representations of the somewhat crude equation of

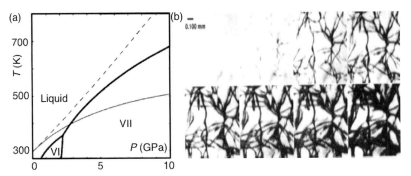

Figure 5.22 Freezing under shock loading. (a) Phase diagram for water showing two particular phases ice VI and ice VIII. The dashed line shows the end states of single shock compression, while the solid line shows an isentropic compression. (b) Image sequence showing nucleation and growth of freezing in shocked water. Frames run L to R; top to bottom. Interframe time 100 ns. Source: reprinted with permission from Dolan, D. H. and Gupta, Y. M. (2004) Nanosecond freezing of water under multiple shock wave compression: optical transmission and imaging measurements, *J. Chem. Phys.*, **121**: 9050–9057. Copyright 2004 American Institute of Physics.

state representations of shocked materials. The most graphic data were achieved using high-speed photography in water where the details of the operating mechanisms could be clearly seen. Application of pressure allows ice VII to form. This phase is more stable than the liquid phase above 2 GPa. A single shock would remain in the liquid phase to high pressure as shown in Figure 5.22(a). However, ringing the shock up within the target approximates isentropic compression and results in a lower temperature loading path shown as the solid line in the figure. Under such conditions, water becomes metastable with respect to ice VII above 2 GPa and changes from liquid to solid during the time of the impulse.

The high-speed sequence in Figure 5.22(b) illustrates features of the heterogeneous process occurring. There is nucleation of solid tendrils at points on the windows containing the sample and changing their material alters the distribution of nucleation points available on a length scale between 10 and 100 μm. Once the tendrils have connected, the form of the network of solid nuclei is fixed and the volume of the solid network simply grows with time. Higher amplitude shocks nucleate more sites whilst kinetics is driven faster, yet the process of freezing is qualitatively the same. The unstable phase of growth to a steady structural state takes of the order of 100 ns and heterogeneous nucleation proceeds in a manner such that the rate achieved agrees with homogeneous nucleation theories that assume continuum behaviour. However, note that the heterogeneity of the phases near the transition means there is no sharp threshold for onset and growth, as might have previously been assumed.

Although such transitions are hypothesised to occur within metals, there are no techniques available at present to directly observe such transformations occurring. However, it is to be expected that the heterogeneity of the processes occurring at defect length scales will dominate the response and determine the structures that result. The new bright imaging sources coming on line in the near future offer the possibility to image such mechanisms occurring, and these facilities will allow illumination of such measurements.

This chapter has focused on the extreme response of metals, yet some non-metals have been used to illustrate processes occurring during transformation. Any discussion of extreme loading would be incomplete without mentioning a holy grail of high-pressure physics: the production of metallic hydrogen. Such a transition is thought to represent the prototypical insulator–metal transition, which must occur at high pressure since the ambient band gap is of the order of 15 eV. Work at LLNL has induced metallisation in hydrogen confined in a sapphire cell with a shock introduced into and then rung up in the target by an impactor travelling at 5.6–7.3 mm μs^{-1}. However, this induced a lower pressure within the material than calculations expected, surprising observers and leading to doubts as to whether the transition had occurred at all; this has resulted in active debate. Present work using other drives is at a critical stage and offers new insight into the physics of the transition. The aim of achieving dynamic fusion in the laboratory represents a key challenge for this community and success will open up new possibilities of energy generation.

5.5.5   Future directions for phase transformation research

The preceding sections have reviewed selected examples of shock-induced phase transformation in (principally) metals that have different microstructures at the ambient state. All classes of structure have model materials where martensitic transformation occurs before freezing or melting in the region between yield and the *finis extremis*. At the present time, the drive to understand material response to pressures at a range of temperatures is driven by the need to create new, physically based equations of state up to this boundary capable of describing material mechanisms in order to predict the response of structures loaded to these levels. In order to reach this point there are several experimental developments that must be achieved to understand where the relevant phase boundaries accurately lie. However, new diagnostics and techniques to probe the material and accurately determine its state in the loading regimes of interest are still needed to answer key questions.

It is necessary to determine the atomic scale rearrangements and their kinetics which may be obtained using time-resolved diffraction experiments under different loading regimes. Further, it is necessary to track the development of the heterogeneous state at the mesoscale where different islands of phase exist near the transition boundary. Coherent imaging from the brightest sources might offer a means of gaining such information. In all transformations it is necessary to determine kinetics by sampling features of the transition at the continuum in order to read across into code models that apply at the engineering scales of interest. One of the most difficult of these challenges will be to directly detect melting in order to understand the relation between static and dynamic regimes. Identifying the mechanisms by which transformation occurs will be key to answering these questions. One result of better defining these will be defining the phase transition rates that clearly govern the timescales for these processes. Although the rearrangement of atoms in a unit cell is very rapid and occurs in less than a nanosecond, the local shear that triggers the process is transmitted by plastic fronts in the metals. Clearly, if the impulse is shorter than the kinetics for a material to phase change, then that transition will not participate in the dynamic process.

A parallel development of novel diagnostics and techniques to directly detect melting would lead to significant advances. There are new classes of these under development including application of ellipsometry, reflectivity, and resistivity amongst others in order to understand response at phase boundaries. In recent years, proton radiography has allowed direct density measurement across phase boundaries that has confirmed the predictions from conservation laws but offers the means to investigate new features of the transformations in the future. In determining properties across the phase diagram, however, the Hugoniot represents only one locus of accessible states. To complete the picture without extrapolation using analytical techniques, off-Hugoniot measurements of the states acquired in pure phases offer the means to better constrain the equations of state derived. Over the past years, techniques for applying ramp loading have been developed in various theatres and use of these new devices along with development of analyses to better understand the deconvolution of the data generated will enable significant advances in understanding. Varying the pressure controllably, along with pre-heating and cooling the specimens before loading, will generate new data across regimes of phase space poorly mapped at present. Along with this, and with parallel advances in better defining in particular high-pressure states in static experiments, it will be possible to achieve better understanding of macroscopic material behaviour across different kinetic regimes.

Along with a bright source one requires recording technologies to be developed that will allow the dynamic loading fields to be imaged or sensed dynamically too, since not only energy thresholds, but also operating kinetics, must be understood to classify the response of materials to extreme loading environments. Typical engineering materials possess a baseline microstructure but also a population of defects within their volumes. It is the understanding of these statistical physical relationships and their effects upon deformation mechanisms and defect storage processes that will drive the development of materials for use under extreme conditions in the future. Thus a combination of measurements with appreciation of their mapping across the length scales will be key to appreciating the new data generated.

From all of these advances will come a new generation of equations of state better able to deliver higher fidelity simulations of dynamic materials response. Such developments have two needs: the generation of better data as outlined above, as well as advances in the analytical tools capable of generating theoretical descriptions of materials from their nuclear and electronic properties. The new numerical formulations should account for the observations noted in the preceding sections, including heterogeneity in the phase mixture present near the transition stress. They also need to account for melt and freeze, particularly in materials near the boundaries of metallic behaviour (like lead and tin) or those of high atomic weight where d- and f-electron bonding becomes significant. The nucleation and kinetics of the atomic rearrangement itself must be reproduced in some temporal manner in the model along with the dependence upon the shear processes driven by the continuum stress state to the volume element that transforms. Finally, and most importantly, the bounds of applicability of a continuum-based model to processes occurring at the smallest length scales need defining in order to determine the regimes over which such models may be applied. Much more work is needed, and new tools must be developed to allow progress to be made, but the foundations for advance over the next decade have been laid.

5.6 Plasticity in compression

The summary above has indicated a hierarchy of microstructural features that determine the response to a step load in the shock regime. These affect two critical parameters that influence observed response at the continuum: the activation threshold for plastic flow to occur and the kinetics with which it can proceed. The temperature dependence of these is a further factor in both of these. The discussion above has also introduced a series of techniques and shown their results upon a representative series of microstructures. They provide two variable parameters which affect the response. The first is pressure that the metal responds to and the second is the pulse duration which determines the coupling with mechanism kinetics. Both of these stimuli lead to higher dislocation density, potentially twinning and phase transformation, and greater stored energy. The metal after loading can exhibit increased yield stress and hardness as a result. Higher pressure leads to a smaller dislocation cell size in recovered targets, which include more twins, an increased number of vacancies and other defects in the microstructure. These observations were in the weak shock regime, however, where recovery techniques can be used. The level of the applied pulse shows two regimes of behaviour relating to the degree of compression exceeding or requiring thermal activation to overcome the Peierls barrier.

The key determinant of the mechanism and its kinetics is the atomic packing of the metal, since this determines the preferred processes and the directions down which deformation occurs. Cubic metals have 12 possible slip systems, whereas the HCP lattice has less. Further, the Peierls stress required to slip planes of atoms further restricts the deformation that an impulse may accommodate by dislocation motion at a given stress level. The critical condition that controls all of these processes is the initial defect population within the material and that fixes both the positions from which slip can begin and also the travel which a dislocation may have before it locks with others or a boundary. If slip is restricted, then a proportion of the strain may be accommodated by twinning and this process is faster than a dislocation travelling through the lattice. However, if the strain accommodated is insufficient, then slip must occur as well to accommodate the applied compression.

All of these plastic processes are reflected in the observed response. The materials with the lowest density of mobile dislocations are BCC metals and they exhibit the slowest rise time in their plastic waves compared to the rapid rise in pure FCC materials where the number of accessible systems is high since the Peierls stress is low. Further, the

FCC materials can achieve the plastic state quickly after step loading on the impact face, whereas BCC metals can only relax to the plastic state after hundreds of nanoseconds. This is reflected in the phenomenon known as precursor decay, which describes the relaxation of the HEL of the metal with distance into the target (time for dislocation processes to act), and which occurs over long distances for BCC metals. Further, it has been shown that the density and interaction of dislocations in these FCC materials result in their strengthening over a 1 μs impulse, which is seen as a higher reload strength. HCP metals have a lower Peierls stress than their BCC cousins but their deformation is restricted by the symmetry of their structure, which means that there are fewer slip systems available to them. On the one hand, this favours accommodation of the compressive strain by twinning as well as slip, and on the other leads to an importance of texture in the response of HCP metals.

The discussion above has reviewed research that has led to the present understanding of the response of metals to a compression pulse. It has not extended to release or spall behaviour, which follow later and thus are *defined* by the effect of the loading which defines the microstructure that the tensile pulse enters and itself moulds. What is clear is that the physical response is dependent upon the defect distribution, in this case responding and evolving under compression, in later sections developing under release of compression and potentially tension.

A sequence of events occurring after an impulsive load is applied to a metal has been described. The form of the pulse considered is a Heaviside step applied at the impact face onto a polycrystalline metal with a population of defects typical of a material lapped flat and smooth. Such a target has a surface prepared to presently available workshop tolerances typical of the experimental work conducted over the past 50 years or so. The impulse could be an instantaneous (or nearly so) pressure pulse from an explosive or laser generated plasma, or the result of impact from a moving mass of a second material propelled at a target. In what follows the term impact will be used but pressure source can be substituted without loss of meaning, although the stress histories applied by a detonation wave, a driven plasma or a moving metal plate are very different.

At the first moment ($t = 0$) impact takes the material into an elastic state. It has strength, but that of a perfect crystal since no defects are activated and no time has passed to define a plastic state. The first atomic plane moves forward, starting compression at the impact face, and this reduced interplanar distance increases the force on the next, transmitting the pulse into the target. The acceleration of the impact plane gives kinetic energy to the atoms and the work done is transferred to heat through the first law. This process takes time, however. Since the degrees of freedom at the atomic length scale are also greater than at the continuum, there is motion set up which establishes equilibrium through the lattice phonon modes excited within the crystal. Such a combination of circumstances has proved rich territory for simulation at the atomic level using molecular dynamics, and work has been published illustrating these processes on several systems and using a variety of interatomic potentials (e.g. Germann *et al.*, 2004; Bringa *et al.*, 2006). The results confirm that classical thermodynamics can be satisfied after some time and that a bulk temperature rise and a Boltzmann distribution of atomic motions behind the shock can be defined, but only after local equilibrium has been established. Such simulations have indicated timescales of the order of tens of picoseconds for such processes.

Although the crystalline packing is one-dimensionally deformed at the continuum scale, with cubic cells becoming tetragonal, the compression also allows the possibility of a reordering of the packing to a denser phase. In many crystals, hydrodynamic compression is known to change the structure of a material to a denser polymorph, from graphite to diamond for instance. Pressure-induced phase transformations are processes so fast that the transformation follows the timescales of the loading that imposes the compression. Recent efforts made to follow them using X-ray diffraction have recorded diffraction patterns of the ambient, intermediate and final phases. The brightness and repetition rate possible with new bright sources becoming available in the near future will allow the detection of events that are limited at present by the lack of photons on the target and the difficulties of applying multiple pulses. The best indication of timescales comes at present from molecular dynamics that suggests times for transformation on the order of 10 ps. Certainly, the driving stimulus will rise at a rate slower than this as it travels any distance beyond the first grain into the target on the impact face since the plasticity processes will take orders of magnitude longer than this.

To summarise, the initial response of an assemblage of atoms determines whether the applied compression is of a magnitude which will allow reordering of the lattice. Thus if the material can respond in this manner, martensitic phase transformation remains the first and the fastest means of responding to the impulse that arrives in a material, although clearly it is only possible to accommodate a fixed proportion of continuum strain in this manner. If greater values are applied, then slip must then operate in addition.

However, nature does not create perfect crystals or atomistically planar surfaces in polycrystalline metals. Such materials are only found in the simulations, with idealised potentials difficult to know accurately at pressures and imposed boundary conditions that simplify calculation. In reality, the material has imperfections within the crystals and consists of multiple grains with boundaries between them. A simple calculation that indicates at least a magnitude for one of these times may be recovered from dividing a typical inter-defect distance by the single crystal elastic wave speed. The modulus is a function of the applied compression so this velocity is greater than the ambient value, since the crystal is perfect and also because of the increased pressure. The grain boundaries are one defect of known dimension, which gives a time of the order of 1 ns for a 10 μm grain size (inter-defect distance). However, it was seen earlier that an annealed material has a dislocation density of approximately 10^{12} to 10^{14} m^{-2}, whilst a cold-worked one has approximately 10^{20} m^{-2}. This gives inter-defect distances of 100 nm, and the corresponding times for equilibration drop markedly to the order of 10 ps. Thus the presence of defects on different length scales allows lateral motion for atoms in the lattice but only when triggered by the downstream influences of one of the defect locations on the impact face. Thus the initial elastic response induces a high stress, which decays as information travels in from defects over a period of tens to hundreds of picoseconds. This initial value must be the ideal strength of the metal at a point near the impact face in the first moments of loading since a region does not feel effects from lattice imperfections until information travels from them. The statistics of the defect population will determine the relaxation of the stress and thus the time taken to establish the plastic (Hugoniot) state in the material. MD simulations have monitored this on similar timescales to the order of magnitude estimates given above.

The passage of dislocations through the lattice not only starts slip but can allow the much faster reordering of the crystal into a twin. One plausible mechanism involves three partial dislocations following each other across a grain. Such a mechanism must operate over a time corresponding to a grain diameter divided by a wave speed characteristic of the plastically deformed lattice, which is lower than that of the perfect crystal. A typical estimate of a typical twinning time might thus be order 1 ns.

The maximum speed that a dislocation can move through the crystal is limited and even if driven at the point at which the Peierls barrier is overcome, it will travel no faster than the shear wave speed (at the relevant pressure). In the weak shock regime it travels more slowly than this and in general cannot travel at a single velocity since the material is inhomogeneous. The dislocations themselves are not ideal but may contain jogs, particularly in materials with higher Peierls stress such as the BCC metals. Thus times for glide are of the order of tens of nanoseconds, and depending upon the dislocation density and available slip systems, dislocations travelling and meeting on different slip planes will result in interaction and locking leading to strain hardening or to recovery if annihilation occurs. These processes take place on timescales of the order of 100 ns.

Finally, interaction and pinning of dislocations at a point also allow the formation of a defect source from which further dislocations can be generated. Such a process takes some time to activate and an estimate might be of the order of 1 μs for such a source to begin operating. Once such times have been reached, thermal transport becomes significant since diffusion timescales are around a thousand times slower than wave transport times. Thus heating and localisation occurring within the metal can give rise to features such as sheared regions in which material approaches melt. Adiabatic shear bands formed in this manner typically take times of the order of 10 μs to form and so are generally not observed in plate impact experiments where loading is generally around half of this time.

5.7   Ramp loading

The discussion above has shown the shock as a pump to probe the state of microstructure in a deforming metal. At the impact face the metal begins to respond to the rapid strain applied and the compression and motion resulting is tracked through the microstructure as it develops at different times (stations) from impact. The compression front launches processes that operate at all the thresholds exceeded by the pulse that propagates. As each threshold is reached, a mechanism with defined kinetics acts to compress the lattice to equalise the strain (and thus the stress state within the solid) and fronts move into the material, each with a faster speed than the last since the solid has higher density and greater sound speed. If the impulse applied does not rise as quickly, compression is applied so that the thresholds for the onset of each mechanism are achieved ever more slowly. Further, if the kinetics are such that each has completed before the next threshold is achieved, then the plastic work done by each has resulted in temperature rises that have had time to dissipate through the target. In this case the material does not rise in temperature at the rapid rates exhibited by shock loading and later time mechanisms proceed more slowly as the rate is generally less in cooler material. Any kinetic models describing these processes must explain experiments that apply mechanisms sequentially as well as forcing them to commence at a single time in parallel as with shock. Thus a ramp loading pulse applied to a target is a validation for the correct formulation of a material model that tracks the response of a metal. In all cases there is a critical rise time beyond which processes that may occur within a particular material volume happen sufficiently late after one another that the global heat generation, integrated from dissipation of the plastic work of each event, achieves a lower final value than in the shock case since conduction has occurred. The global state is not adiabatic, but neither will it be isentropic since there is plastic work dissipated. Thus this process is often described as *quasi-isentropic* or *shockless* loading of the metal.

There is a series of platforms available to apply such loading impulses and each sample deforms with processes occurring at different rates over different scales. The fastest controlled rise is applied by a laser source which induces plasma build up on the impact surface from a *hohlraum* inducing a pulse which rises over *c.* 10 ns and loads a volume element of dimension of the order of 10 μm. The Z pinch platforms magnetically load metals for longer periods, rising over *c.* 100 ns so loading materials over regions of hundreds of microns in the rise. Finally, graded-density impactors launched from guns can provide ramps over a microsecond or more and these probe millimetres of target volume.

Figure 5.23 shows the hypothesised evolution of the microstructure within a metal loaded shocklessly. Figure 5.23(a) shows a series of loading histories typically observed in metals loaded under such conditions. The pulses shown are for a pure BCC metal tantalum loaded to *c.* 12 GPa in the weak shock regime. The developed pulse shows an elastic rise followed by a stress relaxation which precedes a more rapid (more rapid than the driving impulse) plastic following wave which develops as it travels into a shock. A similar pulse is shown for a cold-worked material for comparison. The highest point of the elastic rise in such a trace is known as the isentropic elastic limit (IEL) by analogy with the Hugoniot elastic limit (HEL). It dips to a minimum in the case of the purer material and rises again with the arrival of the faster rising plastic front.

In the annealed metal the IEL shows a higher initial strength, dropping to a lower value after some tens of nanoseconds. Such behaviour is not seen for a pre-worked sample where the initial strength is lower although the plastic front rises in the same manner. Figure 5.23(b) shows a schematic *x–t* plot in which the development of the microstructure is indicated as a grey level with black showing the initial and white the fully developed state in the material; in the case of tantalum at these stress levels this represents the degree of slip within the metal. The elastic front passes through material in the undeformed state until slip is triggered at defects, yielding the metal and

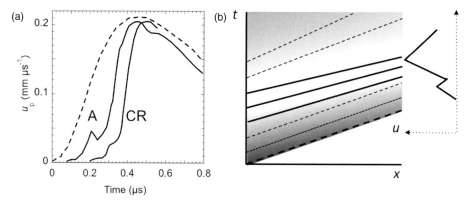

Figure 5.23 (a) Histories from annealed (A) and cold-rolled (CR) tantalum taken using magnetic loading on the target (incident pulse dotted) to induce quasi-static loading on a tantalum target in the weak shock regime (after Asay *et al.*, 2009). (b) *x–t* diagram for characteristics at different (increasing) stress levels in a ramp pulse which converge when densification takes place. The shading refers to the development of the microstructure with black as-received and white developed after substantial slip has occurred. Source: reprinted with permission from Asay, J. R., Ao, T., Vogler, T. J., Davis, J.-P. and Gray III, G. T. (2009) Yield strength of tantalum for shockless compression to 18 GPa, *J. Appl. Phys.*, **106**: 073515. Copyright 2009 American Institute of Physics.

relieving the stress. In the case of the cold-rolled metal the density of defect sites is much higher and slip is restricted since dislocations are soon locked. An equivalent grey scale representation develops the microstructure much faster leading to a light shading much sooner in the process.

Such processes occur with more slowly applied stress drivers and activate the same plasticity mechanisms within the deforming metal. The development of the microstructure that leads the relaxation of the IEL is of the same magnitude at these rates, as seen in shock loading to the same peak stress level.

5.8   Release, spallation and failure

The details of release have been discussed in preceding chapters. Nevertheless, a brief recount of the relevant features of the process in an elastic-perfectly plastic material is useful here. Since materials in which sound speed increases as compression is applied propagate impulses that shock up, the head of the impulse travels at speed U_s, whilst its tail follows at a velocity determined by the bulk sound speed in the compressed material back at ambient pressure where the flow is sonic (Figure 5.24(a)). The picture is complicated in materials with strength by the loading which transits from elastic to plastic at the HEL and then proceeds up to the Hugoniot stress before elastically and then plastically releasing to some residual compression. Figure 5.24(b) shows the Hugoniot in stress–particle velocity space traversed by the material about the hydrostat for an elastic-perfectly plastic material. Figure 5.24(c) shows a stress history for a steel with

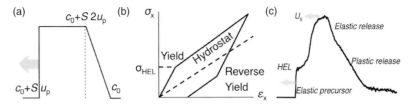

Figure 5.24 (a) Idealised compression history showing shock and release in a material with no strength (hydrodynamic). (b) Hugoniot in stress–strain space for an idealised elastoplastic solid; showing shock and release with final residual strain. (c) Stress history in a loaded steel showing precursors, shock and elastic and plastic release.

an elastic precursor, the rise to the Hugoniot stress and the elastic and plastic release phases. In ideal, homogeneous (fictional) materials, the target has immediately moved to an equilibrium state behind the shock which remains constant for the duration of the pulse so that the releases may be compared with conditions in the virgin material. This is, however, not the case as seen in the preceding sections. One feature worthy of comment before moving on is to note an effect seen in materials where defect storage affects the release of strain to perturb not only the magnitude but also the form of the release.

The *Bauschinger effect* is the continuum result of the mesoscale strain distributions set up in working the metal. Such a phenomenon is found in most polycrystalline metals and is related to dislocation accumulation at barriers. This can result in either local back stresses that can assist the movement of dislocations as release lowers stress, or dislocation sources that can produce opposite slip which annihilates those stored during load, resulting in the same effect. The result of these mechanisms is hysteresis in the loading and unloading on the continuum response and this is observed in the steel above; the unload is much less than twice the HEL, as may be expected from the Hugoniot of Figure 5.24(b).

Figure 5.25 shows an example from the FCC metals copper and its alloy bronze. In both materials the HEL of the unshocked, annealed material is very low, although the alloy has a higher value consistent with the increased Peierls stress in the bronze. As is the case for most FCC materials, the rise of the shock is fast to the Hugoniot stress. The releases, however, show differences in behaviour. The copper shows a regular elastic and then plastic release with a rapid elastic drop and then a convex plastic region as expected from the idealised theory outlined above. Bronze, on the other hand, shows a concave drop from the outset with no clear break between the elastic and plastic regions.

5.8.1   Tensile failure

Failure occurs when bonds are broken within the material allowing a weakened plane to open a failed surface at the continuum. The key feature of a tensile process is that it must first involve some localisation process, which will critically depend upon defect populations within the metal. On the other hand, failure in compression occurs by

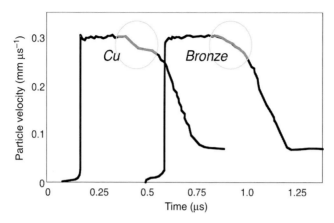

Figure 5.25 The Bauschinger effect (relevant part of history in grey) in shocked copper and bronze.

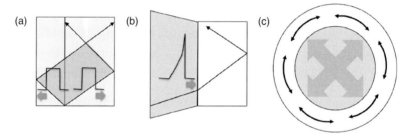

Figure 5.26 Modes of dynamic tensile failure: (a) one-dimensional by wave interaction; (b) one-dimensional wave reflection; (c) two-dimensional expanding ring with circumferential tension. Initial failures in white are release fragments and interact to potential failure further along.

applying shear for a time sufficient that local flow and consequent heating allows the formation of bands of deformation.

Dynamic tensile failure at the continuum frequently involves the shedding of a part of the material and this process is called spallation although the term has other meanings in different areas of science. It is useful to classify the physical mechanisms that give rise to such failure into three groups. These are illustrated in Figure 5.26. The first (a) is a one-dimensional tension resulting from the interaction of two release fans. This has been described previously in Chapter 2. The action of a single release in a loaded region is to restore ambient conditions to the compressed region that it enters. When two releases eat into the same compressed region from opposite sides, then at the point at which they meet, the material must exhibit a region of net tension, and then only at the position where the release fans interact. If the release were instantaneous, such an interaction would not happen since there would be nowhere that net tension existed, but since wave speed is a function of pressure and since the releases disperse, tensile failure may occur over a planar region within a loaded target.

A second failure mechanism (Figure 5.26(b)) results from wave reflection from an interface with a lower impedance material beyond; if the surface is free then the magnitude of that reflection is maximised. In this situation, the rear of an incident pulse meets the head of the reflected one (which has changed phase) and recombines to fix the pressure field near the interface. If that pulse is square then incident and reflected components cancel and the interface sees ambient pressure. If the wave has some structure, however, then the reflected tensile pulse may exceed the incident compression and a net tension then exists which may fail the target at some point within it. The most common application of this occurs in the loading of a metal by an approximately triangular pulse from an adjacent detonating explosive; in this case a scab of metal is ejected from the far side of the loaded metal known as a spall scab.

In the final scenario (Figure 5.26(c)), the metal is failed in a two-dimensional loading mode. In this case a cylindrical metal shell is failed by a hoop stress generated, for instance, by the explosive expansion of the case. In this situation, a line initiation down the centre of the cylinder expands the metal until failure occurs at surface flaws, which then send out further releases which interact to fragment the case.

In what follows, these three modes of tensile loading will be addressed with emphasis placed on the one-dimensional case where mechanisms and their development can be most rigorously investigated. What will become clear is that the science is developed much less in this mode of loading than is the case for compressive loading and, in particular, the analytical framework is largely empirical. This largely results from the simple result that compression homogenises material response whereas tensile stresses localises failure within the material.

5.8.2   Natural phenomena that lead to failure by spallation

Spallation will be used in what follows to describe the planar separation of material parallel to a wave front as a result of dynamic tensile stress components perpendicular to this plane. However, the term has a longer history and is used across science to describe other phenomena or processes. The wave interaction component of the definition is frequently relaxed and in practice spallation is used to describe any process in which fragments of material are ejected from a body due to impact or stress wave interaction. Extension into planetary physics includes meteoritic impacts onto surfaces and the effects of stellar wind on planetary atmosphere, whilst in mining or geology the term refers to pieces of rock breaking off a face due to the internal stresses within the bulk. Even nuclear physics uses the term spallation, although this time it describes a heavy nucleus emitting a large number of nucleons after being hit by a high-energy proton. Finally, anthropology echoes the geological analogy, adopting spallation as the process used to make stone tools such as arrowheads by knapping. In all uses, the term describes a failure and an ejection of material and this is what will be focused on below.

Dynamic tensile failure is a problem in a series of practical applications of which two are shown here for illustration. An important safety concern is the protection of transport from impulsive load. The natural environment imposes a series of insults that

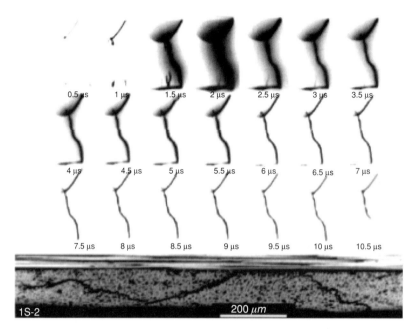

Figure 5.27 Lightning discharge onto a fibre composite panel. The discharge sequence is shown as a negative.

result in impulsive loading that materials must be designed to withstand if they are to be safe and useful for service. The two examples chosen here illustrate electromagnetic and mechanical natural insults upon aircraft: lightning strike onto aircraft fuselage and rain-drop impact onto a window material. These show material response from carbon composites in the first case and from polymers in the second to indicate some features of real failures. The rest of this section will describe the failure processes in metals to return to the theme of the chapter.

Figure 5.27 illustrates a high-speed sequence showing a 100 kA lightning strike (in negative) onto a fibre composite panel at the base. The strike arcs is from an aircraft radome (the curved structure at the top) to a ground panel with a composite plate on it. The offset of the radome to the plate is of the order of 1 m in the sequence at the top. It takes around a microsecond for the discharge path to establish itself and after that time there is current flow for a further 10 μs. The carbon composite panel also shows the effects after recovery. There is a vaporised pit beneath the point of arc attachment and current flows through the carbon fibres to earth from here. The resistive carbon fibres heat up and some vaporise and explode, generating stress waves which travel into the panel, and these fracture the composite layers as shown in the figure. A similar vapour loading of materials is now found when laser illumination of metal foils is used to launch shocks. Again the stress pulse travels forwards, reflects and takes a plane near the rear surface into tension. Such a discharge clearly induces electrical and magnetic effects on an aircraft in flight but the mechanical effects on the fuselage, particularly with the increasing trend toward composite panels, must be considered since such damage will

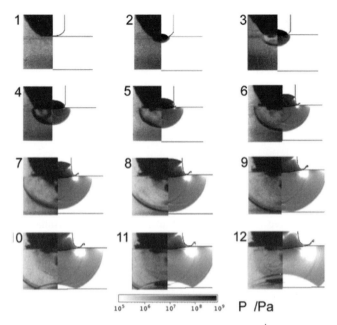

Figure 5.28 Impact of a 3 mm jet travelling at 640 m s^{-1} onto a PMMA target. Impact viewed side on with rear lighting, interframe time 200 ns; exposure time per frame 40 ns. Right-hand side of each frame shows hydrocode simulation with contours of pressure. Source: Bourne, N. K. (2005b) On impacting liquid jets and drops onto polymethylmethacrylate targets, *Proc. R. Soc. Lond.* A., **461**: 1129–1145.

severely weaken components, potentially in an area that appears undamaged and lies away from a surface feature such as an arc attachment pit.

Another application is commonly encountered in structures. The impact of liquid drops onto surfaces is a common practical problem for air transport. Two modes of loading leading to similar results are rain erosion after drop impact and damage due to jets formed when cavities collapse onto surfaces. There have been studies of the interaction of a water jet with various surface materials. The impact of such a jet has been noted to give different damage in brittle and ductile solids. In the latter, surface pits with outer lips were seem. In the former, the damage was considerably less with a central undeformed region, a circular ring of fracture and various subsurface fractured zones noted. The diameters of the central undamaged surface region in impacted brittle materials were found to be closely related to the diameter of the jet head. The result of these analyses also show that the highest pressure was generated at the contact edge of the liquid-jet and this pressure progressively decreased as one moved to the impact axis.

Figure 5.28 shows a high-speed framing sequence with simulation in which a jet impacts a PMMA target. In each case the frame is cut down the axis of the jet to show only the left-hand side of the image. The axisymmetric simulation of the pressure field calculated for each frame is presented on the right-hand side. The scale is logarithmic and becomes darker as pressure increases. The first frame shows the liquid-jet travelling in air before impact. The jet is on the point of impact in frame 1 and a shock travels down into the PMMA in frame 2. The speed of the shock wave is *c*. 3 km s^{-1} and exceeds the speed of the longitudinal elastic wave in the PMMA. By frame 3, the contact edge between the jet and the water surface becomes subsonic and release waves are formed that follow the initial shock wave. A radial jet is visible from frame 5 onwards, moving along the surface of the target travelling at 1.3 km s^{-1} (approximately twice that of the liquid-jet) with a preceding bow shock running ahead.

At the time at which the contact point between water and target becomes subsonic, a surface ring is defined from which a tensile region develops in the PMMA. A toroidal front, visible as two, growing semicircles, propagates from this ring. The original relief motion at the edge is normal to the liquid surface, i.e. directed toward the target. As the inner part of the wave reaches the central axis (frame 4), the fronts interfere and, in frame 5 onwards, the crossing point of this wave may be seen sweeping down the central impact axis. The interaction of the release waves down the central axis of the jet gives rise to a dark ellipsoidal area in the experiment travelling down through the frames. This corresponds to a high-pressure pulse, which traverses this axis. In the sectioned, recovered sample a region of intense shear damage is seen in this area.

Since the jet impacts onto a finite thickness target, the shock reflects in frame 9 from the free rear surface of the target block. This reflected shock and the interaction between it and the oncoming release from the surface, gives rise to tensile failure and the formation of a spall plane near the rear surface – visible as a dark horizontal crack in the figure. This axisymmetric impact gives rise to surface ring damage, a bulk sheared zone with visible banding, and a rear spalled region.

Natural phenomena have the ability to load structures and excite a spectrum of damage modes that result from wave interactions transmitted into the target, and quantifying the material response to such loads is important if materials are to be designed to survive in extreme environments. The ability to predict such failure relies upon understanding, and inclusion into numerical design platforms, of representations of operating failure mechanisms. Understanding these mechanisms requires observation of failure from controlled pulses by direct imaging and sensing of stress histories resulting from the waves generated by these processes.

5.8.3   Failure by spallation

The first studies of note into dynamic tensile failure in materials lie in the work of Bertram Hopkinson at the beginning of the last century (Figure 5.29(a)). Hopkinson studied a range of subject areas, applying basic principles to practical problems. His research had a profound influence on the transformation of aspects of the British effort in the First World War. His subject interests included electromagnetic effects in materials, performance of engines, mechanical properties of materials, explosion and flame and impact phenomena. All of these led to the parallel development of instruments and techniques for measuring relevant states and impulses resulting from experiment and one of these was the mechanical waveguide that bears his name, the Hopkinson bar. At

(a)   (b)

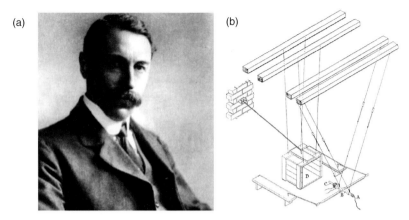

Figure 5.29 (a) Bertram Hopkinson. (b) Hopkinson's adapted ballistic pendulum for the measurement of the pulse shape from detonating gun cotton. Source: after Bertram Hopkinson (1914) A method of measuring the pressure produced in the detonation of high explosives or by the impact of bullets, *Proc. R. Soc. Lond.* A, **89**: 411–413.

the end of August 1918, Hopkinson crashed on a solo air flight to London and was killed at the age of 44. He made many contributions to dynamic materials behaviour, including the realisation that pure metals increased their yield strength in dynamic loading. He also noted the increased yield strength of FCC copper with strain rate, which was not observed in BCC iron.

One of his earlier investigations involved loading metals with explosive, placing gun-cotton (Figure 5.29(b)) on the surface of plate and observing the 'wave action which follows the blow'. This included the failure of scabs of material (spallation) from the rear of the plate:

The results obtained for gun-cotton, though lacking in precision, throw some light on the nature of the 'fracture' which is produced by the detonation of this explosive in contact with a mild steel plate. They indicated that the gun-cotton generated an impulsive force and showed that the pulse length was short since only a small displacement of the steel occurs during its action. Its effect is to give velocity to the parts of the plate with which it is in contact, the remainder being left at rest. (Hopkinson, 1914)

The study of dynamic tensile failure mechanisms in materials in the condensed phase is that of the dynamics of the generation, propagation and interconnection of flaws. In what has preceded this section a range of observations have shown that even in compression, mesoscale regions of tension occur in a metal's substructure concentrated at defects within. In brittle materials where slip is limited and the strain to failure is small such as in ceramics, composites and less ductile metals, the compression phase has dramatic effect on preconditioning the target. In other materials, shear within the substructure may also provide nuclei from which failure might propagate. Even phase transformation is associated with remanent substructural strains which may cause local failure in materials with few slip systems to accommodate the deformation. In what follows these microstructural influences on the observed processes that integrate to give particular forms to continuum failure will be shown for a series of metals, illustrating different behaviours. All of these processes have bulk influence on macroscopic behaviour that results from shear or from tensile loading such as friction in the former case and fracture in the latter.

In this first section the one-dimensional, dynamic tensile failure of metals will be discussed. In all cases failure is over a planar region determined by the interacting release fans. The opening of a free surface occurs here as momentum is trapped in a scab ejected from the rear of the target. The opening of the free surfaces reflects a compression pulse back, known as a pull-back or reload signal. The height of this is proportional to the dynamic tensile strength and a means of recovering this from the measured height was given in the analysis in Chapter 2. This is an idealised description, however, and the real signal recorded has a form that reflects the interaction of releases and reflected compressions with internal structures set up in the loading and unloading period before separation occurs.

The simplest, and most widely adopted, means of modelling such a continuum effect is to allow the target material to fail at a critical pressure. This minimum pressure, P_{min}, gives a means of opening a surface (deleting cells) in a continuum hydrocode and reproducing at least qualitatively the observed features of the signals recorded in experiment. If a fit was required to a particular test, then this parameter may be tuned to return the correct height at least to an observed signal from a measurement station. Although this might reproduce a signal height it is unlikely to capture the shape of the pulse correctly nor predict strengths under different impact conditions on the same metal. Advances beyond P_{min} have used the critical longitudinal tensile stress as a further criterion and fully implemented this to capture full anisotropic yield surfaces to reproduce results down different directions. A further complication is that the threshold itself depends not only on the amplitude of the pulse and the direction of the loading but also on the thickness of the target interrogated. Thus some impulse criteria might seem more appropriate to capture the development of the mesoscale damage. In what follows it is hoped that the complexities of the existing data can be captured to indicate the challenge that exists to reproduce these effects.

5.8.4   One-dimensional failure by interacting releases

As seen previously, the information concerning the tensile deformation processes gathered in a plate impact experiment is gleaned from the reflection of a release front from a free surface and then rebound from an internal failure plane. An overview picture of such an interaction is shown in Figure 5.30. A flyer has impacted from the left and a shock has propagated into the target and back into the impactor as seen previously. Reflection occurs and releases interact in the centre of the target. Since they are fans a zone of material sees a tensile stress which after some time opens a region of damage in the target. The later-time component of the release from the rear surface is now reflected with changed phase from this damaged zone returning to the rear of the target as a compression. The waves bouncing to and fro, elastic and plastic, are shown schematically in the figure to indicate the complex nature of this interaction. A schematic pulse measured

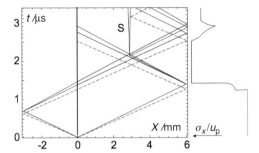

Figure 5.30 Lagrangian X–t diagram for symmetrical impact of copper on copper at 500 m s^{-1}. Elastic wave fronts are dashed, shock and the release fans' heads and tails are solid. A spall plane S opens as indicated reflecting part of the reflected release. Subsequent elastic and plastic wave reflections are indicated in the spall plate.

on the rear of the target in a plane with a low-impedance window is shown to the right and one might monitor the stress histories at this position or indeed record a free-surface velocity history with no window present at all. The relevant wave arrival times can be traced to features on the history.

The relation of the various stresses in the target to a stress history recorded at a low-impedance window was derived in Chapter 2. The derivation presented formulae for the tensile stress required to open the new surface (the spall strength) derived from back-surface signals monitored at a window or a free surface. The window case (where the wave transmits and reflects) gives relations which can be used to deduce the spall strength from either the drop from peak stress or the reload to the pullback stress and represents the most general case. The relations derived were

$$\sigma_{\text{spall}} = \left(\frac{Z_{\text{T}} + Z_{\text{W}}}{2Z_{\text{W}}}\right)(\sigma_{\text{PB}} - \sigma_{\text{min}}),$$

$$\sigma_{\text{spall}} = \frac{1}{2}\left[\left(\frac{Z_{\text{T}}}{Z_{\text{W}}} - 1\right)\sigma_{\text{W}} - \left(\frac{Z_{\text{T}}}{Z_{\text{W}}} + 1\right)\sigma_{\text{min}}\right], \tag{5.5}$$

where Z_{T} and Z_{W} are the impedances of the target and window, σ_{spall} is the spall strength, σ_{W} is the stress at the window (the peak stress of the back-surface signal), σ_{min} is the stress to which release takes the window state and $\sigma_{\text{PB}} - \sigma_{\text{min}}$ is the height of the pull-back signal. It will be seen from the wave diagram that the reload signal consists of information from very different locations. The initial drop is due to a release from the rear of the flyer plate; the reload is from a reflection of a release from the surface where the measurement takes place.

These stress amplitudes can better be understood by reference to the signal illustrated in Figure 5.31, which was gathered from spall in copper. One measure of the strength overcome (in a simple consideration of failure) is to note the height of the reload signal and correct for the transmission from the target to the lower impedance window (or a free surface) using the first formula. However, an equivalent measure is to note the window stress and the minimum after first release and use the second.

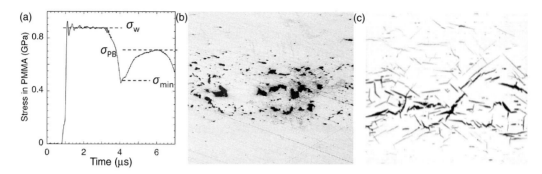

Figure 5.31 Spall in XM copper: (a) stress history at window, where σ_W is stress at PMMA window, σ_{min} is stress after first release and σ_{PB} represents pull-back stress; (b) microstructure of opened plane; (c) failure in ARMCO iron.

In other experiments, a VISAR or another system may be used to monitor free surface or window particle velocity histories for a spall experiment. Here the spall strength is recovered from the measurement using the following:

$$\sigma_{spall} = \frac{1}{2}\rho_0 c_0 \Delta u, \tag{5.6}$$

and the tensile strain rate

$$\dot{\varepsilon} = \frac{1}{2c_0}\frac{\Delta u}{\Delta t}. \tag{5.7}$$

The jump in the particle velocity Δu can be measured from the peak to the base of the unload signal, or the height of the unload signal itself, making the additional assumption of instantaneous failure within the material. The above formulae contain impedances which the reader will have realised vary as one transits from elastic to plastic behaviour. Given that these formulae contain some approximations, it is advisable to use elastic impedance (elastic wave speeds) when the stress is modest and plastic impedance (bulk sound speeds) when in the hydrodynamic regime and to treat the results as indicative recognising error.

Note that these formulae are a simple beginning to interpret details of back-surface histories in these experiments and various efforts have been made to add more features characteristic of the response of the materials probed. Romanchenko and Stepanov (1980) pointed out that failure at a release surface would give rise to a reload signal that was attenuating in its travel to the window surface. Thus more complex analysis is required if accuracy is important. Microstructural development is a further complication that must be accounted for. Nonetheless the formulae presented are used to give an accessible indication of a quantity, and the inaccuracies identified and the details observed discussed in what follows only show that better experimental techniques are required to fully understand the mechanisms operating. In the following descriptions of response it will become clear that it is the reload pulse that gives the key information as to the processes occurring, and in some cases the release from the rear of the target and that of the reload will indicate different thresholds. For this reason it is the pull-back signal, *reflected* from the spall plane, that will be interpreted to yield quantitative measures of failure stress.

5.8.5   Processes in tensile failure

It will be noticed that the station where the pulse is recorded, from which information about conditions in the bulk is deduced, is not close to the failure surface. Deducing tensile failure information from a signal at great distance from where mechanisms act, and based on interpretation of a wave front across which details of spall plane development will be integrated with travel distance, is fraught with danger. This is particularly true for a stress range where a surface has not opened and only local ductile voids have formed, as is typical in metals. Nevertheless, it is hoped that what follows will give some indication of the utility of the technique in not just working back to the stress required to spall the layer using the formulae presented earlier, but also gleaning some information about the mechanisms operating. The signals that are recorded are thus integrated effects of a series of processes occurring at a different location and agreement with simulation is often a validation exercise for empirical failure models rather than a means to derive detailed response. The key to understanding observations of behaviour is to follow the mechanisms developing within the metal. Thus the sections that follow will step through the stages of failure operating in the target as a surface develops. It is not intended that the following is an exhaustive review of all materials investigated over the years.

To start, it is instructive to review a typical result from an experiment on a pure FCC and a pure BCC material which show the classic form for a spall response. The left of Figure 5.31 shows the stress history recorded in a pure (XM) copper subject to a peak stress of *c*. 6 GPa in the target. The HEL of this material is low so that the elastic wave has a small amplitude (as can be seen). The reload pulse rises as expected and indicates a spall strength of *c*. 1.4 GPa in this case. The microstructure of the sectioned target is shown to the right-hand side of the stress pulse. The damage observed consists of a distribution of spherical ductile voids of different sizes distributed about the central position in the target where the heads of the two release fans interact. To either side of this plane, void concentration diminishes since, in the X–t plane, the diamond shaped region representing interaction shows tensile loading down the central position for the maximum time with decreasing load duration either side. This gives a shorter growth time for the voids away from the central plane that accounts for their smaller size. Voids in this material are nucleated and grow from bulk defects primarily at grain boundaries and each grows and interacts with neighbours over time. As the amplitude of the tensile pulse increases, the shear stress behind the shock and the compression of land between the opening voids induces shear localisation between adjacent voids, and examples below will show this behaviour in other metals.

One feature common to all of these locations is that although the signal suggests a complete separation of surfaces within the solid, it is apparent that there is a regime that precedes complete failure over which wave reflection occurs from a region of lower average density but with land between the voids. This is known as incipient spall. At higher stresses voids coalesce and there is a new surface formed which appears rough. At high enough stresses the wave trapped within this spall fragment bounces between two free surfaces and may cause further secondary spall planes within the plug detached. The suite of mechanisms that fail a ductile material in this manner result from nucleation and then the growth of ductile voids and the well-known models of Curran, Seaman and Shockey (1987) describe this process in an integrated fashion.

The right-hand side of Figure 5.31 shows another micrograph from a similar experiment conducted in a pure iron. However, although the observed rear surface history from such an experiment is very similar to that shown for the copper, the recovered microstructure is very different consisting of ductile cracks, primarily again at grain boundaries. This accentuates the statement made earlier that the history at an interface gives the integrated response of continuum mechanical states but not detailed information concerning mesoscale microstructural failure.

It is interesting to return to the velocity history of Figure 5.31(a) and relate its form to that of the signal recorded away from the plane on which the microstructure developed as a result of the loading. The shape and the magnitude of the pulse reflect the operating processes as the plane opens. It would be attractive to find some feature of the signal that indicated the total damage experienced by the solid. Banner found a correlation between the degree of spall damage observed optically in sectioned targets and the ratio of the height of the spall peak to the height of the main shock compression peak (Cochran and Banner, 1977). This may then be related to a damage parameter defined to be the volume of failed material per unit area normal to the strain in a given mass zone of a calculation.

At high driving amplitudes the processes are driven quickly and the free surface separates so that only Newtonian mechanics need to be considering to describe fragment motion. However, in the incipient spallation regime the detailed processes must be understood to describe the response observed in experiment. To track processes through it is logical to chronologically describe the mechanisms occurring and then relate what is observed to each of these in turn. It is the nucleation phase that begins the process and this remains the logical place to start.

5.8.6   Nucleation and growth of damage

Stress concentration around inhomogeneities within the microstructure localises damage to start from defect sites within the structure. Once failure occurs at a point, plastic work can be expended in the metal by the outward growth of damage. Figure 5.31 has shown the two principal forms of incipient damage in metals – ductile voids and ductile fracture. Growth of either of these at a population of sites activated by a tensile pulse will lead to coalescence and formation of a failed plane if the pulse length is sufficient to drive the damage sites to interconnect. However, there is a critical defect size that will concentrate stress before voids can grow from a nucleation site. The growth of failure from that flaw can take one of two forms: a void which opens doing plastic work or a crack that creates new surfaces as it propagates.

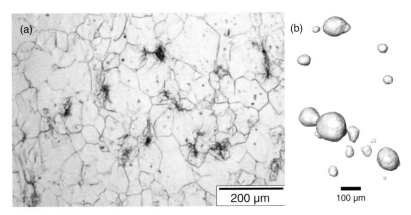

Figure 5.32 Spall in tantalum: (a) ductile failure in the first moments of incipient spallation; (b) tomographic reconstruction of ductile voids at higher stress amplitude.

In either case energy must concentrate and be sufficient to grow from the defect. Thus energy deposition into the opening volume must exceed the surface energy required to grow the void. The reader will recall the arguments of Griffith (1921) in his description of brittle failure where he showed that the new free surface created must be considered in any tally of energy deposition in the crack. A critical condition for propagation must be that the work done at a flaw must exceed the surface energy for it to grow. That condition would have a stress σ acting at a flaw radius r and with a surface energy γ and

$$\frac{4}{3}\pi r^3 \sigma = 4\pi r^2 \gamma , \tag{5.8}$$

which implies a defect scale of the order of $\dfrac{\gamma}{\sigma}$ in an isotropic metal. The surface energy for metals is typically of the order of 1 J m^{-2}, which gives small critical flaw sizes for purely brittle failure. However, Irwin, during the Second World War, showed that the level of energy needed to cause ductile fracture is orders of magnitude higher than the corresponding surface energy, and that plastic work must be taken into account (see Irwin, 1985). The behaviour is not elastic and there are not infinite stress concentrations under these conditions. Thus a realistic energy is of the order of 1000 J m^{-2}, which gives a minimum void size of c. 1 μm.

These arguments show that failure is activated at defect sites and that damage can grow either as a void or through ductile fracture. Which of these dominates is determined by the ease of plastic flow, and this can be judged by the Peierls stress in the materials. Copper slips more easily than iron, which favours void growth as the preferred failure mode. A second determinant is the kinetics of the two processes. Clearly ductile fracture is faster than void growth so that in situations where both are seen to operate the length of the impulse will determine which dominates. The surface energy required to open cracks from defect sites is less than the plastic work required to open voids. If one loads a metal with a pulse that barely exceeds the threshold for tensile failure then only ductile fracture is triggered. Figure 5.32(a) shows an incipient spall region in tantalum. Ductile fracture is the only damage mode excited in this case whereas for greater amplitudes voids are opened, as will be seen below.

The key to nucleation at grain boundaries is that these are often of lower mechanical strength (because of a larger defect population) than within the grains of the metal itself. Further, any impurities within the metal tend to concentrate there. Finally, the lattices are clearly misorientated at grain boundaries, and the level of this misorientation has an effect on the observed tendency to fail, with some boundaries favoured and others not.

Once nucleation has concentrated stress in a region, voids develop within it as plastic work is done on material around the defect site. The nucleation phase is fast and the processes occurring open up free surface quickly; on the other hand, the void growth phase is slower so that these kinetics dominate long processes. Thus there are two distinct phases of deformation which lead to structure in the rise of the reload pulse observed in the stress or particle velocity histories recorded on monitored interfaces. At higher stress levels this rise shows indications of the growth and coalescence processes occurring at the spall plane. Figure 5.32(b) shows a population of ductile voids opened in the same tantalum seen in Figure 5.32(a). The tensile stress is maintained at a level and is held for long enough time that nearly spherical voids are opened around the defect sites. They are spaced in the same manner as the nucleated regions in Figure 5.32(a), indicating that they are a development of the same suite of processes. The voids have a distribution of diameters reflecting the local stress states induced as they grew but none had a diameter less than a micron, reflecting the scale limit of this failure mechanism.

All these comments are of interest, but clearly if the pulse amplitude is high and all processes are driven at the bulk wave speed, then failure and the local kinetics will be overdriven. Features of the incipient region, however, where lower tensile pulses are applied and different behaviours are observed, will be followed below to indicate the individual responses observed.

5.8.7   Pure metals with different crystal structures

It is found that crystal structure exerts two effects upon tensile failure mechanisms observed in spallation: firstly, the Peierls stress controls the ease with which plastic work can be done; secondly, the density and speed of dislocations determine the kinetics of void growth that results. Assuming that the compression pulse is sufficiently long that it applies tension for a period sufficient to grow a void greater than the critical dimension (c. 100 ns) then in pure materials with low Peierls stress (such as pure FCC copper or BCC niobium) void formation is favoured and ductile void populations grow and coalesce to spall the material. The grain size of the material has little effect on the observed strength once the pulse length is long enough to trigger void formation. Indeed the growth of the void is fast in such materials as slip is easy – dislocations are therefore plentiful and their speed is high.

Figure 5.33(a) shows two signals for pure copper spalled by an elastic release and secondly for one that occurs over an extended time. The pull-back signals are in each case identical regardless of the rate of the tensile drive since the void growth mechanism is complete in both cases.

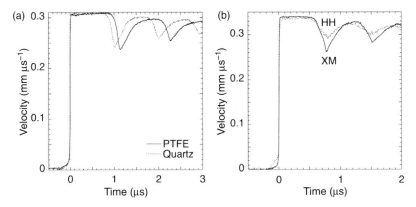

Figure 5.33 (a) Experiments on a pure copper (XM) conducted in a manner that allowed release rates to be changed in each signal. Note that the reload signal is the same in this FCC metal. (b) Two experiments at the same impact velocity on XM and half-hard (HH) copper.

In such materials, the presence of second-phase particles with weak mechanical strength offers an additional population of flaws which can be exploited by the tensile impulse since particle boundaries are in general weaker interfaces than grain boundaries. Figure 5.33(b) shows such a reduction in spall strength for a pure and a half-hard copper; in this alloy carbon particles harden the material and strengthen its response in compression. The presence of an additional population of nucleation sites makes the Hugoniot elastic limit greater in compression but the spall strength less in tension. On compression, dislocations are pinned at the particle boundaries and strengthen the material so preventing slip, whereas in tension they weaken it to spall by providing localisation sites from which failure can be nucleated. It has been observed that when one moves from polycrystalline to single crystal material, the spall strength increases by as much as a factor of around three. It is important to note similarities between these two processes. In the case of copper the void growth mechanism (analogous to cavitation in liquids since ductile voids are formed) is split into two phases: a nucleation and a growth phase. In pure materials, nucleation is made more difficult by the absence of defects from which to grow failure. A larger volume density of smaller voids results in an incipient spall. In materials with second-phase particles, fewer larger cavities are observed. Examples in tantalum below show the same trend.

The pull-back signal reflects the state of the material at the time the release comes to end the compression pulse. In FCC materials hardening can occur as pulse length increases. In the case of the alloy nickel cobalt for instance, the spall strength increases with increasing pulse length applied to the target. A similar result will be shown later for the HCP alloy Ti64. In BCC materials, however, the higher Peierls stress and lower dislocation density mean that, independent of the work done in the compression pulse, increasing the pulse amplitude results in a similar spall strength.

Figure 5.34 shows micrographs of tantalum sectioned to observe the spall plane from the work of Zurek and co-authors (Rivas *et al.*, 2000). The first three SEM pictures

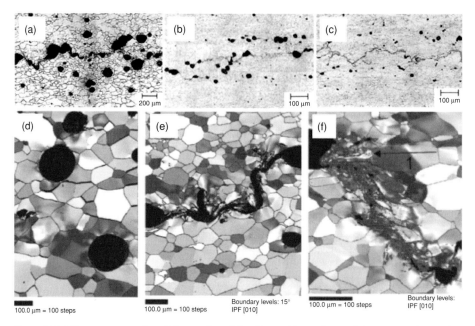

Figure 5.34 Microstructure of spall in tantalum. Metallographic sections of the recovered samples: (a) commerically pure Ta spalled at 252 m s^{-1}; (b) highly pure Ta preshocked and incipiently spalled at 246 m s^{-1}; (c) highly pure Ta incipiently spalled at 246 m s^{-1}. (d) EBSD images of incipient spall low-stress voids. (e) Higher stress, with strain localisation; (f) is a magnified view of (e). Source: from Rivas, J. M., Zurek, A. K., Thissell, W. R., Tonks, D. L. and Hixson, R. S. (2000) Quantitative description of damage evolution in ductile fracture of tantalum, *Metall. Mater. Trans.* A, **31**: 845–851. With kind permission from Springer Science and Business Media.

show incipient spall planes created using three loading paths on the metal shocked to nominally 7.5 GPa and then soft recovered. Figure 5.34(a) shows a commercially pure material; Figure 5.34(b) shows a high-purity material that has been shocked, recovered and then reshocked to observe the effects upon void nucleation; and Figure 5.34(c) shows the high-purity material shocked in its as-received state. All of the micrographs presented show that these materials fail by nucleating ductile voids. The darker lines seen correspond to regions of intense localised shear: mesoscale localisation that is developing within the material prior to failure.

The commercial purity material easily nucleates fewer large voids, some of which in the region of maximum tension have coalesced to form an interconnected fracture surface. The as-received high-purity metal, on the other hand, fails with a large number of more closely spaced, but smaller volume voids. Preshocking this material has increased the defect density and thus the number of nucleation sites and an intermediate response has resulted in this case. In all cases an impulse transfers a fixed total momentum to the metal. Thus the number and spacing of voids is a function of the defect population triggered by the pulse; a large number of smaller cavities or a lesser number of larger ones will in either case integrate to give the same work done and the same total momentum in the pulse.

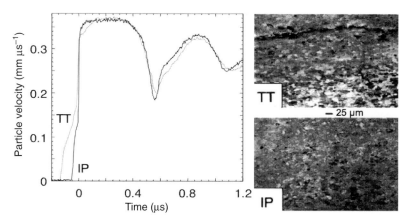

Figure 5.35 Spallation in HCP zirconium: stress histories and recovered microstructures. TT = through-thickness and IP = in-plane loading directions.

Electron backscatter diffraction (EBSD) is a crystallographic technique used to examine the orientation of many crystals which, in a polycrystalline material, can be used to image texture or preferred orientation. At the lowest amplitude, Figure 5.34(d) confirms void nucleation with almost no strain set up in the surrounding crystals. As observed previously, there are no preferred locations for nucleation in pure materials with high grain boundary strengths since cavities are observed within and at grain boundaries. At a higher stress, shear localisation takes place, as can be seen in Figure 5.34(e). At higher stresses still (and greater magnification), the localisation region can be seen to be composed of many smaller crystallites inducing significant strain in the surrounding crystals.

Although tantalum shows ductile void growth under tensile load, other BCC metals such as iron (Figure 5.31(c)), molybdenum and tungsten show failure by fracture. In their case the plastic deformation is more difficult since their Peierls stress is higher than that of tantalum. Thus fracture is favoured over void growth.

Decreasing the number of available slip systems as well as increasing the Peierls barriers results in fracture becoming the favoured deformation mode. The HCP metal consists of crystals with reduced symmetry that results in a limited number of available slip systems and of course Peierls stress is defined for slip in a defined plane. Properties are dependent on the c/a ratio of the unit cell as observed in general for HCP metals. As seen earlier preferred orientation or texture is a term which refers to the development of order in the grain orientations in a polycrystalline aggregate. The decreased ability to slip due to fewer available systems, a higher Peierls barrier and the ordering of microstructure in processing, results in deformation in both compression and in tension showing directional dependence. Thus a key feature for these lower symmetry metals is anisotropy in their response. An illustration of this can be seen in the response of high-purity zirconium which was clock-rolled and then annealed (see Figure 5.35).

Texture effects on fracture can be microscopic, such as the dependence of ease of cleavage on crystallographic orientation, or macroscopic such as the effect of texture-induced

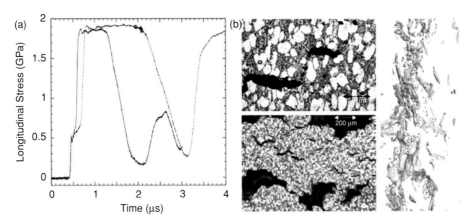

Figure 5.36 (a) Symmetrical impact onto Ti64 targets, 3 mm flyer on 6 mm target and 6 mm flyer on 12 mm target at 550 m s^{-1}. (b) Recovered failed microstructures at two magnifications with tomographic scan of opened voids to the right.

elastic and plastic anisotropy in polycrystalline response. The stress histories show the effects of shock deformation of zirconium in a rolled plate in the through-thickness (TT) plate compared to the in-plane (IP) rolling direction in the plate. Anisotropy was evident at quasi-static rates with flow stresses c. 2.5 times greater in the through-thickness direction. Indeed, HELs of Zr down the two directions are of the order of a factor of two different from one another, as one might expect from the directional elastic properties of the metal. In contrast, there is less effect upon the magnitude of the pull-back signal. It is revealing that the pull-back signal obtained shocking down the in-plane direction is of the same form as noted for the copper and tantalum above, having a rapid rise and then a slower rounded rise to the peak. The in-plane pull-back, on the other hand, shows a smaller initial rise and a lower magnitude.

The right-hand side of Figure 5.35 shows recovered microstructures for the two directions. The damage evolution in both the TT and IP samples contains a population of small (c. 10 microns) nominally spherical ductile voids. There is no separation of the spalled TT or IP samples at this amplitude (approximately five times the HEL) but whereas the damage down the IP direction is almost completely ductile voids, there is some shear fracture in the TT target. The differences in stress magnitude between the two directions reflects the reduced plasticity possible orthogonal to the IP impact direction than orthogonal to the TT one. It is nevertheless the case that the stress required to fail down the two directions in compression is different whilst that required to fail in tension is similar. In both cases, these stresses are found to be less than is the case in quasi-static loading.

Figure 5.36 shows spall in a hexagonal alloy. Ti-6Al-4V is composed primarily of the HCP α phase primary alpha grains within a transformed BCC beta matrix with weak boundaries between the two. The impact velocity was kept the same in the two experiments shown with the peak stress amplitude of c. 7 GPa in the targets, and both pulse length and release rate were nominally doubled at the greater distance as the impact geometry increased in dimension. As seen earlier, the strength of Ti64 remains constant behind the shock front at this stress level in contrast to pure FCC metals where it increases and pure BCC metals where it declines. The stress histories in Figure 5.36(a) were recorded at a PMMA window at the rear of the target and illustrate this behaviour. The HELs of the two metals are indistinguishable from one another even though the stress wave has travelled twice the distance in one case than the other; indeed the compression pulse scales precisely showing rapid equilibration to the yield surface in this alloy. However, the reload signals show different characteristics: the second signal is almost three times the amplitude of the first. Whereas the reload signal height is very different, the drop from the peak of the pulse is similar. Again this shows the simplicity of the assumptions made in the derivation of the formulae used to recover strength when applied in situations where the spall is incipient. Clearly the information contained within the pull-back signal contains details of the dynamic failure mechanisms operating since only this part of the pulse directly samples behaviour after failure has occurred. Indeed the relation of the magnitudes of the drop and the reload to the failure stress work best in a material with lower Peierls stress where failure is dominated by void growth.

Finally, the microstructure of a spall plane in Ti64 at two magnifications and a tomographic reconstruction of the damage is shown in Figure 5.36(b). There is evidence of some ductile void growth but also of ductile fracture in the alpha phase and at boundaries with the beta grains. Quasi-static tensile tests showed failure to be due to a mixture of some fracture near beta phase sections but also a mixture of ductile tearing and some mode I transgranular cracking along with some evidence of shear and/or cleavage in the alpha. Notice that the reload signal shows a rapid linear rise indicating fast fracture as the primary failure mode. Clearly in this case the damage processes are highly time dependent and the spall strength itself increases with distance. Of course the loading has a lower release rate at distance and this in itself will restrict some of the operating mechanisms. This complex suite of observations allows some features of HCP response to be extracted; anisotropic behaviour is observed within grains and fracture is the dominant mechanism that accommodates the applied tensile strain.

5.8.8   Alloys

Metal alloys may be either single phase or may contain second-phase particles at the mesoscale. Single-phase alloys have atoms of different elements within the unit cell and single-phase FCC alloys (like NiCo) show an increase of spall strength with time after first load compared with pure nickel where the spall strength remained constant. This effect is due to an increase of twinning with applied stress which impedes void growth compared with pure nickel.

Macroscale composites such as carbon fibre or glass fibre epoxy are characterised by unzipping at the filler phase where stresses concentrate and the bonding is often weaker than in the binder or filler materials. Two-phase alloys at the mesoscale behave in a similar manner with sites to initiate tensile failure concentrated around inclusions. An

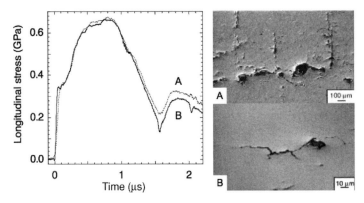

Figure 5.37 Spallation of 1080 steel. Microstructures from targets loaded (A) longitudinally and (B) transverse to the rolling direction.

example is shown in Figure 5.37 where a steel is loaded in plate impact to the same ultimate stress down two directions in a processed steel.

The addition of manganese to the melt in order to remove sulphur and prevent the formation of iron sulphide increases the strength of the metal. Metallurgically it possesses a fully pearlitic microstructure with pronounced MnS stringers formed during hot rolling. The crystallographic texture of the steel was measured using X-ray diffraction and found to be nearly random. However, the microstructure contains manganese sulphide as long straight stringers, aligned with the rolling direction in the processing. The effect of these additions on the compressive behaviour is negligible and the flow stress in directions in-plane and through-thickness with respect to rolling direction is the same. Indeed the stress history to the right of the figure shows that there is no effect upon the HEL caused by shock propagation relative to the rolling direction. However, the tensile response is different down perpendicular directions since the structure is easier to unzip perpendicular to (A) rather than in the direction of the stringers (B).

This material shows extensive microcracking and decohesion along the MnS inclusions and small (*c.* 10 μm) nominally spherical, ductile voids associated with the ends of the stringers. However, there is heterogeneous nucleation of damage normal and orthogonal to them as seen in the micrographs.

A similar behaviour has been observed in other steels, where again the response in compression was isotropic but that in tension was not. Further, the spall in orthogonal directions was also different in the lower stress range but was seen to become similar as the stress amplitude increased. It will be seen from this example and that of the Ti64 that damage is anisotropic and results from nucleation at the mesoscale from an activated population of suitable flaws.

5.8.9   Taylor wave loading

In practical applications, the loading rarely applies a square pulse in the manner discussed in the previous sections. One common impulse, encountered in mining or in

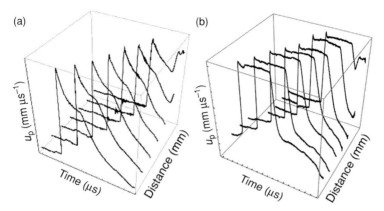

Figure 5.38 Triangular wave (pseudo explosive) loading of ductile metals: (a) particle velocity history of a propagating Taylor wave impulse compared with (b) a square wave impulse on the right.

defence applications, is the pseudo-triangular pulse induced by the explosive loading (Figure 5.38). Such loading in metals reflects and induces a tensile component which fails the target near the rear surface. When failure is complete it results in spallation that launches fragments from the rear surface of loaded plates. Here reflection of a pulse at the rear of a target induces tension directly rather than through the controlled interaction of release waves on a plane within the target as was the case for the square pulse loading described in the previous section. The impulse encountered with explosive loading is often dubbed a Taylor wave since after the rapid Von Neumann spike decays in a target, the pulse follows the form of the Taylor release described earlier: a high stress peak with a rapidly decaying profile which gives a triangular profile to the impulse. Interaction of the later-time lower incoming compressive stress pulse with the reflected tensile peak, gives rise to a spall plane close to the rear surface of the target. The impulse delivered there is generally more localised and of shorter duration than was the case for the square waves where the release fan controls the profile. Further, the impulse at high stress levels is not applied for more than times of the order of tens or hundreds of nanoseconds so that the metal does not have time for slow failure mechanisms to operate in the manner discussed for pure metals with a sustained pulse. Thus, since the pulse is short, comparison between square and Taylor wave impulses depends on a suite of parameters so that simple correlations are difficult; even subnanosecond laser-driven impulses cannot access some of the failure mechanisms excited by the explosive impulse.

Figure 5.39 shows the spall due to a square impulse, as well as that due to a Taylor wave on an FCC stainless steel (SS316L). The material was loaded to three stress levels as seen in the stress histories to the left of the figure. Matched amplitude pulses were launched to load at 6, 10 and 15 GPa and the damage and the spall signals recorded were noted. It will be seen that the highest stress amplitude Taylor wave signal indicated damage with a reload signal height equivalent to the lowest square pulse. In addition, the void diameters in the 15 GPa Taylor pulse were the same and had the same number

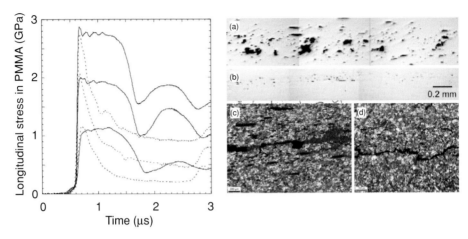

Figure 5.39 Comparison of square wave and Taylor pulse loading of SS316 at three stress levels. Target microstructure shocked to 15 GPa. (a) Square, (b) Taylor wave, (c) EBSD square, (d) EBSD Taylor wave.

density as those in the 6 GPa square pulse. However, although the voids were spread over an area of a millimetre or so for the square pulse, they were localised to around half that in the case of the Taylor wave and were close to the rear surface. Further, the damage created was found within or adjacent to areas of strain localisation.

Such a wave provides a much lower impulse than the square pulse and it neither develops the microstructure at pressure nor allows plastic mechanisms to operate before release occurs. Thus a different response to that observed with an extended pulse occurs, even within materials that respond in a ductile manner such as FCC metals, whilst it is similar to those which deform in a more brittle manner and fail rapidly. To capture such behaviour in models requires both the nature of the failure mechanisms and their kinetics to be included, and such information can only be recovered from experiments that supply loading of the correct amplitude and duration. Finally, an explosively driven pulse loads at a tensile strain rate an order of magnitude higher than that in a square wave where the release has dispersed.

Other short pulse loading experiments have been used to deduce spall strength in a one-dimensional configuration. Flyer plates give rise to tensile loading regions of dimension in the millimetre range, whereas explosive pulses load over hundreds of microns, and laser-driven experiments over a micron or less – the latter two sample regions less than typical inter-defect separations and therefore below the energy threshold for a mechanism like ductile void formation.

5.8.10   Pulse duration and the mechanisms excited in spallation

In the most general case, the failure of a metal under tension follows an integrated suite of mechanistic paths which result in the threshold observed in experiment. To ascribe these

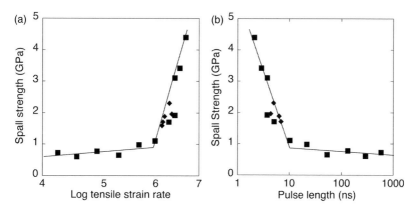

Figure 5.40 Spall strength of pure aluminium as a function of (a) tensile strain rate and (b) pulse length applied in the experiment. Source: after Bourne, N. K. (2011) Materials' physics in extremes: akrology, *Metall. Mater. Trans. A.*, **42**: 2975–2984. With kind permission from Springer Science and Business Media.

results to a single effect is dangerous but evidence of exclusion of particular paths is seen in Figure 5.40. The term strain rate is misleading here since it relates to an average of the applied strain divided by the rise of the pulse at the lowest impulse lengths in conditions where the stress state in the material is not in equilibrium. However, Figure 5.40(a) takes the parameter so defined and shows that as the strain rate increases (which corresponds to both the pulse length and thus spatial region sampled decreasing), the tensile strength increases. The highest values are recorded for the shortest pulses and reflect the sampled zone going below the interdefect distance for the metal under question.

In Figure 5.40(b) the experiments are represented and the strength plotted as a function of the pulse lengths provided by the experiments applying the load. The point at which free surface becomes increasingly difficult to create (corresponding with F, the ratio of the relaxation time for this process compared with the pulse duration) and results in the observed spall strength rising. This relates to the defect population from which failure ensues at the mesoscale on the one part, but also since the trade between energy of the surface created and work done growing the void does not favour cavitation at this length scale (of the order of 1 μm). If the pulse is sufficiently short that it cannot open voids significantly, then the energy required to open a new surface increases bounded by the theoretical shear strength of the solid. As the defect size and volume fraction sampled decrease the strength observed approaches the shear modulus, $G/2\pi$, which, in the case of aluminium, is c. 4.5 GPa (the asymptote to the values recorded at the smallest pulse lengths). At present the shortest impulses are applied with lasers and only these offer the means of accessing the limits of strength in a material.

5.8.11   The effect of substructure on the response to dynamic compression and tension

In the previous sections, the responses of metals to dynamic tensile impulses has been described. The evidence attained to date has shown that response depends critically upon the structure of the material in question, whether it be a pure metal or a single- or multiphase alloy. The underlying substructure at different length scales controls the response. This includes the relevant chemical processing variables such as alloying, second phases, phase stability or interstitial content. It will also depend critically upon the crystal structure which will in turn control related quantities such as the stacking fault energy and Peierls stress at the atomic scale and at larger scale, but also upon mesoscale properties such as the crystallographic texture and the phase distribution.

When a metal is loaded by a shock wave in compression, the state achieved is averaged as the strain field equilibrates given the imposed boundary conditions. The response will clearly also be determined by the physical constraints imposed by boundary conditions upon the loading. This will result in an excited thermodynamic state, which is determined in a particular metal by a series of physical parameters that control the shock process and fix state variables such as temperature, pressure and compression. The pulse will excite a characteristic volume which contains a substructure in metals that will determine an included defect population. In the general case this will consist of a polycrystalline aggregate and each grain may include second-phase particles and likely substructures such as dislocation tangles or cells, twins or remains of a transformed phase. Materials that appear homogeneous at the continuum scale are locally heterogeneous at the microstructural level.

When releases interact or a compression pulse reflects in tension, flow localises at a distribution of weaker nucleation points in the material and triggers damage and relieves strains around them. After loading has completed, the magnitude of the tensile pulse and its duration will have either started partial failure at a population of flaws (incipient spall) or failed the target and launched a scab from its rear. Which one of these results depends upon the amplitude of the pulse controlling the kinetics of the failure. A distribution of critical defects exists for each stress level, but only a subset of these activate mechanisms that lead to failure. Damage can occur by one of two mechanistic pathways. In the first, voids are nucleated, grow and coalesce to form a continuous, failed damage plane, in the second, fracture can separate surfaces more quickly. In both cases momentum is trapped in a scab which flies out to impact or damage downstream. The first mechanism is limited by the kinetics of the plastic processes of void nucleation, growth and coalescence; the second by the crack-opening speed. The crack-opening speed is generally a more rapid process than the former since the process is quasi-two-dimensional rather than fully three-dimensional. The previous section has illustrated a transition in response to dynamic tensile loading as stacking transits from cubic to hexagonal form and then to alloys, as the Peierls stress of the material is increased. It is clear that as the number of available slip systems is reduced, brittle or ductile fracture becomes the favoured failure mechanism and response becomes more dependent upon the direction of the stress field (which is determined by compression on impact and subsequent travel of a shock front) relative to the limited slip directions of the metal. Further, when the time for which tension is held at the defect falls below a critical value, ductile void growth does not occur. The energy required to fail the target is thus partitioned between a surface energy and plastic work term and the former grows quickly as flow processes become more difficult.

The response of all classes of stacking has shown that the failure of metals under tension depends not only on the stress threshold achieved by the pulse but also upon its duration. The impulse applied must reach a critical pulse length and exceed a minimum stress threshold for nucleation of damage to occur. It is the nature of the mechanism by which failure proceeds that determines whether metals will show an increase of strength with pulse length (to give but one example), and this in turn is determined by microstructural features. Ductile metals (such as the FCC group) have low Peierls stress and deform by void nucleation and growth, whereas materials where plastic deformation is more difficult do so by fracture. The flaw morphology may show dislocation and twins at one length scale but also features introduced during processing or heat treatment that give rise to pronounced anisotropy.

With boundary restrictions imposed, anisotropy is observed in the tensile failure that results at one of a suite of possible length scales. The varying response down alternate axes can be due to differences in Peierls stress in the single crystal, which results in different cleavage down different directions. Alternatively it may result from elastic and plastic anisotropy in damage nucleation and growth. In a polycrystalline material, microscopic directionality (grain shape, second phases, etc.) may impose boundary conditions on void growth. Finally, there may exist long-range microstructural constructs such as those in continuous or particulate reinforced composites, laminated composites, and/or functionally graded materials. All of these features of the microstructure result in anisotropy at the continuum that any model at that scale must reflect.

Spallation in single crystal metals is principally controlled by slip and ductile flow with anisotropy leading to aspherical voids; however, in polycrystalline materials, grain boundaries offer lower energy nucleation sites. All these factors give rise to spall strength dependence on a range of features including grain size, flow stress, second-phase particles and inclusions, and phase transformation products. All of these features indicate localisation and nucleation at a population of defects and the recovered surfaces that result from such a loading are dimpled and rough. When material transforms phase and a rarefaction shock can form, the surface formed is smooth, which is consistent with the rapidity of the failure and its small region of loading. The highest spall strengths have been found in single crystal metals where no grain boundaries exist to act as nucleation points for failure and crystal defects are small in extent and widely separated. In continuum materials, however, there are a variety of materials properties that have been shown to affect dynamic tensile failure. These include the grain size, flow stress, second-phase particles and inclusions, and phase transformations which either change parameters for failure in the new phase or revert and leave debris to nucleate damage. Of course as ambient temperature is increased or the stress amplitude climbs heating the metal behind the front, the material approaches melt and the spall strength reduces dramatically.

Finally, the impulse delivered is critical to the appearance of the damage. Square pulse loading at low tensile stress nucleates and fails metals in a different manner to a pulse from an explosive. The factor that controls damage in a metal is the impulse imparted by the load. There is, however, a material defect length scale below which short applied loading pulses do not sample continuum material properties.

What is certain is that much about spall is not intuitive. However, although understanding is still incomplete, it is still possible to follow from quasi-static behaviour and predict to first-order continuum response for tensile loading.

5.8.12   Modelling spallation in metals

The numerical prediction of spallation in ductile metals is a difficult and evolving subject area. The problem is so challenging because of the varied microstructural features observed during the process and the presence of two dominating mechanisms with different kinetics controlling response and summarised above. The simplest approach is to assume that computational cells fail when the pressure falls below a minimum level P_{min}. This approach has some success in that it can show failure will occur at positions that are in reasonable accord with those observed from experiment. However, the value of the pressure assumed has to be carefully fixed if the predictions are to agree with even simple experiments. Clearly there needs to be more physically based modelling to account for the observed features in the flow.

From the early 1970s this was built upon with the introduction of models following a damage parameter D that evolved following particular controlling mechanisms. Through the history of the subject, the principal analytical models produced have followed the development of ductile voids. Solutions have existed for some time for the expansion of ductile voids in materials with strength and integrated models have built upon the work that was done. Observations from plate impact experiments required more complex theory to account for the evolution of D with stress level and pulse length. Key developments came in the SRI nucleation and growth model, which considered void growth and coalescence as well as the effects on the bulk material surrounding resulting from density and modulus variation in materials.

Experimentally it is clear that voids in polycrystalline materials are nucleated at grain boundaries and second-phase particles. Ductile fracture also becomes favoured over ductile void growth as plastic deformation becomes less easy. Thus whilst the most successful models predict situations in which ductile void growth occurs in pure metals, most experimental observations of spallation show some population of blunted cracks. Further, the slow part of the process is the nucleation phase and this, and the transition from nucleation to fracture, is difficult to model at the scales at which these processes occur. Missing kinetics does not allow the transition from ductile fracture to void growth in pure metals as pulse length increases. Workers studying specific systems have produced models that describe their particular metal of interest generally fit to the specifics of its response. Thus no generally applicable model type or even set of material parameters really exists for a pure metal. Thus the difficulties of prediction at the mesoscale are manifest and the challenge of extending to the continuum codes is one of the aims of the present effort.

5.8.13   Metal fragmentation under tensile loads

One key consequence of shear within metal sheets is the break up and fragmentation of containment under dynamic loading. That may include impact, blast or direct explosive loading of structures or of containers, and clearly geometry and the nature of the materials under consideration are key determinants of response. The difference between quasi-static and dynamic fragmentation revolves around the kinetic energy that is present in the latter case. There are several observations of targets that fail in a brittle or a ductile manner by fragmentation. One of the key measurements is of the fragment size and its distribution within the recovered pieces of the target found from trials. Two cases are observed and these are ductile and brittle fracture. In the latter in particular, there are many cracks that have not propagated to create new fragments so that estimates of failure geometries become difficult.

Analyses of fragmentation, including the seminal work of Grady (2006), in general adopt an energy balance approach in which the surface energy of the new free surface created is equated to the kinetic energy of the fragments that result. With such an approach, one needs to be certain that one has identified the mesoscale failure mechanisms and thus has an accurate measure of the new surface created by the dynamic load since the area of new surfaces created is generally higher than that of the fragments. As is the case in many of the controlling mechanisms encountered previously, this is because the continuum stress state is not matched with that at the mesoscale where greater degrees of freedom allow local three-dimensional stress states under global compression that may include significant tensile components. The observed phenomenon is thus critically dependent upon the population of defects that exists within the material at the start of the process. In addition, surfaces, whether they are free outer ones or internal boundaries between different phases, will almost always have a higher density of larger defects than the bulk.

There are several tensile failure mechanisms that might occur depending upon the form of this population and upon the ease of plastic deformation possible within the metal. If spherical bulk defects exist, local tensile strain fields will allow the voids to grow and coalesce or to fail by crack propagation. The fracture process will relieve stresses in the region surrounding the failure and establish a characteristic distance between paths dependent upon interaction mechanisms between adjacent surfaces and the defect state of the material. If the flaw is elliptical then directionality is introduced into the failure. It is frequently the case that mesoscale damage in engineering materials results from a defect population introduced by processing which induces a particular texture. The inclination of the applied load from the dynamic event to the direction of aligned flaws from processing will have a critical effect upon the observed fragmentation. Finally, anisotropy within the material may lead to fracture generated by dislocations nucleated, flowing and then concentrating stress at pile-ups or twin boundaries.

One key problem that has driven consideration of the phenomenon over many years has been the accidental ignition and subsequent detonation of a HE-filled shell stored potentially with many others in an enclosed space. In such a situation the detonation of the explosive causes fragmentation of the surrounding metal case which, propelled

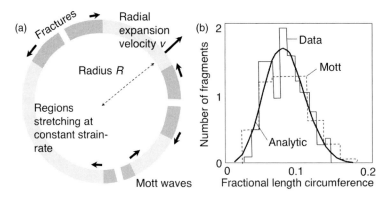

Figure 5.41 (a) The assumptions of Mott theory: failure at flaws under expanding tension leads to release (Mott) waves which interact to further fail a cylinder into a distribution of fragments. (b) Mott fragmentation theory results, showing Mott solution, modern analytical solution, and some sample experimental data points. Source: after Grady (2006).

by products from the detonation, is driven at high velocity into adjacent munitions. These may in turn then sympathetically detonate by the intrusion of fragments from the first exploding shell. N. F. Mott in the Second World War considered the expansion and fragmentation of a cylinder, line-initiated down the central axis (see Mott, 1947). His analysis used a statistical strain-to-fracture model which predicted a fragment size distribution close to that observed in experiment. His intuition took the leap of assuming an initial statistical flaw distribution from which both failure occurs but also propagating waves shield regions by relieving tensile stresses. This has been termed a process of statistical fracture nucleation and growth, or, perhaps more accurately, a process of fracture activation and interaction, and is actively used in its original form to the present day.

The problem he considered was a uniformly expanding and stretching metal ring or cylinder (the Mott cylinder) in which any hardening properties of the plastically stretching metal had saturated (Figure 5.41(a)). The cylinder expands under a constant tensile force and at some point fails at a distribution of flaws. The stress release from points of fracture into the plastically flowing medium drives diffusion waves which prevent further localisation in material adjacent to the failure points but also define a second phase of fracture points when they interact. Working with a deck of cards and graph paper, Mott extracted the one-dimensional fragment length distribution shown in Figure 5.42(b). The data that exist, and that are statistically homogeneous enough to satisfy the theoretical criteria, fit the theory derived quite well; this has several implications for the statements made earlier concerning the energy balance approaches adopted more recently.

There are thus two approaches adopted to describe fracture activation and growth and predict fragmentation in solids. In the case of the expanding cylinder, Mott's statistical strain-to-fracture model has achieved success for the particular physical conditions prevalent to the thin-walled, quasi-one-dimensional state that exists in such a process. It does so by ignoring the fracture energy and predicts an average fragment length proportional to a length scale that is a combination of the flow stress, density and strain rate. Such an approach, however, cannot adequately describe failure mechanics within the bulk of material. In this case, energy-based considerations such as the Grady and Kipp approach should have more success in describing the controlling processes. These methods equate the energy dissipated within the fracture activation process and define the length scale governing the predicted fragment size that contains that fracture energy. The only caveat to these methods is that they must capture all the surfaces opened to correctly apportion the kinetic energy to the failure process which is frequently a difficult task.

The previous three sections have explained dynamic tensile failure by interaction of stress waves within metals leading to subsequent failure zones which nucleate, grow and coalesce to open new fracture surfaces within a material. These processes occur on micron dimensions and over microsecond timescales and show the localisation phase of tensile failure which is the precursor to new free surfaces at the structural scale. Higher length scale processes also act in metals to accommodate compression. Wave processes localise slip at geometrical zones and at the structural scale these adiabatic shear bands represent the means of accommodating large strain in the component.

5.9 Adiabatic shear banding in metals

5.9.1 Introduction

Bands occur in a range of metals loaded in geometries which apply a substantial shear component that fails the metal. This localised deformation is said to be adiabatic when the loading is rapid since the lowering of stress at increasing strain is due in this case to heating in the material which has not the time to diffuse away because of the rapidity of the loading. In fact, the time taken to establish the band is insufficient for the process to be accurately adiabatic, but nevertheless it is fast enough for there not to be significant temperature equilibration in the process.

Adiabatic shear bands (ASBs) have several characteristics that are shared between the different materials that exhibit the phenomenon, including:

(i) the band cuts straight through the metal without regard to microstructure;
(ii) it has a very large aspect ratio with micron thickness and typically millimetre extent;
(iii) there is high strain in the material within the confines of the band;
(iv) the bands are generated by a dynamic nucleation process;
(v) bands are seen to occur in some metals and not in others;
(vi) material within the bands is observed to be harder (or sometimes softer) than the surround on recovery after loading. Thus banding is a process that allows macroscopic slip of structural units and is a prelude to engineering structural response in the metal under load.

ASBs are found in metals of engineering importance, including steels, brass titanium and aluminium alloys, magnesium, uranium and tungsten and its alloys. They are not observed in pure copper or aluminium, however, and some authors have thus ascribed them to materials with high hardness. There are two classes of band observed to form: *deformed bands* and the generally narrower *transformed bands*, where a phase change has occurred on heating. Martensite transformation bands are observed in steels and appear white when etched. But such structures are also observed in titanium and its alloys and also in uranium. This transformed material has different mechanical properties from that initially present in the surrounding matrix resulting in different hardness in the bands. Adiabatic shear is the principal failure mechanism observed in high-rate deformation of amorphous metals. The development of a localised band is favoured in these materials at high strain rates and low temperatures. The discussion below begins with the much wider range of work conducted on polycrystalline metals, but metallic glasses and their response will be returned to at the end of this section.

A breakthrough in understanding was made by Zener and Hollomon (1944) in the testing of steels. They found that the state within the target moved from isothermal to adiabatic as the strain rate increased. In this region the adiabatic stress–strain relation has a negative rather than a positive slope, which makes homogeneous plastic deformation unstable. The downturn in the stress–strain curve by whatever mechanism is at the root of the material instability and leads to localisation. In the dynamic case this is generally localised heating at hot spots within the material that cannot be dissipated by heat conduction and weakens the structure in a feedback loop ending in a failed band. Measurements have been made using infrared detectors amongst other devices and temperatures of the order of 900 °C have been recorded.

Shear bands give a metal an extra mode of plastic deformation. If instability results in a region, small changes in strain, strain rate and temperature can be amplified, relieving adjacent material and producing a localised band of deformation. If the cause of the instability is elevated temperature, then this situation results when heating due to work done by flow in the band exceeds dissipation into the bulk. Thus ASBs form readily in metals of low thermal conductivity where slip is made difficult by some means. If the heating causes a phase change then the region gains new properties and a physical boundary exists within the metal.

Most reports of ASBs result from loading where the nucleation is primarily from surfaces and in this case the band that results cuts across microstructure. However, it is important to distinguish between propagation that occurs rapidly and crosses the microstructure quickly, and the nucleation that occurs slowly at many locations. Although slip bands do occur in metals under dynamic deformation, there is little evidence that they develop into shear bands. However, there are geometries where the nucleation of intense shear will be localised within the bulk of the material and here the microstructure will assume importance. Thus the boundary conditions on deformation will determine the processes and the observations made. Finally, it must be recognised that any analyses of recovered microstructure do not necessarily sample a state entirely due to the deformation in progress during the loading but record the final state achieved after this phase has completed and an unloading and cooling phase has occurred. Comments will be made below that support the view that many of the observations made occur in the release phase of the process. Ultimately, in order to attempt

Figure 5.42 Sheared compression targets after loading in a compression split Hopkinson bar: (a) Ti; (b) Ti-6Al-4V. Source: after Gray III, personal comm.

to counter this mode of failure, it is necessary to change the thermal characteristics of the alloy or composite to control localisation and subsequent failure if effects are to be controlled.

5.9.2   Techniques used to study shear banding

There have been several techniques developed to study shear response in materials; however, there are few available for dynamic studies and a subset of these will be mentioned here, although the following is by no means an exhaustive review.

One integrated test in which adiabatic shear has been extensively observed is in ballistic impact in a range of geometries. Studies of these effects have frequently shown that bands nucleate from geometrical features (such as machining marks) on the surface of, or at macroscopic interfaces within targets in preference to microstructural defects in the bulk. This mode of localised deformation sets up slip bands in fully dense metals and forms plastic hinges which allow failure to collapse structures. It is thus a longer timescale and larger length scale phenomenon than the dislocation-driven slip considered above. Thus in order to study the localisation event, a suitably long loading pulse must be applied under well-defined conditions to a designed target geometry. This has evolved two classes of test which have dominated reported experiment: the Hopkinson pressure bar in various geometries and the contained, imploding cylinder test.

Shear bands have been seen extensively in compression tests within Hopkinson bars. Recovered, sheared compression samples are shown in Figure 5.42, showing clearly how the microstructure has slipped to accommodate the axial applied strain. Bands are clearly seen at 45° to the loading axis in both titanium and the titanium alloy Ti64 which have formed during the late-time response of the compression samples; slip is reduced in HCP metals. Unfortunately it is difficult to use this simple geometry to gain quantitative information on the shearing process. However, various developments using the Hopkinson bar principle have been developed to study the phenomenon in detail. One geometry obviously suited to study this mode of deformation is the torsional bar. In early adaptations of the technique by Duffy, a thin-walled cylinder was twisted and observed to deform and localise using high-speed photography and monitored incident and transmitted wave pulses to gain quantitative data. Typical results for a steel are shown in Figure 5.43.

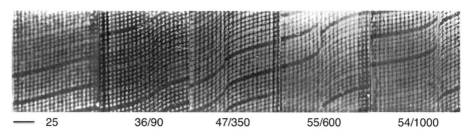

| —— 25 | 36/90 | 47/350 | 55/600 | 54/1000 |

Figure 5.43 Shear bands in the torsional Hopkinson bar. Each snapshot shows the average shear strain (%) applied and, when appropriate, the local shear strain at the localisation. The scale bar to the left is 500 μm long. Source: reprinted from Marchand, A. and Duffy, J. (1988) An experimental study of the formation process of adiabatic shear bands in a structural steel, *J. Mech. Phys. Solids*, **36**: 251–283. Copyright 1988, with permission from Elsevier.

In such a bar, a torsional pulse was launched through down the rod to interact with the target by the sudden release of stored torque at one end. The target was a thin-walled tube with integral hexagonal flanges which fitted into sockets in the bar ends. Measurements of the stress and strain in the specimen were inferred from records of the strain in the incident and transmitter bars. Figure 5.43 shows photographs of a grid pattern etched onto a sheared target in five tests. Localisation may be seen to occur over an 80 μs period in which the shear stress starts to drop as the material fails. Although providing quantitative information on the deformation of the structure, it only yields qualitative indications on that of the processes occurring, since clearly such a thin-walled cylinder suffers three-dimensional structural deformation as well as material failure at the mesoscale. Thus a direct relation to bulk material properties of the steel from which the cylinder is constructed using such a geometry will be complex.

A more controlled loading geometry in which deformation is within the bulk of the metal is achieved using a compression bar with top-hat geometry specimens (Figure 5.44(a)). This technique has been used on a range of materials to induce shear banding in a controlled manner to quantitatively assess material response. The test forces localised shear by dint of the geometry of the specimen rather than creating instability in the material flow field. The hat-shaped specimen has a circular cross-section and a narrow overlap region that isolates the deformation. The displacement is controlled either by a steel ring which limits movement of the narrow end of the specimen or by fixing the duration of the loading pulse by varying the striker bar length.

Figure 5.44(b) shows results from such a test on a stainless steel. Localisation of the deformation and formation of the shear bands can be observed clearly in the sequence in Frame 4, 41 ms after loading begins, whereas deformation occurring in the bulk is homogeneous before then. As the band forms, the stress–strain response exhibits the classic signature of instability with the shear stress decreasing as strain increases. A key observation in these interrupted tests is that the microstructural defect concentration must reach a critical level before the band can develop and allow plastic flow to focus in a defined area.

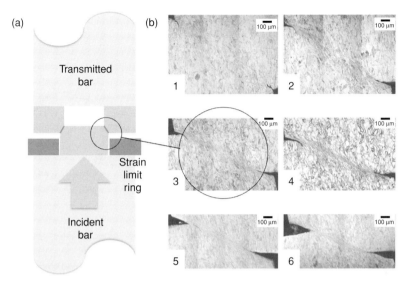

Figure 5.44 (a) Top-hat geometry specimens in compression in the SHPB (after Gray III *et al.*, personal comm.). (b) Postmortem observation of shear band forming in top-hat targets in SS316 loaded in Frame 1 for 25.1 ms; Frame 2 for 31.0 ms; Frame 3 for 36.1 ms; Frame 4 for 41.2 ms; Frame 5 for 51.5 ms; and Frame 6 for 61.8 ms. Source: Xue, Q. and Gray III, G. T. (2006) Development of adiabatic shear bands in annealed 316L stainless steel: part I. Correlation between evolving microstructure and mechanical behaviour, *Metall. Mater. Trans. A.*, **37**(8): 2435–2446. With kind permission from Springer Science and Business Media.

Whereas the geometries described above are used primarily to observe the evolution of single bands using predetermined localisation nuclei, the final geometry described is one of the few that allows nucleation and development of a population of bands. This thick-wall cylinder technique forces the collapse of a cylindrical cavity by the action of explosive flow behind a slow detonation front travelling down a charge located on the external wall of the cylinder (Figure 5.45(a)). The detonation sweeps down the axis with of the cylinder driving the thick walls in to fill the cavity at the centre. The explosives are carefully chosen to ensure there are no asymmetries in the collapse of the cylinder and this ensures that the initial phase is symmetric with homogeneous radial plastic deformation. At some point a population of shear bands is nucleated within the collapsing cylinder that develop over time during the collapse. As flow becomes favoured in particular bands, others become shielded and further deformation is arrested in their growth. The recovered and sectioned samples after the tests typically show spiral shear bands starting at $45°$ about the radial direction which is the direction of maximum shear stress in this geometry (Figure 5.46(b)).

The stress field develops as bands travel through the metal so that at later times (greater radii from the centre) they have adopted directions aligned to the flow in the homogeneous region of the deformation. Of course such a test is a valuable validation for developed models providing a controlled loading in a 2D geometry. However, hydrocode submodels or analytical descriptions have not yet reached the state where they can routinely handle

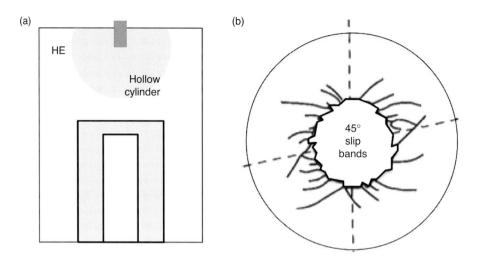

Figure 5.45 (a) Thick cylinder collapse geometry due to Nesterenko and Bondar (1994). Detonation about a hollow tube collapses the cylinder. (b) Nucleation of the bands occurs at the inner interface and the final state achieved in a test is shown schematically in the figure.

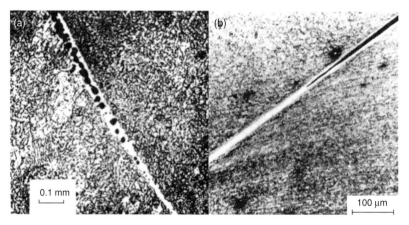

Figure 5.46 (a) Macroscopic ductile voids in a shear band in DU. (b) Brittle fracture down an ASB in steel. Sources: after Raftenberg, M. N. (2001) A shear banding model for penetration calculations, *Int. J. Imp. Engng.*, **25**(2): 123–146 and reprinted with permission from Elsevier; and Walley, S. M. (2007) Shear localization: a historical review, *Metall. Mater. Trans. A*, **38**(11), 2629–2654. With kind permission from Springer Science and Business Media.

these geometries, particularly such a multidimensional, defect-dependent test that has localisation such as occurs here. Nevertheless it affords a valuable link to the real situations found in loaded components in dynamic applications.

5.9.3   Nucleation of adiabatic shear bands

The nucleation phase has different effects upon observed response according to boundaries on which the compression acts to apply the load. There are thus a series of observations that have been made under varying loading modes that describe microstructural effects that tie to operating mechanisms with associated kinetics activated during the operating process.

A series of experiments by Hutchings impacted targets with varying sized spheres. Adjusting the diameter of these projectiles allowed bands of different number density and thickness to be formed. In particular, there was a diameter of projectile (0.4 mm) below which no shear bands at all resulted. This corresponds to a compression period of *c.* 150 ns, although the process continued beyond this time and thus may not be adiabatic. If the bands in dynamic loading are due to heating then clearly there is a nucleation time involved in the process within which the temperature moves from a homogeneous value behind the compression front to localised high-temperature zones within it. Further, it is a coupled mechanical and thermal process. These integrated mechanisms mean that analytic forms and simple dimensional scaling cannot work in such a situation since the timescales of operating processes and those of the loading become near the same ($F = 1$).

In the case of the imploding cylinder experiments, there is evidence that inhomogeneities within the microstructure nucleate bands since populations observed on recovery developed around defects whilst when purity was increased, the cylinders broke into fewer fragments.

One of the key features in determining nucleation times is the ability of the metal to conduct heat away from local regions of intense plastic work into the bulk. In heat transfer analysis, thermal diffusivity is a useful measure that describes this transport (defined to be the ratio of thermal conductivity to volumetric heat capacity). It has the SI unit of m^2 s^{-1} and is given by

$$\alpha = \frac{k}{\rho c_{p,}} \tag{5.9}$$

where k is the thermal conductivity, ρ is the density and c_p is the specific heat capacity.

Materials with high thermal diffusivity are capable of rapidly adjusting their temperature to that of their surroundings since they are able to conduct heat quickly in comparison to their thermal bulk. To maintain a homogeneous temperature during a rapid dynamic process would be easier for materials with high diffusivity values. Thus localisation by adiabatic shear banding is favoured in materials with low values of this quantity. Table 5.2 gives a selection of thermal diffusivities for common metals with selected non-metals included for comparison. However, there are two routes to fail the structure – slip and fracture. Some materials deform in a brittle manner before plastic work and localisation can occur. In the general case shear banding and ductile fracture are favoured in metals with fewer slip systems but ductile response (as seen in Figure 5.42).

Work has been done loading a series of the materials listed in the table and much has been published particularly describing the appearance of the bands, or the fracture surfaces that result. There are several pertinent features worthy of note which indicate some features of the nucleation process. There is evidence of void formation in the bands and these voids appear to be at two different length scales. Microscopic features suggest that a void process preceded band formation. Larger voids (sometimes associated with fracture that has propagated after) suggest that the hot and softened zone has failed in this manner in the later stages when tensile forces have acted. When fast plastic deformation occurs, defects are generated, travel and interact with one another and as seen earlier this

Table 5.2 Thermal diffusivities of metals with some reference materials for comparison

Material	Thermal diffusivity $(\times 10^{-6} \text{ m}^2 \text{ s}^{-1})$
Diamond	1207.2
Cu	111.6
Al	86.4
Al6061-T6	69.0
W	63.0
Fe	22.0
Ta	21.4
Ni	14.9
U	12.1
Aluminium oxide (polycrystalline)	12.0
Carbon steel (1%)	11.7
Zr	8.9
Ti-6Al-4V	2.9
Common brick	0.5
Window glass	0.3
Wood (yellow pine)	0.1

may lead to bulk defects from which further failure can proceed. Once such sites grow and coalesce, they will relieve strain fields around themselves leading to local, quickly evolving mesoscopic deformation with high local flow leading to high temperatures. Indeed, there is suggestion that in some cases the surfaces of material either side of the band have rubbed past one another with further frictional contact after formation.

As the deforming metal within the shear bands gets hotter clearly the metal softens but additionally there is clear evidence in particular metals that thermal thresholds have been crossed where the loaded material may also undergo temperature-induced phase transformation. Such transformed bands have been observed in steels and in titanium, as mentioned earlier; in steels white bands result since a martensitic region remains after unloading.

5.9.4   Analysis

Various attempts to pose the problem in simple terms have been attempted over the years and these have resulted in several analytical approaches. These can be summarised into two groups. The first (due to Grady and Kipp, 1993) follows the development of the bands using a diffusion of momentum argument similar to that assumed in their analysis of Mott fragmentation (see next section). The second due to Wright and Ockendon takes a perturbation approach (see Wright, 2002). Both theories are one-dimensional and describe steady-state conditions. Further work by various authors has extended these to a greater number of dimensions and has included particular material response to match experimental data considered.

The analytical methodology is sketched out here in order to give the reader a feel for the operating mechanisms and the interplay between the properties assumed to be of importance. One key simplification, found true in shear driven by expanding fronts from surfaces, is that the band cuts through the microstructure regardless of grains or boundaries. If this is the case then it may be assumed that band formation is controlled by bulk response and classical conservation laws may then be applied for mass, energy and for linear and angular momentum. Further to that, one needs to impose thermodynamic constraints so that the second law is obeyed. Beyond this, one needs to represent heat flux, generally through Fourier's law. Clearly the development of any theory depends critically upon the constitutive equations that describe the conversion of work into deformation and heating. Here the selection of laws determines the complexity and the applicability of the results obtained. As seen previously there are a number of constitutive descriptions that might be applied to this problem but representation of effects such as work hardening will be critical to the development of instabilities in this case.

Predictions from these works have derived quantities such as band width and spacing which can be used to compare their predictions with experiment. Grady and Kipp assumed that shear localisation would cause unloading of the strain field in adjacent material thus preventing further banding from occurring. Wright and Ockendon gained their prediction from the influences of heat conduction and inertial effects. Interestingly each of these theories predicts spacing within an order of magnitude of the experimental result obtained using the collapsing cylinder technique.

To give some idea of useful quantities taken from solutions, some simple results are summarised. Firstly, the estimated width of a band, δ, can be shown to be

$$\delta = \sqrt{\frac{k\Delta\theta}{\tau\lambda}},$$ (5.10)

where τ is the shear stress, λ is the strain rate, k is the thermal conductivity and $\Delta\theta$ is the temperature rise. Using this result and observations from the literature one can deduce two useful parameters that will clearly vary from material to material but which serve as rules of thumb in considering the phenomenon. The development of a band from the onset of dynamic loading to full development takes of the order of 10 μs. Further, the zone affected by the rapid heating is of the order of 10 μm in width. Of course, since this is a thermal diffusion phenomenon, the width of the band is not well defined but gives an indication of the effected zone dimensions. Further, transformed regions in recovered microstructures are not necessarily representative of the fully deforming band whilst it was active, although these numbers represent useful rules of thumb for the reader. Clearly this treatment is insufficient to describe the process in detail. For instance, higher strain rates are observed to produce narrower bands. Nevertheless it gives a useful indication of the order of the dimension to be expected in the process.

The spacing L between bands can also be estimated from

$$L = 2\pi \left[\frac{m^3 kc}{\lambda^3 a^2 \tau}\right]^{\frac{1}{4}},$$ (5.11)

where c is the specific heat, m is the strain rate sensitivity and a is the thermal sensitivity. These typically take values of a of the order of 10^{-3} and m of the order of 10^{-2} for steel.

This gives a spacing of millimetres at low strain rates, reducing to hundreds of microns at 10^4 s^{-1}. These values, derived from assumptions of developed bands in a continuum, approach the limiting value obtained in ball impact in the Hutching's work described above.

5.9.5   Features of shear bands in metals

The presence of a heated, softened and thus weakened region within the bulk of a deforming solid provides the optimum sites for further failure mechanisms to operate. The shear band is thus frequently observed to contain features within it that have not necessarily evolved at the same time as the plastic flow occurred in the compression phase of the loading. The damage takes a ductile or brittle form depending on the material and its state within the band after plastic deformation has occurred and is observed to take one of two forms: voids or short cracks that appear within the shear band (Figure 5.46). Their presence is an indication that, at the mesoscale, there appears to be an apparent tensile component which acts upon the band nucleating failure sites from which these features can develop; their formation and growth depends upon the incident impulse length (the strain applied).

A further question is posed by the observation that in one class of bands, the feature is not responsive to the details of the local microstructure; they cut across grains regardless of preferred orientations of the components present in the material suggesting that they have a surface origin. Furthermore, they are observed in amorphous metals. Thus, in these cases the development of the band is controlled by an average of the thermo-mechanical properties of the metal rather than the grain level details of the microstructure. In this class of bands, surface defects such as machining marks, scratches, etc. appear to nucleate and the bands follow paths of maximum shear stress that are unrelated to the microstructure. Again this emphasises the structural failure mode that shear bands provide when operating at the macroscopic scale in materials.

Of course for the case of experiments where bulk populations of bands are observed, such as in the cylinder compression tests, there is some effect of microstructure with the suggestion of a population of excited shear trajectories shielding the rest with only a few going on to propagate further. In these cases there is also evidence at the tip of propagating bands that microstructure is having an influence particularly when the band propagation is not overdriven.

The microstructural features of shear banding in polycrystalline materials have been discussed above, with comment made upon the evidence relating to the effects of defect populations upon the resulting response. There is a class of metals where such considerations are less relevant and study of the phenomenon is in some ways simplified. On the other hand, the properties of the metals within which the banding forms are less studied than their polycrystalline cousins, although that situation is changing rapidly at present. The key breakthrough in producing metallic glasses in bulk form came when several families of multicomponent alloys were discovered that contained many different elements so that upon cooling the constituent atoms could not coordinate before mobility was stopped. By this means bulk amorphous metals were produced that exhibited

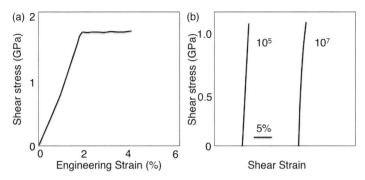

Figure 5.47 Vitreloy-1 (41.2% Zr, 13.8% Ti, 12.5% Cu, 10% Ni and 22.5% Be) loaded (a) in compression and (b) in shear at two different strain rates. Source: reprinted from Dai, L. H. and Bai, Y. L. (2008) Basic mechanical behaviors and mechanics of shear banding in BMGs, *Int. J. Imp. Engng.*, **35**: 704–716 with permission from Elsevier.

high strength and hardness, high fracture toughness and improved wear resistance as a direct result of their disordered structure. Such amorphous metals derived their strength directly from their lack of defects (such as dislocations) that limit the strength of crystalline alloys. However, the lack of significant room-temperature plasticity has limited the application of bulk metallic glasses (BMGs) as engineering materials.

Recent work has started to present a coherent view of their properties. The behaviours may be classified into two groups. The first occur at low stresses and high temperatures and in this regime homogeneous flow occurs across the entire bulk. At high stresses and low temperatures, however, a second mode of deformation is observed in which localisation occurs and the flow is inhomogeneous. This occurs below the glass transition temperature of the metal and here shear bands dominate the response particularly to impulsive loading.

Under quasi-static loading, metallic glasses do show some (admittedly limited) plasticity in compression. However, the behaviour under tension and shear is quasi-brittle. Typical results for one of the first commercial BMGs produced when loaded quasi-statically are shown in Figure 5.47. The behaviour in compression is almost elastic-perfectly plastic whereas that in tension and shear is almost entirely elastic regardless of the strain rate tested at.

Localisation leads to failure by fracture similar in morphology to that observed in silicate glasses, yet the metallic glasses are not brittle in the same manner. One means of classifying ductility is through the fracture toughness (K_{IC}). Vitreloy-1 has a K_{IC} of 80 MPa m$^{1/2}$ that may be regarded as typical for a ductile BMG and is of the same order as that for polycrystalline metals. A polycrystalline ceramic like alumina has a K_{IC} of 3–5 MPa m$^{1/2}$, whereas brittle silica glasses have K_{IC} values of 0.6–0.9 MPa m$^{1/2}$. Thus metallic and silicate glasses have very different natures as they fail.

Although such differences in fracture toughness are found, recovered fracture surfaces do show characteristic morphologies in common with silicate glasses displaying mirror, mist and hackle zones, for instance, which are related to the speed of crack propagation. However, there are also liquid droplets and melted belts which indicate a degree of

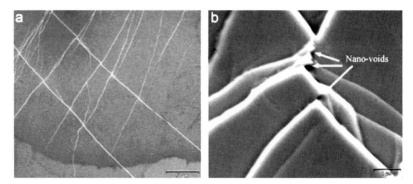

Figure 5.48 (a) Multiple shear bands and (b) nanoscale voids at shear band interaction sites. Source: reprinted from Dai, L. H. and Bai, Y. L. (2008) Basic mechanical behaviors and mechanics of shear banding in BMGs, *Int. J. Imp. Engng.* **35**: 704–716 with permission from Elsevier.

adiabatic heating in the shear fracture zones. Figure 5.48 shows the occurrence of a family of shear bands formed in compression of a BMG. Evidence of plasticity is found in the form of nanoscale voids at the intersection of these trajectories but at a length scale lower than that of the deformation bands at engineering scales. Whereas a population of bands is seen in compression, very few or (with increasing load) only a single band dominates in tension.

The dynamic deformation behaviour of metallic glasses is dominated by adiabatic shear since they represent a class of microstructure where other forms of defect-based plasticity are not present. This limitation in response is both useful in reducing the density of defects and thus increasing strength, but also limits the tensile and shear performance. Whereas this lack of ductility may be regarded as a weakness, combination with other phases in a composite may potentially yield a much better material for future design requirements.

5.9.6   Shear banding in energetic crystals

One of the principal mechanisms responsible for the accidental ignition of energetic materials concerns shear bands that propagate within explosive crystals. If deformation occurs by localisation into a heated band as described in previous sections then such crystals have an extra degree of freedom with which to respond and that is by chemical reaction. In this case localisation of plastic work is said to occur at hot spots and when chemical reaction ensues further heating and potentially catastrophic runaway reaction can occur (Figure 5.49) (see Chapter 8). The key advance made in formulating this picture was that localisation allows the reaction threshold to be overcome in a small volume of material that would not otherwise be reached if the mechanical work done was converted to heating through the bulk. This behaviour has been investigated in many classes of explosive since the pioneering work of Bowden and Yoffe in 1958 yet the concept is still an important research area today (see Bowden and Yoffe, 1985).

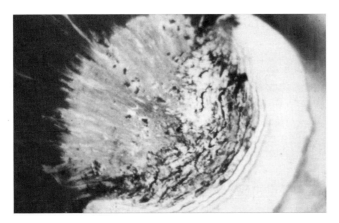

Figure 5.49 Heat-sensitive film discoloured under a disc of PETN loaded under a falling weight. Darkened linear regions are interpreted as shear banding. Some emulsion is removed by flow of the products from violent reaction to the top left. Source: after Field, J. E., Bourne, N. K., Palmer, S. J. P. and Walley, S. M. (1992) Hot-spot ignition mechanisms for explosives and propellants, *Phil. Trans. R. Soc. Lond. A*, **339**: 269–283.

This is because it is a qualitative description of events since the resulting response may range from localised burning to full detonation and this is determined by a range of boundary conditions imposed by other factors some of which can be controlled and some which cannot. These critically include the confinement of the reacting crystal that determines the pressure that builds around the burning site and consequently the rate of reaction that occurs within the explosive. In a deforming composite of many energetic crystals, the temperature field at the mesoscale will be highly inhomogeneous with many hot spots defining a population of reacting sites. However, the conditions at a subset of these will be sufficient to cause runaway reaction and this may propagate outwards to the rest of the explosive. These hot spots are termed critical. In a series of elegant experiments the physical conditions for criticality were determined. Three factors were observed to be of key importance for ignition under ambient conditions. These were the dimension of the heated site, the duration for which the heating was maintained and the temperature that was attained for this period. Clearly the chemical and physical properties of the energetic material itself have a key role to play in this, but over the large range of materials tested a common series of thresholds were attained. The critical parameters for these variables were determined to be:

(i) size between 0.1 and 10 μm;
(ii) time duration between 10 μs and 1 ms;
(iii) temperature greater than 700 K.

In summary, the localisation of uniform deformation leads to hot spots within an energetic composite. Reaction may start at many of these sites throughout the material where one or two of these conditions is satisfied. However, although many such regions will exist, only a few will have all the conditions required to lead to the formation of critical hot spots. It is the identification of the conditions leading to one of these critical ignition sites, in a particular explosive and geometry and under a specified thermal or mechanical load, that drives safety concerns in governments throughout the world concerned with the safe transport and storage of energetic materials. This will be returned to again in Chapter 9 when explosives will be discussed as a material class.

Clearly the flow within a shear band localises heat production within slip planes of micron-dimension. That explosives are observed to initiate at such bands shows that temperatures reach in excess of 700 K in these regions. In metals this results in melting, thermal phase transitions and enhanced slip whilst compressive loading persists.

5.10   Metallic response under dynamic compression

The response of metals at the continuum is always driven by processes occurring within the substructure. Every target recovered and examined contains a rich assemblage of microstructural behaviour at the mesoscale that must be packaged into a form recognisable as measurements taken at an observer's length scale. Ordering of the response of this complex suite of components at each scale requires integration into a lesser set of assemblages each typical of a region. By these means, bulk response can be explained in terms of lower scale structural behaviour. The length scales considered here begin with the atom, but dynamics at the mesoscale has the principal control upon continuum response.

Clearly the structure contains defects: vacancies within the lattice; grain boundary structures; as well as the components themselves from which materials are formed. Metals that have already been heavily worked have a high density of dislocations locked into a structure within grains that reduces slip distances for dislocation travel. In classifying the response of the targets loaded in the experiments described here, one is considering an assemblage of pure annealed metal grains in which defects are concentrated around particular scales; these are at the microscale, in planar defects within lattices, and at the grain (meso-)scale where boundaries form between ordered sub-structures and frequently where voids or second-phase particles exist. Each element possesses a different bond strength which changes as atomic number increases and induces a different packing which results in planes that may be slipped past each other with different degrees of ease. FCC materials have closer packed slip planes and dislocation motion requires less energy than is the case for their BCC counterparts. Equally defects are easier to form since the bonding is weaker, leading to higher densities of flaws within the lattice. HCP metals have fewer systems and slip is thus more difficult. As pressure increases, defect concentration reduces and the response of the metal under load becomes simpler as the microstructure homogenises. The discussion below will assume that grains have five activated independent slip systems within the crystal to simplify the arguments presented.

Plasticity is accomplished by twinning, phase transformation or slip in loaded metals. Twinning and phase change are sub-nanosecond shuffle rearrangements that accommodate a fixed proportion of the applied strain in the impulse. Both processes are fast (nanosecond) and result in the metal quickly attaining a strengthened microstructure with more boundaries to pin slip. At the highest stresses twinning is triggered in many metals, but in the weak shock regime the over-riding effect is for dislocation interactions that rapidly achieve a stable microstructure and stress-state equilibrium under the applied strain. More plasticity may be accommodated by slip. Dislocations travel at the shear wave speed at maximum in these metals. However, the wave speed is a function of pressure, increasing as the target is compressed. Assume for simplicity that its variation with pressure is small in the weak shock regime. Differences in the Peierls stress (the force required to move a dislocation on a slip plane) are reflected in the different substructure of metals. The metals with higher Peierls stresses have greater resistance to defect motion in the lattice and show a radically reduced ability to cross slip and store additional defects. BCC metals possess high Peierls stress, which restricts dislocation motion as the energy for thermal activation is decreased and further fewer nuclei are activated from which slip can commence as loading begins. These considerations mean that BCC metals have fewer operative slip systems than FCC crystals in this regime since there are no close-packed planes in BCC crystals, which means that a greater stress is required to initiate slip. HCP metals also possess fewer operative slip systems. The structure and the shock-loading direction thus determine the population of glissile dislocations and this results in more planar slip.

This review of the processes occurring within a lattice when applied compression has started to densify the metal has indicated three phases to its response. In the first, localisation around defects initiates planes which allow the second process, flow, to occur. This phase continues until terminated by the slip plane interacting with another surface (a boundary or itself) to lock the accommodation of compressive strain and harden the metal (localisation, flow and interaction; see below).

5.10.1   Volume elements for stress-state equilibrium

There is a range of operative scales which map out regimes of behaviour. These are principally defined by the nature of defects that occur within that size range. The microscale defines atomic to grain dimensions, the mesoscale from grain to component size, the structural from component size and up. In each of these scales it is possible to define volume elements that sample sufficient defects that their effect over that volume is the same from element to element.

The concept of a representative volume element (RVE) was introduced many decades ago by Hill (1963) to describe particularly inhomogeneous composite microstructures. There it was defined as a sample of a microstructure that is 'structurally entirely typical of the whole mixture on average' and contained 'a sufficient number of inclusions for the apparent overall moduli to be effectively independent of the surface values of traction and displacement, so long as these values are macroscopically uniform'. Other researchers have modified the RVE concept as, for instance, the 'smallest material volume element of the composite for which the usually spatially constant "overall modulus" macroscopic constitutive representation is a sufficiently accurate model to represent mean constitutive response'. An alternative approach takes the statistical nature inherent in composite microstructures to capture the characteristic of a macroscopic composite structure. In what follows a volume will be defined to include a representative volume of defects that yield a uniaxial strain response in shock across the volume.

A metal volume element, sampled at a scale less than a grain dimension, recovers mechanical properties which average the point and line defects within it. To recover average properties within such a volume element, measured over a volume or on a surface, requires one to sample a lattice with a sufficient population of activated inhomogeneities within it. Let us define the element size so that the statistics of that defect population within is just large enough that the population within it is sufficient to make the summed properties within each element the same. Take this number to be of the order of ten defect separations in each direction to give a homogeneous stress state within after some defined period for equilibration after an element is loaded. This means that one must construct a volume element that contains of the order of a thousand defects and this volume will be referred to as a microstructural unit (MSU) in what follows for these applications. Within each MSU there will be three phases of response to compression that follow one another in time. Densification leads first to *localisation*, then to *flow* once bands are defined and finally to an *interaction* phase when flow must terminate.

The defects referred to here are viewed as vacancies within a metal that provide a surface to expand or to initiate new flaws that aid deformation. The defect population does not show a smooth increase of dimension as length scale increases. There are vacancies of sizes concentrated around the atomic dimension and then another population around the grain size, which is of the order of a thousand times greater. Response at greater scales again requires consideration of structural components that have free surfaces that act to define subunits and again these are of the order of a thousand times greater in size in most materials. One then enters the macroscale and the realm of impact engineering to determine structural response. The defect population is controlled by processing and metallurgy has grown up over the centuries exploiting this ability to design defect distributions into metals at the mesoscale. Since this chapter has considered the response observed in laboratory scale targets, it will be most relevant to consider microstructural units at the mesoscale. The term MSU in what follows will refer then to the mesoscale (unless otherwise stated) and thus it will be typically ten grains in each direction since the primary defects are at grain boundaries which will correspond to a side length of *c*. 100 μm. This will be crossed by a longitudinal wave in *c*. 20 ns and dislocations may be driven to boundaries where they will lock over such times.

In FCC metals where the dislocation density is higher and slip can occur more quickly the plasticity is completed faster as dislocations interact with one another within grains and lock so that microstructure becomes set until release occurs. The defect locking scale is then of the order of 100 nm. However, whilst this gives fast stress equilibration within a grain, there is a further factor since grains are single crystals and there is an assembly of orientations at the grain scale which must be equilibrated to give a stable state. This suggests an MSU of dimension *c*. 10 μm for FCC metals.

In contrast slip is slower in pure BCC metals and the defect density is also lower, so that locking and stress equilibration takes longer to occur. It is the development and settling of the microstructure that defines the stability of the stress state in these metals and this equilibrium that determines the time taken to display the continuum response.

Unless a shock can generate a high density of defects at high amplitude it is unlikely that defects will interact and lock in BCC metals during typical experiments in the weak shock regime. The combination of lower defect concentration and slower dislocation propagation times makes the response in BCC metals a factor of around ten slower than in FCC materials. In the case of BCC tantalum, the microscale MSU is of the order of 10 μm, which means that it is approximately a grain dimension where the next scale of defects is found. Thus for BCC materials the mesoscale MSU is taken as 10 grains (order 100 μm) and this will be taken in what follows to be typical.

Thus the transit across an MSU represents the time required to generate a steady wave behind which the stress state is stable. However, it represents rather more: a dimension that defines an area over which a measurement, e.g. of stress, should be taken to infer a parameter that can be applied meaningfully over the defect length scale over which the measurement takes place. Also it defines a volume element, a target dimension, that should be sampled to determine properties relevant to understanding the response due to a particular impulse.

A typical continuum target is of thickness c. 10 mm. At this dimension one is sampling 100 MSU in a time of c. 2 μs in the case of a FCC metal. The velocity of waves is slower in the case of BCC metals and in such targets dislocations only lock at grain boundaries and then not in all cases. This means that a 10 mm target may consists of 10 MSU in the BCC case. The stress state is locked into such a unit when the microstructure has developed and defined a strength stable within it until waves from boundaries at a higher length scale in the hierarchy arrive to take a larger element to a new stress. The time for stress equilibration is of the order of 20 ns in a pure crystal. In the case of BCC metals, however, it is more like 200 ns. Further, in HCP metals, where the slip systems are reduced but the Peierls stress is relatively low, the stress equilibrium will lie between these two values. Here the MSU is defined by activated defects under the compression pulse.

5.10.2 Weak and strong shock regimes

A shock launches an impulse which drives compression into each unit in the target. As a wave passes through, deformation is triggered as each threshold is exceeded by the driver. This begins microstructural change in each MSU by phase change, twinning or slip, and as the shock becomes stronger more and more defects are seeded into the flow. Focusing on the slip process first, the mechanism completes ever faster as stress increases and dislocations glide and eventually lock to define a stable microstructure. This is represented in Figure 5.50(a), which shows a schematic x–t plot with grey level contours indicating microstructural development as a scalar variable. At the lower stress levels (weak shocks) shown the elastic wave (black) precedes the shock front and the process continues behind the shock until the deformation achieves a stable stress state with microstructure developed (white). However, it may not achieve a stable state since release arrives before the structure has completed the deformation process. This is because either the operating kinetics was not rapid enough (deformation is assumed to cease when the shear stress is removed) or the stress levels were too low.

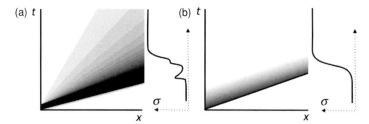

Figure 5.50 A representation of metallic response in (a) the weak and (b) the strong shock regimes. In (a) the elastic wave (black) precedes the shock front (grey). The shading is a representation of the development of substructure behind the disturbance with black the as-received structure and white the fully developed structure to accommodate the applied strain. The processes take longer as the wave front develops in these metals – weak shock regime. (b) The shock front contains the full development of microstructure – strong shock regime.

As the kinetics of the plastic processes gets faster, the plane at which the stress state is stable moves closer to the front until microstructure develops within a defined plastic front, achieving sonic conditions to the rear. The high pressure drives slip faster and faster until ultimately a stress-equilibrated microstructure is achieved behind the shock. As the pressure rises and compression occurs the defects are driven more and more rapidly to the point at which the defect generation and storage processes become limited to within the shock front itself. At this point the defect density that can be created by a shock has saturated and the Weak Shock Limit (WSL) has been reached. This is a single overdriven shock (Figure 5.50(b)) and the stress state behind it is finally stable at this point and the strain rate is thus zero in this region; the wave is finally steady and observations such as the Swegle–Grady relation can be applied. Microstructure recovered from targets loaded in this regime shows no pulse length dependence and the developed structure flows behind the front with no further processes taking place. Once at this value, all plastic work is complete within the shock and the microstructure behind is not developing. This overdrive pressure represents the transition at which the ultimate strength of the plastically worked and defect-locked solid is reached, since beyond this point the processes cannot drive a greater defect density to collapse the structure. However, as compression continues the Hugoniot steepens parallel to the hydrostat until the density reaches the limits at which strength changes nature – the *finis extremis*.

5.10.3   Thermodynamic paths within the shock front

Finally, moving to the continuum scale it is possible to track the thermodynamic paths taken by the metal on the transit to the steady state. Figure 5.51(a) shows the principal Hugoniot in the σ–v plane as a solid curve and Figure 5.51(b) shows strength τ as a function of the longitudinal stress achieved σ.

The locus of states is shown at different times as a series of dashed trajectories: the curves show the moment of impact (t_0) and two intermediate times (t_1, t_2) before the final solid curve at which a stable stress state is achieved (t_3). Each dashed line represents a

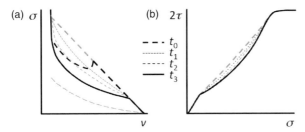

Figure 5.51 (a) Principal Hugoniot of an elastic plastic (solid). The linear dotted line shows an extension of elastic line to the overdrive point ($t = t_0$) – the start of the *strong shock* regime. Dotted curves give loci of states of partial development of the microstructure under plastic work to the stress times t_0, t_1, etc. The lower dashed curve is the hydrodynamic curve. Heavy black dashed curve is the locus of a typical weak shock history. (b) Shear strength as a function of impact stress. Each curve represents different fields in an evolving microstructure developing during compression.

different locus of responses in the development of the microstructure. The solid curve is locus of the stable deformation states after the stress state has equilibrated. The path taken by a stress state in the weak shock regime rises initially along the elastic curve. As time progresses the deformation mechanisms act to deform the crystal and take the structure to new states at later times. The stable stress state curve will always be reached if the driver has acted for long enough time. However, short duration loading will leave the microstructure in a state where a stable stress state has not yet been achieved and a partially developed microstructure is locked in by release, becoming the starting state for any subsequent reload.

At stresses below the overdrive point but above the steady HEL for the material, the state on the impact face rises along the elastic trajectory and then relaxes to a final value through slip, with microstructure developing during the stress reduction. Molecular dynamics simulations suggest this time to be of the order of 100 ps in a metal. Elastic compressions are experienced at volume elements within the bulk at lower peak stresses since relaxation by previous defect interactions has occurred. The impulse starts deformation on the impact face and as the wave sweeps further into the target these unsteady effects are not transmitted and the elastic front reaches a value consistent with the equilibrium volume element and the impulse of the loading received. This phenomenon of HEL overshoot known as elastic precursor decay (as noted previously in this chapter) is seen extensively in BCC metals where stress relaxation is slow compared with that in FCC targets where the process is rapid. It is manifested in the evolving stress state as the time required to gain stress-state stability in the MSU for that structure.

A sample path is shown as a heavy dashed line in the figure for a position away from the impact face of the target. The stress rises to a value consistent with the deformation kinetics operating up to that time. As the microstructure develops, the stress relaxes toward the Rayleigh line and is swept up to the Hugoniot stress thereafter in the shock. The strength, the offset of the loading state and the hydrodynamic response thus develop over time. This is illustrated in Figure 5.51(b). The strength itself takes an initial elastic state (on the impact face) and decays to a final locus over time as shown in the figure. For a range of stress amplitudes in the shock, the strength is observed to increase. However, as seen in the figure, at a certain point the value plateaus at a maximum stress – the theoretical strength.

It is of interest to tie down this critical stress within the material. It is the point at which the stress state behind the shock becomes steady and the deformation processes become confined within the front. This stress is that at which the shock speed becomes that of the elastic wave speed in the material; for copper this value is at c. 40 GPa, in tantalum at c. 70 GPa. Thus the stress state is not steady behind the front until the wave is in the strong shock regime and it is only at stresses greater than these that the microstructure behind the front does not develop and defects are locked. Thus a key threshold in shock loading is the *shock-overdrive stress* or the Weak Shock Limit (WSL), which defines the regime where the approximations of the Hugoniot derivation become precisely applicable. In the regime below this where the shock is weak, the microstructure is still developing behind the shock to some degree and the absolute strength determined depends upon the loading time of the shock impulse used to drive the impact. This is discussed further in Chapter 9.

The shock must sweep a MSU and be supported if a steady equation of state is to be probed from the flow, since the conditions assumed in the Rankine–Hugoniot relations assume a steady wave in the solid. In the strong shock regime the rise of the shock is of width one MSU but since their volume is tiny, defect populations contribute little to the equation of state. Yet the strength measured must depend upon the volume element sampled since the strength determined depends on the volume element sampled and the time for which one waits for the integral measurement to be taken. Thus the measurement of strength depends critically on the methods adopted and the MSU volume sampled.

5.10.4   Ramp wave compression

The situation in which the loading front rises in amplitude more slowly than its head can cross an MSU is worthy of further examination. In this case the pulse loads in a ramp waveform and can be launched either by a variable impedance plate or by a magnetic pulse. In a single shock impulse the MSU is swept by a wave driving plastic work within it, and effects are completed within the shock pulse with consequent effects in the next unit beyond. In a ramped waveform each unit now sees not the full stress jump but some part of it. This means that the work done by the stress increment that crosses the unit is now less than that generated by a shock and thus the temperature generated is consequently less. Therefore a critical condition will exist for a given microstructure which will require that the rise time of the compression pulse must be such that the plastic work generated in compressing it equals the time to dissipate heat from the unit if it is to remain in an isothermal state. A balance between heat generation and dissipation will exist for any pulse that is not a shock and a full range of responses will be evident between the adiabatic conditions in the shock and the isothermal conditions in a critical ramp. Thus there may exist some loading rate at which static and dynamic high pressure will show equivalent results if the processes that define the final state are the same in each case. Further, the static and dynamic strengths for a volume element will show the same features since the defects and boundaries that trap deformation are the same in each case. Only temperature and stress state will change.

Of course the rise of a pulse sampling microscale dimensions within a grain can be very much faster in this region since the micro-MSU is much smaller in dimension than that at the mesoscale. Thus shockless loading at one length scale becomes shock loading at a greater one since the defect sizes are much larger and plastic work done in compressing them lead to heating, which is greater at the increased scale. Moving from lattice to grain scales multiplies a pure metal's MSU by a notional difference in length scale that varies by three orders of magnitude. Thus a laser experiment providing a *shockless* load might rise in 5 ns, a Z pinch experiment in 100 ns and a shock in 5 μs, and each is sampling a state which cannot lie on the equilibrium Hugoniot for a volume element for that length scale.

5.11 The response of metals to dynamic tension

The response of metals under dynamic tensile loading in this interpretation is paradoxically easier to address than that under compression. A means of treating the reaction of inhomogeneities to the loading pulse has been discussed above with a global stress state defined over a volume by three-dimensional wave interactions within transmitting information between local defects activated in the pulse. By the time waves reflect and interact to take a MSU into tension, the compression front has steadied and a global stress state is defined. The front has a defined rise time activating a statistical distribution of flaws at a particular stress level. Again if the stress state is above the WSL then the structure has no volume defects and spall occurs in more homogeneous regimes. It is still the case, however, that in relieving pressure at a spall plane the wave is generally taking a material volume into tension much more slowly than a shock loaded it in compression since release fronts disperse from a reflection surface. In the weak shock regime there will be a series of effects at play.

In the following discussion, the assumed failure process will be ductile cavitation within the metal. Ductile fracture is also observed in some metals, or even a combination of fracture and cavitation but such dual mechanism processes will be considered further later. The flaws within the microstructure respond to a tensile pulse through a two-step process of void nucleation and growth. The nucleation will be easiest at the largest flaws not, as in compression, an average of a population communicating with one another to fix a global state, but this time by the action of defects found in the tails of the flaw size-distribution under load. Once cavitation begins, the effect of wave interactions with competing sites around will result in a particular location being favoured. Figure 5.34 shows (a) the first nucleation sites and (b) the developed void field in spalled tantalum. One can see that the density of nucleation sites is greater than the final number of developed voids. Again the mechanisms for such interactions will be sub-MSU wave interactions and for the same reasons that the MSU was defined to fix global uniaxial compression response, the unit also defines the development of a void in uniaxial tensile loading.

Hence a failure site grows within an MSU and at long enough pulse duration (since nucleation and growth each have kinetics) these sites will coalesce to form a connected surface isolating momentum within a new material component at the laboratory scale. The limiting impulse seen by the metal is the tensile stress in the reflected loading pulse. And the shortest time over which that stress can act is found in the rise of a rarefaction shock and that time will be the transit time of the MSU under load. In general, stresses and times will be less than these values, yet it is the applied impulse that determines the response; the stress and time duration of the pulse.

In the weak shock regime the interplay of the kinetics of nucleation and growth with the partial development of the microstructure will result in a rich range of response to tensile pulses. The fastest (laser driven and Taylor wave) will not extend loading for long enough to grow voids. Equally longer tensile pulses with slower rises may not activate more than a few flaws growing so slowly that the damage reduces and the strength appears greater (Figure 5.39).

In more complex materials a second failure mechanism is also found with different, faster kinetics. Ductile fracture operates at greater speed but has an energy similar to that of void failure in these metals under load. In complex alloys a combination of both mechanisms occurs but again stress relief within a MSU will confine sites to a single location within one in the incipient phase. Further, strength within the surrounding matrix will affect both nucleation thresholds and growth rates and, in the case of an anisotropic material, either through crystal anisotropy or texture, different responses down different directions will be observed in the weak shock regime, as seen in the HCP metals above (e.g. Figure 5.36). The structure of the reload signal (faster rise then slower development) reflects these effects. So the transition from ductile fracture to void grown and the thresholds for slip and twinning are microstructure sensitive but depend critically on the defect populations and their behaviour under load to seed the change required to accommodate the applied strain.

Thus the MSU defines a region required for steadiness that could otherwise be expressed as the minimum unit for a global equilibrium in the metal under compression or tensile loading. Across it, the thermodynamic states for that length scale and the strength of the element are defined. Within it the local state averaging selects a single flaw to develop and fail the material under tension. The Hugoniot conditions apply to states behind fronts where the strain rate has dropped to zero and this requires the pulse length to exceed the length scale of the microstructure to yield a set of parameters that can be used to construct models valid in that region. As the pressure amplitude of the loading pulse increases further defects are generated and the volume of the MSU decreases. At the WSL the material strength is overcome and the MSU volume shrinks to that of the unit cell. The Freya defines the applicability and the regime of use of a particular test, but also the scale over which the model that results from measurements from it applies. These constraints additionally show the regimes where scaling can be applied in the representation of materials physics in simulation.

5.12   Final comments

The understanding of the impulsive loading of metals requires linkage of the continuum scale response recorded in tests to the microstructural observations from recovery experiments. The continuum state histories drive the application of a metal in its operational environment but to understand materials deformation and to construct a physical description of a metal's response requires information on the component-, meso- and microscale processes that occur. The available data indicate that aspects of processes occurring at the microscale, such as phase transformation thresholds, are well understood and can be accurately modelled with advanced quantum mechanical techniques even when the electron states are complex. Processes leading to plastic deformation in the weak shock regime are, however, dependent upon defect populations, their activation and subsequent transport and interaction, and these processes occur over times that map primarily to the mesoscale. The evidence presented shows an evolving microstructure and this must be described by analytical models that also evolve if the response is to be captured at the continuum. Existing descriptions derived using continuum tests and quasi-static loading will function well for the regime that they were fit to. However, these models cannot be adapted to describe transient loading and the next generation of constitutive models must account for the operating physics if they are to be useful in this class of problems.

The application of load to a material or structure proceeds in three stages as the material yields to accommodate the applied strain. In the first moments instabilities are initiated which open transit planes through the microstructure to allow flow once deformation commences. Then slip occurs to allow the movement of the driving piston into the bulk. If the strain has not been accommodated then eventually these slip processes lock against one another or onto a second interface, locking the deformation and providing greater resistance to compression. This phase is commonly described as hardening. Localisation, flow and interaction (LFI) occur within metals at the microscale, mesoscale and (as will be seen later) at the component and at the structural scale too. These classes of response are the top-level processes by which condensed matter (whether that be single crystal or a population of engineering structures) responds to an applied mechanical force and each has a timescale and an energy budget that accompanies the accommodation of the strain. The times and distances that bound the dimensions of this behaviour are the defect length scales and deformation speeds. Once compression at a lower one has been exhausted, higher length scale units with larger-acting mechanisms must be accessed until the driving compression has been arrested. Strength is the bulk response of a volume element at each scale, a function of the processes acting in each unit at the corresponding time, and as both time and scale increase, the strength of the volume elements accessed decreases as each defect scale is crossed. The physical models fit to one length and timescale will not fit to those in a different scale and so analytic simplifications must be better defined to the stress, length and the time range within which they operate if invalid extrapolation or interpolation is not to result.

Thus the range of theatres available to the experimental community must be selected with a mind to the scales that they can access. Fundamental physics operating within the unit cell may be studied using pulses introduced by lasers or DACs probing perfect crystal volume elements at the microscale. This provides data pertinent to the construction of equations of state for pure materials and illuminates the limits of strength within solids. Launchers (Z pinch, gun or HE loading) access the micro- and mesoscale where the grain-scale physics resides that fixes the defect structures that control material strength at the continuum. At the scale beyond that are the devices of impact engineering that inform the behaviour of stuctures for use in the field.

In the weak shock regime, at pressures less than the WSL, the inhomogeneity of nature results in assemblages that deform as units accessing key physics tied to defects within their volumes. It is the appreciation of these statistical physical relationships and their effects upon response that will drive the development of materials for use under extreme conditions and understanding of the structures and their behaviour under load.

5.13   Selected reading

Metals under dynamic loads

Bourne, N. K., Gray III, G. T. and Millett, J. C. F. (2009) On the shock response of cubic metals, *J. Appl. Phys.*, 106, 091301.

Davison, L. and Graham R. A. (1979) Shock compression of solids, *Phys. Rep.*, 55(4), 255–379.

Gray III, G. T. (2000) Classic split-Hopkinson pressure bar testing, in *ASM Handbook, Vol. 8: Mechanical Testing and Evaluation*. Materials Park, OH: ASM International, pp. 462–476.

Energy balance in shock

Grady, D. E. (2010) Structured shock waves and fourth power law, *J. Appl. Phys.*, 107: 013506.

Swegle J. W. and Grady, D. E. (1985) Shock viscosity and the prediction of shock wave rise times, *J. Appl. Phys.*, 58: 692.

Phase transformations

Duvall, G. E. and Graham R. A. (1977) Phase transitions under shock-wave loading, *Rev. Mod. Phys.*, 49(3): 523–580.

Release, spallation and failure

Grady D. E. (2006) *Fragmentation of Rings and Shells: The Legacy of N. F. Mott*. New York: Springer.

Meyers, M. A. and Aimone, C. T. (1983) Dynamic fracture (spalling) of metals, *Prog. Mater. Sci.*: 1–96.

Adiabatic shear

Bai, Y. L. and Dodd, B. (1992) *Adiabatic Shear Localization*. Oxford: Pergamon Press.
Duffy, J, Campbell, J. D. and Hawley, R. H. (1971) On the use of a torsional split Hopkinson bar to study rate effects in 1100–0 aluminum, *J. Appl. Mech.*, 38(1): 83–92.
Wright, T. W. (2002) *The Physics and Mathematics of Adiabatic Shear Bands*. Cambridge: Cambridge University Press.
Xue, Q., Nesterenko, V. F. and Meyers, M. A. (2003) Evaluation of the collapsing thick-walled cylinder technique for shear-band spacing, *Int. J. Impact Engng.*, 28: 257–280.
Zener, C. and Hollomon, J.H . (1944) Effect of strain rate upon plastic flow of steel, *J. Appl. Phys.*, 15: 22–32.

Representative volume elements and microstructural units

Hill, R. (1963) Elastic properties of reinforced solids: some theoretical principles, *J. Mech. Phys. Solids*, 11: 357–372.

6 Brittle materials

6.1 Introduction

To obtain high strength often suggests inorganic and non-metallic materials where hardness provides resistance against high thermal or mechanical loads. In a book concerned with dynamic extremes, these stimuli include materials propelled to impact at high velocity. The design paradigm requires a material to deliver an easily formed structural component capable of resisting impulsive loads of high amplitude and arbitrary duration. At the present time the reality is that these aspirations are only partially met. The response of these materials to idealised one-dimensional loading under shock is not yet understood in full detail and fully three-dimensional loading is only described empirically. Nevertheless the response of glasses and ceramics to dynamic loading has been investigated by the impact community over the past 30 years so that at least a library of data exists. In that time much has been learnt but vital questions remain unresolved, particularly understanding contact, penetration, fragmentation, inelastic behaviour and failure that are encountered in the response of a brittle material to impulsive loading.

Even the qualitative understanding of the response of brittle materials to dynamic loading has not been reflected in advances in constitutive models for them. This results from an incomplete knowledge of operating mechanisms that are consequently not reflected in global models. Further, there is a wide range of microstructures represented in this grouping, ranging from amorphous silicate glasses to polycrystalline ceramics containing both crystalline and amorphous phases. Clearly to construct adequate models for such heterogeneous materials to work on numerical platforms requires a macroscale description of behaviour, yet at present even subscale approaches have not described the processes operating in these heterogeneous media where the phases interact at the mesoscale. It is scale that remains the key frontier that bridges the continuum to microscale behaviour, and it is the mesoscale where the defects that control failure in the bulk are found.

Amorphous glasses have a Hugoniot elastic limit of c. 4 GPa but the polycrystalline materials are characterised by higher elastic limits (6–20 GPa). All by definition show low spall strengths (0.3–0.6 GPa). Typically a reflection of these quantities may be found in the ratio of the hardness to the Young's modulus, H/E, which is c. 10^{-3} for metals, 10^{-2} in alloys but only 10^{-1} in ceramics. One feature of composites with phases formed from constituents with this combination of properties is that pre-existing defects are activated and new damage is introduced, even when shocked to relatively low stresses (about half the Hugoniot elastic limit; HEL). These materials possess few slip systems and localised, often covalent bonding, and this makes plastic flow difficult at the crack tip favouring fracture which implies brittleness for the bulk. The fracture toughness (K_{IC}) is therefore low, typically a tenth or less that in metals. The tensile strength, σ_T, may then be expressed as

$$\sigma_T = \frac{K_{1C}}{\sqrt{\pi a}}, \qquad (6.1)$$

where a is the length of the largest flaw.

Various failure criteria, which have been used in numerical codes, have been constructed, fitted and variously modified in an attempt to account for these behaviours but even so modelling damage and the eventual failure of brittle materials continues to be a grand challenge for the theoretical and computational physics communities, even under quasi-static conditions. This has been due in part to difficulties in studying these complex phenomena under relevant conditions and at the length scales required. The time to failure of bulk volumes in this class of materials is not fast as is the case for metals since their strength is greater and inelastic processes are slower. Thus common present-day platforms have not isolated mechanisms in the same manner as they have been able to for metals.

The brittle targets under consideration here include inorganic and non-metallic materials including oxides such as alumina and zirconia but also carbides, borides, nitrides and silicides. Polycrystalline ceramics have a crystalline and a glassy phase bonded at high temperature and pressure to create composite microstructures. They can thus be ionically or covalently bonded and the crystalline phase has a complex unit cell with limited slip systems or is amorphous so that plasticity can have no part in response. Even in crystals where it is possible, dislocation motion is difficult with only a small population of defects nucleated and high Peierls barriers, so that slip can only occur down a few directions under high loads. Thus in crystalline phases, twinning is favoured as the primary deformation mechanism and of course this can only accommodate limited strain, whereas in amorphous materials the response is generally fracture. Further, in polycrystalline ceramics the component crystals are also generally anisotropic and have high melting points. All of these features mean that fracture is the primary deformation mechanism observed in brittle materials and control of its initiation and propagation is the best means of optimising material resistance to failure and penetration. One means of retaining the benefits of ceramics whilst easing the difficulty of working with them is to use these materials as components of macroscale composites, and these have importance in withstanding high impulsive loads. In addition to high hardness, the binding phase may also possess other properties such as some ductility, large bulk moduli, high melting temperatures, chemical inertness, high thermal conductivity, etc., which makes them highly desirable for industrial applications. This gives this class of materials excellent compressive strength and thermal resistance, but means that they are difficult to work and form which makes construction of components challenging and gives them a significantly higher cost than forming ductile metals.

Quasi-statically and dynamically this results in a range of properties for brittle solids that show clear distinctions between this class of materials and metals. A summary

Table 6.1 Properties of representative brittle materials: density ρ, Young's modulus E, shear modulus μ, Poisson's ratio ν, longitudinal wave speed c_L, shear wave speed c_S and fracture toughness K_{IC}. Each of these materials defines a class and these properties vary between members of that class. Alumina, for instance, has a range of microstructures with a factor of two or more difference in strength between them. Thus these should be taken as typical not as absolute values

	Borosilicate	Soda-lime	DEDF	Gabbro	Al_2O_3
ρ (±0.05 g cm^{-3})	2.23	2.49	5.18	2.88	3.80
E (GPa)	73.1	73.3	52.8	84.8	346
μ (GPa)	30.4	29.8	21.1	35.5	140
Poisson's ratio ν	0.20	0.23	0.25	0.27	0.23
c_L (±0.01 mm μs^{-1})	5.56	5.84	3.49	6.21	10.30
c_S (±0.01 mm μs^{-1})	3.45	3.46	2.02	3.51	6.07
K_{IC} (MPa m$^{1/2}$)	0.8	0.7	–	c. 1	3.4
HEL (±0.5 GPa)	4.0	4.0	4.0	–	7.7
	SiC	**TiB$_2$**	**B$_4$C**	**AlN**	**WC**
ρ (±0.05 g cm^{-3})	3.16	4.48	2.52	3.23	15.0
E (GPa)	422	522	448	315	634
μ (GPa)	181	238	191	127	262
Poisson's ratio ν	0.16	0.09	0.18	0.24	0.21
c_L (±0.01 mm μs^{-1})	11.94	10.91	13.9	10.72	6.89
c_S (±0.01 mm μs^{-1})	7.57	7.31	8.7	6.27	4.18
K_{IC} (MPa m$^{1/2}$)	3.8	5.5	–	–	–
HEL (±0.5 GPa)	13.5	13–17	16.0	9.5	7.0

for a range of glasses, geological materials and polycrystalline ceramics is presented in Table 6.1 with the link to dynamic properties presented through the HEL in the lower row of the table.

6.2   Brittle failure

As seen above, brittle materials are strong in compression. They do not, however, typically achieve their theoretical shear strength. As given earlier this may be approximated as

$$\sigma_{Th} = \frac{\mu}{2\pi}, \tag{6.2}$$

where μ is the shear modulus for the material. For the glasses in Table 6.1 above, the strength one calculates is typically c. 4.5 GPa, which is around 25% greater than the HEL of the material (discussed later). For polycrystalline ceramics, this is typically more like twice or three times the HEL and again, the meaning of the observed elastic limits in these materials will be reviewed below. In all cases, the covalent or ionic bonding in brittle solids gives these materials greater compressive strength than metals.

In crystalline solids loaded in the weak shock regime the passage of a shock front induces the generation and storage of defects behind the front as seen in earlier sections on metals. These may include dislocations, point defects, deformation twins and stacking faults, as well as phase transformation products. However, for materials where the number of slip systems is limited, there is little possibility to store defects in this manner. Indeed in some brittle materials, the structure is amorphous. Thus the principal response to high strain rate loading in brittle materials is failure by fracture.

Cracking can occur in a polycrystalline ceramic in various manners. An initial flaw within the microstructure may concentrate stresses which, unrelieved by plastic flow, will initiate collapse under load to form cracks emanating from that point. Alternatively there may be a pre-existing fracture within the microstructure, for instance at a grain boundary that depending upon its orientation to the applied load may allow the crack a means to propagate from the edges of the pre-existing site further in the direction of the loading applied. Of course there may also be anisotropy introduced into a polycrystalline material by the processing route itself. In such a case, residual stresses will exist at grain boundaries and any flaws or fractures at these points may propagate as load is applied by a dynamic impulse. Thus residual stresses and cracks are a continual limit on performance that must thus be minimised in the processing of polycrystalline ceramic materials as they are synthesised.

In the final stages of loading a population of pre-existing flaws may receive an impulse which causes them to increase their size and potentially join with one another, to fail the material down a population of intersecting fracture surfaces. This might be seen particularly in dynamic tensile (spall) failure as was observed in some metals at low temperature or high strain rate where failure occurred down grain boundaries and cracks were visible within the incipient spall regions (as seen earlier). Populations of flaws loaded in tension fail at the largest of the defects. A similar target loaded in compression, however, fails at a stress that depends on the average flaw size within it.

Several distinct modes of failure are observed in ductile and brittle solids. The first occurs when a crack crosses the microstructure without regard to grain boundaries or other microstructural features. In polycrystalline materials, if the fracture crosses grains, the behaviour is said to be *trans-granular*. On the other hand, the grain boundary region is generally weaker than its bulk. It is generally disordered and contains defects and thus brittle failure is more likely to follow grain boundaries. In such cases, failure is said to be brittle *intergranular*. If there is a possibility that the material may fail at defects within the grains themselves then that failure may propagate from each defect site and join one failed region to the next. Such a population, developed and coalesced, gives a ductile failure surface characteristic of metal spall planes. Such a fractured path is said to be *ductile transgranular*.

Observations of the effect of flaw distributions on the strength of brittle materials were first published in 1921 by Griffith. He noticed that the theoretical strength of bonds within a material was around a hundred times the tensile stress required to fracture the material in experiment. The discrepancy between these strengths was subsequently resolved in terms of flaw distributions within the material under test. He conjectured that the cohesive strength is only reached in the material at the tip of a sharp crack. The propagation of fracture was also analysed by Griffith. His analysis was based on energy balances within a cracked target using a simple approach based on previous work by Inglis in 1913.

Orowan (1934) adapted the concepts of Inglis considering the stress field around an elliptical flaw within a uniaxially strained body under tension. This approach allows calculation of the stress concentrations at the crack tips at the sharp ends of the ellipse, and shows them to be much greater than that on other surfaces. Thus using this argument it was possible to show that the local stress necessary to propagate a fracture was less in a flawed material than it would be in one containing only a few active defects. Griffith arrived at this argument by using an energy approach. Each advance of the fracture front occurs when more elastic energy is released by the crack extension that is needed to create new surfaces. This allowed him to derive a relation for the critical stress necessary for failure.

Griffith solved the problem in two dimensions for a plate stressed biaxially. He considered the stress on a crack oriented at an arbitrary angle to one of the stresses and showed that fracture occurs when the maximum stress reaches the theoretical strength. Over the intervening years the analysis has been extended to account for more complex loading and more realistic fracture geometries. In the early 1960s, McLintock and Walsh (1962) considered a changed loading in compression rather than tension and constructed an analogous analysis. They hypothesised that cracks closed under compression transmitting normal and tangential stresses. This altered the failure conditions in the model. Further, they modified the criterion to consider friction on crack faces. With all of these modifications, they produced a modified Griffith criterion which described the brittle response and delivered stress predictions commensurate with experiment. In particular the modified Griffith criterion described the compressive strain as a function of the confining pressure and this modified criterion has application to the situation found in dynamic loading where the material experiences load under inertial confinement. This will be referred to in later sections; however, these simple concepts and beginnings are the limits on material response that extreme loading explores.

6.3   Crack speeds and fracture propagation

When brittle materials fail in macroscopic compression, it is generally the case that the microstructure fails by fracture at microscopic defects initiated by mesoscale tension. In the case of glasses this is driven by local tensile stress regions at inhomogeneities such as scratches on surfaces or at voids or pre-existing cracks within the bulk. When a fracture opens it does so in response to forces applied during loading. There are three idealised crack loading modes: I, II and III (see Figure 6.1). A mode I fracture opens in response to a tensile stress normal to the plane of the crack. In mode II a shear stress acts parallel to the plane of fracture, perpendicular to the front. A mode III crack is a tearing motion where shear stresses act parallel to its planes.

Each of these modes can propagate fracture at defined speeds that are limited by the velocity at which a disturbance can propagate on a free surface on a material, called the

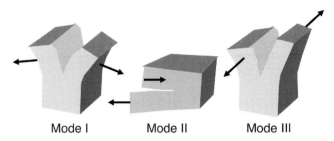

Mode I   Mode II   Mode III

Figure 6.1 Crack opening modes in fracture.

Rayleigh wave speed. As seen earlier, this is close to the shear wave speed in materials (see Section 2.2). However, this work considers materials under compression at stresses beyond yield and as cracks propagate under dynamic loading the energy deposited at the crack tip may result in branching. This bifurcation has been found not to lower the velocity of cracks and its occurrence depends upon the stress intensity at its tip. Clearly in the dynamic case, fracture is driven into the material at such a rate that there are complex interactions between the suites of propagating cracks that affect the bulk properties observed in experiment. Although derived to describe lone cracks, these concepts provide a basis on which to consider the behaviours observed under applied dynamic compression in what follows where macroscopic failure occurs as a population travels together to fail the material.

6.4   Material classes and their structures

6.4.1   Glasses

Glasses heated above their transition temperature attain a viscous state. This means that flow becomes easier in molten glasses allowing them to be worked in a variety of industrial processes. The viscosity of molten glass is measured in poise ($1 \text{ P} = 10^{-1} \text{ Pa s} = 1 \text{ g cm}^{-1} \text{ s}^{-1}$). In the production of glasses they are heated so that they can be worked in the range 10^4 to 10^7 P; on cooling they pass back through their glass transition temperature (T_g) to form the amorphous solid discussed here. The glass transition temperature, T_g, is defined to be that temperature at which the viscosity reaches 10^{17} P, which indicates that the glass transition represents a threshold below which the material may be regarded as stable to subsequent deformation (its state is still not fixed since even at T_g a sheet of window glass left for 10 000 years would deform perceptibly under its own weight). A second important temperature threshold for the behaviour of glasses is the annealing point. At this temperature (10^{13} P) internal stresses within the microstructure can be relaxed in a short time – typically 15 minutes. The final threshold is the strain point (10^{14} P) which is the maximum viscosity from which the glass can cool rapidly to room temperature without fracture. Each of these quantities describes mechanical behaviour as a function of temperature and is key in the production and working of glasses.

Table 6.2 Oxide additives to form four silicate glasses. The highest silica content is found in borosilicate glass (Pyrex) and the proportion decreases through float glass, DEDF (type D Extra-Dense Flint) and LACA

Additive	Pyrex	Soda lime	DEDF	LACA
SiO_2	80.6	72.6	27.3	18.42
Al_2O_3	2.2	1.0		
Na_2O	4.2	13		
K_2O		0.6	1.5	
PbO			71.0	
MgO	0.05	3.94		
CaO	0.1	8.4		19.84
Fe_2O_3	0.05	0.11		
La_2O_3				33.28
As_2O_3			0.1	0.1
BaO				1.01
B_2O_3	12.6			25.34
ZrO_2				2.01
ρ (kg μ^{-3})	2230	2490	5180	3520

Once in a viscous state the glass can be formed by a variety of different methods. Typically these may be by simple pressing, blow-moulding where molten glass is inserted into a mould and distributed in a thin layer across the inside by application of a jet of gas, or simply by rolling. One of the better-known processes, used in the production of window glass, is the float process. A tank holding the molten silicate is fed under gravity on to a molten tin bath. The glass floats on the surface of the molten tin forming a thin, uniform layer. After leaving the tin bath the sheet passes on to rollers on which it cools and solidifies with the viscosity continually increasing as it travels. Driving the rollers allows constant traction on the sheet drawing it from the molten bath and allowing it to be sliced when solidified into sections of suitable dimension to be removed, stacked and stored for future use. The details of this process whilst interesting also have relevance here since they explain the defect structure within the solid which will be referred to later. The drawing process of the viscous liquid entrains air bubbles into the material and these act as nuclei for failure during load.

The constituents added to the glass ultimately affect key properties as well as allowing different forms of processing to be undertaken on the material. All silicate glasses have properties related to their crystalline counterpart, quartz. Pure silica is difficult to form into amorphous material. However, silica crystallites may be fused to make usable components of pure composition. In general the working temperature ranges are altered by the addition of network modifiers to the molten glass which have the effect of lowering the temperature required to produce usable viscosity material for large-scale processing, but also give the resulting solids a range of useful mechanical properties. The range of oxide additives in three types of glass is summarised in Table 6.2. To obtain a working range of viscosity (10^4–10^8 P) in fused silica requires temperatures

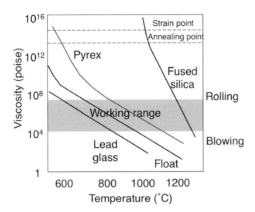

Figure 6.2 The viscosity of glass at various pressures.

between around 1300 and 1500°C (see Figure 6.2). The addition of B_2O_3 to form borosilicate glass (or Pyrex) brings the temperature range down to between 800 and 1200 °C. Common window glass, also known as float or (because of its composition) soda-lime glass, can be worked between 700 and 1000 °C. Addition of lead oxide and other heavy metals forms lead glasses, found in crystal tableware, lowers working temperatures to between round 600 and 800 °C. The range of temperatures quoted in the preceding descriptions maps to various working methods available to the manufacturer. The lower end of temperatures quoted are appropriate if the working of the glasses is restricted to rolling operations, whilst at the highest temperature quoted more complex operations where flow is necessary become possible such as blowing into suitable moulds.

Although altering the processing properties of the glasses produced, the addition of oxides changed the microstructure. Materials with high silica content such as Pyrex have a structure which is open and this largely determines its low density. The network here is bridged by boron atoms from the added B_2O_3. The addition of sodium and potassium oxides to form soda-lime glass partially fills the open microstructure, increasing density and lowering compressibility compared with borosilicate. The alkali metal oxides do not, however, fill the free volume available in the network. With the large amount of lead oxide in a glass such as DEDF the material is highly filled and the density reflects the elements added, increasing to more than twice that of borosilicate. The reduction of compressibility with the addition of interstitial oxides to the microstructure has a consequent effect upon the properties observed under dynamic loading to high pressures. The structure is believed to consist of planar arrays of alternating silica-rich and lead-rich regions giving some order to the microstructure. This has the effect of lowering elastic strength and increased density making it easier to work for fine cut glass and giving it a response more typical of a metal than its more open-structured counterparts.

The previous chapters have emphasised the controlling effect of defect nature and population on the dynamic properties of a material; their size and separation fix the strength and the kinetics of localisation, flow and interaction that it sustains under load.

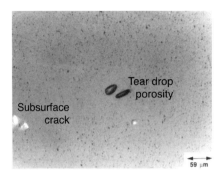

Figure 6.3 Optical image showing the microstructure of soda-lime glass.

Optical examination of commercially produced soda-lime glass shows two classes of defect present within the microstructure (Figure 6.3): pores and microcracks within the bulk. The first of these corresponds to a population of entrained air bubbles in the low-viscosity sheet as it flows across the bath. This results in the solidified material after processing with ellipsoidal pores within the structure. These teardrops are of order 100 μm along the major axis and 50 along the minor, but a target typically contains a normal distribution of such flaws. The volume density of defects across the entire volume taken through the block produced is, however, low. A second class of defects found in the amorphous microstructure consists of planar cracks within the bulk of the material. Such cracks are typically tens of microns across and of similar separation and volume densities as the pores. Both classes of defect are separated by distances of the order of millimetres and thus their small size and low volume density results in no optical degradation of the glass sheet when in place. In fact the reader is unlikely to be aware of these defects in window glass despite many years of looking through the material in question since they lie below the resolution that would perturb the transparency of a typical sheet. Nevertheless they do provide a population of nuclei at which local strains can concentrate and from which failure can propagate through the bulk material under compressive load. Their dimension couples particularly well to the length scales of the shock fronts introduced into the material on impact.

6.4.2   Polycrystalline ceramics

To appreciate the response of polycrystalline ceramics it is useful to view their production and how that results in the microstructures that respond to dynamic loads. These hard materials form with both ionic and covalent bonding. Some ceramics contain metals such alumina (Al_2O_3), titanium diboride (TiB_2), aluminium nitride (AlN) and zirconia (ZrO_2) and are ionic with some covalent nature. Others include the non-metals silicon, boron and carbon, which are essentially covalent, with carbon bonding to silicon or boron. The responses of silicon carbide (SiC) and boron carbide (B_4C) are discussed below but the material of choice for strength and hardness would clearly be diamond.

   The ionic ceramics form crystal structures which are close to those observed within the metals that form them. For instance, alumina, revisited in more detail below, forms a distorted HCP lattice. The covalent ceramics form more complex structures still. Silicon carbide forms a relatively simple structure which results in its properties being essentially isotropic. However, there are several allotropes of SiC which may be present in the polycrystalline materials. Boron carbide on the other hand is more unusual. Its complexity results in a wide range of lattice structures and B_4C may thus not be an accurate reflection of the exact composition of a region. This means that the form of a boron carbide material is not fixed as one moves through the microstructure as local phases with different composition may be found. In both ionic and covalent ceramics, the complex unit cells make slip limited and this results in high hardness but also brittleness. Further, the crystals themselves are anisotropic which has consequences for the behaviour of polycrystalline targets.

Synthesis of a polycrystalline ceramic is an attempt to produce a bulk material with properties as close as possible to those of a single crystal. The usual production route takes a powder and attempts to achieve a result with as near to full density and with as close to 100% of the constituent major phase as is feasible. In reality, this is never achieved, with the resulting material a composite with the major phase dominant but with subspecies and other processing features remaining. At the beginning of the process one starts with a powder which is mixed with a binder or additives to speed the next phases of production. At this time, some means of consolidation must be adopted which presses the composition (die pressing, slip casting, plastic forming or injection moulding). At this point, binder is burnt out to leave a shaped powder form ready for densification. This allows some control in achieving a particular geometrical form to the resulting compact which can then be consolidated to give a near net shape component.

The next phase of forming requires sintering. Snow is compressed into a snow-ball by applying pressure and the same principle is adopted with the shaped powder to form an engineering ceramic. Powder particles are bonded together resulting in shrinkage, densification and grain growth. There are various processes used, including liquid phase and high-pressure sintering as well as reaction bonding. High-pressure sintering produces the highest density materials with properties closest to those of single crystals of the majority phase but at the cost of processing time and complexity.

All sintering processes occur at high temperatures of around two thirds of the melting point (1000–2000 °C) where rapid diffusion occurs and at high pressures in the range 100–200 MPa. At such temperatures, particles bond to form necks which grow and thus densify the powder. Reactive additives frequently form low melting point glasses which remain within the composition in some quantity after processing. Further, such a route inevitably leaves a degree of porosity which hot isostatic pressing (HIPing) can reduce significantly. Other routes, such as liquid phase sintering, can lead to full density but result in decreased strength and high-temperature performance. Another, reaction bonding in the silicon-based ceramics, leads to high porosity. These defects are key to the poor properties found under load. Thus high-performance materials for dynamic applications are generally HIPed as will be seen below.

A schematic of the resulting composite microstructure is shown in Figure 6.4. A material is shown with features within the bulk which interact with a pressure pulse

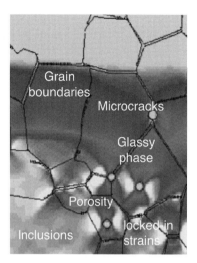

Figure 6.4 Schematic of polycrystalline ceramic microstructure (under shock loading) showing defects at the mesoscale and shades of pressure. The sketch is c. 5 µm across its base. Source: after Bourne, N. K. (2006a) Impact on alumina. I. Response at the mesoscale, *Proc. R. Soc. Lond. A*, **462**(2074): 3061–3080.

travelling up from the base. The ceramic consists of an assemblage of grains bonded to one another. A glassy second phase sits at some grain boundaries which weakens these regions. Further, some of the boundaries themselves are not bonded at all but consist of microcracked regions where failure can nucleate; the sintering process always leaves some degree of porosity. There are also pores at triple points where grains touch as well as within their bulk. The heating, cooling and consolidation processes further leave some residual strain around these features which acts to mediate the effects of the loading pulse when it arrives. Finally, the grains may contain inclusions of different chemical but also physical properties. These will further localise strain in the lattice but also act as points where mechanical forces concentrate on loading. As will be seen by the contours of pressure in the figure, a dynamic pulse induces a fully three-dimensional stress field at the mesoscale in the ceramic.

These features are found in all polycrystalline materials but each different class has its own nature and behaviour. The reader can find texts elsewhere on a multitude of ceramic microstructures, but only one is focused on here to illustrate the observed behaviour. That is alumina, one of the most commonly used materials.

6.4.3   Case study: alumina

Under the names corundum, sapphire and ruby, alumina is found both in nature and across society in various applications. It is also one of the most commonly used polycrystalline ceramics consisting of sapphire crystals bonded with a glass phase and sintered as discussed above. With varying degrees of high pressure applied for longer times, the purity and density can be pushed towards that of sapphire.

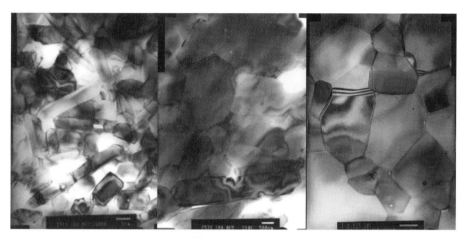

Figure 6.5 Left to right: transmission electron micrograph (TEM) images of alumina ceramics of 88.0, 97.5 and 99.9% alumina by weight, respectively.

Figure 6.5 shows three alumina ceramics of 88.0, 97.5 and 99.9% purity by weight, respectively. The 880 was pressureless-sintered and the 999 was hot, isostatically pressed; the 975 was also HIPed, but to a lesser extent. This process has several consequences including reducing porosity and growth of the grains. The porosity in the microstructures consists of spherical pores within the grains and triangular ones at boundaries (as seen in these polished transmission electron micrograph (TEM) sections). The circular voids have been attributed to spherical inclusions introduced during sintering; the triangular ones to incomplete bonding at the triple points during hot pressing. Some of the flaws visible in the micrographs result from preferential etching during the ion milling process used to form the TEM foils, so they must be taken as qualitative indications. Further, it is known that remnant residual stress fields may be present within the grains after production and that this may drive further damage during loading. All this being said it is clear that such a ceramic offers a complex three-dimensional composite to a shock at the mesoscale regardless of the planarity and precision of the loading at the laboratory scale.

The 880 alumina consists of grains of varying geometry and includes some long and narrow ones that show crude alignment. Further, there is a variation of their size with some large grains evident. In contrast, the micrograph of the 999 alumina shows that the grains are well equiaxed and similarly sized. There is much less porosity, although there is still evidence of intragranular spherical voids and openings at triple points.

As noted earlier the HIPed material possesses a uniform equiaxed grain structure, with the pressureless-sintered material having the greatest grain size variation. All three aluminas contain intragranular spherical voids and inclusions. There is also evidence of dislocation networks within the alumina grains seen in Figure 6.6, particularly within those containing porosity. There may also be free dislocations, indicative of residual strains that might become mobile as the shock passes through. What is clear from these microstructures is that it is the properties of the sapphire crystals which must dominate

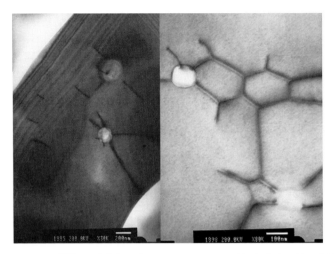

Figure 6.6 TEMs of dislocation structures around included voids within sapphire grains within polycrystalline aluminas 999 and 975.

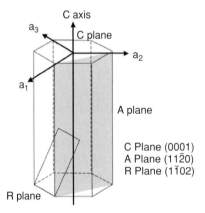

C Plane (0001)
A Plane (11$\bar{2}$0)
R Plane (1$\bar{1}$02)

Figure 6.7 The sapphire unit cell.

the response of the materials as they are loaded. Yet the response to impact must be dominated by fracture as the ceramic undergoes failure to accommodate the applied strains.

Sapphire is a rhombohedral (space group $R\bar{3}c$) crystal. A consequence of this is that fewer slip systems are available than would be the case, for instance, in a hexagonal metal. The unit cell is shown in Figure 6.7. Pyramidal and basal slips allow homogeneous deformation of alumina polycrystals. Further, two other systems will also satisfy the von Mises criterion. However, the critically resolved shear stress necessary to achieve slip on these additional systems is appreciably higher than that needed to slip on the basal plane. This results in sapphire behaving as an anisotropic material, with material properties (including wave speeds) dependent on the relation of the loading and the crystallographic orientations. The elastic properties reflect this asymmetry with a range of moduli and

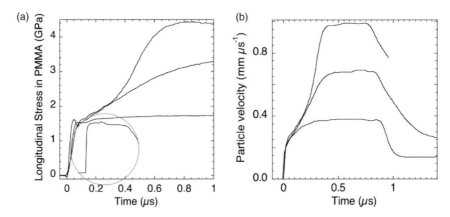

Figure 6.8 (a) Three histories of a DEDF glass shocked over a stress range. The insert shows trace at close to HEL on timescale c. five times those of other traces to show the relaxation of Hugoniot stress in this region. (b) AD995 alumina shocked over a stress range that shows characteristic behaviour of these ceramics in one-dimensional shock loading.

Poisson's ratios according to the loading direction. According to the general form of Hooke's law,

$$\sigma = C\varepsilon, \tag{6.3}$$

where C is a 6×6 elastic constant matrix with elements $\{c_{ij}\}$ in Voigt notation. The single crystal elastic constants for sapphire, c_{ij}, are

$$\{c_{11}, c_{33}, c_{44}, c_{12}, c_{13}, c_{14}\} = \{497, 501, 147, 163, 116, 220\}.$$

Thus, in a polycrystalline alumina, basal slip will always be initiated first, unless other planes are oriented so that the resolved shear stress on the basal plane is small, and this will lead to stress concentrations which will result in intergranular cracking. For this to be avoided, pyramidal slip must be activated by high shock temperatures or high confining pressures. Further, material parameters are temperature dependent even in polycrystalline materials. At the stress levels that are typical of ballistic regimes (where the Hugoniot elastic limit is a significant fraction of the peak stress amplitude), shock temperatures will be low compared with the melting point of sapphire and the shock pressure will be the principal means available to suppress microcracking.

To begin reviewing dynamic properties, as in previous chapters, compression loading will be addressed first with the aim of explaining the experimental observations from a mechanistic viewpoint. One of these is the complex wave shape observed as a step-loading wave at the contact face that disperses in plate impact experiments.

6.5   Shock loading of glasses and polycrystalline ceramics

Figure 6.8(a) shows loading histories for the filled glass DEDF, and Figure 6.8(b) the polycrystalline alumina AD995 at different stress levels. In all cases a step impulse was driven from the impact face by the impact of a projectile. The dispersion of the pulse shows characteristic features of the response of these brittle material classes. The glass histories show shock propagation through the lead glass DEDF, which has lead oxide fill almost filling the amorphous silicate network and resists densification as a compression front passes through it. The instantaneous loading on the impact face thus rises quickly to the elastic limit and then shows a concave rise at higher stresses before reaching the Hugoniot stress in the material. The trace inserted in a bubble shows a further shot on which the temporal scale is much longer. Here the stress rises to just above the HEL of the material and after a microsecond shows a dip on the peak of the pulse. Clearly failure is occurring in the glass after the elastic front has passed.

Figure 6.8(b) shows the response of the armour alumina, AD995. Here the pulse has a very rapid (sub-nanosecond) elastic rise and then an S-shaped form at the higher stresses at it rises to the peak shock stress. Figure 6.8(a) and (b) are on similar temporal scales showing that the rise of the elastic pulse in the alumina in (b) is faster than the rise in the glass although the form of the shock after the elastic rise is similar in the two materials. In the loading of glass at just above 4 GPa the pulse reaches a flat peak stress level but then stress relaxation from an initial value to a lower one over half a microsecond or so occurs.

Three wave profiles for the 99.5% of full density alumina, AD995, are shown in Figure 6.8(b). The histories show typical form for such materials with a rapid rise to a first elastic limit, then a convex region to a point of inflexion, followed by a concave section rising (at the highest stress amplitudes) to the peak of the shock. When shocked up to *c*. 1.5 times the HEL, the pulse shows a rapid rise to the first Hugoniot elastic limit, followed by a convex ramping region until the Hugoniot stress is reached. At higher stresses, the pulse shows a rounded, convex form, which then rises more sharply to the peak stress as did the glass. The mechanisms controlling the processes defining the elastic limit are discussed below and it will be shown that it is the composite behaviour of the ceramic that leads to two values bounding the elastic threshold. In the past there has been confusion as to which point to take for the elastic limit; in some work the HEL is defined to lie between the first break in slope of the fast elastic rise and the peak stress achieved, whilst in others it is described as the break at the top of the first elastic rise. Certainly the latter threshold shows features seen in metals such as elastic precursor decay discussed in the last chapter. There are several other observations of the compressive behaviour of ceramics that have not been fully explained including issues of grain size and porosity dependence of the elastic limit (the HEL was observed to decrease with increasing grain size, while the 'plastic' wave rise time increased with increasing grain size) and the answers to these puzzles are best explained by studying behaviour in the mesoscale flow.

The spall strength in glasses reaches zero once the elastic limit has been reached and failure has propagated to the station where the measurement is taking place. In the composite ceramic it can actually increase as porosity (and thus defects) declines and the proportion of alumina increases. However, this was observed to occur only over a range of stresses below half the elastic limit of the material and above that the ceramic had none. In all cases the mechanisms responsible for the composite behaviour may be collected to explain these effects; the ceramic contains ductile grains bonded with a brittle interlayer. Thus there are features of the response that depend upon the grain structure within the sample including their size and conformation and under loading the plasticity within them. Secondly, there are strong effects due to the closure of porosity and fracture within the glassy binder phase.

Glasses illustrate the response of a material where failure is purely by fracture; ceramics marry limited plasticity with a binder that cracks to allow the material to accommodate strain. In the following sections, these observations and their explanation will be highlighted first for amorphous glasses and then for polycrystalline ceramics since the behaviour of the first sets the backdrop for describing deformation in the second. As before 1D shock loading will be used as the test bed to describe the operating mechanisms excited as the impact drives into these brittle materials before applying the materials response to explain their performance in other environments.

6.6 Dynamic compression of glasses

The simplest brittle behaviour is found in amorphous glasses so discussion of these is a prerequisite for considering the more complex behaviour of composite polycrystalline ceramics. As previously, there will be no attempt to exhaustively review the response of a large range of materials; rather a few will be focused on to represent the behaviour of the whole class.

The behaviour of amorphous glasses is characteristically driven by compression of their network structures. The free volume within the silicate network contained, and the failure that occurs therein, determines the behaviour observed under load. The various different glasses differ primarily in the nature of the added oxide, and the degree to which the network is filled as seen above. Three glasses will be followed here: borosilicate (better known by its trade name Pyrex), soda-lime (float) glass and DEDF, a lead-oxide-filled material. The first has an open structure, the second a partially filled one and the last is near fully filled. They will be referred to as BS, SL and DEDF in what follows.

Table 6.3 shows typical quasi-static and elastic properties for the glasses and the most striking difference between them is that density varies from 2.2 to 5.3 – nearly a factor of 2.5. As one would expect, this has a dominating effect upon the equation of state. The elastic impedance of the three materials follows the same trend: $Z_0 = 18.1$ for DEDF, 14.5 for soda-lime and 12.5 for borosilicate. Thus the materials order themselves with the Hugoniot of DEDF lying above that of SL and BS the lowest of the three.

6.6.1 Equation of state response of glasses

Figure 6.9 shows typical stress histories taken at some distance into tiles of the three glasses: (a) DEDF; (b) soda-lime; and (c) borosilicate. The rise of the signal is rapid to the HEL in DEDF; so rapid that the elastic front can be seen bouncing in the epoxy layer around the gauge (Figure 6.9(a). In Figure 6.9(b) the histories in SL glass show a different

Table 6.3 Properties of glasses discussed in the text

	Borosilicate (BS)	Soda-lime (SL)	DEDF
ρ (± 0.05 g cm^{-3})	2.23	2.49	5.18
E (GPa)	73.1	73.3	52.8
μ (GPa)	30.4	29.8	21.1
Poisson's ratio, ν	0.20	0.23	0.25
c_L (± 0.01 mm μs^{-1})	5.56	5.84	3.49
c_S (± 0.01 mm μs^{-1})	3.45	3.46	2.02
K_{1C} (MPa m$^{1/2}$)	0.8	0.7	–
HEL (± 0.5 GPa)	4.0	4.0	4.0

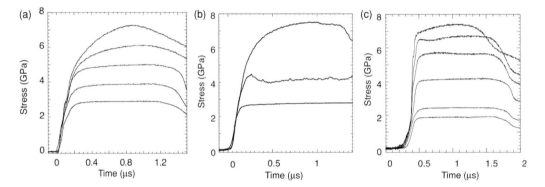

Figure 6.9 (a) DEDF embedded gauges at 12 mm from impact. (b) SL glass with embedded gauges at 10 mm. Note relaxation at c. 4 GPa. (c) BS glass, embedded gauges at 5 mm. Relaxation occurs on higher parts of pulse.

behaviour. The rise of the pulse is slower but the upper trace shows the same convex rounded form. The central trace, just above 4 GPa, shows a characteristic drop from its initial value by 4% of its peak value at this stress. The open structured borosilicate glass (Figure 6.9(c)) shows two regions: a ramped precursor and a more rapid rise thereafter. This slow-rising section is characteristic of materials that are of open structure and are compressing to a denser form. However, after some time the network has compressed and the stress starts to rise again rapidly. Note again at an intermediate value above the HEL there is a dip in stress at the peak of the pulse which does not occur at higher stresses.

These histories are characteristic of the behaviours observed for open and filled glasses. This dip on the longitudinal stress coincides with failure in the target and is connected with other observations of delayed failure discussed later in the chapter. However, the locus of peak stresses recorded in such experiments also informs behaviour and some features of the Hugoniot are shown in Figure 6.10(a).

Figure 6.10 presents the principal Hugoniots for the three glasses. The dotted lines represent loci of elastic responses in the three materials. The points were taken at target thickness of 10 mm in each case and represent values of the stress where the

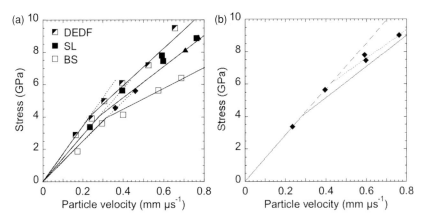

Figure 6.10 (a) Hugoniots for the three glasses. (b) Hugoniot for soda-lime glass taken at different positions into the target. Error on each point *c*. 2%.

wave has settled to a steady peak. The Hugoniot of the higher density DEDF glass lies above that of SL which in its turn lies above borosilicate as expected given their densities and impedances. In Figure 6.10(b) the same data are presented along with other points derived in the same manner from experiments conducted at different distances from the impact face of soda-lime glass. Now the Hugoniot is seen to evolve with thickness of tile (and hence over time) from a purely elastic response on the impact face. The large dashed extension of the elastic line extends upwards and includes a state determined in thin (1 mm thick) tiles of the target. This non-steady behaviour in the response of these materials is driven by increases in stress and defect generation behind the wave front and so the greater compressions converge towards the steady-state Hugoniot as stress increases, eventually lying on the same states above *c*. 10 GPa.

Thus the Hugoniot for glass evolves with time at points on the impact face and thus also as a function of distance from it as a wave runs into the target. Further, the elastic limit (the HEL) decreases markedly with distance and the shear stress is relaxed by failure. Both of these features indicate a slow kinetics in the failure mechanism from an initial elastic state on the impact face through the development of inelastic deformation in the solid as one moves through the tile. This is equivalent to that seen in metals but on very different timescales. In the last chapter the relaxation from the elastic state to the Hugoniot on the impact face was estimated to be *c*. 300 ps from molecular dynamics simulations of the behaviour. In this regime defects were tens of nanometres apart and communicated in grains of single crystal density at elevated wave speeds so that sub-nanosecond timescales were not unreasonable for stress-state equilibration. In the case of glasses the defects are millimetres apart and communication is at some fraction of the shear wave speed in this region of stress space. Consequently failure times are of the order of a fraction of a microsecond at this length scale. It is this transposition of deformation kinetics by three orders of magnitude from one class of materials to the next that differentiates metallic from brittle behaviour.

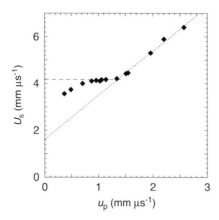

Figure 6.11 Plate impact data on a Russian glass similar to soda lime. The dashed line represents the bulk sound-speed in the intact glass. From Dremin, A.N. and Adadurov, G.A. (1964) Behaviour of a glass at dynamic loading, *Fiz. Tverd. Tela*, **6**(6): 1757–1764.

Further, these observations make such considerations vital in considering brittle response whilst being a negligible complication for ductile metals.

Figure 6.11 shows the results of other experiments on a glass similar in composition and properties to SL in which the relation between the shock velocity U_s and the particle velocity u_p was measured. The behaviour is linear at values of $\{U_s, u_p\}$ above $\{4,1.25\}$ which corresponds to a pressure of c. 12 GPa in the glass target. Up to that level the curve follows a path which asymptotes towards the bulk sound speed of the intact glass, c_0, calculated as

$$c_0 = \sqrt{c_L^2 - \frac{4}{3}c_S^2},$$ (6.4)

from the measured values of the *intact* longitudinal and shear wave speeds, c_L and c_S. At stresses accessing states above this region (where the stress state is unsteady) the material follows linear behaviour as expected.

In the experiments conducted to produce the gauge records shown above, a metal flyer plate was fired at a stationary glass target with a sensor embedded between glass blocks. However, a similar experiment could be done where a glass flying plate impacted a gauged metal target with sensors embedded and if this were done at the same impact velocity as in the first case one would expect to record the same stress levels in each target. Results from such an experiment are presented in Figure 6.12. In (a) the two gauge histories are displayed for such an experiment and clearly the stresses recorded in the two plates are different from one another with the stress in the glass target less than that in the metal. Figure 6.12(b) shows the steady-state stress levels plotted in the Hugoniot plane. Up to 4 GPa the stress levels follow each other but above this level indicate two states with the stress in the copper lying above that in the glass. The copper on glass curve follows the Hugoniot curve discussed above, but the stress in the copper takes a value determined by the elastic impedance of the glass. This behaviour does not continue indefinitely. At around 10 GPa the stress in the metal drops onto the

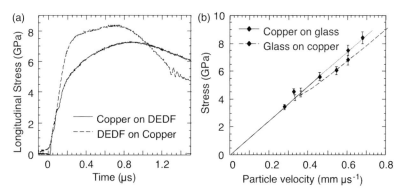

Figure 6.12 Converse loading experiments on the filled glass DEDF. In the first a glass target was impacted by a copper plate; in the second a copper target was impacted by a glass flyer at the same impact velocity. The stresses are not the same in glass and metal plate as the failure process does not arrive at the sensor location until a later time. Source: reprinted with permission from Millett, J. C. F., Bourne, N. K. and Rosenberg, Z. (1998) Observations of the Hugoniot curves for glasses as measured by embedded stress gauges, *J. Appl. Phys.*, **84**(2): 739–741. Copyright 1998, American Institute of Physics.

Hugoniot curve. There is thus a region of behaviour between *c*. 4 and 10 GPa where the material shows a dual behaviour which is the result of delayed failure on these timescales.

In equation of state experiments like those conducted above it is always a hope to be able to measure the HEL of the material as well as the peak stress recorded at different compressions. However, in the open structured materials such as soda-lime and borosilicate glass the elastic limit is frequently obscured by features that result from compaction of the microstructure to a fully dense form. This will be commented upon later for the case of porous compacts where this behaviour is especially prevalent. However, as seen above in DEDF glass, the silicate network has been almost fully filled so that it is possible to see clear breaks in slope between elastic and inelastic compression in the material. One can identify a similar feature for float glass where despite the ramping shock front a slight break in slope is noticed at *c*. 4 GPa. Histories recorded in open-network glasses showed an elastic-inelastic transition in behaviour equivalent to that observed in a filled material. Then this threshold is the failure stress of the silicate network in these materials and represents a compressive elastic limit for all glasses.

Figure 6.13 shows stress histories taken for tiles of different thickness in DEDF and soda-lime glass. The behaviour of the filled and the partially filled glass noted above is clearly seen in the two groups of histories. Those in the filled glass are more easily interpreted since there is a rapid initial elastic rise followed by a break to a convex ascent to the peak Hugoniot stress. The stress at the break between the elastic and inelastic fronts is seen to decrease with distance as indicated by the dotted curve. This decay of the elastic precursor occurs over a distance of 16 mm and asymptotes to a final stress level which can be converted back to indicate an elastic limit of *c*. 4 GPa.

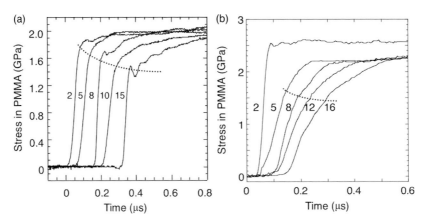

Figure 6.13 Back-surface stress histories in (a) DEDF and (b) soda-lime glass shocked to *c*. 6 GPa at different tile thicknesses. In both cases the impact conditions were similar. Source: reprinted with permission from Bourne, N. K. and Millett, J. C. F. (2001) Decay of the elastic precursor in a filled glass, *J. Appl. Phys.*, **89**(10): 5368–5371. Copyright 2001, American Institute of Physics.

In Figure 6.13(b) a similar suite of histories for soda-lime glass show the dominant effect of the open microstructure on the response of the glass. The compression of the microstructure ramps the shock with high stress levels travelling slower than lower ones in this regime. There is a break on the shock rise visible on the more dispersed histories which corresponds to the equivalent transition from elastic to inelastic deformation seen in the DEDF and is also at *c*. 4 GPa. The trace recorded at 2 mm does not show a clear break and its peak stress is higher than those at greater target depths. This corresponds to an elastic response from the glass as noted earlier for the Hugoniot and for embedded gauges which show a relaxation at close to this level.

As with all materials, glasses exhibit precursor decay in their response, particularly so since the failure time for these materials takes so long in comparison with ductile metals. Thus these observations indicate that there is a transition from elastic to inelastic deformation in glass at *c*. 4 GPa and that the process responsible takes time to establish itself, which can be estimated by noting that the distances travelled imply delays of several microseconds.

The process defining the end of elastic behaviour in an amorphous glass is fracture and it is this that defines the limit in the experiments conducted above. The theoretical shear strength of glass in these experiments has been calculated above (from Eq. (6.2)) at around 4.5 GPa, close to these measured values. The interesting feature of this process is that fracture is a volume additive process and proceeds at a limiting velocity from discrete initiation points. In the case of slip in metal plasticity the process is initiated from inhomogeneities at the microscale such as vacancies, whereas in the case of fracture in these materials the sites are on length scales of mesoscale dimension on the surface of the glass. All the features of the times and dimensions are scaled by 10^3, making glasses perfect to illustrate inelastic behaviour in a manner that may be read over to crystalline and polymer materials.

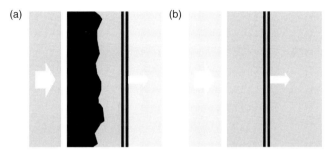

Figure 6.14 Failure waves in glass: (a) asymmetic impact (flyer high Z and target glass); (b) reverse impact to (a) (flyer glass and target high Z).

On exceeding the strength of the material, the inelastic process by which glass responds is fracture. Once a crack is initiated at a flaw on the impact surface, the tip travels at a speed determined by the stress level that accelerates to the Rayleigh wave speed in the material in the limiting case with the shear wave speed increasing with pressure. At the highest stresses this asymptotes to the shear wave speed in the intact glass. At the point at which the inelastic fracture front is travelling with the shock (which the reader will see is c. 4.2 mm μs^{-1} in Figure 6.11) and the wave is steady, the value of the particle velocity can be determined and the stress calculated. For this condition $U_s = c_s + u_p$, at which point

$$\sigma_x = \rho_0 c_L u_p, \tag{6.5}$$

which occurs at c. 12 GPa in these glasses.

Of course glasses are also to some extent open-structured and the pressure behind the front is driving densification as well as fracture. Further, cracking, unlike plasticity, is a volume additive process in itself so that the microstructure itself is compressing rapidly as well.

Other workers have found seemingly anomalous behaviour at around this threshold. For instance, Gibbons and Ahrens (1971) recovered shocked glass and measured its refractive index. They reported change beginning above 4 GPa with densification complete by 12 GPa.

The nature of the failed glass also has consequences for its behaviour in the region defined by stresses above the fracture limit and below that at which the failure occurs at the front itself. Clearly a fractured mass has no tensile strength, unlike metals. This means that release entering a tile will disperse material rapidly. Two opposing scenarios are seen when materials of different impedance are impacted in this region and these are sketched in Figure 6.14. If a higher impedance metal impacts glass in this stress regime, release takes failed material to a lower stress and higher particle velocity, disconnecting the target from the impactor. When a glass flyer impacts a high-impedance anvil, there is no such process and the glass remains compressed on the target until release can arrive. (Further indications of this behaviour were seen in Figure 6.12(a).) The two upper embedded histories at c. 6 and 7 GPa do not show the flat peak of lower stress

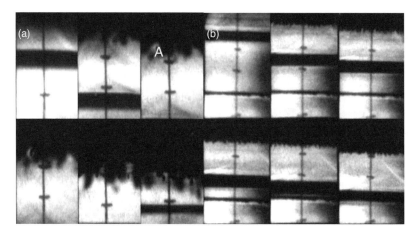

Figure 6.15 High-speed photography of (a) soda-lime and (b) borosilicate glass shocked from above in each case by impactor out of shot. The shock strength is lower than the Weak Shock Limit (WSL). Frames run left to right, top to bottom. (a) 0, 1.4, 2.0, 2.2, 2.8, 3.2 µs; (b) 0, 1.2, 1.4, 1.8, 2.2, 2.6 µs after impact. Source: reprinted with permission from Bourne, N. K., Rosenberg, Z. and Field, J. E. (1995) High-speed photography of compressive failure waves in glasses, *J. Appl. Phys.*, **78**(6): 3736–3739. Copyright 1995 American Institute of Physics.

elastic shocks. Instead the release from the failed material is catching the shock and the stress drops slowly and without the expected rapid wave arrival.

Thus a regime exists where two targets can support different stresses for a time dependent upon the speed at which the failure process can travel. And although this time is of the order of microseconds in the case of glass, and can thus be measured with ease, the same features must take place in all shock processes at the length scales of the defects contained. Thus in metals, where typical distances (in a pure metal at least) might be 100 nm, that time will be of the order of 100 ps. Thus brittle materials offer a window into the evolution of the unsteady stress state in the localisation phase of stress state as it transits to a steady value. In order to better understand these behaviours it is instructive to image operating processes in brittle materials using high-speed photography. These may be extended to opaque materials in terms of the lessons and behaviours observed in the stages of flow development that ensue.

The failure process described above can now be imaged in the glasses under shock. Figure 6.15(a) shows a copper flyer (out of picture) impacting soda-lime, and Figure 6.15(b) onto BS glass where failure is initiated at the impact face. In both cases a shock is driven from the top of the picture down the blocks and visualised from the side. The morphology of the fracture fronts is different in each case. In the soda-lime glass to the left, the shock proceeds down the block as the dark, flat bar passing downward. The fracture front follows behind with discrete cracks visible, some coming out of the focal plane of the camera and appearing somewhat blurred. Individual inhomogeneities crossed by the shock (one labelled A in the figure) start failure themselves as it interacts with them. These fronts were first detected by wave reflections by Kanel when they were known as *fracture waves*. As the inelastic flow mechanism in all brittle materials,

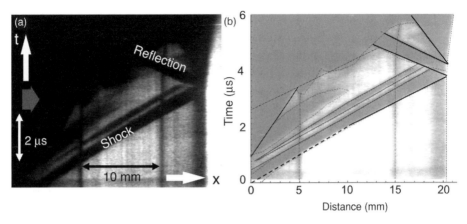

Figure 6.16 Streak image down the impact axis for a shock in soda-lime glass at a stress below the WSL. (a) The image shows vertical lines (markers on the target) 10 mm apart for scale. (b) Superposed time and distance axes.

however, they are better dubbed *failure waves* since similar fronts observed in the relaxation zone in ceramics have some ductile component.

The image in Figure 6.15(b) shows the same experiment in borosilicate glass. This time there is a sheet of PMMA and an interface visible in the picture. The morphology of the failure front is very different to that in the soda-lime case. The fractures are much smaller in dimension leading to a less diffuse failure front. There is again an indication that the fracture paths are at some angle to the shock travel direction.

To assess the failure quantitatively, it is instructive to image an axis in the target using streak photography. Figure 6.16 shows such a sequence for soda-lime glass shocked to a stress of *c.* 5 GPa; above the 4 GPa threshold for inelastic behaviour but below the overdrive condition discussed earlier. In Figure 6.16(a) the image recorded is displayed directly whilst in Figure 6.16(b) scale and measurement markers are superposed in order to clarify the operating waves. The shock progresses across the target through the sequence, reflecting after 3.5 μs or so and accelerating the rear surface. A fracture front propagates back at this time. Two vertical scale lines ruled onto the target fix distance, but can be seen to be refracted and deflected by the moving and compressed glass behind both shock and release.

The front travels at a velocity which settles to the faster elastic wave speed but is initially some 20% slower than this. Further, the shock width broadens with travel distance. The schlieren imaging displays the differential of the compression and so the development of a second parallel dark region behind the shock front denotes the relaxation in stress noted earlier with compression in the front. The fracture itself follows at a slower speed (around 2 mm μs^{-1}) which even correcting for the particle velocity is only *c.* a half of the shear wave speed. It is worth noting that there is a delay between shock breakout and the fracture leaving the impact face. This incubation time is a further complication that accompanies analysis coupled with compression of a partially filled structure in soda-lime glass.

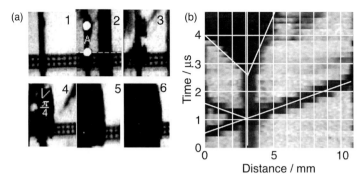

Figure 6.17 High-speed photographic sequence showing the impact of a 10 mm thick copper coverplate travelling at 503 m s^{-1} onto a soda-lime glass laminate. (a) Framing sequence with 600 ns between frames; (b) pseudo-streak down marked dashed axis. Source: from Bourne, N. K. and Millett, J. C. F. (2000) Shock-induced interfacial failure in glass laminates, *Proc. R. Soc. Lond. A.*, **456**: 2673–2688.

The delay results from the development of tensile driving stresses at inhomogeneities from the interaction of the metal flyer with the glass surface. Here local peaks and troughs will interact with the driving plate to produce local tensile strains which drive the cracks into the target. Differences in impedance and in material wave speeds will accentuate these effects and act to drive more fracture since surface strain states will be fully three-dimensional. This may be minimised by matching material properties at the impact face, with a symmetrical impact being the limiting best case. Symmetrical impacts on ceramics stopped fracture fronts from propagating into the bulk. Nevertheless, imperfections on the surface interacting with the shock will give rise to failure. An illustration of this is shown in Figure 6.17. In this case a glass block has two grooves scratched across it with a diamond tool. The surface is then made an internal interface by bonding a second soda-lime block to the first.

Figure 6.17(a) shows the interface in frame 1 to the left of the picture. The grid is a mm scale added to make quantitative measurements from the sequence. In frame 2 a shock has crossed the interface and it is seen moving to the right in that frame. The locations of the two scores are shown on the interface as two white disc markers. From frame 3 onwards, a wedge of black, fractured material can be seen propagating out from the two source sites. The fracture cone makes an angle of 45° to the impact axis as can be seen in frame 4. By frames 5 and 6 the two damage sites have coalesced and a single black front propagates outwards behind the shock.

Along the lower of the cracks in frame 2 is marked a horizontal, dotted white line. This corresponds to a one-dimensional line of sight which is imaged as a streak sequence in Figure 6.17(b). The shock can be seen entering the target and crossing the interface unaffected through the sequence. A wave travels back from the interface itself reflecting imperfections in contact at this location. In the upper region the damage can be seen propagating away from the damaged site introduced into the tile. However, the crack does not run immediately but delays for *c.* 2 µs before propagation begins. This reflects the necessary build up of driving strains at the crack tip required to start propagation

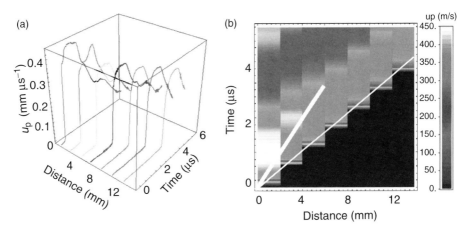

Figure 6.18 Impact onto the lead glass DEDF. A copper flyer has impacted a block as shown in the previous photograph and each element of the particle velocity has been tracked through the target at 2 mm intervals from the impact face. (a) Particle velocity histories at each station. (b) Lagrangian X–t diagram for the impact with the velocity magnitude indicated as grayscale.

into the tile. Clearly if the stresses were increased to beyond those required to drive fracture at the shock velocity, then failure would be occur as the wave swept through the target. This delay between the arrival of a pulse to the fracture of the target allows the glass to support stresses for some time before the failure of the bulk. Although this is seen and measured at laboratory scales and times in glasses, similar failure kinetics operates in metals but at scales and times three orders of magnitude smaller so that at present this cannot be resolved.

A further variable of interest to track as the failure process occurs is the particle velocity behind the shock u_p. This may be measured at a series of stations by embedding a series of gauge loops into a glass tile, applying a magnetic field perpendicular to the impact plane and monitoring the induced EMF as described previously. A typical response from the lead glass DEDF is shown in Figure 6.18. The evolution of the pulse is clear to see. A higher particle velocity component of the pulse is seen riding atop the main part of the wave which has dispersed by the third gauge element, caught by the release. The result of the pulse continues onward through the tile and is reminiscent of the stress traces of Figure 6.8(a) that show similar form. The particle velocity level recorded at the first two sensors corresponds to the elastic response of the glass until the failure front penetrates from the impact face, relieving stress and accelerating the fractured material as discussed above. There is a break on the elastic rise which can be seen to be decaying at each sensor until asymptoting at c. 0.2 mm μs^{-1}, which is the c. 4 GPa threshold for elastic response discussed above. In Figure 6.18(b) a Lagrangian X–t diagram for the impact is shown. The shock can clearly be seen crossing the target with the release to lower levels catching from the rear. On the first two gauge locations the lighter higher level can be seen travelling slower than the shock. It corresponds to the fracture wave discussed above taking the DEDF to higher particle velocity as shown in the Hugoniots and discussed in relation to the asymmetric impact of glass on metal.

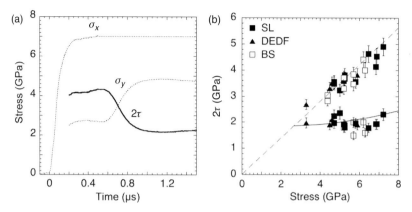

Figure 6.19 (a) Components of the stress field measured at a station 3 mm from the impact face in soda lime glass below the WSL. (b) Strength data in the initial elastic and the final inelastic response for the three glasses followed through this section. Source: after Bourne, N. K., Millett, J. C. F. and Field, J. E. (1999) On the strength of shocked glasses, *Proc. R. Soc. Lond. A*, **455**(1984): 1275–1282.

This has been presented previously in Chapter 4 when the operation of strain sensors was described. Figure 4.8 showed a series of strain traces (and the reader will remember that strain is u_p/U_s) recorded at increasing pulse amplitude in soda-lime glass. These had precisely the same features and incidentally also revealed that above a threshold the sensor returned to a residual strain which corresponded with the failed state of the glass sampled. This not only shows that both sensors are sampling details of the failure accurately but also that one can directly sample the jump from elastic to inelastic behaviour with state sensors in brittle materials.

The particle velocity histories complement the data obtained on the stress levels in targets to provide a complete picture of the macroscopic mechanical state within the material during the processes operating and provides a window into the first moments of failure in an impacted solid. Whether brittle or ductile, the defect sizes and density at the micro- or mesoscale determine not only the strength of the material under compression but the hydrodynamic response and the timescales over which it develops.

6.6.2   Strength response of glasses

If the longitudinal and lateral stress components are measured in a glass block as it is loaded, the effects of the propagating fracture front become apparent in the embedded stress histories and can be correlated with the photographic and other evidence presented earlier. Figure 6.19(a) shows the recorded stress levels at a sensor location some 3 mm from an impact face and shows stress components in the direction of, and perpendicular to the shock travel. Twice the shear stress at this location is plotted as the difference between the stresses in the other two histories. Although these histories were obtained from soda-lime glass, similar behaviour is also observed in DEDF, borosilicate and other glasses that have been examined in this stress regime.

The longitudinal stress jumps to *c.* 7 GPa and ramps upward at this gauge location. The lateral stress rises to *c.* 3 GPa for 0.5 μs but then jumps again to 5 GPa thereafter coincident with the arrival of the fracture front. The shear stress follows a trajectory starting at just above 4 GPa and then after 0.5 μs drops to *c.* 2 GPa. The failure front has halved the strength of the glass taking it from intact to comminuted material at this stage.

At each stress level between *c.* 4 and 10 GPa the failure wave is accelerated and the stress ahead leaps to the elastic value whilst once it has passed strength behind fails to *c.* 2 GPa. The locus of the unfailed and failed strengths is shown in Figure 6.19(b) for the three glasses. The shear stress ahead of the front takes the elastic value as expected until failure reaches the gauge location as cracks arrive. It then drops to the failed value of the shear stress which defines the inelastic branch of the Hugoniot. The shear stress levels are the same for the three glasses which is remarkable given differences in some of the bulk properties of the three materials. Density, for instance, varies from 2.2 to 5.2 amongst these three glasses yet the strengths are governed by the silicate network that bonds the materials together and takes a common value.

This dual behaviour occurs in the range 4–9 GPa for all glasses tested, both natural and manmade, even if they contain oxide fillers. However, Gibbons and Ahrens note that fused silica shows elastic behaviour up to 8 GPa which might be expected for such a crystalline composite.

Glasses deform in a manner that shows them clearly different to metals. The inelastic deformation mechanisms in the two material classes operate with kinetics that limit their spread to a velocity of around the shear wave speed for dislocation motion or for fracture. This limiting speed is clearly a function of pressure and since the shock is subsonic the process of deformation beyond the elastic limit is one that takes the material to achieve sonic conditions behind the shock front. The density of defects from which the two processes initiate is very different, however, with separations of hundreds of nanometres in the case of metals or hundreds of microns in the case of brittle flaws. This defines three orders of magnitude difference in the typical deformation times determined from the kinetics for the two processes. Thus relaxation from the elastic to the plastic state appears instantaneous in the case of metals but takes time for glasses that equate with macroscopic processes of significance that occur in the weak shock regime where the failure lags the shock. Similar features are found in the polycrystalline composites dubbed ceramics considered next, which contain amorphous glasses as a binder phase in their microstructure.

To bridge the gap between amorphous microstructures comprising large silica networks to composite structures containing many different phases it is instructive to consider the behaviour of a one-dimensional, purely brittle solid. A glass laminate, containing many thin layers, might be considered as an analogue to a multicomponent brittle solid. If shocked in one dimension it gives an overview of the effect of multiple interfaces with a binder on the response of a true 3D composite. Figure 6.20(a) shows a schematic of the microstructure that results when a series of 1 mm soda-lime glass plates are bonded to one another and then shocked. A wave travelling across the stack encounters

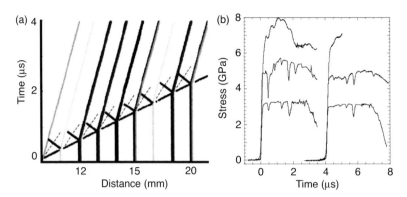

Figure 6.20 (a) Distance–time plot for an impact onto a glass laminate. The plates are each 1.1 mm thick and are bonded with thin (0.1 mm) layers of epoxy. (b) Interfaces in a soda lime glass laminate (the first, third, fifth and seventh). The lower stress level is in the range before failure fronts are observed. The upper two are in this region. Note the rounded front that appears at the first location at the highest stress level. Source: reprinted with permission from Bourne, N. K. (2005c) The shock response of float-glass laminates, *J. Appl. Phys.*, **98**(6): 063515. Copyright 2005, American Institute of Physics.

internal boundaries within the composite triggering failure at each interface. As it does so fracture is initiated in the plate in front and by release at the lower impedance boundary. However, internal interfaces delay the failure of the stack as seen in the example above. Thus the bulk response is different from that of a monolith. Figure 6.20(b) shows pulses transmitted through 1 and 3 tiles in the laminate (sensors at 1.1 and 3.3 mm from impact). Below the 4 GPa transition in behaviour, the pulse travels unperturbed. Local drops are seen corresponding to releases from the 100 µm glue layers found between the tiles. As the stress increases beyond this level, the pulse attenuates, and the apparent elastic limit decays with distance through the target asymptoting at greater distance to a steady value.

Multiple interfaces arrest failure at each component of the stack and act to delay the failure travelling forward relative to that observed in a monolith. Although the ultimate Hugoniot stress for failed material is reached, the kinetics for the failure process are arrested thus increasing the time for which it remains intact and the distance travelled by a wave before attenuation takes place. By this means, interfaces play a role in controlling the times to failure in the layered targets and laminates accommodate failure more successfully than monoliths would in their place.

6.7   Dynamic compression of polycrystalline ceramics

As shown above, a polycrystalline ceramic is a composite composed of crystalline and amorphous components and seeded with a population of flaws left in the material as a result of its processing. Because of important applications, the dynamic response of strong ceramics to shock wave loading has been addressed extensively over the past 30 years, principally by empirical ballistic testing. In that time much has been learnt but vital questions still remain unresolved. The continuum response of polycrystalline ceramics has been extensively tested, since they combine high compressive strength with low mass. For instance, these properties are of interest in developing armours to respond to projectile impact where high elastic limits and spall (dynamic tensile) strengths are known to give better ballistic performance. However, there are many features of response that are unexplained by continuum processes. Rather, behaviour is controlled by mechanisms operating at the mesoscale that, in polycrystalline ceramics, defines a length scale of importance of the order of the grain size, which in common examples is around one micron.

Applications abound for materials that display such strengths. A typical steel has an HEL in the region of a GPa with optimum armour steels around 2 GPa. However, the Hugoniot elastic limits (HELs) of polycrystalline ceramics range from 6 to 20 GPa. For instance, boron carbide has an HEL of *c.* 16 GPa and silicon carbide one of 15.5 GPa. Titanium diboride and aluminium nitride have also attracted attention since they show several thresholds when loaded in compression. Polycrystalline ceramics also have high impedance (many around that of copper) due to their high wave speeds, and the rise of the elastic pulse in a pure material is generally short (less than 1 ns, a measurement time limited by the fastest velocity interferometry measurement of the free-surface velocity). Finally, glass boundaries within the tile fail to allow comminution of the target under pressure. This failure from surfaces (failure waves) in polycrystalline ceramics in several materials has been observed and operates in an analogous manner to that seen earlier for glasses.

A summary of relevant properties found in experimental shock investigations and that remain for explanation include:

 (i)　the high HEL of shock-loaded ceramics (6–20 GPa);
 (ii)　the observation that the shear strength increases with shock pressure;
(iii)　that the form of the measured stress and particle velocity histories of these materials show a rapid elastic rise followed by a convex curved region when loaded to about 1.5 times the HEL, and that at higher speeds turns back to give a very rapid rise at higher amplitudes to form an S shape;
(iv)　that the materials show low spall strengths which vary with position (0.3–0.6 GPa);
 (v)　that surface failure can be driven into the bulk by the propagating wave.

The rest of this section hopes to explain each issue in depth in what follows. One feature of their response is that pre-existing defects are activated and new damage is introduced, even when shocked to relatively low stresses, about half the Hugoniot elastic limit (HEL). Various failure criteria, which are used in numerical codes, have been tried in an attempt to account for their advantages and discrepancies but as yet no single approach has had success in determining the response required over a broad range of impact conditions.

At present, engineering models for ceramics are generally empirical. The simplest are plasticity-based adaptations of metals' models, but the most widely used continuum extension of elastoplastic models is the Johnson–Holmquist description that has evolved into a series of forms to cover a variety of responses. In these models the state within a

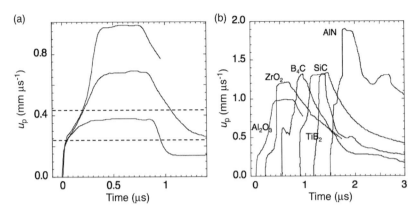

Figure 6.21 Response of ceramics under one-dimensional shock loading below the WSL. (a) Alumina loaded to three different stress amplitudes, the two levels are upper and lower yield (see below). (b) A variety of polycrystalline ceramics with different features in their response. Source: after Grady (1997).

material element, brought to a yield surface, is then moved to the failure curve which lies below the envelope and imposes a decrease in strength. A second route is to explicitly assume that fracture mechanics operates on the materials and consider failure criteria similar to that of Griffith for brittle solids. Further, in recent years, advances in numerical treatments have had some success in overcoming the weaknesses of regular, polygonal meshes which confine failure to non-physical fracture paths.

What follows will review ceramic response to shock by considering the compression of alumina. Lessons for other systems will be drawn from this example as the discussion develops. It is hoped that this will allow the reader to draw out general principles rather than producing an exhaustive review of all experimental work. As previously, the section will start with one-dimensional shock compression to start the discussion of the micromechanical behaviour.

The response of aluminas to impact has three regimes of behaviour. Figure 6.21(a) shows a series of traces for AD995 which illustrate these with the dotted line boundaries indicated. The elastic regime is indicated by a rapidly rising part of the history which reaches a break at the first boundary (the lower HEL). This represents a lower threshold for the material. Above this region the pulse shows a convex region up to a second threshold or point of inflexion (the upper HEL), after which it curves upwards in a concave manner to the maximum stress as in other shocked materials. This characteristic form to the pulse is repeated in other ceramic materials and so explanation must be sought in the response of the microstructure under load.

Figure 6.21(b) shows histories for all of the ceramics considered here. The quasi-static data for the materials have been presented at the beginning of the chapter in Table 6.1 and the shock constants and HEL in Table 6.4. The history for AD995 presented earlier is seen to be typical of many of the others in major respects, although each material has its own peculiarities. The most obviously different response comes from

Table 6.4 Shock data for sapphire and selected polycrystalline ceramics. Note that these properties refer to particular batches of different purity and porosity and are given as representative values for the reader

	ρ	c_L	c_S	c_0	S	HEL
Sapphire*	3.98	10.56*	6.24*	8.74	0.96	12/22*
AD995	3.88	10.6	6.25	7.71	1.27	6.71
ZrO$_2$	4.51	5.88	3.35	2.0	1.67	5.4
B$_4$C	2.51	13.9	8.74	9.06	0.91	16
SiCB	3.23	12.18	7.74	8.0	1.0	15.7
TiB$_2$	4.48	10.9	7.32	8.48	0.85	4.7/15
AlN	3.23	10.7	6.27	8.5	1.0	9.4

* Sapphire is anisotropic; average values are given. Wave speed ranges from 10.6 to 11.2 mm μs^{-1} on the r to the a and c crystallographic directions (see Figure 6.7) and HEL is a function of direction with min/max shown.

AlN which displays a martensitic phase transformation. The other materials all show a response which has an elastic and an inelastic phase in this regime of behaviour. The two materials of note that display distinct responses are boron carbide and titanium diboride. Boron nitride has a more angular aspect to its shock profile. The elastic limit has a sharp peak with a drop behind and a jagged rise to the peak. It is similarly angular on release. Titanium diboride also shows an interesting form displaying two cusps in its profile. The release behaviour is symptomatic of differences in the rise.

A series of such shots allows the construction of Hugoniots for the materials. Clearly each differs in magnitude of elastic limit and density so that the curves spread a range across the σ–u_p plane. The vital constants for the shock behaviour are reproduced in Table 6.4 and the reader is referred to the standard compendia for the curves themselves.

6.7.1 Recovery of shock-loaded alumina

To understand the form of the continuum response requires some better knowledge of the controlling micromechanisms at work within the shocked target. Relevant embedded mesoscale sensors are not yet capable of this and so it is necessary to proceed (as with metals) with the microscopic examination of shock-recovered targets and deduction of operating mechanisms. To do so in brittle ceramics has proved very difficult, however, and only a few works have done more than comment on the state of fractured material. It has been possible to recover shock-loaded AD995 alumina and examine its response microscopically and these results will be discussed here.

Figure 6.22 shows materials analysis from the recovered targets. Figure 6.22(a) shows a snapshot from a simulation of a momentum-trapped recovery fixture to the left and an optical photograph of a stained section through a recovered target to the right. These fractures correspond to releases from the fixture during the recovery process. The

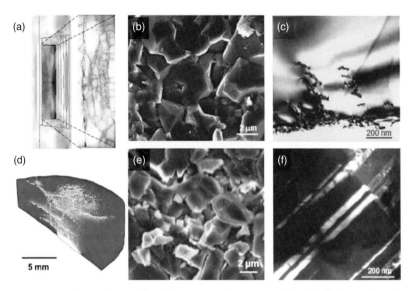

Figure 6.22 Observations of the microstructural response of the shocked samples of AD995 at 6.0 and 7.8 GPa. (a) Optical view with positions of interfaces in the recovery fixture superposed. (b) SEM micrograph and (c) TEM taken from the sample shocked below the HEL.(d) Tomograph of half the recovered target. (e) SEM and (f) TEM for the specimen loaded above the steady value of 6.7 GPa. Source: reprinted with permission from Bourne, N. K., Millett, J. C. F., Chen, M. MacCauley, J. W. and Dandekar, D. P. (2007) On the Hugoniot elastic limit in polycrystalline alumina, *J. Appl. Phys.*, **102**: 073514. Copyright 2007, American Institute of Physics.

recovered discs could be examined non-destructively using X-ray computed tomography (XCT) and sectioned to perform optical and electron microscopies. At the mesoscale, shock deformation features were dominant. The failure mechanisms were investigated using scanning (SEM) and transmission electron microscopy (TEM). Two samples that had been subjected to shock-induced stresses below and above the HEL were separately sectioned and imaged. SEM observations show mesoscale details and suggest a transition in fracture behaviour from intergranular dominated fracture below, to cleavage dominated fracture above the HEL. Below there is some intergranular fracture observed. Cracks propagate down grain boundaries nucleated from voids within the microstructure. Above the grains have been crossed by transgranular fracture leading to smaller particles more closely packed. The TEM images show microscale features. Below the HEL one can see dislocations at boundaries. These are few in number and assumed to lie in the basal and prismatic directions. Above there is twinning observed within some of the grains. Thus crossing the HEL threshold corresponds primarily to the appearance of twins within the crystals. Plastic deformation occurs within the grains above the HEL for the alumina studied here and the complex unit cell and dearth of slip systems favours twinning as the preferred plastic deformation mechanism. However, twinning can accommodate only small strains within the material. Fracture is the only mechanism available to take these loads once twinning has been exhausted and it is the results of communition that are commonly observed in less well-controlled experiments.

All of this emphasises that alumina yields plastically at the lower HEL of the material but that fracture is occurring in tandem across the microstructure from within the elastic regime to the highest stresses applied.

6.7.2   Mesoscale response

This section will address compression loading with the hope of explaining experimental observations of continuum response. One of these is the complex wave shape observed as a step-loading wave at the contact face that disperses in plate impact experiments. One means of amplifying behaviour is to employ mesoscale modelling to amplify mechanisms and behaviours. In particular it has been successfully adopted for materials where localisation allows small regions to achieve critical thresholds that are not reached using continuum numerical descriptions. Thus numerical simulations are shown here to reveal fluctuating stress states and localisations of energy that evolve in three-dimensional analysis to a steady stress state with appropriate equation of state and elastic-plastic material strength descriptions. Real microstructures are input in simulations to connect with continuum measurements on materials examined.

In the previous chapter the equilibration of material states was integrated over a microstructural unit (MSU) containing a number of defects. The stress state in such a cell took time to reach a steady state and this was illustrated to depend upon details of the defect distribution and the metal structure. Similar considerations can be used here to view a similar unit constructed from the SEM of the 99.9% pure alumina in Figure 6.5 containing defects population critical to continuum response in these materials at the mesoscale. Figure 6.23 shows pressure contour plots for the simulation of impact of a MSU of this alumina onto a rigid lower boundary at the three velocities. The flow and shock structure are clear to see in this microstructure since the high density of the alumina composite, the better equiaxed grains and lower porosity make the flow steady over shorter times than with lower density materials. The grain and void morphology fixes the front curvature and rise. Interestingly, the elastic front is more uniform than the following plastic one as a result of the more complex, time-dependent defect interactions that affect the latter's rise. There are no tensile stresses created within the grains, even around inhomogeneities, and so it is plasticity within the alumina that determines the yield within the composite, just as was seen in the recovered material. At higher velocities, conditions are more uniform behind the shock front, and the reader will observe that the shading becomes quickly regular, showing that behaviour is thus less sensitive to these effects. The left-hand side of the MSU contains five void sites. Two of these are representations of defects within grains, while two are regions at grain triple points, where no glass phase has reached. A dry joint (in the lower right-hand corner) is a grain boundary into which no glass has intruded, leaving open free surfaces. This last flaw is analogous to what remains after an intergranular fracture that might result from processing, or from damage during some previous impact event. At the lowest impact velocity none of the voids close in the timescales of the presented frames. However, local wave focusing and release occur, leading to inhomogeneous pressure fields around them. Thus, the net

500 m s⁻¹

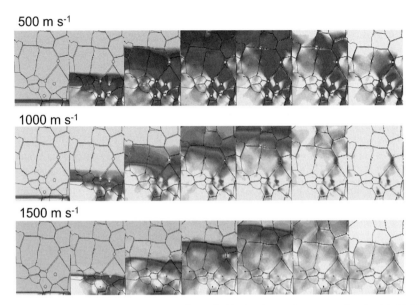

Figure 6.23 Simulations of shock loading of 999 alumina at three stress levels: (a) 20 GPa; (b) 40 GPa; (c) 60 GPa. Contours of pressure are plotted. The first frame is 50 ps after impact and subsequent frames are 200 ps apart. In all cases the alumina is below the WSL. The snapshot is 10 μm high. Source: after Bourne, N. K. (2006a) Impact on alumina. I. Response at the mesoscale, *Proc. R. Soc. Lond. A*, **462**(2074): 3061–3080.

effect of defect sites on the right-hand side of the microstructural cell is to slow the shock, leading to curvature over several micrometres.

This inhomogeneity at the mesoscale illustrates the unsteady nature of flow in the weak shock region behind a propagating front. However, as the amplitude increases towards the strong shock threshold, stress state stability is achieved more quickly and the inelastic phase of the loading becomes confined within the shock front. Homogeneous behaviour and uniaxial strain are only found in steady regions in the flow and these are only established here immediately behind the front when the shock transits to the strong shock regime. However, in materials with such high elastic limits (that in boron carbide is *c.* 20 GPa) the strong shock regime is above 100 GPa and such prodigious pressures are accessed in only a few specialist applications. This means that brittle materials generally respond in the weak shock regime and assuming stress state stability will result in unphysical descriptions of their behaviour.

Of course the principal feature of ceramic properties is their high elastic limits. Such simulations as those conducted here on alumina, along with the recovery experiments shown earlier, confirm that the lower HEL of the material is determined by plasticity within the grains. In these simulations there were no tensile stresses created within the alumina grains, even around inhomogeneities, and thus it is plasticity that determines the yield within the composite. There is no indication that the glass binder phase reaches conditions where it fails under compression. Rather, the composite microstructure and

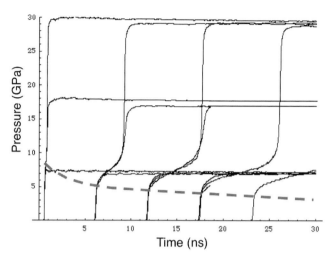

Figure 6.24 Station histories for the simulations of Figure 6.23 at distances 10, 110, 210, 310 and 410 mm from the impact face. After Bourne, N. K. (2006b) Impact on alumina. II. Linking the mesoscale to the continuum. *Proc. R. Soc. Lond. A*, **462**(2075): 3213–3231.

the release effects of other phases control the yield. Porosity has a major influence at this scale, since voids disperse shock. When favourably oriented, they may also cause tensile zones. Thus the shocked region near the front is one of local wave focusing and release, leading to inhomogeneous pressure fields acting for some distance behind. The pores do not collapse at lower impact velocities, since the alumina grains are within their elastic region. However, glass boundaries within the tile may fail allowing comminution of the ceramic to occur. This has completed by the upper HEL. These simulations have not included fracture models and are only intended to illustrate the gross features of response. In reality this will dominate around the collapsing defects and increase the communtion observed in the ceramic under load.

In such materials then, where the tensile strength is low, the roughness of the impact faces in targets used for experiment is a key seed for the observed behaviour. Thus, a roughened surface may lead to failure and causes a delay after the initial impact time (as the surface features are deformed) before the failure front advances.

The link from the mesoscale to the continuum requires some means of extrapolating the simulation to run down many rows of the unit structure shown above. Extension to the continuum scale then allows comparison with experiment to verify features observed. The state is followed down the length of the column and the result averaged across the mesh boundary at the head of the cell and the histories are shown in Figure 6.24. The response is seen to reproduce the features of the histories observed in experiment (Figure 6.26). The pulse rises quickly and the first break in the rise shows decay and finally achieves a pressure close to the expected measured value at a station millimetres from impact. A convex region then rises to a second threshold which decays more weakly at the five stations. Finally, the measured value of the peak pressure near the impact face is higher than that within the bulk since the Hugoniot itself develops with propagation into the target.

When compared with continuum sensor records on the same material the derived measurements of elastic limit and of the peak state achieved agree well with continuum measurements. Note that the input variables for the calculations are the numerical descriptions of individual phases. These derived simulated strengths and wave speeds at continuum scales are entirely the result of the mesoscale geometry and the development of the defect related damage within the composite. Of course these ceramics are normally regarded as continuum materials rather than composite mixtures. However, a means of introducing the properties of individual component phases into a larger scale representation is always useful.

The composite nature of the ceramic suggests that one needs to investigate a means of deriving the continuum values for the shock parameters of the ceramic from those of the individual phases. This requires the derivation of suitable mixture rules applicable to the material. A first means of deriving the parameter c_0 (the intercept at zero pressure of the shock velocity particle relation) is to equate it with the bulk sound speed of the material through the standard formula derived in Chapter 2. This yields a value of 6.6 mm μs^{-1} which is clearly too low for such a ceramic. A linear mixture rule for the calculation of c_0 and S from the data yields

$$c_0 = \alpha c_0^{\text{Alumina}} + \beta c_0^{\text{SL}}, S = \alpha S^{\text{Alumina}} + \beta S^{\text{SL}}, \tag{6.6}$$

where α and β are the volume fractions of alumina and glass, respectively. Application of such a rule yields values too high for the bulk sound speed to explain the results. Thus, a reduced c_0 is calculated where

$$\frac{1}{c_0} = \frac{\alpha}{c_0^{\text{Alumina}}} + \frac{\beta}{c_0^{\text{SL}}} + \frac{\gamma}{2u_{\text{p}}}, S = \alpha S^{\text{Alumina}} + \beta S^{\text{SL}}, \tag{6.7}$$

and that has been shown to successfully describe the behaviour when the response is hydrodynamic. Here, γ describes the volume fraction of porosity within the microstructure.

6.7.3   The Hugoniot elastic limit

Stress histories for aluminas are shown in Figure 6.25(a) and (b) for shock experiments in 88% and 99.9% pure materials respectively. In the first set of histories, the rapid rise and concave then convex curvatures to peak stress are very evident. These correspond to two elastic limits in these composite materials called lower and upper here. These pulses correspond to the lower two of the simulations in Figure 6.24. In Figure 6.25(b) a series of stress histories in the elastic and plastic regions is also shown. Since the sensor sits on a PMMA backing window, a reflected release occurs when the elastic wave first arrives, which appears as a small reload signal on the peak plateau of the history. There must be another front travelling within the target for there to be such a reflection before release arrives and this must indicate further failure occurring within the material under shock.

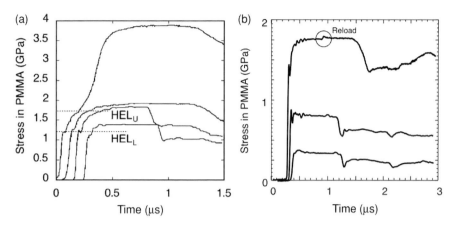

Figure 6.25 (a) 880 alumina driven to reach overdriven response showing two yield thresholds: lower L and upper U. (b) 999 alumina rising above yield in alumina grains. Reload signal from failure front on peak. Source: reprinted with permission from Bourne, N. K., Millett, J. C. F., Chen, M. MacCauley, J. W. and Dandekar, D. P. (2007) On the Hugoniot elastic limit in polycrystalline alumina, *J. Appl. Phys.*, **102**: 073514. Copyright 2007, American Institute of Physics.

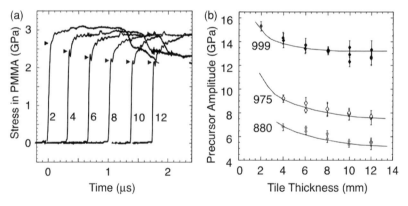

Figure 6.26 (a) Precursor decay in 999 alumina at travel distances indicated on each history. (b) Magnitude of the lower yield stress as a function of distance travelled into the target showing the limit to be non-steady in each case, settling more slowly for the lower purity aluminas. Source: reprinted with permission from Murray, N. H., Bourne, N. K. and Rosenberg, Z. (1998) The dynamic compressive strength of aluminas, *J. Appl. Phys.*, **84**(9): 4866–4871. Copyright 1998, American Institute of Physics.

The histories in the preceding figures show the development of the lower threshold which represents the onset of plasticity in the alumina grains. Figure 6.26(a) shows the evolution of this elastic limit with travel of the wave through the target in a 99.9% pure alumina. This threshold develops with travel distance as seen previously for the glasses over a distance of 10 mm or so. This reflects the kinetics of the integrated inelastic slip processes within the microstructure incorporating plasticity but also intergranular fracture that accompanies reaching this limit. The complex unit cell in alumina favours twinning and limits processes within the grains leading to the high elastic limit when the grain has failed down the hard deformation direction. Just as in the case of the other materials considered in this book, the transit from an elastic state on the impact face to an inelastic one at depth is a time-dependent process that evolves as one samples at stations through the target.

In Figure 6.26(b) the lower elastic limit is plotted as a function of distance for three different purities of alumina from 88 to 99.9%. Firstly it will be apparent that the ceramics closer to theoretical maximum density (TMD; so sapphire in this case) show higher elastic limits. In all cases there is precursor decay. However, the phenomenon is more prominent and occurs over the larger distances for the lower purity material indicating that the kinetics for the steadying of the deformation processes are slower in the lower purity materials. The observations made in this section have shown that data from continuum compression experiments and one-dimensional shock recovery allow one to evaluate the micromechanical behaviours operative during the shock process in a polycrystalline ceramic.

The discussion above has noted two operative thresholds in stress histories as well as microstructural observations of alumina behaviour under shock. The composite contains a randomly oriented array of anisotropic sapphire grains bonded with glass phase and containing a population of pores. The lower yield point in a shock experiment corresponds to deformation down the easy direction and the upper to the hard, z-axis of the alumina grain. In a polycrystalline target this means that between these stresses an assemblage of elastically deformed, within a matrix of plastically deformed grains will exist favouring fracture at the weakly bound grain boundaries. Further, twinning in the grains is favoured over dislocation motion given the few slip systems present in the complex unit cell and so fracture across grains and down twin boundaries is also observed. Thus, not all grains are activated at the lower HEL. The alumina unit cell is anisotropic, and HEL values from 12 to 22 GPa have been recorded down different directions in single crystal sapphire. Thus, the lower HEL represents the onset of plasticity in suitable oriented grains and the stress levels accessed by the shock will transform the easy and then at higher stresses the hard direction in the material until all are compressing in the upper state.

The convex part of the pulse, from the first break from the elastic rise to the second point of inflexion on the rising pulse, corresponds to the mixed response region resulting from grain anisotropy in an assemblage of sapphire crystals and thus, at higher applied stresses all of them have deformed. This picture sees the response of the alumina as having three regimes: elastic; mixed response; and a regime in which the continuum applied stress has exceeded the yield stress in grains of all orientations at the mesoscale and where the grain boundaries have failed. These regimes have thresholds separating one from another, which are termed the lower and upper HEL values (HEL_L and HEL_U). In the mixed response region, in which the flow has some grains behaving elastically and some grains behaving plastically, it has been shown that the relation between shock and particle velocities becomes non-linear. This results from the different wave speeds and Poisson's ratios in the assembly of grains deforming at these stress levels and of course this anisotropic strain state will act to pull apart grains and fail boundaries in the flow.

The stress fields and motion at the mesoscale are fully three-dimensional, as seen above. Continuum boundary conditions apply preserving one-dimensional strain but at a length scale which averages over ten or more grains. Thus an individual, anisotropic sapphire crystal will not be constrained in a uniaxial strain field but in general a mixed one, and for many grains, a uniaxial stress field. Thus, the yield stress can be related to that at the HEL through the relation between uniaxial stress and strain giving HEL_U as 12.5 GPa and HEL_L as 6.9 GPa given Poisson's ratio for sapphire of 0.3. These values give upper limits for the two shock thresholds in a polycrystalline alumina where the grains can relax laterally to attain a 1D stress state. When other phases of lower strength are present and other processes contribute to failure these thresholds will be lowered. But for a pure alumina such as the 99.5% pure material AD995, this value for HEL_L is close to that measured. Thus consideration of the deformation of the mesoscale features in the microstructure is a vital component in understanding the observed continuum behaviour in polycrystalline materials and further examples of this will be seen when one considers failure processes occurring under shock. In such materials multiple phases clearly determine failure times; consideration of atomic scale slip within the grains that is more important in metals is less dominant an effect although clearly it determines the elastic limits. Thus different materials have failure that is dominant at the composite scale and it is the corresponding times for mesoscale processes that determine their response to impulsive loads.

Mesoscale simulation has illustrated some of the interplay between operating mechanisms within the composite. Futher a simple composite model explains both the observations of structure and thresholds measured but also the characteristic shape of the stress pulses as well as features of the measured response. At every scale a series of processes operates with varying kinetics to accommodate the applied strain at the macroscale and the transition from unsteady to steady processes and establishing stress-state equilibrium takes many microseconds in brittles as opposed to tens of ns in metals. Further, the strong shock regime where stress state stability is ensured behind the shock is less accessible for this class of materials compared with loading of metals. Strength in compression, however, is countered by weakness in tension and the next sections probe the failure of ceramics during dynamic loading.

6.7.4 Strength and delayed failure

All of the ceramics considered here contain an assemblage of grains bound in some manner with a weaker second phase. Moreover that phase is brittle so that local tension will fail this boundary and this will lead to a distribution of fragments after unloading. Fragment size has been looked at by Grady in an attempt to assess the deposition of energy into the target; balancing the surface energy to the kinetic energy imparted by the pulse. He noted several features of the fragment distribution, for instance that the mean grain size of the fragments increases with pulse length. In the final account, the internal damage was an order of magnitude greater than surface area of fragments indicating that considerable work was being done in processes that occur alongside brittle fracture of the microstructure.

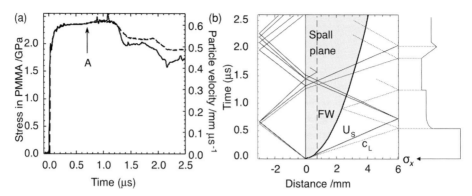

Figure 6.27 The impact of a copper flyer of thickness 3 mm and a 99.9% pure alumina target of thickness 6 mm at 778 m s^{-1}. (a) VISAR (dashed) and gauge (solid) histories mounted in a back-surface configuration. There is a reload signal appearing at A (indicated). (b) X–t diagram with elastic and shock fronts, marked with their velocities, c_L and U_s, and the failure front FW, which is decaying in velocity with time as the root of distance. A schematic of the stress trace is shown to the right of the figure. Source: after Bourne, N. K., Rosenberg, Z. and Field, J. E. (1999) Failure zones in polycrystalline aluminas, *Proc. R. Soc. Lond. A.*, **455**(1984): 1267–1274.

There is a suite of micromechanisms operative that can allow energy dissipation. These include not only fracture but potentially phase transformation (such as in AlN), dislocation motion or more likely twinning in these materials dissipating plastic work, activation of interfacial defects in the second phases and, after failure, sliding of new surfaces with consequent friction. Finally, collapse of any porosity with grains, or at grain boundaries. All of these available nucleation sites act as triggers for fracture or plastic flow at sites throughout the material and new defects can be generated by the passage of the pulse. Any recovered target will exhibit features of all of these mechanisms in action but the fracture processes do not show a threshold; rather crack density increases monatomically as loading amplitude increases, whereas plasticity activates at defined thresholds (HEL$_L$ completing by HEL$_U$) so that they are evident in the stress histories recorded in plate impact experiments.

If there are sufficient flaws on the surface of a ceramic tile, impact will induce local release and trigger fracture in the surface region. However, the failure is dissipated quickly by plastic flow in the grains relieving the strain at the crack tips and blunting at pores between grains etc. The failed region that remains has lower impedance and will reflect waves evident to suitably placed sensors. Figure 6.27(a) shows simultaneous velocity and stress histories taken at a low-impedance window on the rear face of a ceramic target. If the flyer plate is sufficiently thin such that full release does not arrive first, then its reflection at this lower impedance, failed region at the impact surface returns the wave as a compression pulse arriving at a time indicated by the marker A in the figure. After this time, the particle velocity and the stress recorded at the window also cease to scale in the same manner with one another showing that the release processes in this region further alter the impedance of the target.

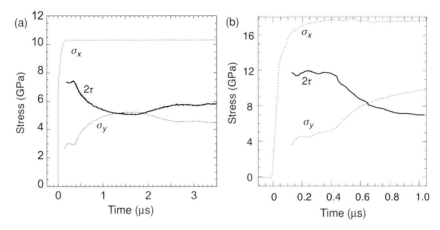

Figure 6.28 Longitudinal and lateral stress histories (dashed) and shear stress history (solid) for (a) alumina, AD995 and (b) SiC showing delayed failure from the elastic to the inelastic state.

Figure 6.27(b) shows a detailed X–t diagram for the impact indicating the travel of waves in the target. The speed of the front decays into the tile as the travel of the fracture front dissipates energy so that there is a parabolic form for the trajectory labelled FW. Reflection of the initial shock first from the window and then from the failed region returns the wave to the sensors as a compressive load. The shaded region thus represents a part of the target with lower compressive and tensile strength than that ahead of the failure front. This will also affect later waves returned from planes such as those induced in the experiment shown and where the spall strength can be seen to be negligible. Further comment on this will be made below. A second feature of note concerns the activation of flaws at the target surface. The experiment shown was conducted with a copper flyer and a ceramic target, thus with different impedances and sound speeds in the two materials at the impact face. When the same experiment is conducted symmetrically, with flyer and target the same, failure can be suppressed since there is much less surface strain and creation of local release to initiate fracture. As is the case with glass, it is fracture that causes bulk shifts in impedance. But plasticity in the grains mediates its effect, altering the kinetics of the process and localising failure to the impact surface in the case of polycrystalline materials.

The tensile strength of ceramics across the failure front is mirrored in a further material variable, the shear stress across the boundary. Like glass, it too must change behind the front taking the ceramic from an initially high strength to a lower value in the failed material. This can be followed at the continuum, as with the glass, with sensors to measure the longitudinal and lateral components of the stress field. Such records are shown in Figure 6.28 for (a) alumina and (b) SiC, but other ceramics show similar behaviour. In the geometry of the experiment, the tile is sectioned and a sensor introduced into the cut. In a brittle solid, the cut acts as an ideal flaw to initiate failure into the surrounding tile. The longitudinal and lateral stresses rise quickly to the HEL but then more slowly to the Hugoniot stress. Near the impact face the stress remains high for around 500 ns before decaying to a lower value. Again, the material can display an elastic strength for some time before it returns to an inelastic state. The damaged material on the other hand has a failed strength of c. 5 GPa at this stress level. Similar behaviour is observed in the SiC with the material maintaining strength for a time before decaying to a failed value.

These measurements are a window into the failed state of the ceramic which it adopts in any but the most idealised of loading states. The one-dimensional continuum strain field of plate impact can suppress fracture and show the earliest stages of compressive failure as a boundary condition on later behaviour. The sensitivity to the small lateral strains introduced by asymmetry in impedance between impactor and target shows both the very small strains to failure of the material (which are well documented) and also the dependence upon localised states at the mesoscale. Brittle behaviour is dominated by fracture instigated from mesoscale flaws in a similar manner to explosives, where localised high-temperature states control bulk ignition. In both cases continuum descriptions that integrate over the volume of the target cannot capture the key mechanistic details that control behaviour and subscale models are necessary to reproduce response. In that respect appreciation for the behaviour of brittle materials, and particularly their mathematical description, are in a state that is removed from that existing for metals since the initiation and path to steadiness in that case is many times, in some structures orders of magnitude, faster compared with the impulses of interest in impact applications. This is not true for ceramics where the processes are evolving. Nevertheless controlled, quantitative measurements of the states of failed material can be used directly to connect with failed states in practical loading situations as will be shown below.

In that respect, the deviatoric behaviour can be summarised in a plot detailing initial and failed strengths in the manner as shown for glass above (Figure 6.19(b)). Figure 6.29 shows the initial and failed strengths measured for the range of brittle solids discussed above. The known values of the longitudinal and lateral stresses are used to calculate the shear strength of tested glass, alumina, SiC, B_4C and TiB_2. One glass (borosilicate) is included but as has been seen, this represents the class since they lie on the same trajectory. All of the materials have two states lying on one of two curves representing an intact strength ahead of the fracture front, and a failed one behind it. The intact state lies on the elastic trajectory for the material, which, in the space presented here, is a line of a slope relating to its Poisson's ratio. Clearly different materials have different values of v, and so elastic response defines a sector with limits given by Poisson's ratio 0.09 (TiB_2) and 0.25 (glass). The locus of states of deformed materials, the failure curves, lie to the right of the elastic region and rank the strengths of the materials behind the propagating shock. The glass is the weakest solid presented here whilst TiB_2 is the strongest.

The crystalline materials exhibit greater strength than their amorphous counterparts as expected. The materials and their means of preparation do change many of their hydrodynamic properties yet within groups of common crystalline phase, initial strengths are similar. Straight lines are added to the figure to indicate trends but more detail can be seen in the papers in which this work was documented. Clearly, the failed strength of the materials tested does not, in general, remain constant as a function of the impact stress.

Table 6.5 Failed strengths of materials taken at the point at which the failure curve intersects the elastic line.

Material	Density (± 0.05 g cm^{-3})	HEL (± 0.5 GPa)	$2\times$ failure strength (± 0.2 GPa)
SL glass	2.49	4.0	1.9
Alumina	3.80	7.6	5.3
B$_4$C	2.51	16.0	7.1
SiC	3.16	15.7	11.4
TiB$_2$	4.48	15.0	13.0

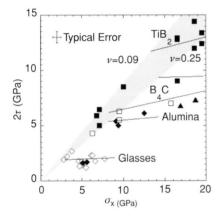

Figure 6.29 Deviatoric response at 2 mm from the impact face for the classes of material discussed. The grey shaded region represents a suite of elastic response trajectories for the materials and has boundaries between $\nu = 0.09$ (TiB$_2$) and $\nu = 0.25$ (glasses). Source: reprinted from Bourne, N. K. (2008) The relation of failure under 1D shock to the ballistic performance of brittle materials, *Int. J. Imp. Engng.*, **35**: 674–683. Copyright 2008, with permission from Elsevier.

Different processing routes modify material behaviour and in some cases the failure curve then shows hardening, and in some cases softening behaviour. The initial failed state on the elastic curve represents a common comparison position to judge between different ceramics and this is used below to compare with other performance measures. Some typical values for these strengths are given in Table 6.5.

6.7.5 Behaviour in tension

The spall strength of brittle solids shows equivalent results to those obtained quasi-statically. The fracture mode is flaw dependent and proceeds with great speed. In bulk solids that have not been carefully prepared, the value is thus of the order of a tenth of the yield stress whereas in metals it is comparable. The failure wave comminutes material so that it cannot support any tensile load. In ceramics, the material exists

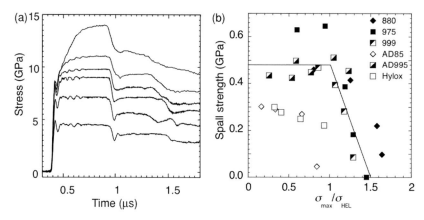

Figure 6.30 (a) Symmetric impact at various stress levels below and above HEL$_L$ for 97.5 pure alumina. The height of the spall signal reduces as one approaches 1.5 times this stress. (b) Variation of the spall strength in six aluminas of varying content showing the trend of spall strength reduced to zero at 1.5 times HEL$_L$. Source: from Bourne, N. K., Rosenberg, Z. and Field, J. E. (1999) Failure zones in polycrystalline aluminas, *Proc. R. Soc. Lond. A.*, **455**(1984): 1267–1274.

predominantly as plastically deforming grains yet failure still occurs in the brittle phase at grain boundaries by fracture. Thus behind the surface failure front the spall strength is zero as shown earlier for glasses. Ahead of it, the material shows finite strength but only up to about one and a half times HEL$_L$.

Figure 6.30(a) shows a series of experiments on a 97.5% pure alumina where the impact induced peak stress levels of up to twice the HEL of the material. The spall plane was induced to lie in the region away from the failure zone at the impact surface. The spall stress remains finite up to the HEL and then decreases to zero over subsequent experiments to higher peak stress. This behaviour is found over a range of aluminas from 85% to 99.9% pure materials. Figure 6.30(b) shows the spall strength as a function of the impact stress for six materials in this purity range. The absolute value of the spall strength is reproduced, whilst the magnitude of the stress is normalised by the HEL of the material to account for the different strengths of the ceramics. The spall strength has a constant value of *c.* 0.5 GPa until the HEL is reached and beyond this threshold the value then drops. There are two groupings of values; the first for close to full density materials and a second for lower purity aluminas where the spall strength starts around half that of the purer materials and then decays to zero more rapidly.

In the case of ceramics close to TMD, the insight given by the recovery on the AD995 suggests that the onset of plasticity, in suitably oriented grains, creates some flowing and some rigid grains in proximity, which builds local stresses at grain boundaries that fail some, thus allowing easy cleavage in a tensile pulse. Increasing stress results in ever-greater plastic deformation in more grains until at 1.5 the HEL, interfacial bonding is all but destroyed leading to zero spall strength when a tensile pulse arrives.

6.7.6   The elastic limit in other polycrystalline ceramics

A picture has been presented of the behaviour in compression and tension of glasses and polycrystalline ceramics. In particular, a key feature of their response has been shown to be failure by fracture and plasticity within the grains. Evidence has been presented in some detail for the glasses and from alumina's response to shock to illustrate operating behaviours. Most other polycrystalline ceramics behave in a similar manner under compression. However, there is one that shows different response and it is instructive to look at yield criteria applied across a range of brittle solids to determine whether these correspond with a ductile or brittle response.

The previous sections have shown a suite of mechanisms operating in materials under compression composed of different crystals and phases with defects built within their microstructure. The two mechanisms that have controlled response are those of plasticity in crystalline domains and microfracture between grains. Alumina has been shown to reach its elastic limit by onset of plasticity down favourable directions within the grains. Although microfracture occurs at levels below and above this threshold, it is this that denotes the lower HEL of the ceramic. However, not all brittle materials respond in this manner. Glass for instance displays inelastic behaviour above 4 GPa and the mechanism that yields this response is fracture in this case. The steady HEL of the material must obey a yield criterion describing the relation between the strength of the material and the one-dimensional yield stress. For the two cases of ductile and brittle yield, in one-dimensional strain, the criteria are the shear stress at yield given for ductile materials as

$$2\tau = \sigma_x - \sigma_y = \frac{(1-2v)}{1-v}\sigma_{\text{HEL}}, \tag{6.8}$$

which represents the Tresca or von Mises criteria in one-dimensional strain.

If the response is brittle, the strength at yield, τ, is related to the stress at yield (the steady HEL) through the Griffith criteria in one-dimension

$$2\tau = \sigma_x - \sigma_y = \frac{(1-2v)^2}{(1-v)}\sigma_{\text{HEL}}. \tag{6.9}$$

These two expressions give a means of calculating a value for the strength at yield from the recorded HEL of the material assuming different types of failure mechanisms in each case. The values for these quantities are well documented for the five ceramics and for comparison a steel is included since clearly it follows ductile yield.

Figure 6.31 shows pairs of points calculated for the strength of five ceramics and a metal at yield using the brittle and ductile yield criteria described above. Each material has an HEL_L measured from steady values reached during loading which is used in Eqs. (6.8) and (6.9) to yield values for ductile and brittle strength at the HEL. In the figure, the two points for B_4C are circled. Brittle yield (denoted by a square symbol) is at $c.$ 8 GPa while ductile yield (denoted by a diamond) corresponds to a shear stress of $c.$ 13 GPa. Each of the other materials has a pair of values for this quantity measured where the failed strength intersects the elastic line for each material. Thus the calculated and measured values can be compared one with the other and if there is correspondence,

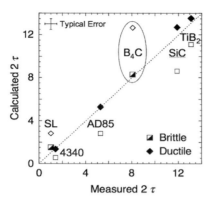

Figure 6.31 Measured shear stress plotted against calculated values for two yield criteria: ductile (Tresca/von Mises) and brittle (Griffith). The correlation for all of the materials shown is with ductile yield save for glass and boron carbide which fail in a brittle manner.

each material would lie on the dotted diagonal shown. For each of the materials, one of the pair lies on or close to this trajectory. In the case of B_4C it is the square symbol corresponding to brittle yield. In the case of the steel 4340 it is the diamond symbol denoting ductile behaviour. The point nearest the correspondence curve is given a solid symbol whilst that away from this curve is open.

The six materials show examples of both ductile and brittle behaviour. Only two have strengths consistent with brittle yield; SL glass and boron carbide. The rest (4340 steel, alumina, SiC and TiB_2) show ductile behaviour. The titanium diboride point is interesting because it displays stress histories with two cusps on the rise for the shock; the datapoint presented is that calculated for the upper one in the material. This will be returned to below. Even given that there is error on the measured failed strengths and on the HELs, the agreement between the measured and calculated values is good. Further, the separation between the ductile and brittle pairs is sufficient to make the assignment of mechanism unequivocal in all cases. Although the behaviours of glass and steel at the lowest values are comforting sanity checks on the method, it is valuable to show ductile processes in alumina (for HEL_L) where the recovery experiments have shown plasticity operating within the grains. Although unit cells are complex similar behaviours are observed in SiC and TiB_2. However, the complex unit cell and the variation in composition inherent in B_4C makes it reasonable that the behaviour is entirely brittle with slip very difficult.

The fascinating behaviour of all the other ceramics cannot be covered here in the detail it deserves. A few highlights are given in these remaining paragraphs to illustrate the rich detail available in other materials. First consider the response of TiB_2. Histories for the loading have been given in Figure 6.21(b) and the material shows two cusps on its compression Hugoniot corresponding to two features on the rise of the shock. Recovery experiments on TiB_2 have shown that the stress at the first cusp is a threshold for the propagation of intergranular microcracks and that at the second corresponds to intragranular fracture and plasticity. The material also fractures in a failure wave in shock experiments. Spall strength measurements also support a failure wave and microcracks leading to reduced spall strengths behind the shock. In this particular case the Hugoniot shows features of the composite mesoscale response characteristic of polycrystalline ceramics with the same suite of underlying mechanisms and responses.

The kinetics of operating processes determine the shape of the stress history observed by embedded or external sensors and two questions are posed by these observations. The first concerns the mechanism by which cusps on a Hugoniot are generated. Extending this one might ask why other polycrystalline or amorphous brittle materials do not show such behaviour and exhibit structured Hugoniots. One observation is that fracture must discontinuously increase the internal volume and a threshold for such behaviour might give rise to a Hugoniot cusp. This is consistent with alumina and SiC not displaying features, since they have been observed as not having a lower failure-wave threshold in similar experiments.

Obviously suitable ceramic microstructures may undergo martensitic phase transformation under shock. A classic example is shown amongst the histories in Figure 6.21(b), the Wurtzite $B4$ to NaCl $B1$ phase transformation in AlN which occurs at c. 22 GPa. The change is additionally accompanied by a large volume change of c. 20%. These thresholds and volume changes indicate that a large amount of energy is liberated in such a transformation. The phase transition wave does not rise monatomically as one might expect for such a front. Rather it shows a convex region before rapid rise. This behaviour suggests complex kinetics in the process and may be a region over which transition occurs depending on the orientation of crystals and anisotropy in their strengths as seen earlier in the HEL mixed response region for alumina.

The strength of brittle crystals is a function of the difficulty in slip down complex unit cells. A composite constructed from many such anisotropic components glued with a brittle glass phase and containing flaws and pores affords a rich tapestry of behaviours which result from this complex mix. Yet most originate from the bonding at the mesoscale where failure occurs in these ceramics and it is there that one must look for guidance to explain bulk response observed at the continuum.

6.8   Ballistics

The response of brittle materials to shock loading and their subsequent failure has implications for a number of practical scenarios involving impact for both engineering structures but also in nature. At present, the most successful models of penetration of armour constructed for brittle materials are semi-empirical since not all of the mechanisms operating during impact have been correctly identified and described. These include wave propagation occurring in the initial stages as well as the interaction of the geometrical boundaries during the penetration process.

The processes that occur behind a shock front, where both pressure and temperature are elevated and the material is driven at great speed, have been enumerated above. The consequences of these conditions give rise to an interplay between material response and the travelling wave fronts. It is very evident in brittle solids that it is not only the magnitude of the impulsive load that determines behaviour, but also the kinetics of

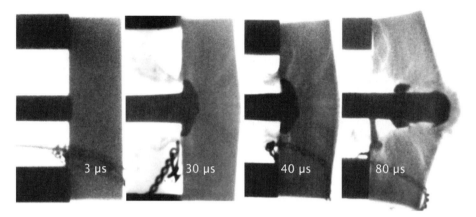

Figure 6.32 An X-ray sequence showing a steel rod 10 mm in diameter hitting a block of BS glass at a velocity of about 540 m s^{-1}. The first image was taken 3 µs after impact, the second 30 µs later. By the third frame, the rod has eroded a pit in the glass and the block begins to fail. In the final frame, 80 µs after impact, a disc of metal sheared off by the glass can be seen left behind in the surface crater. Source: reprinted from Bourne, N. K. (2005a) On the impact and penetration of soda-lime glass, *Int. J. Imp. Engng.*, **32**: 65–79. Copyright 2005 with permission from Elsevier.

the response of the material or structure that it contacts. In the shock-induced loading of high-strength glasses and ceramics, effects can be slow compared with the failure of the bulk which results in inelastic behaviour in the solid. The practical consequence of this is that a material like glass can stop an incoming projectile where a metal plate will not, even though the impact turns the glass into a pile of small fragments afterwards.

Consider what happens when a steel rod is fired into a block of borosilicate glass at a speed of 540 m s^{-1} (Figure 6.32). Within a few microseconds of the impact, the rounded nose of the rod is blunted as it flows across the glass. X-ray images reveal that after 30 µs the rod becomes shorter and slows down before the block fails. By this time, the shock wave has moved away from the impact site, so that it no longer plays a direct role in the events. However, the interaction of the wave with the glass has lowered its strength behind failure fronts, allowing the rod to penetrate its surface and thus determine the length of time for which the block can remain intact. At some point, the flowing steel erodes a pit in the glass. Cracks creep outwards parallel to the surface as chips fly back from the face, and a faint cone of failed material can be seen to form at the rear surface of the glass. Meanwhile, the steel is forced to flow back on itself away from the impact site.

Eventually, the glass shears off a disc of metal. This sharpens the rod, thus allowing it to move more easily through the remaining failed material. To maximise protection, the key to defeating the fragments is to tame the initial shock and thus slow the break up of the target block. Cracks in a brittle material like glass propagate at many kilometres a second, which might suggest that such a material would not support the impact stress. Yet when the loading pulse travels faster than the fractures can follow, parts of the material can retain their strength until the cracks arrive. Thus the right material can

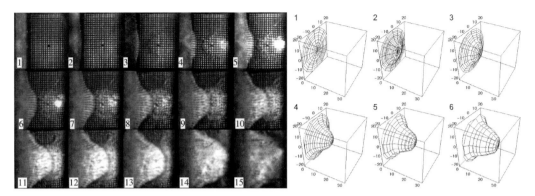

Figure 6.33 Penetration of a round-nosed rod through a 10 mm thick soda-lime block at 540 m s^{-1}. The first 12 frames are 5 μs apart and the last 3 are 20 μs. Source: reprinted from Bourne, N. K. (2005a) On the impact and penetration of soda-lime glass, *Int. J. Imp. Engng.*, **32**: 65–79. Copyright 2005 with permission from Elsevier.

protect vehicles, buildings or people if it is selected with a mind to its behaviour when a shock sweeps through it.

Ballistics has proceeded over the past 50 years examining behaviour using experiments that primarily involved flash X-ray imaging such as shown above and recovery of targets and (if possible) projectile after impact. In order to better understand the integrated response of the material to all the operating processes better-instrumented experiments are necessary to test response and to validate new materials models in hydrocodes. In previous sections, a simple introduction to basic concepts has been presented, but it is the goal of this book to describe the materials response to dynamic events and full details of terminal ballistics are better related in specialist texts. Nevertheless, the intimate relation of materials failure to the response of the targets in such loading has not always been appreciated so two pertinent examples are given below.

Experiments characterising ballistic impact must control parameters such as rod launch, pitch and yore. If a steel projectile similar to the previous impact is launched at a plate, the most sensitive measures of the interaction are seen at the rear surface. The photograph shows side and end-on views of the impact with the deformation transformed to 3D surfaces to the right-hand side. Such experiments can be performed with round and square nosed rods. Varying target thickness and nose geometry illustrates some general features. The penetration of the thinner tiles is dominated by the exit phase with material highly comminuted at the rear surface (Figure 6.33). Thicker tiles show more localised deformation with a divergent entry phase which allows surface deformation of rod followed by an exit localised to the impact axis. The impact of a penetrator with a flat nose gives more uniform deformation and penetration, and less radial flow.

The observed behaviours show the importance of the initial wave effects upon target behaviour. Firstly they precondition the target and determine subsequent performance, and secondly they define the form of the exit phase for the impactor. Nose geometry has a role in setting up conditions for preconditioning the target but as the target increases in

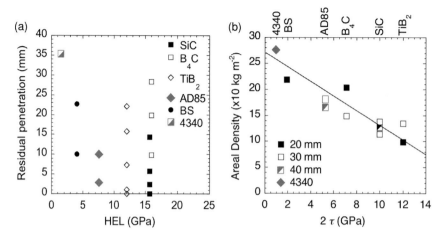

Figure 6.34 (a) Residual penetration vs. HEL for the six materials. (b) Areal density vs. strength in the failed state for 4340 steel and armour ceramics subject to normal impact at 1750 m s^{-1}. Grey points indicate metal and alumina targets discussed earlier. Source: reprinted from Bourne, N. K. (2008) The relation of failure under 1D shock to the ballistic performance of brittle materials, *Int. J. Imp. Engng.*, **35**: 674–683. Copyright 2008, with permission from Elsevier.

thickness, the entry phase becomes less important. The ultimate test for an impacting rod is in achieving a depth of penetration into a semi-infinite target. To convert kinetic energy in the projectile into removing a volume of target material represents a test of materials performance to characterise either impactor or target. In particular, such experiments are key indicators of which parameters determine target performance. In one series of experiments a tungsten rod was fired at fixed velocity at a series of targets; one a steel and the other brittle solids.

Figure 6.34(a) shows penetration data for three thicknesses of five ceramics placed onto a steel semi-infinite witness block and laterally confined, and impacted with the same projectile. The curves show the depth of penetration recorded in ceramic and steel, converted (in Figure 6.34(b)) to areal density, ρ_A, to mediate for the different densities encountered between the different ceramics

$$\rho_A = \rho_C t + \rho_{4340} d, \tag{6.10}$$

where t represents the thickness of ceramic plate (ρ_C), and d represents the penetration distance into 4340 steel (ρ_{4340}). There is an additional point where no ceramic plate was added to the block and impact was allowed to occur directly upon it.

In the plot one point (top left) was obtained by the rod impacting a monolith consisting just of the semi-infinite 4340 steel backing block whilst in other cases the facing tile was penetrated too. Figure 6.34(a) shows the correlation between penetration depth and yield strength (the HEL). Clearly there is scatter and little obvious dependence discernible. However, in Figure 6.34(b) there is a clear correlation between failed strength and areal density. The failed strength of the material is taken to be that measured at the take of the failure curve in the plot of Figure 6.30. Considering that there is a spread of velocities in these ballistic experiments, that a range of processes operate in the flow around a projectile through a comminuted ceramic, and that the waves are not divergent in plate impact, the relation is strong and indicates the value of creating quantitative measurements of the properties of a failed state. It is interesting to note that the material with the highest HEL, B_4C, does not have the best performance as might be expected on the basis of purely its HEL since beyond this elastic value strength rapidly falls away relative to the other ceramics.

6.9   The response of brittle solids to dynamic compression

The preceding sections have presented continuum measurements of stress histories, high-speed imaging and mesoscale modeling to illustrate the behaviours of the class of brittle materials including silicate glasses and polycrystalline ceramics. Near the impact face of both metals and brittles the strength decays from an elastic to an inelastic state with kinetics dependent upon operating mechanisms. Clearly this occurs in the weak and strong shock regimes but the response in the weak shock regime will dominate discussion here.

The discussion of the response of metals concentrated on the later time flow and interaction phase of the response since the localisation phase was transient and very fast in such loadings. It will be recalled that there is a transitory surface state that occurs on the impact face before the metal achieves state equilibrium and a stable peak stress. This relaxation zone transits the stress in the metal from a purely elastic state in the first few atomic planes to a plastic one some distance into the bulk by shear from defect sites which locally deform plastically and eventually communicate with one another to define the plastic state for the bulk. With inter-defect spacings of the order of 10 nm and communication at the shear wave speed this gives times for stress state stability of the order of a nanosecond over a localised zone of dimension c. 10 nm. Molecular dynamics simulations have confirmed this and calculated times of c. 500 ps for such equilibration. In metals this non-equilibrium relaxation zone is generally insignificant (10 nm) when compared with the dimensions of impacting bodies or the scale of transmitted impulses. The relaxation zone defines the region in which the localisation processes act to establish a plastic state and in metals cannot interact with processes occurring at the laboratory scale which occur on timescales and length scales thousands of times greater. In the case of brittle failure the response presented shows similar top-level behaviour yet with very different coupling to the continuum. Again the initial stages of deformation will relax the elastic stress in the material by cracks driven from defects in the bulk. Brittle fracture is the mechanism by which the material forms slip planes and the defects are voids or internal fracture zones within the solid.

In the case of BCC metals, high Peierls barriers to slip slow relaxation to a stable plastic state in c. 500 ns at a stress of 10 GPa. In the case of glasses the material holds its elastic strength for a similar time before it starts to decay to its inelastic state by the interconnection of microcracks. This failure front is seen before overtake occurs and illustrates release of the initial elastic compression by crack propagation from flaws on the impact face taking the material to its final failed inelastic end-state. Eventually failure occurs within a single shock front at *c*. 10 GPa in a glass. The defects are now of the order of 1 mm apart and fracture propagates at slowly increasing speeds reaching the Rayleigh wave speed (close to that of shear waves) as a maximum. This gives a time of the order of 1 μs for the equilibration of the Hugoniot stress and defines a relaxation zone on the front of the target of several millimetres in dimension. This is now at laboratory scales and the time for interaction of objects with brittle materials can be all, or at least occupies a significant part, of the localisation processes in the response.

A second difference in behaviour concerns the nature of the mechanism by which slip occurs in ductile and brittle solids. The reader will remember the definition of a microstructural unit (MSU) in a solid as the volume element which integrates a population of defects sufficient that after a relaxation period the element is in a stable mechanical state. Such a unit exists at a length scale below the state sampled by the observer to consider response. In the case of a purely brittle amorphous solid (such as a filled glass) the mesoscale contains the defect population that defines that length scale on which laboratory/structural scale observations depend. After the localisation phase microstructural units move into a flow phase where in the case of metals dislocations propagate to accommodate the applied strain by slip and eventually move into a locking phase where the microstructure can harden by interaction and accumulation of dislocation debris or interaction with grain boundaries. In the case of brittle solids fracture is the dominant flow mechanism and whilst it can be arrested by interaction with an interface (laminated glass works by this means) the scale of the relaxation zone generally means that the interface is likely to be the free surface of a sample or component in which case the material will release and fragments will separate giving zero strength to the target. In common experience impacted glasses produce an assembly of fragments whilst impacted metal is dented but intact. Thus in only the most specialised situations will there be any interaction between fracture surfaces that causes the material to retain strength in a flow after the localisation phase is complete. There is no equivalent to dislocation hardening processes in purely brittle response in for instance a dense glass where fracture surfaces operate as the slip planes.

The final class of materials of importance is composite polycrystalline ceramics. They have an anisotropic ductile phase bonded by a low volume of a glass binder which further contains a population of flaws at the mesoscale which result from material processing. The crystal phases are hard, inorganic materials with few slip systems and complex unit cells. This, and the anisotropy of the material bed, result in a little plastic strain being accommodated in twinning by the crystals but rather the vast majority leads to fracture between grains and down twin boundaries. Again the surface has a relaxation zone but this time it is activated at of the order of ten times the stresses at which metals respond because of the much greater elastic strength and further is of dimension millimetres. This time the failure processes are slower than in glasses with the inter-granular fracture process more difficult. When the elastic strength is overcome, the cracks may free grains which flow against one another and can exhibit greater resistance whilst under

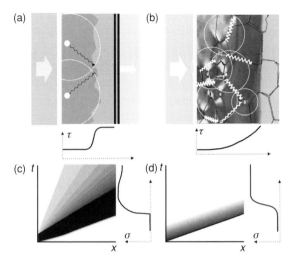

Figure 6.35 Shock deformation of a brittle solid below the WSL. (a) Pure brittle response. A surface relaxation zone exists whilst slip planes at the mesoscale are formed by defects which fracture. The strength in compression at late time is elastic behind the shock and lower behind a failure front. (b) Failure of a composite polycrystalline ceramic. Cracks propagate between grains and down twin boundaries from mesoscale flaws. The statistical defect population produces an integrated strength which decays with distance behind the front. Strengths are sketched beneath each schematic. (c) The weak shock regime. The elastic wave (black) precedes the shock front (grey). The shading represents the development of microstructure from black, undeformed to white, the fully formed inelastic state. The processes take longer times as the wave front develops in these brittle materials as the kinetics of deformation processes within them is slower. (d) The strong shock regime. The shock front contains the full development of microstructure. Schematic stress pulses are shown to the right of each.

compression as stress increases resulting in materials that can resist penetration after the relaxation zone has failed. Thus a locking phase is possible in ceramics.

These summaries illustrate the response of brittle solids in the weak shock regime to compression impulses on the impact face of a solid. Such a solid is one that accommodates compressive strain by slip down fracture planes at the mesoscale; defined to that at which defects are discernible and lead to an unstable state in response. The model for mechanical loading discussed for metals in the last chapter can be extended to brittle solids in the same manner here. The microstructural unit for brittle solids defines the onset of compressive failure and contains an assembly of crack nucleation sites within the solid. In glasses defects are hundreds of microns apart and to cross a MSU takes a few 100 ns (Figure 6.35(a)). Further, the defect population on interfaces drives the response of the solid as that within the bulk is orders of magnitude further spaced. Thus failure is driven from interfaces. The low density of defects in the bulk and the slow travel time of cracks means that the elastic state may be retained for hundreds of nanoseconds, compared with hundreds of picoseconds in metals. The failure time is governed by how long it takes such a unit to fail by fracture which will be of the order of a microsecond since communition and then consolidation must occur, which are both slow processes.

In the case of polycrystalline ceramics the defect density is higher and the elastic state is relieved by intergranular fracture (Figure 6.35(b)). However, the inelastic state still takes time to stabilise since the alumina grains must flow and bed and this process is again slow. The alumina ceramic AD995 is a composite having grains, within which slip systems are limited, and containing a brittle intergranular glass phase. It has shown itself capable of retaining its elastic strength for around 500 ns before relaxing begins to a failed state as with glass. Shock and recovery of AD995 alumina has shown evidence of twinning in the grains above the lower elastic limit of the composite ceramic and trans- and intergranular fracture within the microstructure in this range as discussed above. Again the high strength of the grains and the difficulty of preparation of hard solids means that a surface layer has a markedly higher defect population than within the bulk. Nevertheless the failure processes are still brittle and fracture surfaces are held under compression defining the strength for further flow. In the cases of both glasses and ceramics the tensile strength of the material is zero whilst in the case of metals it is not.

Figure 6.35 also shows schematic $x–t$ plots for shock propagation in a brittle solid; in (c) weak shock and (d) strong shock loading. As with metals, the weak shock regime is one in which the stress state is unsteady as it settles by defects communicating to fail the material and takes it inelastically to the Hugoniot for the solid. Fracture runs at speeds some fraction of the shear wave speed and further defects may be of the order of a millimetre apart in these materials. Thus the elastic state can persist for up to a microsecond before relaxing through a series of states as comminution of the material occurs. In the case of metals, however, defects are on a scale of microns apart and thus the relaxation of the elastic state is less than a nanosecond and not seen using conventional sensors within solids. Once failure has started the material cracks and fractured fragments flow until the comminutiae has attained a stable stress state on the Hugoniot and failure is complete through the front. The material is then in the strong shock regime (d) and the shock from then on travels at a speed faster than one at the intact strength of the material.

The overall picture painted here will be familiar to the reader from the previous chapter. However, the $x–t$ plots, and particularly those for the weak shock regime, look similar to metals but contain very different temporal evolutions in the two classes of material. The dislocation is a weakened plane at the microscale within the metal which accommodates strain by forming and reforming bonds with atomic dimensions between the planes. The crack is a free surface at the mesoscale and the two planes that then move to accommodate the strain do not have any bond between them. Thus the plasticity processes in brittle solids are taking place at a length scale far removed from the atomic dimension. Crystalline solids have a level of response to accommodate strain at a length scale lower than that for brittle solids. The material will access the scale that fails the material most easily as ever but compression selects defects of importance to respond to the acting piston. Once the length scale at which matter must respond has been defined the timescale is fixed by the communication of a plastic stress state over a representative defect population. It is that time which an experimental tool for investigating response must match to yield data from which a predictive model can be constructed and in materials such as these in many cases such a tool does not yet exist.

6.10   Final comments

As ever with this expanding field, new experimental tools and data evaluation techniques must be developed to illuminate the effects occurring within these deforming solids. Also there are key lessons to be learned from high-speed quantitative imaging so that comparison between technique and sensors can be used to validate mechanisms in material behaviour.

Micromechanics controls the conditioning of the impact zone ahead of an incoming penetrator defining the entry phase and the density of nucleation sites and nature of fracture in the projectile's path. At the stresses encountered in ballistic impact the material is in the weak shock regime and failure is delayed with the stress state unsteady for some time behind the shock. This process describes the phenomenon dwell observed in ballistic impact experiments. Further, the steady penetration depth into a semi-infinite block of damaging ceramic scales with the failed strength of the materials independent of whether the targets are metals, brittle glasses or polycrystalline ceramics. Of course once a released state is present at the exit of the penetrator, the strength drops ahead of it as confinement is lost.

To conclude, the ultimate strength of strong ceramics controls the conditions that define the subsequent response of the material since the relaxation zone will not yet be in a steady failed state. This is a function of the point at which they undergo inelastic flow. Failure in these materials is principally by micro-fracture between the grains which is a function of the density of defects activated by plasticity mechanisms within them and in the amorphous intergranular phase. Future work must completely define the mechanisms by which particular materials operate when subjected to load. Understanding the kinetics at work in these states will allow design of the materials themselves as well as better structures for all our protection in the future.

6.11   Selected reading

Fracture

Cox, B. N., Gao, H., Gross, D. and Rittel, D. (2005) Modern topics and challenges in dynamic fracture, *J. Mech. Phys. Solids*, 53: 565–596.

Freund, L. B. (1990) *Dynamic Fracture Mechanics*. Cambridge Monographs on Mechanics and Applied Mathematics. Cambridge: Cambridge University Press.

Griffith, A. A. (1921) The phenomena of rupture and flow in solids, *Philos. Trans. R. Soc. London A.*, 221: 163–198.

Lawn, B. R. (1993) *Fracture of Brittle Solids*, 2nd edition. Cambridge: Cambridge University Press.

Brittle solids under dynamic loads

Bourne, N. K., Millett, J. C. F. and Field, J. E. (1999) On the strength of shocked glasses, *Proc. R. Soc. Lond. A*, 455: 1275–1282.

Bourne, N. K., Millett, J. C. F., Rosenberg, Z. and Murray, N. H. (1998) On the shock induced failure of brittle solids, *J. Mech. Phys. Solids* 46(10): 1887–1908.

Chen, M. W., McCauley, J. W., Dandekar, D. P. and Bourne, N. K. (2006) Dynamic plasticity and failure of high-purity alumina under shock loading, *Nature Mater.*, 5: 614–618.

Grady, D. E. (1997) Shock-wave compression of brittle solids, *Mech. Mater.*, 29: 181–203.

Grady, D. E. and Kipp, M. E. (1993) Dynamic fracture and fragmentation, in *High-pressure Shock Compression of Solids*, eds. Asay, J. R. and Shahinpoor, M. New York: Springer-Verlag, pp. 265–322.

Kanel, G. I., Rasorenov, S. V., Fortov, V. E. and Abasehov, M. M. (1991) The fracture of glass under high pressure impulsive loading, *High. Press. Res.*, 6: 225–232.

Ballistics

Bourne, N. K. (2008) The relation of failure under 1D shock to the ballistic performance of brittle materials. *Int. J. Imp. Engng.*, 35: 674–683.

Rosenberg, Z. and Dekel, E. (2012) *Terminal Ballistics*. London: Springer.

7   Polymers

7.1   Introduction

The man-made and the natural environment surrounds man with a wide array of plastics
and polymers both natural and man-made. A polymer is a molecule which contains
repeated units of a particular chemical base segment, and there are many types found
across organic chemistry. A plastic is a term that covers a wide range of mostly synthetic but also some natural organic products that can be moulded or extruded. Thus
while all plastics are organic polymers, not all polymers are plastics and in general
may need to be modified with other additives to form useful materials. In everyday parlance plastic and polymer are terms often used interchangeably but in fact
there are many other types of molecules, both biological and inorganic, that are also
polymeric.

The word polymer has ancient Greek roots, compounded from *poly* (meaning *many*)
and *meros* (meaning *parts* or *units*), whilst *plastic* has a root that indicates a solid that is
malleable being easily shaped or moulded. Natural plastics may originate in biological
systems such as tar and shellac, tortoise shell and horns, as well as tree saps that produce
amber and latex. These plastics may be processed with heat and pressure into a host of
different products. If natural polymers are chemically modified then other plastics result
and during the 1800s these processes produced such materials as vulcanised rubber and
celluloid. In 1909 a semi-synthetic polymer was produced called Bakelite, soon followed
by fibres such as rayon (1911). The ability to work and particularly to cast or mould
them to component shapes made plastics increasingly dominant for manufacturing.
However, restrictions on supply of natural materials during the Second World War led
to the modern predominance of synthetic plastics. This time period saw development of
nylon, acrylic, neoprene, polyethylene and many more to replace the natural products
that could no longer be imported. Post-war the plastics business has developed into one
of the fastest growing industries in the world.

In the past 50 years manufacturers have grasped upon the adaptability of polymeric
materials to realise new possibilities in design and production. Applications have used
pure polymer materials but also employed them as composites with other material classes
such as fibre or second-phase particles embedded within. Their strength and flexibility
has allowed them not only to be cast but also to be drawn in a manner that optimises
material's microstructure to benefit from the inherent strengths of the polymer chain.
Thus as applications of polymers and polymer matrix composites grow, they are placed

under increasingly more extreme conditions in harsh environments such as in space or high-temperature conditions in production environments on earth.

In particular, increasing use of polymers and polymer matrix composites in structural engineering applications thus makes it critical to understand their mechanical response under high pressures and strain rates. In particular, the aerospace, defence and automotive industries require an understanding of the shock response to address safety concerns for vehicles and structures. Polymers have received relatively less attention at high pressure and strain rate than other classes of material. The early work addressed microstructural development under static pressure with little emphasis on dynamic response. However, a few key studies have been conducted on which the present discussion builds.

Work describing the shock loading of polymers indicated from the first behaviour different in nature from that observed in metals. For example, Champion, in studying polytetrafluroethylene (PTFE), noted a change in slope in the shock velocity, U_s–particle velocity, u_p curve, corresponding to an impact stress of about 0.5 GPa, and giving a volume change of 2.2%. However, the first comprehensive overview of the shock response of thermoplastics, thermosets and rubbers was only done relatively recently by Carter and Marsh (1977, republished 1995). More modern work has built on these studies and will be reviewed in the sections below.

7.2   Classification and key behaviours

7.2.1   General comments

Polymers are a class of amorphous solids with varying degrees of order to their microstructure in the bulk controlled by the thermomechanical history of their synthesis. Quasi-statically they display a number of different regimes accessed as temperature rises and in which they behave very differently. In the glassy regime polymers exhibit high stiffness (large modulus) that drops across the glass transition as the microstructure expands and becomes a rubber. At higher temperature response again enters a viscous regime where it starts to flow with a final phase at higher values still where they eventually decompose. Thus the mechanical properties of polymers change rapidly with temperature but also interestingly with the time that the materials spend under load. Here the weaker bonding found in bulk polymers can be easily overcome by the forces exerted under load, rearranging microstructure by analogy with what happens under heat. This makes their response markedly different from the metals and ceramics considered earlier, but of course with behaviour intimately linked to the microstructure of the classes of plastic found in applications.

The polymer chains themselves may adopt amorphous or ordered conformations according to the chemical nature of the repeat unit. Polymer crystallinity must also be understood, since the mechanical properties of crystalline polymers are different from those of amorphous ones. However, crystallinity here is not of a close-packed nature familiar from metals or ionic materials; rather it refers to an ordered arrangement of polymer chains within a spherulite, which is again not fully dense. However, as one may expect by analogy, polymer crystals are both stiffer and stronger than amorphous networks and this may be used to advantage to produce strong fibres. In particular, it is common in polymeric processing to engineer semicrystalline materials where the microstructure consists of crystalline domains embedded within an amorphous matrix that is much less ordered. This is particularly true for the class of thermoplastics, so extensively used in modern manufacturing. Such solids therefore display different response to loading at quasi-static rates than under shock.

In metals, ionic solids or ceramics the atomic structure, comprising repeating unit cells with the component atoms of the material, is key. The crystalline solid contains regular arrays and thus may be described with reference to bonds within that fundamental volume. In a polymer, the molecular building block which forms the chain, the monomer from which it is made, takes an equivalent role but provides only short range order. These are known as repeat units. A huge variety of common polymers exist and the monomer units for a selection used here to illustrate behaviour are shown in Figure 7.1. The chains themselves and their arrangement under different environments determine statistical limits upon observed behaviours. However, their complex arrangement make analysis more difficult.

The simple polymer chain consists of repeat units of the monomer down its length. The average length of such a chain can be represented by analogy with a random walk in arbitrary forward directions and thus the length of the chain is $\sqrt{(n\lambda)}$ where n is the number of the repeats and λ is the length of the copied unit of the chain. An example is shown in Figure 7.2(a). In some monomer units, however, it is possible to arrange these units in different manners which are distinct from one another in atomic arrangement. One of these conformations occurs as a result of rotation around the chain. When atoms are coplanar, the unit adopts a higher energy to that when they are apart. This results from the tetrahedral arrangement of bonds around the carbon atom. If there are side groups along the chain, their geometrical arrangement with respect to one another will alter the energetic state of the unit with their proximity driving the arrangement into a higher energy conformation.

In these arrangements the different energy states are given specific nomenclature which is concerned only with the geometry of the chain. Looking down a carbon–carbon bond in a monomer unit shows two extremes of arrangement accessed by rotations around the bond. These rotational isomers all have different energy states which depend only on their conformation. In a simple organic molecule where the side groups are on the same side the conformation is called *cis* and has a greater energy than that where they are staggered (*trans*). In polymer chains, however, the chains on either side of an individual bond are more complex and take on a theoretically infinite number of conformations. Where the backbone continues down the same side of the bond this arrangement is called *gauche* whereas if it passes down one side it is dubbed *trans*. The *gauche* state has higher energy and entropy than the *trans* simply as a result of its geometrical configuration, although the difference is small for simple structures. In dynamic loading, particularly shock, transient temperature rise will quickly take the chain to states in which this barrier is overcome, however, so that its significance is marginal in such situations (Figure 7.2(b)). It is the inter-chain distance δ which controls

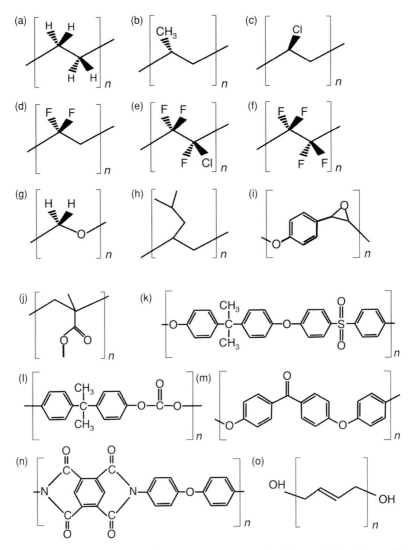

Figure 7.1 Polymers discussed below with monomer units shown: (a) PE; (b) PP; (c) PVC; (d) PVDF; (e) PCTFE; (f) PTFE; (g) POM; (h) P4M1P; (i) epoxy; (j) PMMA; (k) PS; (l) PC; (m) PEEK; (n) PI; (o) HTPB. The sketches mix notation to show bonds in the simpler molecules whilst adopting the normal convention of representing C atoms at the breaks in slope of lines with H atoms not shown in the more complex ones.

the dynamic behaviour of the plastic at the smallest length scale and this is determined by the confining pressure and the Van der Waals' forces between chains.

7.2.2   Crystallisation of polymers

In the general case the chain will contain random arrangement of molecules and side groups and this will drive the entropy of the system. In amorphous polymers the material

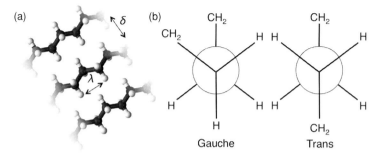

Figure 7.2 Polymers chain configurations: (a) simple polyethylene chains, λ = length of unit, δ = interchain distance; (b) rotational isomers of the rotated repeat units down the chain. Trans is the lowest energy conformation.

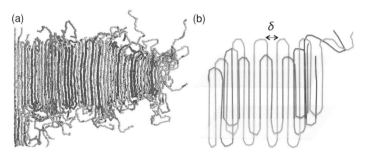

Figure 7.3 Polymers chain stacking: (a) chain folds on itself held by Van de Waals bonding; (b) three-dimensional lamellae formation with stacked polymer folds.

adopts the maximum entropy conformation (given by the Boltzmann distribution), and the chains are arrayed randomly throughout the material. When there is some degree of crystallinity the entropy is lowered. It is thus important to determine how much crystallinity exists within a polymer target in order to attribute conformation changes during loading to microstructural rearrangements that have occurred and there are two main techniques that are capable of delivering such information. One is X-ray crystallography (as with more regularly bonded solids) whilst the other is differential scanning calorimetry (DSC). Interestingly these two techniques can deliver different results from one another.

In semi-crystalline polymers, the material contains crystalline regions embedded within an amorphous matrix. No polymer is completely crystalline, but careful manufacture can attain structures that have a high percentage of crystalline phase. For instance, high-density polyethylene is 80% crystalline. The polymer crystal is made up from one-dimensional sequences of folded chains where the repeat distance between is given by the chain spacing (Figure 7.3(a). The chains fold one upon the other to form sheets dubbed lamellae that are of the order of microns in extent and c. 10 nm in thickness (Figure 7.3(b)). In a melt there are always heterogeneities such as dust or foreign particles which act as nucleation points. The crystals grow from these regions within the structure in helical strands that radiate out from a nucleation point to form regions that

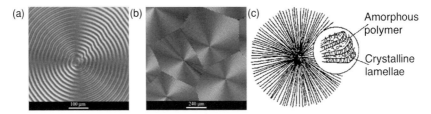

Figure 7.4 Spherulite structures with cross-polarised light microscopy. (a) Spherulite in poly-3-hydroxy butyrate (PHB), scale 100 μm. (b) Spherulites in PHB, scale 240 μm. (c) Schematic of microstructure within spherulite. Source: images copyright DoITPoMS Micrograph Library.

appear circular in pictures of the microstructure. These structures are know as *spherulites* and are formed of helical strands stacking out to occupy regions of the microstructure (Figure 7.4). The lamellae grow out radially interspersed with amorphous material, and have characteristic appearance viewed in section with polarised light showing a cross form on the image (Figure 7.4(a) and (b)). The maltese cross is simply oriented to that of the crossed polar filters used to image the pattern but indicates the order in the polymer chains beneath interacting with the incoming beam to filter the transmitted light.

Despite the helical structure determined by the spiral growth of individual crystalline domains, and the circular cross section of growing spherulites, they pack together to form polyhedra as they impinge upon one another at a higher length scale with grain-like morphology and with planar boundaries. The appearance is thus similar to an etched metal at this length scale. In semi-crystalline plastics the lamellae define the mesoscale microstructure of the polymer and have a new population of defects which have a population at this scale.

7.2.3   Microstructure

7.2.3.1   Thermoplastics and thermosets

The comments above show that plastics adopt one of two classes of microstructure; they are either entirely amorphous or contain a proportion of their volume which is partially crystalline. Clearly the amorphous materials have no long-range order and consist of a three-dimensional extended network structure. The second class of materials consists of simple linear chains that may be partially ordered into crystalline structures. The latter group are dubbed *thermoplastics* whilst the former are called *thermosets*.

The picture of the polymer chain painted above, packed into a spaghetti-like microstructure of intertwined backbones, is typical of the group of thermoplastics of which simple polymer backbones such as polyethylene (PE) or polysulphone (PS) are members. The polymer chains are generally a linkage of carbon–carbon bonds (sometimes with other atoms present such as oxygen) either single or double bonds according to the chemistry of the monomer. The polymer chains themselves may either order into the zig-zag structures seen in spherulites where they are relatively closely packed with therefore greater strength, or, if they are free chains randomly intertwined, then the forces between will be the electrostatic Van der Waals' or London attractions. The Van der Waals' force describes the integrated intermolecular attractive and repulsive forces. The attractive components result from electrostatic interaction between adjacent charged molecules, polarisation and correlations in the fluctuating polarisations of nearby atoms. The repulsive component prevents the collapse of molecules as they are compressed. In the case of polymer chains, these forces are a consequence of dipoles formed as a result of the electronic state of adjacent atoms on chains in close proximity to one another. This has the result that the forces between chains are at least an order of magnitude smaller than those along the carbon backbone.

In these materials heating the material is a means of allowing chains to overcome these weak electrostatic attractions and to slide over one another. By this means melting occurs. Plastics may be linear chains or additionally have side groups dangling from them. The geometrical arrangement of these groups has further effect upon microstructure. This property is termed *tacticity* and describes the relative stereochemistry of centres in neighbouring units within a macromolecule. If all side groups are positioned on the same side of the chain the polymer is said to be *isotactic*. However, it may be possible for the groups to alternate regularly in which case the chain is said to be *syndiotactic*. Finally, the groups may alternate randomly in which case the structure is said to be *atactic*. If the chains are folded upon each other and form crystalline structures then regularity is required in the stereochemistry and so all but the atactic structures are able to form crystalline microstructures.

7.2.3.2 Amorphous polymers

Three-dimensional networks may be formed by cross-linking the polymer chains during synthesis of the plastics. Epoxies, polyesters and resins are examples of this class of material. This cross-linkage of the polymer chains means that the bulk has no fixed melting point but rather a region exists over which bonds are broken in the material. Cross-linking may be achieved during fabrication using chemicals, heat, or radiation and this process is called curing or vulcanisation. Once cured, *thermosets* cannot be remelted or reformed into another shape so giving them a fundamentally different nature to the thermoplastics discussed above. In general they exist (under ambient conditions) below their glass transition temperature.

Elastomers are also amorphous polymers but they exist primarily above their glass transition temperature so that much more segmental motion is possible. The response of elastomers reflects a linear geometry with a small amount of cross-linking as will be seen in the examples that follow.

Clearly the behaviour of these materials is critically dependent on the transition in behaviour that occurs as the amorphous microstructure heats. The response to an impulse is thus determined by how the rising pulse couples with the length scales of the various microstructural elements. At the base level is the polymer chain itself with covalent carbon–carbon bonds of the order of 1.5 Angstroms. At greater scales the chains themselves are more weakly bound by the dipole interactions between them. In polyethylene this distance is of the order of 5 Angstroms. Finally, there are crystalline and amorphous regions, showing great directionality in bonding and properties within the bulk material as it is loaded. These crystalline regions are at the mesoscale and are of the order of 1 μm in size. The interaction distance of a pulse front rising in a nanosecond is of the order of a micron in typical polymers and this gives an indication that the front interacts with the mesoscale structure as it propagates from the impact face through the target.

In all that has been said above, the properties have been related to the morphology that arises from the base monomer chemistry. However, the materials used in engineering are composites that contain structures on a greater scale. For a given base monomer a variety of manufacturing processes can be employed to produce a selection of grades or classes that can be sold under either industry-wide nomenclature or under a brand name. Moreover, a number of additives and fillers are commonly added to industrial polymers to improve specific properties, mitigate aging processes, or simply reduce cost by addition of inexpensive additives such as talc or Wollastonite. These differences can significantly change polymer properties, making it important to understand the specific polymer pedigree. Such composites are more correctly refered to as plastics.

7.2.4   Glass transition

Despite some regions of order in thermoplastics particularly, all polymers exhibit some proportion of their structure which is in the amorphous state and which, when heated, does not exhibit a sharp melting point. Thus even highly crystalline polymers do not have a well-defined melting point since the amorphous phase exhibits a glass transition. Amorphous materials have free volume which is lost as temperature decreases from the melt. Solidification may thus be regarded as an increase in viscosity during cooling until free volume is removed and the polymer becomes glassy. For an amorphous plastic such as PMMA this occurs at around $100\,^\circ$C. The glass exists in a metastable state, and entropy and density depend on the cooling history. Thus the glass transition is a dynamic phenomenon with the time for the process to occur and the temperature history adopted interchangeable quantities. Thermodynamically it must be described by a statistical threshold such that the free energy exceeds the activation energy required to move a statistically significant proportion of microstructural elements. The transition thus also acts as a flow criterion and so T_g must depend on the conformation of the chains and the molecular composition of the monomer unit. Thus architecture of the chain, and the addition of large side groups or creation of suitable copolymer compositions, can be used to control T_g for particular applications. For instance, heating a fabric above the transition with a heavy iron crosses the glass–rubber transition allowing alignment of the polymer chains in cloth under the weight applied.

A large number of dynamic polymer properties obey the time–temperature superposition principle which results from this. This behaviour has a direct bearing upon the properties observed in dynamic loading since here mechanical work done on the chains, rather then the application of heat, results in rearrangement of the polymer microstructure. Thus, above the glass transition temperature the properties are dominated by entropy, whilst at lower temperatures the orientation of the chains within the

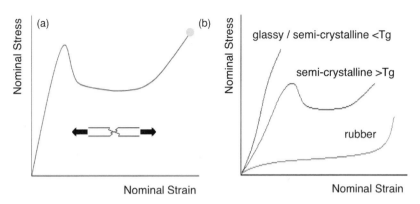

Figure 7.5 (a) Typical polymer stress–strain curve. (b) Effect of temperature above and below T_g and rubber behaviours. Images copyright DoITPoMS Micrograph Library.

structure maintains the strength of the polymer. As temperature increases with mechanical work, stretching of the bonds reduces entropy but since there is a region of behaviour with order at a range of scales, substantial strain recovery is possible on unloading. Thus whilst the name might suggest otherwise, plastics are not 'plastic' in the sense one might read over from metallic behaviour since deformation is not irreversible.

7.2.5 Polymer stress–strain curve

Stress–strain curves show the response of a material to an applied (usually tensile) stress. They allow important information such as a material's elastic modulus and yield stress to be determined for continuum conditions. Accurate knowledge of these parameters is paramount in engineering design. Plastics, however, show a different type of response to metals and brittle materials. Although all matter shows time dependence in its response, the strength and statistical nature of the interchain bonding in polymers, and the presence of other phases in plastics renders their response to fast loading more delayed. Thus even in tests regarded as quasi-static, the response to deformation may be inherently time-dependent which makes their behaviour very different from other classes of material considered above. This is seen not only in the loading phase but in the unloading phase as well where significant recovery may occur over hours or days after the test. Machining a material to a particular tolerance for instance gives a stable result for polycarbonate, whereas the sample moves significantly over days after in the case of PMMA. Thus attempts have been made to account for this dependence in representations of the constitutive response of the material. From an engineering perspective the mechanical response has significant dependence on time (normally represented as strain rate) and temperature. However, it is clear that there are a range of mechanisms operating with different kinetics in the response of plastics and that fitting a law to one regime is insufficient to describe any general response.

Typically a polymer shows three regions of behaviour in (for example) a simple tensile test (see Figure 7.5). There is an elastic region where the stress rises initially at least linearly to some peak after which necking commences and the stress drops back to a lower value. This yield behaviour occurs at around 5–10% strain; much higher than seen in a metal for instance. It may be that brittle failure may occur at this stage and the process halts here. In other plastics the neck is elongated in the second region, in which the polymer chains are gradually unwound from one another, increasing strain and allowing the material to extend at constant stress. This behaviour is called cold drawing and extends the microstructure along the sample loaded until the necking process is completed down the sample. In the final stage of loading, there is a strain-hardening stage since the strength of the bonding down the chain length means that the polymer is now significantly stronger than it was before, as a result of chain alignment down the loading axis. This process continues until no further alignment is possible and fracture occurs.

As shown in the discussion above, the mechanical response of polymer materials couples with the processes that allow the chains to unwind from one another and align during the loading. As temperature is lowered or loading time is decreased, then there is less opportunity for this to occur. A thermoset below its glass transition temperature will behave in a brittle manner whereas a semi-crystalline polymer or a rubber above its glass transition will behave in the manner shown above (Figure 7.5(b)). Following that, the same behaviour at higher strain rate will fail the polymer in a brittle manner too, since now the loading rate will not allow the reorganisation of the polymer chains parallel to the direction of the applied stress. Thus temperature and time show equivalent effects which must be captured in a model to describe the behaviour observed in dynamic experiments. Further, the picture is richer still since there is a great spectrum of behaviour across different polymer chemistry and microstructural architecture. In shock, the structure is essentially frozen at the largest scales since there is little time for kinetics to occur and small applied strains to drive rearrangement. It is with this background in mind that the following discussion of response in shock loading will be constructed.

7.2.6   Yield in polymers and the effect of pressure

Earlier discussion has shown that polymers have strength determined by microstructural rearrangements at three distinct length scales. This behaviour ensures that determination of strength by traditional means renders a series of values with some statistical spread fixed by variations in the packing of the chains within the structure. In thermoplastics with crystalline domains, strength can vary by a significant factor down a chain length or across a grain. Thus anisotropy can be a strong factor at the mesoscale giving rise to such behaviours as variable wave speed and packing density.

Polymers also show strength in compression which is greater than that observed in tension. This means that polymer yield surfaces are not symmetric. Yield stresses are also shifted to higher values in the compression quadrant. This results from the molecular arrangements within the solid under ambient conditions. Further, free volume present in the microstructure means that mechanical properties are sensitive to pressure. Typical results for such behaviour are shown below for pressure (Figure 7.6).

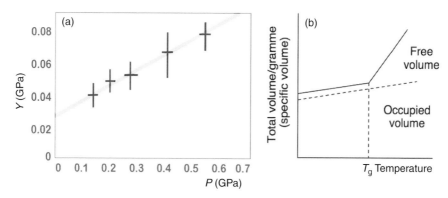

Figure 7.6 (a) Schematic effects of pressure upon yield stress. (b) Free volume as function of temperature (strain rate).

The behaviour observed when load and temperature cycles are superposed gives rise to complex behaviours, as seen above, which have effects that stabilise at times beyond that of the loading itself. When a polymer is loaded, unloaded and then annealed for instance, the material appears to exhibit the same features as an aged microstructure has.

To understand this behaviour it is necessary to consider the nature of the bonding within the polymer. The polymer chain consists of a series of covalent bonds that hold the material in geometries adopted by the tetrahedral conformation of carbon, forming the basis of this class of structures. These chains are modified but other elements, principally in side groups, interact with one another in a secondary manner. These ionic interactions are generally over greater distance and are in the form of weaker Van der Waals' forces. Of course in amorphous rubbers or thermosets there are cross-links sculpting a three-dimensional network of such bonds. Thus over long times and larger temperatures, statistical materials like polymers will tend to their equilibrium states. However, this book is concerned with transitory and extreme loads where the material is placed under very different conditions and where the chances of reaching equilibrium are at best severely restricted or where densification has reached a new state. There is a large raft of phenomena to explore in this arena as the following sections will illustrate.

7.3　High strain rate properties

The response of polymers loaded with ever-faster rising impulses shows different regimes of behaviour. Testing at intermediate rates from quasi-static to Hopkinson bars shows strengthening over a broad range of strain rates up to 10^3 s^{-1} of the same order as seen in metals. However, above this value there are different regimes apparent. The first group of microstructures shows rate sensitivity decreasing above 10^3 s^{-1} whilst a second shows constant rate sensitivity. A final group shows strength increasing above 10^3 s^{-1}. This behaviour is shown in Figure 7.7 for three polymers. Clearly the response is governed

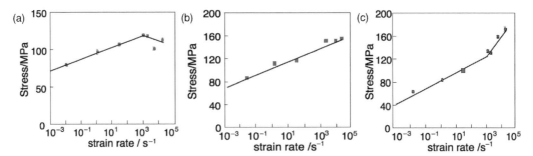

Figure 7.7 (a) Yield stress vs. strain rate for PES. (b) Yield stress vs. strain rate for dry nylon 6, a partially crystalline thermoplastic. (c) Yield stress vs. strain rate for PVC, an amorphous thermoplastic. Source: after Walley, S. M., Field, J. E., Pope, P. H. and Safford, N. A. (1989) A study of the rapid deformation behaviour of a range of polymers, *Phil. Trans. R. Soc. Lond. A*, **328**: 1–33.

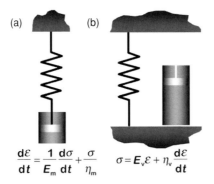

$$\frac{d\varepsilon}{dt} = \frac{1}{E_m}\frac{d\sigma}{dt} + \frac{\sigma}{\eta_m} \qquad \sigma = E_v\varepsilon + \eta_v\frac{d\varepsilon}{dt}$$

Figure 7.8 The spring and dashpot representation of polymer response: (a) Maxwell element; (b) Kelvin-Voigt element. The form of the constitutive relation connecting stress σ and strain ε for each element is shown beneath, where E is the elastic modulus and η and the assumed viscosity.

by the nature of the microstructure existing within these polymer materials. However, with existing data, it appears that behaviour and structure appear uncorrelated.

Mechanical response in polymers has been traditionally represented mathematically using an analogue. The Maxwell and Kelvin-Voigt models represent viscoelastic behaviour using mechanical components such as springs to provide resistance to motion in phase with it and which acts to resist displacement, and dashpots which resist motion via viscous friction so providing a resulting resistive force proportional to velocity which absorbs energy. A combination of springs and dashpots (see Figure 7.8) in series and parallel with one another add time dependence to the behaviour of the strength model for the polymer and allow delayed response to step loads and stress relaxations to be reproduced. In the most modern representations of polymer behaviour under shock loading many springs and dashpots must be employed to fit to the required response. This formulation allows good agreement with experiment but the fitting employed must be sensitive to the range of loadings to which it may be applied after the model has been constructed.

7.4   Dynamic compression of polymers

Whereas mechanical data exist in some quantity for the response of polymers subject to quasi-static conditions and laboratory rates, this is not the case as the loading becomes more rapid and the amplitude increases. In the shock regime this is especially true, with only one extensive study completed (using explosive lenses by Carter and Marsh) and written up into an openly accessible report by Fritz and Sheffield. This means of loading only subjected targets to high stress levels and Taylor wave loading and so little detail was seen for lower amplitudes where the shock was not overdriven. Modern techniques using new tools such as PDV with more established ones such as particle velocity gauges, and encapsulating sensors into the flow to observe bulk states evolving with the applied load, open up a much richer picture into the observed behaviour. Nevertheless, the pioneering work of Carter and Marsh is still the largest definitive body of work that exists for shock data to high pressure. The following text will refer to and base comment on the data they collected extensively throughout the pages below. However, their deductions still largely hold true due to the paucity of work done in this area, although more detail is slowly becoming available as time goes on. It thus seems appropriate to quote the authors' abstract to that work summarising their findings.

The Hugoniot equations of state of a large number of representative polymers have been obtained. Two aspects of the results are particularly striking (Carter and Marsh, 1977):

(i) The $U_s–u_p$ Hugoniots of all the polymers extrapolate to bulk sound velocities higher than the ultrasonic values, an indication of a rapidly varying rate of change of compressibility in this region. This is attributed both to the two-dimensional nature of polymer compression and to the form of the inter-chain interaction potential.

(ii) A relatively high-pressure transformation (in the range 20–30 GPa), characterised by a change in slope of the $U_s–u_p$ Hugoniot and sometimes by a large volume change as well, is observed for all of the polymers. This transformation is probably associated with pressure-induced cross bonding. In particular, for those polymers which contain rings in their monomer structure and which display the largest volume change at transformation, it is proposed that carbon–carbon covalent bonds along chains are broken and tetragonal bonds between chains are formed in a manner analogous to the graphite–diamond transformation.

The equation of state of polymers is then an obvious place to start a review of properties. As in previous sections, several key materials will be highlighted and followed in detail through their equation of state and strength properties. It will be important to try and contrast in the discussion the responses of thermoplastics, thermosets and rubbers and of course copolymers containing a combination of these units within. First, however, we will present a table of a series of polymers with some critical properties to aid the reader in comparing the various materials in what follows (Table 7.1).

The above range of materials is far from an exhaustive selection but includes all the polymers that are discussed later in the text. This gives some means of assessing the

Table 7.1 Dynamic properties for a series of polymers

Name (abbreviation used in the text)	ρ (g cm^{-3})	c_L (mm μs^{-1})	c_s (mm μs^{-1})	c_0 (mm μs^{-1})	S
Acetal: polyoxymethylene (POM)	1.41	2.29	0.79	2.50	2.02
Hydroxyterminated polybutadiene (HTPB)	0.93	1.46	1.00	1.53	2.84
Poly-(4-methyl-1-pentene) (P4M1P)	0.83	2.19	1.08	2.07	1.59
Polyamide (Nylon 6-6) (PA)	1.14	2.54	1.08	2.67	1.69
Polycarbonate (PC)	1.20	2.19	0.89	1.87	1.37
Polychlorotrifluoroethylene (PCTFE*)	2.13	1.74	0.77	2.05	1.66
Polyethylene (PE)	0.95	2.46	1.01	2.86	1.57
Polyetheretherketone (PEEK)	1.30	2.47	1.06	2.52	1.71
Polypropylene (PP)	0.90	2.58	1.26	2.86	1.49
Polysulphone (PS)	1.24	2.25	0.93	2.35	1.55
Polyurethane (PU)	1.27	2.39	1.03	2.54	1.57
Polymethylmethacrylate (PMMA)	1.19	2.72	1.36	2.6	1.52
Polyvinylchloride (PVC)	1.38	2.29	1.08	2.33	1.50
Polyvinylidenefluoride (PVDF)	1.77	2.10	0.85	2.58	1.58
Teflon Carter & Marsh (PTFE High P)	2.15	1.29	0.71	1.84	1.71
Teflon Low P regime (PTFE Low P)	2.15	1.23	0.41	1.14	2.43
Thermoplasticpolyurethane (TPU, Estane)	1.14	1.75	0.75	2.43	1.65

*a component of Kel F.

effect of microstructure and chemical composition on the observed dynamic response of the materials. The first necessity is to find a means of assessing the elastic properties of the material. Certainly an obvious measure to begin with is by looking at the longitudinal wave speed in the material. In metals and brittle materials ultrasonic transducers can be used to measure the longitudinal (c_L) and transverse (c_s) wave speeds in the material. Typically such transducers have a dimension which is on the order of 10 or 20 mm which clearly averages response over the microstructure encompassed in the measurement. If that is the case, the isotropic relation for wave speed can be used and the bulk sound speed (c_0) can be calculated through the relation

$$c_0^2 = c_L^2 - (4/3)c_s^2. \tag{7.1}$$

Clearly, this may be inappropriate in the case of polymers for several reasons. Firstly the moduli of polymers are critically dependent on frequency so that wave speed too can alter with the transducer selected and secondly there is critical temperature dependence found for many polymer properties as seen earlier. Finally, thermoplastics in particular are very anisotropic (difference of more than a factor of ten between bond strengths down and between chains) so that crystalline phases will show similar anisotropic response.

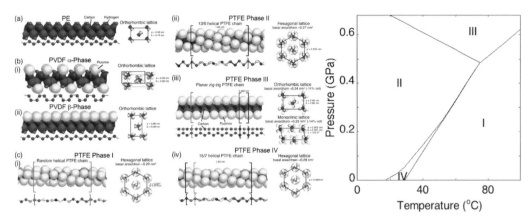

Figure 7.9 The crystal structures of (a) polyethylene (PE), (b) the two most common polymorphs of polyvinylidenefluoride (PVDF) and (c) the four phases of polytetrafluorethylene (PTFE). (d) The pressure–temperature phase diagram. Source: reprinted with permission from Bourne, N. K., Millett, J. C. F., Brown, E. N. and Gray III, G. T. (2007) The effect of halogenation on the shock properties of semi-crystalline thermo-plastics, *J. Appl. Phys.*, **102**: 063510. Copyright 2007 American Institute of Physics.

Thus materials with high crystallinity, and in cases where the size of the crystalline domains is large in comparison to the sensor, must be treated with caution and the wave speed confirmed down different directions to compare results and to ensure that isotropic relations can reliably be used.

As an example of the high-pressure response of polymers, a thermoplastic has been chosen to illustrate the complexity found in the low-pressure response. A commonly used material is polytetrafluoroethylene (PTFE, tradename Teflon), which is a widely employed fluoropolymer for engineering applications. The material is semicrystalline under ambient conditions with its linear chains adopting several, complex phases within crystalline domains near room temperature and ambient pressure. The phase behaviour of PTFE is shown in Figure 7.9 where the conformation of the chain can be directly compared with the simpler polyethylene (a). Under laboratory conditions at room temperature, a pressure-induced phase transition has been reported in PTFE at 0.50–0.65 GPa (Phase II–III). Phase II PTFE consists of a helical conformation with a 13 atom repeat unit and a well-ordered hexagonal packing of the helical chains, while in phase III the helical conformation gives way to a planar zig-zag and the chain packing takes on an orthorhombic or monoclinic lattice structure. Recent work using a high-pressure diamond-anvil cell and near-infrared Raman (NIR) spectroscopy suggests the transition occurs at 0.65 GPa and exhibits around ±0.05 GPa of hysteresis. The phase transition results in a 13% local volume change within the crystalline domains and a considerable reduction in compressibility. PTFE also exhibits two atmospheric pressure, crystalline transitions at 19 °C and 30 °C.

Figure 7.9 shows that in the impact stress regime this, and indeed polymers in general, have rich and varied response giving rise to a series of possible behaviours which are potentially accessible by a dynamic event which raises both the temperature and

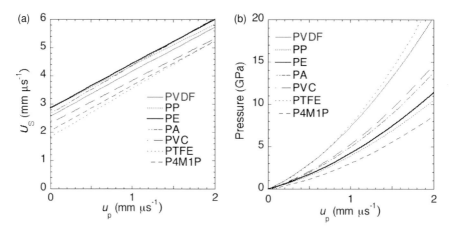

Figure 7.10 Schematic low-pressure polymer Hugoniots for a selection of thermoplastics constructed from measured c_0 and S. (a) Shock velocity U_s v. particle velocity u_p; (b) pressure vs. particle velocity. The materials' Hugoniots scale in this space (from the lowest upwards): poly-(4-methyl-1-pentene), polypropylene, polyethylene, polyamide, polyvinylchloride, PVDF and PTFE. PE is represented as a heavier line for reference.

the pressure. This offers the possibility of accessing states down this loading path not reached by other means which suggests that these materials have important practical uses which are as yet unexploited.

7.4.1   Equation of state

Determination of the low-pressure Hugoniots for a range of thermoplastics shows several behaviours placing the relations obtained for polymer behaviour into a series of defined groupings. In what follows, the c_0 and S quoted in the literature are used to construct Hugoniots using the conservation laws. This allows easy comparison of material properties but densification in these materials means that the Hugoniots have more structure as the lowest stresses as will be seen in later sections. A number of thermoplastics is illustrated in Figure 7.10 to show the observed behaviour: the shock velocity–particle velocity response is shown in Figure 7.10 (a) whilst the pressure–particle velocity behaviour is shown in Figure 7.10 (b). It should be said at the outset that thermoplastics have a range of polymorphic phase transformations which will be commented on in more detail in the discussion below. Other thermoplastics too have a series of these which will determine their observed behaviour at these low pressures. The transitions in thermoplastics involve repacking of the carbon chains and typically occur at a few GPa consistent with particle velocities of tens or low hundreds of m s^{-1}. It will be seen that the effects of these upon the measured shock constants are subtle in this regime, and so data taken from Carter and Marsh where the material is loaded to higher levels is used to compare response between different monomer chemical units.

The simplest microstructure, to which other behaviour is referenced in what follows, is that of polyethylene (PE) and it is plotted as a heavier solid line. It will be seen that it lies in the mid region of the thermoplastics chosen for this comparison. The U_s–u_p data show variation only in the bulk sound speed of the material that can be seen to change across the range of microstructures presented. This relates to the variation of the modulus of the material and also to the changes in density that occur due both to chemical content and packing of the chains. The slopes of these lines are very similar for this range of plastics. It has been hypothesised that the slope is a function of the pressure dependence of the bulk modulus of the material and normalising for density shows a range of values for this quantity. The value of the bulk sound speed constant, c_0, obtained experimentally varies by 30% and this reflects the microstructural response under pressure that these materials experience. Unlike simple metals, a bulk sound speed calculated from ultrasonic measurements of c_L and c_s does not correspond to the intercept of the U_s–u_p plot as mentioned earlier and it generally lies above this value. This behaviour shows the collapse of the microstructure with the modest bonding found between polymer chains.

A better discriminant of base-unit form is found comparing the pressure–particle velocity response. There is now a spread of curves as seen in the figure and again PE lies near the centre of this grouping. Below it is another simple thermoplastic, polypropylene (PP). The methyl group on the carbon backbone may be expected to have two effects upon the response, making the chain more bulky and providing steric hindrance to flow in the loaded polymer. This results in a lower density that is seen in this phase space but will also reduce Van der Waals' interaction between the chains.

Poly-(4-methyl-1-pentene) (P4M1P) is a semicrystalline polyolefin where the presence of an even larger group off the chain accentuates both of these features. It is the only semicrystalline polymer in which the density of the amorphous component is higher than that of the crystalline fraction at room temperature. It's low density and large side groups rank the Hugoniot of this hydrocarbon below any of those considered here.

The other materials shown all have other elements introduced into the carbon backbone or into side groups. The polyamide (PA), nylon 6-6 lies above PE by some amount. Above that again lies polyvinylchloride (PVC), teflon (PTFE) and the partially fluorinated plastic PVDF. In each case a heavier atom has been added to the chain and the density is increased through the series. Further, the electronegativity of the four elements increases down the series nitrogen, chlorine, oxygen and fluorine and this is reflected in increased strength of the carbon-ligand bond and decreased Van der Waals' forces between chains. This combination of density and force ensures that Hugoniots lie above their hydrocarbon neighbours. In pressure–particle velocity space these carbon backbone thermoplastics have Hugoniots that simply scale with density with PTFE at the highest pressures and P4MP1 at the lowest.

Work on Teflon itself, obtained from various commercial and pedigree sources, has shown there to be no significant effect passing across materials of controlled production and purity through to commercially pure targets at least within the experimental scatter (Figure 7.11). Figure 7.11(a) shows stress–particle velocity Hugoniots for the range of materials; a best-fit line is superposed onto the data. There is little difference between the various production routes plotted in this plane. However, this low–pressure region contains the phase transformation from phase II to phase III which occurs

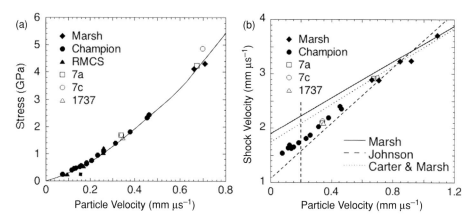

Figure 7.11 Hugoniot for teflon (PTFE): (a) stress–particle velocity; (b) U_s–u_p. Phase II to phase III transition occurs at $c.\ 0.19$ mm s^{-1}. The phase II to III boundary is shown as a vertical dashed line.

at 0.7 GPa and at a particle velocity of $c.\ 0.19$ mm s^{-1} (note the dashed line in the figure and the phase diagram of Figure 7.10). Other measurements have tracked this transition from the release of the state via a rarefaction shock and these will be discussed below.

The dependence of the shock wave speed on the induced particle velocity is shown in Figure 7.11(b). There are a series of best-fit lines through the data which are gathered for a range of commercial materials and specially produced plastics presented here which show a linear response for all the materials above a particle velocity of $c.\ 0.2$ mm μs^{-1} (the phase transformation). The intercept on the shock velocity axis c_0 is less than the longitudinal sound speed c_L as noted previously. It is generally the case that polymers don't display linear behaviour so that quoted values for shock constants generally depend upon the velocity range studied. The reader is cautioned (as previously with low-stress data for glasses) in using databook values for c_0 and S without checking their range of applicability. Three lines are plotted on the curve using fits to literature constants for PTFE. It is clear that these derived curves can only be used in a defined range since the response of polymers is more complex in the low-stress regime than the elastic-perfectly plastic approximation that describes metal response.

The equation of state measurements reflect the integrated effect of the density of the materials and their strength and in this region the hydrostatic behaviour is dominated by Van der Waals' forces between interacting polymer chains; larger for fluorinated ones than the hydrocarbon chain in PE. PTFE exhibits the largest repulsion and has the strongest covalent bonds, which is reflected in the increasing melting points of the materials as one moves from PE to PTFE. Carter and Marsh saw that the low-pressure data can be approximated assuming that only a two-dimensional force is present between chains in shocked thermoplastics. Agreement can be obtained between shock, diamond anvil and calculations for thermoplastics when the geometry of the as-received material is known. Such an analysis follows from a simple model of a polymer based on a

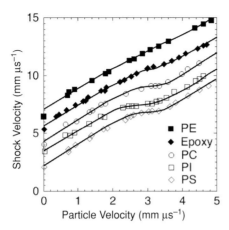

Figure 7.12 $U_s–u_p$ for five polymers to high pressure. Transition occurs at 20–30 GPa for each polymer. The shock velocity is displaced for each curve to illustrate behaviours more clearly. Thus PS is in correct position, PI + 1, PC + 2, Epoxy + 3, PE + 4 mm μs^{-1}. Source: after Carter, W. J. and Marsh, S. P. (1977/1995) *Hugoniot Equation of State of Polymers*. Los Alamos, NM: Los Alamos National Laboratory.

vibrating chain in a rigid cylinder. The force field, ϕ, was represented by a law of the form

$$\phi = A \exp \left[-\frac{r}{r_0} \right] - \frac{B}{r^6}, \qquad (7.2)$$

where A and B are constants and r represents displacement. This potential reproduces low-pressure behaviour for the observed Hugoniot data.

Nearly all polymers show further high-temperature (2000 K) and high-pressure transformations at 20–30 GPa for which the volume change is also large (Figure 7.12). Several authors have speculated upon the nature of this transformation including Carter and Marsh themselves (1977/1995). As has been stated above, in the low-pressure regime the anisotropy of polymer microstructure is very marked. Low-pressure transformations in this regime are due to rotations and rearrangements of chains one against the other subject to the weaker Van der Waals' and London inter-chain forces. But the covalent bonding down the chain length is an order of magnitude greater in strength. However, low-pressure transformations and closer packing of chains with pressure will result in the magnitude of the inter-chain forces approaching that of those down their length. In this regime the atoms between chains begin to interact in a manner similar to that observed down them. When this degree of compression has been achieved, the nature of the observed behaviour approaches the situation seen typically in loading of other solids and is more isotropic than at ambient pressure. Thus the material compresses via the interaction of carbon atoms with one another and covalent bonding in structure (such as the delocalised benzene rings) is eventually overridden, making anisotropy less important. The nature of this process causes significant volume change and the transformation that results is analogous to the carbon–diamond one where covalent bonds are broken in favour of tetragonal ones between planes.

There is a fundamental shift in behaviour and properties across this phase boundary and the resulting microstructures may be expected to have markedly different properties to those of the parent polymer. It also represents the Weak Shock Limit (WSL), the start of an overdriven strong shock regime in plastics familiar in other classes of material where the response is isotropic and defects are locked.

7.4.2   Polymethylmethacrylate (PMMA)

Although the aforementioned materials have interesting properties, one stands out as of particular importance to workers in the shock field. This may be because it has been studied in more detail than all the others, and its use as a transparent window in a variety of experiments where imaging of an opaque surface beneath is required. Polymethylmethacrylate (PMMA) is a transparent thermoplastic sold under many trade names across the world including Perspex, Plexiglass and Lucite. The material was developed in 1928 and has been used as a transparent window along with polycarbonate (PC, Lexan) in many applications where structures are subject to impact. In 1936 production of acrylic safety glass began and during World War II this glass was used for submarine periscopes, windows and for canopies and gun turrets in aircraft.

The polymer has a combination of attractive physical properties compared with other plastics and further it is environmentally stable and is additionally soft and easy to fabricate into complex shapes for use in a range of experiments. It has a density of 1.15–1.19 g cm^{-3}, less than half the density of even a borosilicate glass. Of course, its dynamic tensile strength, although higher than that of glass or polystyrene, is lower than that of polycarbonate and other engineering polymers. However, it is found not to shatter but instead breaks up into larger pieces and this fracture behaviour will be noted again below.

Interest in the shock properties of PMMA came with the development of the velocity system for any reflector (VISAR) by Lyn Barker in the late 1960s. Use of a free surface to measure velocity allows a history of the different wave arrivals, but after the first moments the subsequent oncoming waves experience interactions with reflected releases from the surface. One solution to mediate this is to use a transparent window attached to the target material through which the reflecting interface may be viewed, and PMMA is one typical choice for this application. In order to interpret the frequency shifts introduced, the window's change of refractive index must be calibrated against the applied pressure and thus many experiments have been conducted to accurately record its optical and mechanical response.

Figure 7.13(a) shows particle-velocity histories for shots at different impact velocities. Each history shows rapid (<1 ns) risetimes but with a rounded top. The rounding appears at higher particle velocity levels (> 0.1 mm μs^{-1}) showing the delay introduced to stress increments above this level. Introducing a ramp wave graphically shows this behaviour. Input waves at two amplitudes are shown in Figure 7.13(b). Those below 0.1 mm μs^{-1} show shocking up with steepening of the waves whilst above this level there is less effect. This indicates that free volume is lost within the bulk of the polymer as the compression proceeds. The polymer backbone contains oxygen, with adjacent carbonyl groups as

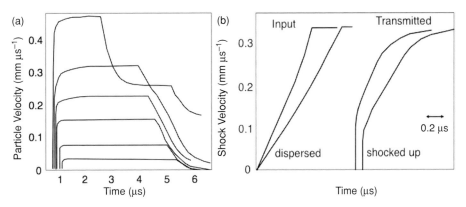

Figure 7.13 (a) Rear surface particle-velocity histories for impacts on PMMA. (b) Ramp loading input and transmitted pulses. Source: reprinted with permission from Barker, L. M. and Hollenbach, R. E. (1970) Shock-wave studies of PMMA, fused silica, and sapphire, *J. Appl. Phys.*, **41**: 4208–4226. Copyright 1970 American Institute of Physics.

well as methyl additions, and these conspire to allow fast initial compression before the chains approach each other sufficiently closely that electrostatic repulsion occurs resulting in an apparent stiffening of the solid.

These observations are reflected in the behaviour observed at two stress levels in the elastic and in the compressed phase of response. Figure 7.14 shows impacts on PMMA at *c*. (a) 150 and (b) 450 m s^{-1}. The particle velocity histories shown to the left track the propagation and development of the shock through the material at a series of stations through the target. They are shown to the right, contour coded to indicate the propagation of the pulse through the sensor. The upper figure shows elastic response from the material. The pulse travels over 12 mm at a velocity close to the bulk wave speed in the material and with release travelling at close to that same velocity. The pulse is flat topped save the overshoot at the start which is a sensor effect as discussed in Chapter 4. In Figure 7.14(b), however, the behaviour is different with rounding of the pulse evident as it develops through the target and as may be seen to right of the figure, where the release is catching the shock markedly whilst the front travels now at a faster speed than in the first case. These two snapshots show the beginning of the collapse of the microstructure, overcoming the interchain repulsive forces to compress the material.

This behaviour is reflected in the equation of state measurements for the polymer. Figure 7.15 shows the Hugoniot and U_s–u_p curves for PMMA. The Hugoniot in stress–particle velocity space shows close to linear behaviour below around 0.2 GPa. Above this value, however, the behaviour is more complex. Here the response is non-linear up to $u_p = 0.4$ mm μs^{-1}, above which the material behaves in an elastic-perfectly plastic manner. This non-linear form reflects the compaction processes occurring within the microstructure. Thus the behaviour has been described as pseudo-viscoelastic below 0.7 GPa and elasto-plastic above this value although this terminology is deceptive as the mechanisms are very different. The response found in the material originally studied has been compared with data in various programmes on different grades of the polymer and

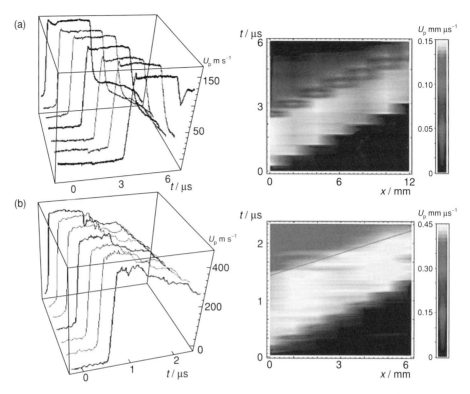

Figure 7.14 Two PV gauge experiments in PMMA at u_p of (a) 150 and (b) 450 m s^{-1}. Traces at different Lagrangian positions to the left and reconstructed X–t to the right.

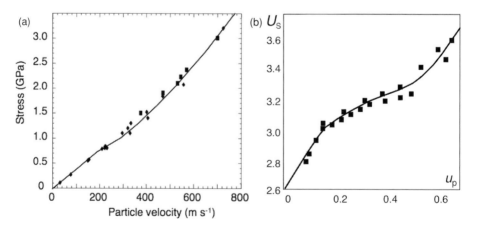

Figure 7.15 Hugoniots for PMMA: (a) P–u_p; (b) U_s–u_p. Source: reprinted with permission from Barker, L. M. and Hollenbach, R. E. (1970) Shock-wave studies of PMMA, fused silica, and sapphire, *J. Appl. Phys.*, **41**: 4208–4226. Copyright 1970 American Institute of Physics.

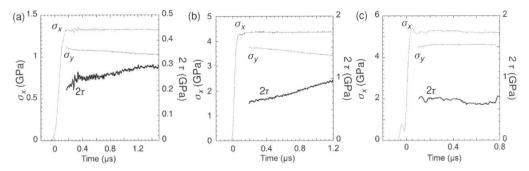

Figure 7.16 Sample histories for longitudinal, lateral and shear stress for three thermoplastics: (a) PE; (b) POM; (c) PTFE. Sensors observing the bulk response are 4 mm from the impact face.

has shown that there is negligible difference in the recorded behaviour with pedigree and across manufacturers.

The picture displayed here is typical of the observed response for a whole series of polymer microstructures. The weak Van der Waals' interchain forces are easily overcome by even modest compression. The resulting densification and stiffening of the bonding structure means that an elastic region does not truly exist comparable with that seen in the undeformed state. In contrast what is observed is a regime in which after initial fast densification, the material exhibits properties that appear to have a behaviour akin to elastic response in other material classes, but which result from compression at impact. Thus what is dubbed a bulk sound speed is faster than the as-received longitudinal wave speed and this state persists until a second threshold is reached beyond which compression can occur to a state where the directional nature of the chain structure is overcome. PMMA overcomes this second threshold between the two velocity levels illustrated in the example above in Figure 7.15 and this can also be seen in the equation of state data published for the material above a particle velocity of 0.2 mm μs^{-1}.

7.4.3   Strength and failure

The reader will notice from previous chapters that the strength histories of compressed materials are a sensitive indicator of the nature of the microstructure under study. The evolution of the shear stress during impact in a range of polymers has been determined as previously in the lower pressure regime with embedded gauges. These histories reveal behaviours which are grouped according the chemical structure of the monomer units as will be seen in the examples chosen below. In order to focus on the behaviours observed different structural types will be allocated a champion to allow comparison between different classes but the general features of response can be applied to similar structural elements. Response will be compared back to polyethylene since it represents the baseline thermoplastic in all these discussions.

Figure 7.16 shows the development of the stress field within polymers compressed by impact in the low-pressure regime. In each case sensors were embedded in the flow, a distance from the impact face to observe the development of the stress field behind the shock front. The three thermoplastics chosen for this comparison illustrate different behaviours in their response to the applied load; they include polyethylene (PE), polyoxymethylene (POM; acetal) and Teflon (PTFE). Starting with a hydrogenated carbon chain each example adds atoms to the monomer unit of increasing electronegativity in each case moving from PE to acetal and Teflon. Whereas PE and PTFE have a carbon backbone, POM has divalent oxygen atoms there whilst PTFE has substituted hydrogen atoms for monovalent fluorine around the chain.

The longitudinal stress rises rapidly in all cases until reaching a flat plateau, whilst the lateral stress approaches a stable value in each example over a different time. In each case the difference between the two values determines twice the shear stress behind the front that is represented by a solid line in each component of the figure. The shear stress in polyethylene rises behind the shock and POM, over the microseconds of the impulse applied. This behaviour contrasts with that observed in PTFE, where the shear stress rapidly achieves a steady value which it retains for the duration of the pulse even at the higher stress level. The addition of an oxygen into the chain mediates the rate of increase of the shear stress over time and controls forces felt between the chains as they are forced closer but fluorine resists interchain interaction at these stress levels.

No material can respond instantaneously to a load at some station within a target and polymers are no exception. In the low-stress regime compression forces components of the molecule together against the electronic conformations that the polymer chains adopt. These electronic interactions, and their development over time, condition the material to resist an impulsive load and determine the threshold beyond which and the rate at which stress and strain fields develop within a component after loading. Thus kinetics in polymers reflects the electronic environment around elements of the backbone, and for thermoplastics in the low-pressure regime, such as those applied for the three polymers illustrated above, increasing electronegativity makes the interaction distances closer but the potential barrier higher for compression to occur behind the front. Looking at other plastics confirms that the electronic environment around the polymer chain resists densification as the material loads. Finally, the shear stress either equilibrates rapidly and appears constant if below the interaction threshold, or rises behind the shock front as the microstructure responds to the load in the low-pressure regime for all cases where data are available, showing a strengthening of the polymer over the duration of the pulse with pressure.

To look in more detail at further aspects of the observed behaviour, consider the effects of adding both phenyl and carbonyl groups to the chain. Polycarbonate is a clear plastic used in a variety of applications. The effect of these side groups is to allow its packing to lie between POM, where the electron concentration is within the polymer chain, and PMMA where the oxygen atom is further away within a side group. Polycarbonate has a weaker electric field around the chain than is the case for POM but shows similar behaviour. Figure 7.17 shows experiments carried out on PC at two stress levels (*c.* 1.5 and 3 GPa) in the low-pressure regime. The sensors map the stress fields at two locations, 2 and 6 mm from the impact face, and in each case it will be seen that whilst the longitudinal stress follows a similar history, the lateral stress shows a different response. This behaviour occurs since in this experiment the impacting flyer plate was

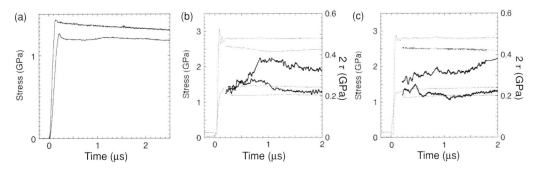

Figure 7.17 Impact on polycarbonate at two stress levels *c.* 1.5 and 3 GPa. Metal flyers launched at targets with lateral and longitudinal sensors at 2 mm and 6 mm from the impact face. (a) Lateral stress histories for the sensor 2 mm from the impact face. In both cases the longitudinal stress induced was 1.5 GPa. One symmetric (PC flyer), one asymmetric impact (metal flyer). To allow details of the history to be observed more easily, the symmetrical impact stress has been multiplied by a factor of 1.1 so that the histories lie distinct from each other for clarity. Stress components (dotted) and shear stress (bold) developed at (b) the 2 mm station, (c) the 6 mm one.

of a different material to the polycarbonate and there was not continuum flow in a zone near the impact face. This means that this region sees an evolving stress field compared with that seen at later time within the bulk of the target.

Figure 7.17(a) shows the two lateral stress histories for the shots in which the sensor was positioned 2 mm from the impact face side by side. In both cases the longitudinal stress induced was 1.5 GPa but the two experiments represent different loading conditions; in one case, an asymmetric impact with a metal flyer plate and in the second, a symmetrical impact where the impactor and target were both polycarbonate. The two stress histories show different response over the first microsecond or so but lie one upon the other thereafter. It can be seen that there is a perturbation in the case of the metal flyer for the first microsecond, where the shear stress behind the front relaxes to that seen in the symmetrical case. In that case the stress falls monotonically as is also seen for the asymmetric case for sensors placed at 4 and 6 mm from the impact plane.

The region near the impact face sees an unsteady response for a period of a microsecond or so before the lateral stress levels (Figure 7.17(b)). This does not correspond to any wave transit across the target or within the impactor but to equilibration of the microstructure to the lateral strain that exists at the impact face (in classical terms where the different Poisson's ratios of the materials can allow lateral motion). In the bulk of the material (Figure 7.17(c)) the lateral stress drops monotonically as seen for PE and POM above, showing the shear stress rising behind the front as both these polymers show resistance to flow. This behaviour indicates orbitals interacting more as they move closer, increasing Van de Waals' forces between them and impeding motion between chains. In addition, the shear stress rises more rapidly at higher stress as the impulse compresses microstructure more rapidly to achieve high density.

This chapter has principally discussed the response of thermoplastics in this low-pressure phase of polymer behaviour. Figure 7.18 shows the response of a thermoset;

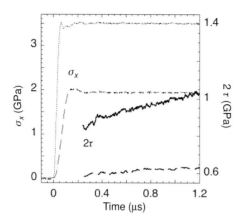

Figure 7.18 Response of the thermoset epoxy. Two stress levels 3.5 (solid) and 2 GPa (dashed) showing evolution of the shear stress behind the shock.

epoxy. The material has now an isotropic network structure and so crosslinks dominate the strength rather than weaker Van der Waals' forces observed earlier. As shock amplitude increases the material responds with shear stress rising behind the front as the pressure in this range increases. The figure shows two stress levels with the impulses displayed along with the two corresponding shear stress histories. At lower amplitudes the lateral stress is constant as the pulse develops but at these higher ones it falls and as can be seen this epoxy strengthens in an almost linear manner over the microsecond or so of the experiment. As the pulse amplitude is increased, the rate of rise increases and over the stress range sampled it does not asymptote to a plateau where a constant strength is observed although this will eventually happen with increased shock amplitude.

As seen throughout this book, the flow near an impact surface is only one-dimensional at the continuum scale and then only where flyer and target are of the same material and the flow at the continuum is symmetrical. In the previous chapter the dominating features of a surface relaxation zone were noticed as brittle materials attained an inelastic state by microfracture driven from the impact face; such a zone also exists in metals but equilibration by slip is much faster. There is always such a region where there are materials of different impedance where surface lateral strain must exist and in which assumptions of one-dimensional flow are restricted. This region is greater in polymers than is the case for metals and brittle solids as can be seen above, but in both the other classes of materials there are features of the behaviour that are due to this asymmetry in the flow. In metals this is seen in variation of microhardness in material in the hundreds of microns near the impact face, whilst in brittle ceramics, fracture waves are initiated when the impact is asymmetric that are not present in the symmetric case. However, in polymers the surface zone is of a different nature since in this case the response is dominated by densification of the initial open structure; interaction of side groups and Van de Waals' attraction in thermoplastics or cross-linking bonds in thermosets. Over the loading times that can be studied on plate-impact platforms the response in the low-pressure regime is unsteady as the microstructures collapse under the stresses applied.

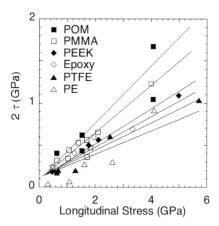

Figure 7.19 Strength of six polymers 4 mm from impact face as a function of longitudinal stress induced at impact.

In thermoplastics this compresses Van der Waals' bonds between chains with increased stress required to do so as constituents become more electronegative or increase steric hindrance. Faster pulses again see a snapshot even further distant from steady response. Eventually the stress becomes high enough that these forces are quickly overcome in the front and the stress state then becomes steady and the behaviour will be akin to that of metals discussed earlier. This densification zone defines the response of plastics to dynamic loads and shapes their individual behaviours, especially in the low-pressure regime.

Having seen that the polymers respond with different kinetics to input impulses, it only remains to note the effect upon the shear strength observed from such measurements in a range of polymers. Figure 7.19 shows a representation of hardening (although perhaps a better word is stiffening) behaviour for the shock strength of the polymers discussed above in this low-pressure regime and for the first microsecond or so of loading. The simple hydrocarbon PE has lower strength whilst the strongest materials are the polymers containing oxygen; POM, PEEK, PMMA and epoxy. The monomer structures of PEEK and epoxy are similar and these polymers show similar mechanical strengths. However, they represent two different groupings – thermoplastics and thermosets – and the strength behaviour does not appear to be sensitive to the cross-linking. Bringing the carbon chains closer to one another increases Van der Waals' forces and strengthens the microscale polymer structure as pressure increases. The behaviour is reflected across the range of simple monomers illustrated. A further feature is the steric hindrance provided by attached side groups seen in the cases of PMMA. Other hydrocarbons show similar trends as molecular architecture becomes more complex.

Along with the thermoplastic hydrocarbons, the thermoplastic PTFE with its fully fluorinated chain shows low strength. Here the electronegativity of fluorine dominates the electronic structure and favours repulsion between adjacent chains. Van der Waals' attractive forces are less important in interactions of the fluorinated chains as the repulsive components dominate at low pressure. This is less true for other molecules showing the different effects of the electronic nature of the included atoms as electronegativity changes between polymer monomers containing nitrogen, oxygen, chlorine and fluorine.

The strength of polymers is controlled by electronic and steric interactions which in combination act to define the regime that these materials inhabit during dynamic loading. In the low-pressure region, the molecular spacing and conformation is key in defining the development of the material's strength as pressure increases. However, in all the cases investigated, polymers show increasing shear strength with pressure. A change in behaviour is found in almost all plastics when the pressure reaches the point at which Van der Waals' repulsions or cross-linking bonds are overcome and atoms interact in three-dimensions rather than within the chain structures defined above. At this point fully developed isotropic carbon structures form which respond to strong shock in a wave similar to other solids but in a manner very different to that in the low-pressure regime. This threshold may be regarded as phase transformation. In the low-pressure regime crystallinity also has an effect and average spherulite size has been observed to change in recovered, shock-loaded targets. Simple engineering arguments (borrowing a concept such as Hall-Petch for instance which equates smaller grain size with greater strength) would suggest that this alone would result in increased strength for the shocked over the unshocked material. However, the mechanisms for inelastic flow here are dominated by densification and not by slip, although strengthening may result in either case.

Finally, it has been noted that the weak, inter-chain Van der Waals' bond strength of thermoplastics and inter-chain cross-linking bonds in thermosets has an effect not only upon the ease of compression but also upon the kinetics with which they can respond. They have much lower density, more open structures than closely packed metals for instance, so that a stage of densification must occur in polymers before simple shock relations may be used in the manner of other solids; this transition point is the transition from the low-pressure densification regime to strong shock behaviour that occurs between 20 and 30 GPa for these plastics. The transmission of information is much slower than in metals or brittle solids and the effects of phenomena such as the asymmetric loading found at interfaces are observed at the continuum scale in macroscopic sensors in polymers whereas they are not in metals or brittle solids. What is clear is that simple assumptions about isotropy or the stress states in the flow are not valid in the low-pressure regime. Inhomogeneous flow must occur in these cases at the microscopic and mesoscopic length scales. Such observations are reflected as well in the more complex response occurring in multidimensional loading such as occurs in integrated experiments such as the Taylor test. Examples of these will be shown below to illustrate the range of observations introduced as complexity and dimensionality increase.

7.4.4   Behaviour in tension

The response of polymers to dynamic tensile loading is not yet sufficiently well understood to attribute mechanisms unequivocally to material response. There have also been relatively few published results in this area compared with the much fuller dataset that

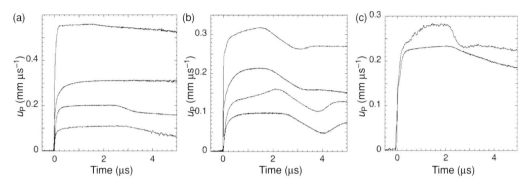

Figure 7.20 (a) 2.5 mm PTFE flyer onto 5 mm PTFE targets; (b) 2.5 mm Kel-F-800^{TM} flyer onto 5 mm Kel-F-800^{TM} targets; (c) 2.5 mm estane^{TM} flyer onto 5 mm estane^{TM} targets.

exists for other material classes. As in so many areas, the simple concepts borrowed from the elastic-perfectly plastic construct that are applied so readily across the literature are woefully inadequate to explain the observed results. Polymers nevertheless have many important applications in a series of arenas in the modern world where it is vital to properly understand their rapid tensile response; particularly since they form the binder phase of so many composites important to the engineering community for a vast range of products. There are data in the literature on a few thermoplastics, thermosets and rubbers loaded with a range of pulse lengths and strain rates. This section will focus upon PTFE, PMMA, estane and the plastic Kel-F (which is a composite of fluorinated polymers), illustrate the principal features of response.

Symmetrical spall experiments with microsecond pulses conducted using PTFE are shown in Figure 7.20(a). The pulses are expected to be *c.* 4 μs long (assuming elastic wave speeds) but in fact each is much less than that showing the dominating effects of densification on the microstructure increasing release speeds. In the highest velocity trace, the pulse flattens and then a gradual decline from the peak impact velocity begins. Whereas the two lowest pulses show some release at the end of the compression phase, there is less evidence of this in the upper two cases. Here other mechanisms are operating. In none of the histories is there any evidence for a pull-back signal with any appreciable amplitude. In conventional interpretations of spall this height is taken to be an indication of the tensile strength of the material, suggesting that in this case it possesses none. The low strength of PTFE with its fluorinated, repulsive molecular chains reflects this behaviour.

In reality these histories and those observed elsewhere show that conventional interpretation of high rate failure is inadequate in the cases shown here. This is because the kinetics of inelastic response in polymers is much slower than those observed in metals and brittle solids. In the case of PTFE there is a further complication in that low-pressure phase transformations which act to further densify the structure are present as well. A pressure-induced phase transition in PTFE occurs at *c.* 0.5 GPa (phase II–III) which divides the range of histories presented between the lower and upper two traces. This transition, its kinetics and the interaction between release and transformation in the polymer spherulites may explain the lack of any features in the reload pulse.

The experiments conducted on Kel-F-800^{TM} are presented in Figure 7.20(b). Kel-F-800^{TM} (also known as FK-800) is an amorphous, 3:1 copolymer of chlorotrifluoro-ethylene (CTFE) and vinylidene fluoride and thus a composite. The lower velocity histories show reload signals that suggest spall strengths of around 50 MPa but at higher velocities these reduce. Paradoxically, at the highest impact velocity, there appears to be some spall strength indicating that alternative damage mechanisms may operate or that as in PTFE low-pressure transitions may have occurred in one of the phases. There is a different kinetics and different operating mechanisms in this case.

Figure 7.20(c) shows two traces for a polyurethane with the lower velocity showing a flat-topped compressive pulse but the signal does not reload. At a higher velocity where ramping can be seen on the top of the pulse, there is a reload signal present although in recovered targets there was little or no visual evidence of damage evolution or fracture. This series of symmetrical spall experiments has shown that PTFE displays no measurable spall strength even at the lowest velocity tested, Kel-F-800^{TM} displays decreasing spall strength with increasing stress as expected, and spallation occurs in the polyurethane, but at a higher stress than either PTFE or Kel-F-800^{TM}. All three polymers exhibited unexpected structure in their pulse shapes.

Symmetrical spall experiments on PMMA have shown a similar range of behaviours. Clearly the processes involved in the nucleation and propagation of damage are critically dependent upon the form of the loading impulse and how it couples both to the mesoscale and atomic microstructures of the polymer. Further, individual active processes have different operating kinetics propagating failure. Reload signals from microsecond pulse lengths are of similar magnitude to those discussed above for the fluorinated polymers and are of the order of 50 MPa. All thermoplastics have open structures and take time to both densify and thus also to release. Thus it is of no surprise that with such short impulses, the stress state in the material has not equilibrated either in compression or tension and that damage and failure occurs at later times than the impulses applied can sample.

A one-dimensionally recovered PMMA target shows key features of the failure process in a brittle polymer. To observe the response of PMMA to spallation, a target was loaded and recovered in a specially developed recovery fixture to trap lateral momentum. The recovered sample plate is shown in Figure 7.21(a). It is of the same lateral dimensions as previously found indicating the success of the recovery method for polymers. Placing the target between crossed-polarising films showed no lateral strains remaining within the target after load–unload and recovery. The disc thickness had, however, compressed by the expected plastic strain given by

$$\varepsilon_{\text{res}} = \frac{4}{3}\ln\left(\frac{V}{V_0}\right) = \frac{4}{3}\ln\left(\frac{U_s}{U_s - u_p}\right). \tag{7.3}$$

The sample was also designed to allow spall to enter the target without rear cover plates present to trap the tensile pulse. The impact face, however, was covered with a sacrificial plate which was fractured, indicating some lateral strain at the impact face.

(a)   (b)

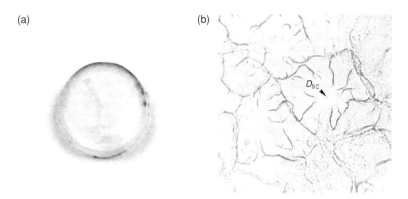

Figure 7.21 Shock-recovered sample of PMMA: (a) recovered target; (b) spall plane. A spall crack, D_{SC}, is labelled and there is a flaw visible at its centre.

Figure 7.21(b) shows the spall plane viewed through the target and down the shock direction. It consists of an area with ductile fractures which appear to initiate at flaws within the sample. One such nucleus can be seen arrowed in the figure. It is surrounded by a circular region and further out cracks are initiated which travel and interact with damage from other zones at a greater distance again. There is a large population of these flaws visible as discs throughout the volume and these indicate active nuclei from which failure has been initiated within the polymer.

The recovered target when placed under cross-polars shows no fringes due to stress birefringence indicating no residual strain due to the loading and confirming the success of the technique. The crack appears in a penny-shaped geometry and shows brittle fracture occurring within the PMMA as it fails. Thus although the complex form of the reload signal shows that the fracture process is clearly slow compared with that seen in brittle metals or glasses, the form of the damage suggests ductile fracture rather than the nucleation and growth of voids as the operating mechanism in PMMA. This is at first glance surprising given the weak bonding at the molecular level within polymers yet the response rates of microstructure to the compression phase are such that the microstructure cannot respond to rapid loading within the microsecond impulses applied.

Nanosecond pulses have also been applied to PC by irradiating the surface with a laser pulse. These impulses were sufficient to induce facture near the rear surface of the target comparable with damage seen when loaded with a Taylor wave from an explosive driver. Other work on PMMA has estimated the tensile strength in such loading at c. 1 GPa. This contrasts sharply with the measured strengths for microsecond pulses and consequently lower strain rates where the strength is an order of magnitude or more less that these values. Clearly the coupling of mechanisms to these impulses is very different and account must be taken of this when trying to compare results across loading regimes when the deformation mechanisms and their kinetics are as yet not adequately known.

As seen earlier, different duration impulses probe volume elements at different scales. Damage induced with a laser impulse probes microscale defects at the level of the chain structures within the loaded structure. However, with a microsecond pulse the impulse probes defects at the mesoscale and opens a different form of damage. The kinetics of bond strain in polymers and the possibility that rearrangements at the chain and at the spherulite level can occur with relative ease means that, whilst short-lived impulse may pull an area apart, suitable relaxation can equally bring areas back into proximity where the weak Van der Waals' attractions can once again operate. In order to rupture polymers sufficient strain in the tensile impulse must be present to yield damaged material after the interaction and it is often the case that in plastics the strain to failure may not be achieved before the pulse has released back to an ambient state.

7.5   Shock polarisation of polymers

Polymers are good insulators and are used as dielectric materials in a whole range of applications throughout the electronics industry. If some conduction is required, plastics can be synthesised with dopants to control the electrical behaviour. Further, over the last 20 years plastic electronics has become one of the forefront disciplines in modern physics opening a new world of electronic behaviour for these materials. Pure polymers are known to show polarisation at pressure and display a piezoelectric response when shocked. This manifests itself in several observed behaviours. The intense mechanical loading induces localised strain fields in regions within the polymer that activate and nucleate a range of defects which are associated with electrical charge concentrations distributed inhomogeneously behind the front in the weak shock regime. Such an effect is observed in all classes of polymer and manifests itself as a polarisation field which increases in strength with shock amplitude. This field induces voltages generated across the shock front which can be measured with suitable techniques (see below). However, the limited study of these effects has observed behaviours which appear counter-intuitive; one of these is the generation of an induced field across the front which changes direction as shock amplitude mounts at thresholds which do not appear to relate to the mechanical state within the material. The coupling of electrical and mechanical effects is of course fundamental to the compression of the material and the interplay of electronic and mechanical properties is thus a fascinating area of study that has not received the attention it deserves. This close relation has important implications in polymers and ionic solids where charge separation may occur as seen below.

As an example of the observed behaviour it is instructive to look at the properties of PMMA, whose mechanical properties have been considered at length above. Shear strength measurements show an initial increase to 7.5 GPa before dropping to near zero. Similarly temperature measurements show a marked increase in temperature at 2.0 GPa as the material is shocked. These data were explained in terms of the onset of a shock-induced, exothermic reaction, although there has been no other evidence for this presented. However, a similar threshold was noted for a dynamic electrical polarisation of PMMA. Thus both polarisation and conductivity might be expected to alter at a mechanical threshold in behaviour in this region.

To differentiate polarisation from resistance change behind the shock, a differential circuit was designed which was adapted from methods used to determine the conductivity of sapphire under shock. The measurement removes voltages produced in the induced shock field allowing determination of those induced by resistance changes. Some examples of this were shown for KCl in earlier chapters. PMMA targets did not conduct up to a value of 11 GPa in the shock. Further, induced electrical fields across the shock front were measured at all levels studied, but no changes of field direction with increasing stress were observed in the range considered although their size increased by an order of magnitude.

These electrical phenomena have further physical effects. Light emission is observed from many shocked dielectrics and may be connected with brittle response since fracture (which separates charge) is known to lead to fracto-emission. Bulk effects from defects with electrons trapped in vacancies may also contribute. Certainly light emission is a frequent complication for high-speed photographic studies of impact phenomena and this shows the intimate connection between mechanical and electrical effects. Such observations are still to gain the cadre of data necessary to fully define the operating mechanisms and as such this offers a rich vein of research for the future.

7.6   PTFE, PEEK and PMMA Taylor cylinder impact

The previous sections have shown the compression response of polymers in densification, where the microstructure strengthens, and strong shock regimes. The experiment of choice for tracking the integrated response of a material under load has in modern times been the impact of a right cylinder onto a rigid boundary or onto itself; the Taylor test. As seen in previous chapters, in its modern form, the impact of a right cylinder onto a rigid anvil is most often used as a validation test for models introduced into hydrocodes, since it has the potential to fully exercise constitutive models for the material, combining axisymmetric geometry (and, thus, little mesh complexity), development of histories of radii and of plastic waves, and easy recovery of impacted samples. When impact is asymmetric, however, it is necessary to carefully lubricate and align the cylinder with the target to ensure useable results since the stress state in the impact zone is critical and friction will blur behaviour. Examples are shown in what follows of a series of experiments conducted on selected polymers over a wide range of impact velocities which illustrate many of the observed responses found in other polymers.

The results of two experiments on commercial, off-the-shelf (COTS) PTFE are given in Figure 7.22. Two sequences show impacts of nominally 10 mm diameter, 50 mm long cylinders at velocities of 125 and 140 m s^{-1}. In both cases, samples hit rigid, hardened-steel anvils which were polished and lubricated with MoS$_2$ grease to ensure zero friction at the impact surface. There is an abrupt ductile–brittle transition in behaviour between (a) and (b). A threshold occurs at c. 130 m s^{-1} where the recovered rod length becomes markedly shortened over a small velocity range corresponding to the onset of fracture

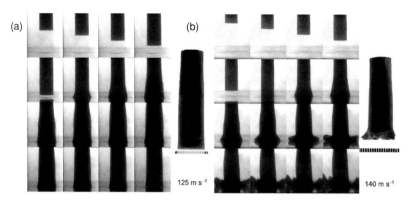

Figure 7.22 Taylor impact of PTFE onto a hardened steel anvil; impacts of nominally 10 mm diameter, 50 mm long cylinders at velocities of (a) 125 and (b) 140 m s^{-1}. Source: reprinted with permission from Bourne, N. K., Brown, E. N., Millett, J. C. F. and Gray III, G. T. (2008) Shock, release and Taylor impact of the semicrystalline thermoplastic polytetrafluoroethylene, *J. Appl. Phys.*, **103**: 074902. Copyright 2008 American Institute of Physics.

and development of petals which hinge outward to absorb the forward momentum of the rod and accommodate strain. This is a similar threshold to that observed for pedigree PTFE which was produced using controlled composition and preparation.

Of course, PTFE has a low-pressure phase transition and this occurs across this boundary in behaviours. In the pedigree and in the COTS material the transition is only occurring within the crystalline domains, and these make up only *c*. 50% of the material. It is unclear what effect the surrounding amorphous PTFE might have. Observations have shown changed mechanical properties after transformation including altered moduli and increases in crystallinity. Thus PTFE has shown several features in response observed in recovered targets that are a result of the phase change and which have implications for the interpretation of such an integrated test.

Taylor tests on other fluorinated polymers such as PEEK, polychlorotrifluoroethylene (PCTFE) and PE show different behaviours. These semicrystalline polymers show no phase transformation but globally similar fracture behaviour, and zones of increased radial strain as observed in PTFE.

Figure 7.23 shows sections of four PEEK Taylor cylinders after impact. The loaded end of each recovered cylinder has a strongly concave nature, showing that considerable relaxation has occurred after impact. Discolouration immediately under the impact face can be seen in all cases and in this region material is significantly darker than the bulk of the cylinder. Clearly the polymer is undergoing rapid deformation as the impact face flows across the anvil and sees high pressures, temperatures and large lateral strains not present in other parts of the rod. This change in colour corresponds to regions where the percentage crystallinity of PEEK is decreased relative to the original spherulite morphology in the as-received sample. It thus shows the complex cycling of the mesoscale microstructure under load and recovery that occurs under the extreme conditions at the impact face that drives the irreversible changes in material properties that result from such loading.

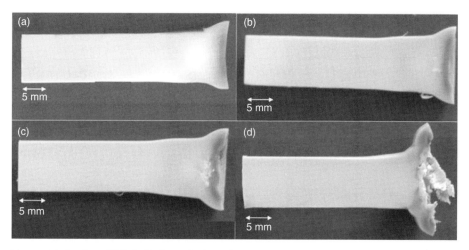

Figure 7.23 Sectioned PEEK Taylor cylinders after impact. Impact velocities: (a) 247 m s^{-1}; (b) 276 m s^{-1}; (c) 303 m s^{-1}; (d) 349 m s^{-1}. Source: reprinted from Millett, J. C. F., Bourne, N. K. and Stevens, G. S. (2006) Taylor impact of polyether ether ketone, *Int. J. Imp. Engng.*, **32**(7): 1086–1094. Copyright 2006 with permission from Elsevier.

Again the results of Taylor tests on these thermoplastics are critically dependent on the operating mechanisms at very different length scales with the response conditioned by both chain and mesoscale morphology of the material as it deforms. The modern adaptation of the Taylor impact is a fully integrated experiment that highlights continuum behaviour, which has origins that reflect this multiscale response. There are distinct regions of deformation that are controlled by the compression and viscoplastic flow then followed by release and fracture within the polymer, in a region within one diameter from the impact face. This region, and only a central conical portion of it, experiences the high impact stress. Simple, one-dimensional strain experiments only reveal a portion of the behaviour that results from the ostensibly two-dimensional loading that is assumed to occur within the rods. In reality the loading is fully three-dimensional at scales below c. 10 μm since most polymers contain a crystalline phase so that there is an interplay between this and the surrounding amorphous phase as well as the fact that at lower length scales the material is highly anisotropic with local moduli varying by orders of magnitude. Thus multiaxial loading in the Taylor geometry, and the composite nature of the microstructure at the mesoscale, results in a range of observed operating mechanisms which favour varying compression behaviour but also fracture between transforming crystallites and the amorphous matrix. This test is a complex validation experiment for material models, but great care must be taken to specify the quantities for comparison given the range of operating mechanisms and length scales available for diagnosis.

Simple interpretations of behaviour are to be cautioned against, however. The complexity of the failure in polymers makes prediction of their response difficult with the present state of knowledge of materials properties. There are suites of mechanisms for all types of behaviour that are as yet not fully investigated both in states of compression

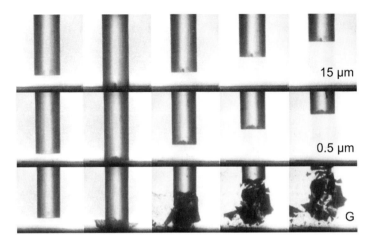

15 µm

0.5 µm

G

Figure 7.24 PMMA 10 mm diameter PMMA rod impacting a hardened steel anvil at 75 m s^{-1} with 80 µs between each frame. The top set of the three shows a rod impacting onto the anvil with 15 µm grit paper on the surface, the middle with a 0.5 µm paper, and the bottom set with a well-greased surface (G).

as well as tension. The well-known thermoplastic PMMA shows complex behaviour in all aspects of its response. PMMA is an uncrystallised polymer whose glass transition temperature ranges from 110 up to 135 °C so that at ambient temperature it is hard, rigid and brittle with little elongation. In compression, the U_s–u_p relation exhibits the S shaped form noted through the literature. In tension, its spall behaviour has also been noted to be anomalous. It is known to be a brittle material and so it is not surprising that its failure should also show interesting properties when it is loaded under multiaxial and continuously varying forces.

Figure 7.24 shows three sequences for the impact of a 10 mm diameter PMMA cylinder travelling at 75 m s^{-1} onto a polished, hardened steel anvil. Each is of five frames and they are presented in lines of images stacked one upon the other. The top set of the three shows a rod impacting onto the anvil with 15 µm polishing paper on the surface. The rod can be seen in all three cases approaching the surface in the first frame. Impact in frame 2 propagates a small crack back down the loading axis of the cylinder and this travels around 1 mm into the rod. The cylinder rebounds in subsequent frames. The middle sequence shows impact at the same velocity onto a 0.5 µm grit paper. There is a wider zone of damage now with a family of cracks across the impact face propagating back into the rod as above and to a similar depth below it. The lower sequence shows impact onto a polished and well-greased anvil; in fact a thin halo of grease can be seen erupting around the cylinder nose at impact showing the lubricated region on this face. In subsequent frames a dark region of fractured material can be seen extending one diameter back into the rod. This region corresponds to several propagating cracks which travel back and comminute the PMMA due to the lateral strain at the impact face causing tensile release. The fragments of plastic can be seen flying apart in subsequent frames under the influence of the extra strain lateral motion at impact that was allowed by greasing the anvil.

This behaviour illustrates the brittle nature of failure in this polymer and the effect of very slight alterations to the boundary conditions for the loading. At 67 m s^{-1}, even in the case where the impact surface was well lubricated, brittle failure of the rod in the manner observed above was not seen. Thus it can be seen that by using varied roughness surfaces, there was sufficient restriction of lateral motion at the impact face to prevent brittle fracture from propagating. Clearly the roughness introduced couples to the mesoscale microstructure within the material. By introducing local three-dimensional strain fields on that scale, propagation of fracture from surface flaws is suppressed where in the case of a lubricated surface it is not. The Taylor test again shows its use as a means of obtaining an overall snapshot of material response whilst highlighting critical features of behaviour that can result in important mechanical effects. However, since failure is dramatically altered by slight variations in boundary conditions, this example is a telling illustration that care must taken to ensure that the experiment is tightly controlled if useful quantitative data are to be extracted from it. Taylor introduced a simple catch-all to screen materials (metals) for use at in impact under the strains of war-time drivers and national need. He did not advocate it as a means of deriving detailed materials behaviour in one experiment and this illustrates the necessity to remember that in the modern era. No one platform will capture all salient features of a material's dynamic behaviour. It is a suite of controlled measurements in different loading modes that will clarify the apparent complexities in material behaviour. Like any of the other tests discussed in the preceding section and chapters, the Taylor test has a series of phases and responses that the material must follow. The first stages have a localisation phase followed by an extended period where strain develops. In the localisation phase the material begins to flow across the surface and then fails under circumferential tensile strain into cracks which propagate back to allow the material to hinge. Once formed, the material can adopt a second failure phase where it accommodates the applied strain required. Shock, release and failure occupy those first moments and it is the time dependence of polymer shock densification and failure that is crucial to achieve the stage where the stress state is stable and strain can develop. The test graphically illustrates the response of a structure to dynamic compression.

7.7 The response of plastics to dynamic compression

The response of polymers and plastics (a composite of polymers) to dynamic impulse shows a behaviour different to that of metals and ceramics but understandable in terms of the nature of the packing of the carbon chains. The response of fully dense metals or ceramics is dominated in the first moments of load by the creation of localised regions to accommodate macroscopic flow within the microstructure. The elastic phase of compression of the material is small before plastic response ensues. In contrast polymers densify to large strain against weaker interchain Van der Waals forces before materials begin to exhibit a steady state behind the shock. This typically starts at strains of over 10% which transient loading frequently may not reach. Thus stress state stability can only be fully achieved once the polymer has been loaded to a compression at which the carbon atoms interact at maximum density. This is seen as a transition to a strong shock regime in the material. At stresses lower than this level the material is caught in a state of compaction and the amplitude and duration of the pulse determines the state which is reached in that process. At low stresses (near the yield stress) the polymer under load can take tens or hundreds of microseconds to equilibrate the stress state and this makes data acquired from devices such as the split Hopkinson bar only valid at large strains. Thus the shock again allows a snapshot of the response as time develops and this makes it a window into the processes which define the *densification* region below the Weak Shock Limit. However, they may not be complete before the impulse ends at stresses below the WSL.

The locus of shock states in this lower stress regime tracks these processes. As with amorphous glasses the Hugoniot appears initially convex for polymers, showing that even highly crystalline thermosets below their glass transition have free volume in an environment of several GPa, particularly since they are heated as well. Thus polymers show ramped rises on their shock fronts at later stations away from the impact face with no evidence of elastic waves over the microseconds of their loading pulses. This is because these materials are so compressible that the speed at which waves transmit in the shocked flow (the bulk sound speed c_0) is greater in plastics than the longitudinal wave speed c_L for ambient elastic waves and the strengthening that is seen happening at later times doesn't operate in the first moments. Whereas plasticity processes develop rather slowly in some metal structures in the weak shock regime, the densification processes in all plastics take time. Their kinetics are best viewed from development of the shear stress behind the front, but the complexity of the conformations the chains may adopt to accommodate the strain is seen in the response of PTFE discussed earlier. The strong shock regime is reached when rearrangement occurs to maximum density within the front and the strain rate behind becomes zero and this transition region occurs most simply for the hydrocarbon PE but over a wider stress range for polymers with side groups and/or electronegative atoms within the chain. Across all of this stress range, the compression of the open structure against the Van der Waals' bonding or cross-linking in the cells does work that is converted to heat within the lattice. With this type of bonding it is no surprise that the strain input over a particular time is directly analogous to heating to an equivalent temperature since densification is the mechanism operating most directly here with more complex rearrangements at the atomic level at much lower energy than those in rearrangement of crystalline lattices.

Thus in the low-stress region a picture of response may be constructed. The microseconds of present arrangements see the microstructure collapsing to a fully dense form with a number of possible structural rearrangements to accommodate the applied compression possible (phase transformations) in some polymers. The weak, interchain bonding is easily overcome, which has several consequences and ensures that the elastic wave speed of the uncompressed material is almost always less than the bulk sound speed extrapolated from shock data. Under modest compressions the weak Van der Waals' bonds are overcome and this stiffens the structure and increases the density in the low-stress region. This behaviour is reflected in the response of PTFE and PMMA discussed in detail here but also in the whole range of polymers studied under shock.

The strength increases behind the shock and this results from closer proximity of the molecular components with strong attractive bonding and steric entangling which resists further strain as time develops. The strength of the repulsion reflects the electronegativity of the included atoms, with sulphur showing less effect than oxygen and with fluorine supplying the greatest force. At these pressures cross-linkage between chains shows negligible difference to the behaviours observed with thermoplastics that have none and which illustrates that these different configurations, so important under ambient conditions, are homogenised as pressure increases.

Thus polymers move through the three stages of compression in a rather different manner to metals and brittle materials. They remain in a state of stress which takes time to equilibrate while locally the chains rearrange three-dimensionally to accommodate the strain. Once fully densified they can flow as all material classes must and in the strong shock regime their response is similar in form. However, the densification regime is one in which chains are working to achieve a stable final state and this may take tens of microseconds, beyond the impulse characteristics of modern test platforms at relevant stresses. The rearrangement of chains into the new conformation consistent with the compression applied by a shock and the consequent temperature induced within the densification regime takes times that may extend to hundreds of microseconds near the yield stress of the polymer ranging down to the rise of the shock at the Weak Shock Limit.

In previous chapters a representative volume element was defined over which defects within the structure were averaged to yield a stable stress state in the volume. In the case of amorphous polymers, the inhomogeneities in packing will be dominated by the folded chain lengths of the plastic at the smallest scales, and by amorphous and crystalline regions in thermoplastics. In both of these, texture introduced by processing will add defects potentially at a higher scale again. In the pedigree materials used in laboratory trials targets may not contain the large-scale inhomogeneities that are found within production materials; thus care must be taken to understand the correct defect scale when reviewing performance particularly under tensile loading. Certainly a microstructural unit (MSU) of size 100 μm will include a statistical distribution of spherulites and amorphous regions in a typical partially crystalline polymer to recover uniaxial strain loading at the macroscale. In a three-dimensional network amorphous polymer the cell might be several orders of magnitude smaller.

A piston compresses the microstructure on the impact face to a defined strain and this starts rearrangement of the microstructure to new conformation that equilibrates lateral stresses and defines a new strength for the compressed polymer tangle. For the MSU discussed this gives a time for equilibration that is of the order of microseconds since the reptation speeds are some fraction of the particle velocity which is itself only hundreds of metres per second in the densification regime. Further, the target does not reach maximum density even if stress-state equilibrium is achieved. Of course as pressure increases sound speeds increase and the temperature rises, and this hastens the processes until the shock to full density can occur within the front itself.

On release, the dispersed reflection of a wave already ramped during transmission through a bed meets a diffuse microstructure consisting of closely packed and entangled chains all raised in temperature. The nature of failure is thus dependent both on pulse amplitude and on the duration of the impulse but also on the dimensions of the target which will determine its dispersion through the compressed material. Defects at the mesoscale including processing flaws will seed failure as ever and there may be differences between crystalline and amorphous plastics in the flow. Thus defects exist at the macro- or the spherulite scale from which a new surface may open and in the case of PMMA the failure sites appear similar to those in the early period of growth in metals (Figure 5.32(a)) and to those in brittle materials. However, the time required to complete these processes is much longer than is the case for metals or brittle solids and so continuum behaviour is only defined for long pulse drives in these polymers.

The timescales of processes operating within the polymer at the target length scale (tens of millimetres) take milliseconds to complete because of slow sound speeds and sluggish kinetics coupled with large MSU sizes. Stress-state equilibrium thus takes equivalent times to achieve before continuum properties can be measured. This regime applies directly to loading in split Hopkinson pressure bars (SHPBs) where the transition between the wave-dominated mesoscale regime and a stress equilibrated structural response takes place. There is a danger here that applying too short a pulse in tests, particularly in amorphous materials above the glass transition temperature (rubbers), may mean that the target does not achieve equilibrium before the loading pulse ends.

Finally, the Taylor impact described above represents a laboratory scale test which illustrates some of the processes in the densification and flow phases of deformation under compression. The examples shown above for PEEK, PTFE and PMMA illustrate the three flow behaviours possible in these tests: flow, localised failure, hinging and fracture. The kinetics of these processes in the two loading phases play against the duration of the pulse applied which is determined by the double transit time of an elastic wave down the cylinder axis to fix the rebound phase of its motion. In each class of material, the defects in the microstructure seed inelasticity in the initial densification in shock followed by later localisation and failure. In the case of PEEK, tensile hoop strains during the flow on the surface play against strengthening mechanisms observed within the microstructure to prevent failure in the loading time available. If the cylinder velocity is increased, eventually the localisation phase of the process will continue until the whole cylinder has failed into fragments and the centre of mass of the system will obey Newton's law as required. In PMMA, lateral strain on the impact face soon starts fracture and fragmentation, and the momentum is transferred to the lateral flow of failed material. PTFE shows both densification and then plastic flow followed by fracture as the tensile hoop stresses increase. The plastic then fails and hinged struts are formed which move outward on the anvil to accommodate the strain applied in the pulse. It is the bar's length which determines the pulse length and its velocity the high, transient impact stresses which drive the kinetics of operating densification in the zone that fails to accommodate the strain. The transient stress state in the region a diameter from the impact face is resolved by failure as the material relaxes strength to allow flow across the surface to arrest the incoming projectile.

7.8 Final comments

The response of polymers again emphasises the need to combine results from all the existing platforms used over the last years for materials testing with potentially new techniques that can match the pulses applied to the slow kinetics that operate as these materials densify and then flow under extreme loads. Advances in new tools and diagnostics can illuminate their response dramatically and with their low density the materials can be penetrated by X-ray more easily than for instance metals under load.

Micromechanics dominates polymer response yet mechanisms remain to be fully understood. Any length scale has three phases of behaviour: localisation, flow and interaction as discussed in earlier chapters and, at least in the densification regime, polymer stress state stability is frequently not achieved in existing laboratory tests. Thus integrated existing loading techniques developed for metals (such as Taylor impact) reveal the transit to stress state stability but over milliseconds for polymers where microseconds were sufficient in metals. Thus constitutive models for use in hydrocodes are at present top-level, semi-empirical fits to behaviour that have severe limitations in their applicability since they borrow concepts based on viscoplasticity and are fit to data in which the stress state is rarely steady. Future development of theory and experiment must happen in tandem if models are to improve their accuracy and applicability in the future.

The region below 30 GPa is a rich undiscovered country for polymer behaviour and the details of subtleties in chemical composition of the monomer unit and the transitions in packing and flow during compression remain a fascinating tranche of subject areas for future research. It is kinetics and chain behaviour that must be tracked to illustrate response in order to fully reconcile observed states with the microstructures that are loaded. Their complexity has mediated their use in dynamic environments, but greater understanding will make polymers materials of importance for the structures of the future.

7.9 Selected reading

General background

Bourne, N. K. and Gray III, G. T. (2005) Soft-recovery of shocked polymers and composites, *J. Phys D. Appl. Phys.*, 38: 3690–3694.

Mills, N. J. (1993) *Plastics: Microstructure and Engineering Applications*. London: Edward Arnold.

Shock compression

Bourne, N. K., Brown, E. N., Millett, J. C. F. and Gray III, G. T. (2008) Shock, release and Taylor impact of the semicrystalline thermoplastic polytetrafluoroethylene. *J. Appl. Phys.*, 103: 074902.

Bourne, N. K., Millett, J. C. F., Brown, E. N and Gray III, G. T. (2007) The effect of halogenation on the shock properties of semi-crystalline thermo-plastics. *J. Appl. Phys.*, 102: 063510.

Carter, W. J. and Marsh, S. P. (1995; republished by J. N. Fritz and S. A. Sheffield from a report put together in 1977) *Hugoniot Equation of State of Polymers*. Los Alamos, NM: Los Alamos National Laboratory.

Zerilli, F. J. and Armstrong, R. W. (2007) A constitutive equation for the dynamic deformation behavior of polymers, *J. Mater. Sci.* 42: 4562–4574.

8   Energetic materials

8.1   Introduction

This chapter will detail the response of a class of materials dubbed energetic to signify that they can break bonds and react under load. These substances contain a large amount of stored chemical energy that can be released if appropriate thermal thresholds are exceeded. Such materials combine a fuel and an oxidiser; fuels are typically carbon or hydrogen, oxidisers are oxygen or a halogen like chlorine, for example. Combining hydrogen and oxygen to form water liberates $13\,260$ J kg^{-1} and burning petrol with oxygen (air) $30\,000$ J kg^{-1}. Yet the high explosive TNT liberates only 4080 J kg^{-1}, less than 15% of the amount liberated by petrol. The difference is that fuel alone burns only where oxygen is present; a spillage will burn for minutes with oxygen from air, for example. Yet a TNT molecule contains oxygen within it and can liberate energy in the microseconds the reaction front takes to transit the molecule and break bonds. Therefore the difference between these fuels lies in the power that the molecule supplies in the form in which the material exists on ignition. Energetic materials may be solids, liquids or gases, but condensed-phase materials will be followed here as earlier in the book. Further, they need not necessarily be organic. There is increasing need for higher performance, lighter weight and safer composites which use reacting metals as well as more conventional materials and using new material morphologies which have increased surface areas, such as mixtures of nano-materials or designed nano-composites. However, the principal energetics used at the present time include a range of elements that react with oxygen and these will be discussed in what follows.

Energetic materials include explosives, pyrotechnics, propellants (gunpowders or rocket fuels) and fuels (diesel and gasoline). In most cases, thresholds are breached at local sites within the material and reaction spreads quickly under the inertial confinement, driving a mechanical wave ahead. The associated rate-limiting chemical kinetics is in this case faster than thermal conduction competing to quench propagation from a site. If the thermal excursion is triggered by an impulse, then mechanical work must be done to raise the temperature above the barrier. In the limiting case, the response of an energetic to shock will proceed from a deformation stage to ignition and then finally to a combustion of some kind behind an inert front. This chapter will describe each of these phases and the evidence for, and results of, each one.

There are several different classes of explosive of different phase but also ranked according to their performance when ignited. Equally, it is often convenient to refer to homogeneous and heterogeneous forms of microstructure and reaction state. However, throughout preceding sections of this book the statistical nature of defects and second phases within real solids has been emphasised. These same considerations should be tabled here (and even extended to liquids too) since no real explosives will be found without defects in structure and composition. Nevertheless, theory has been developed for homogeneous and heterogeneous explosives and some consideration of when these terms may or may not apply is necessary. Clearly liquid high explosives or gas mixtures (such as hydrogen and oxygen) may be regarded as homogeneous. Carefully grown single crystals of energetic materials may be as well. However, there must be physical constraints coming at a particular scale. The key in understanding ignition comes in defining a scale below which defect size can cause propagating reaction without consideration of an inhomogeneous medium and this idea will be expanded in what follows. The concepts embraced here have importance in present materials since the engineering explosives of the modern world tend to be designed composites whose formulation is defined by the need to counter particular threats whilst maintaining performance when required. The counter example is that of imperfect, flaw-ridden materials used in terrorist encounters where the dangers of accidental ignition lie in manufacturing and then transporting or disposing of these materials once they have been made.

In what follows, an account will be given of ignition points followed through the suite of available reaction routes leading to full detonation in a charge in the ultimate case. As with many areas of science, the route to explaining the operating physics largely follows that taken in historical development so that a brief digression into the background development of the field highlights many of the concepts required.

The earliest use of energetic compositions was recorded in the far East. The four wonders of the ancient Chinese world were magnetism (second century BC), paper (first century AD), printing on paper (ninth century AD) and gunpowder (ninth century AD). Alchemists were searching for an elixir of eternal life but instead found a composition they named *huo yao*; *huo* meaning fire and *yao* meaning medicine. Its effect was more violent and spectacular than first expected since:

Of the composition of *huo yao*, sulphur is pure yang (male) and saltpetre is pure yin (female). When these substances come together the result is noise and change. (Wu Jing Zong Yao (1044 AD))

The effect and its value were soon realised and the first explosive, gunpowder, was tracked across the world over the next 100 years. The route west via the middle east reached Britain by the thirteenth century and its western roots are often attributed to Roger Bacon. Production facilities came later to support the English crown starting around the fourteenth century. Over this period the recipe for three ingredients seems to have settled close to the chemical optimum used today (Table 8.1).

Gunpowder (or for the last few hundred years black powder) from the earliest times:

... consisteth of three essential ingredients

Brimstone: whose office is to catch fire and flame of a sudden and convey it to the other two,

Table 8.1 Gunpowder composition development over time

Constituent	Bacon, *Opus Majus* (1242) (%)	Middle East (1260) (%)	British government (1635) (%)	Modern (after 1780) (%)
Potassium nitrate	38	75	75	75
Charcoal	31	16	12.5	15
Sulphur	31	9	12.5	10

Charcoal: pulverised which continueth the fire and quencheth the flame which otherwise would consume the strength thereof,

Saltpetre: which causeth a windy exhalation and driveth forth bullet.
(Dr Thomas Fuller, The History of the Worthies of England 1662)

Thus it is a mixture of potassium nitrate (the crucial oxidant), charcoal and sulphur. Fortuitously sulphur also lowers the temperature of the reaction between the nitrate and carbon, making the mixture easier to light and reducing erosion on barrels. In this form it has been widely used as a propellant in guns and as a pyrotechnic in fireworks.

It was first important militarily in the thirteenth and fourteenth centuries and spawned the advent of guns into warfare, but from the sixteenth century it was also used in engineering for dredging (1550) and tunnelling (1605). The operation and purity of the compositions used were fraught will uncertainties, however. A petard was an efficient means of undermining defences yet variation in powder chemistry or ignition threshold frequently left the user 'hoist by his own'. As Hamlet says:

For 'tis the sport to have the engineer
Hoist with his own petard: and 't shall go hard
But I will delve one yard below their mines,
And blow them at the moon: O, 'tis most sweet,
When in one line two crafts directly meet.
This man shall set me packing:
I'll lug the guts into the neighbour room.
            (*Hamlet*, Act III, Scene 4)

Of course explosives in the hands of insurgents were always an issue and terrorist actions came to prominence in 1605 when the Gunpowder plot brought the propellant to public notice. However, gunpowder burns rather than detonating and so it (along with other low explosives used alongside) was less potent than the materials developed in the coming years.

The reaction considered liberates carbon monoxide, carbon dioxide and nitrogen thus

$$4KNO_3 + 7C + S \rightarrow 3CO + 3CO_2 + 2N_2 + K_2CO_3 + K_2S \qquad (8.1)$$

Since only 43% is liberated as gas, the precipitated solids thus retain heat making the material inefficient as an agent to do work after expansion.

Sulphuric and nitric acids were discovered $c.1800$ and the process of nitration soon after. The modern age was born with the development of nitroglycerin (NG) by Sobrero in 1846 which ushered in the era of high explosives. Nobel industrialised its manufacture in the 1860s and made composites adding *kieselguhr* in 1867 and nitrocellulose in 1875. These explosives were sold under the well-known epithet dynamite and the age of commercial blasting was born at this time. Dynamite was primarily a commercial material and spawned explosives used in blasting, excavating, mining, tunnelling and demolition.

Nobel was the first of a series of figures who contributed key concepts to the science of explosives. A selection of members of that fraternity are shown in Figure 8.2(a). They include Chapman (1899) and Jouguet (1906, 1917) who developed the first theory of detonation; Nobel who encouraged explosive engineering with the development of dynamite (1867); Zeldovich (1940), von Neumann (1942) and Döring (1943) who developed the next significant improvement to the Chapman–Jouguet model for detonation; and finally, Bowden and Yoffe (1985) who developed the concepts of ignition and the first models of hot spots in solids. Through these people explosive science has prospered to the present day where the materials are used in engineering as precision sources for the loading of materials and structures.

In the run up to the First World War, the main formulations fielded by the UK were picrates and TNT (trinitrotoluene), with PETN used by the German army. Between the conflicts, the powerful formulations RDX and HMX were developed and in the Second World War these, along with mixtures (RDX/TNT, Comp B, and others), were used in various applications. Torpex (RDX, TNT, Al), for instance, was a composition used in torpedoes since the aluminium component lengthens the underwater impulse generated after detonation by reaction with seawater at later times. Since then these formulations have remained as the principal components for military use since systems with RDX and HMX are widely produced and documented and their chemical performance is close to optimal. The most recent new formulation (produced in 1987) is CL20, a nitroamine explosive developed primarily for use in propellants. It has a better oxidiser-to-fuel ratio than HMX and RDX and is thus superior to conventional high-energy propellants and explosives in performance. The principal formulations contain key groups holding fuel within the molecule. These include the $-NO_2$ group containing oxygen for combustion. Key variants are:

$-O-NO_2$ nitrate: e.g. nitroglycerine,

$-C-NO_2$ nitro: e.g. TNT,

$-N-NO_2$ nitramine: e.g. RDX.

The molecular structures of important explosives are shown in Figure 8.1. Before passing on, there is one further class of reactions which liberate heat in a manner analogous to the fuel–oxidiser reaction of explosives. The thermite undergoes an oxidation–reduction reaction between a metal powder and a metal oxide. An example might be copper oxide or iron oxide and aluminium. In this case the reactants, stable at room temperature, burn with an intense exothermic reaction after ignition and stable alumina results. The

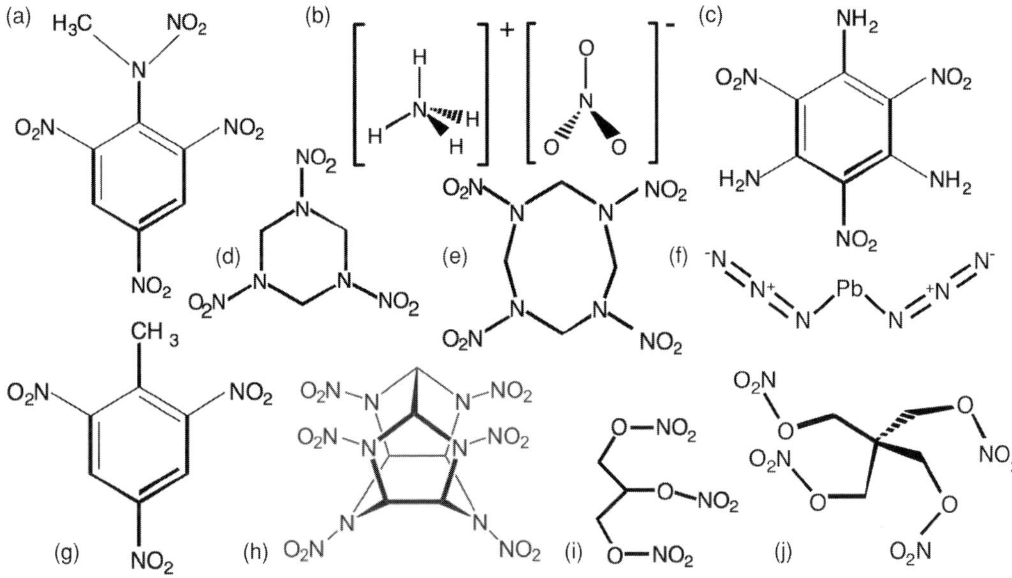

Figure 8.1 Selected explosive compositions: (a) 2,4,6-trinitrophenylmethyl-nitramine (TETRYL); (b) ammonium nitrate (crystalline); (c) triaminotrinitrobenzene (TATB); (d) RDX; (e) HMX (Octagon; High Melting eXplosive, His Majesty's eXplosive, or High-velocity Military eXplosive); (f) lead azide (crystalline); (g) trinitrotoluene (TNT); (h) CL20; (i) nitroglycerine (NG); (j) pentaerythritol tetranitrate (PETN).

pyrotechnic burn is often accompanied by a coloured flame so that such mixtures are often used in fireworks, but the high temperature is important in several industrial applications – for instance in the explosive welding of metals. Some modern explosives have incorporated nanometric metal additives to increase surface area and liberate late-time energy (with the metal contributing close to the reaction zone) as well as new thermite mixtures which react at fast rates.

In the modern world, energetic materials have assumed an important part of engineering capability and a series of devices have been developed for use in a range of applications. The ability to shock, fragment and move materials makes explosives the means of choice for blast and fragmentation used to excavate on the surface or at depth in mines, to create tunnels or even to create smooth walls using directed charges (smooth blasting). They may be used to clear large structures under precise control, to demolish buildings or to bring down chimneys with precision using laid sequences of blast or directed cutting charges. The latter may also be used underwater to cut parts of industrial sea platforms or sever pipework and cables. Indeed the short impulses from detonation may be used to work and form the metal under load as well as merely to push it with precision. In these applications forming or pipe-closure are simple uses, but energetic welding, cladding, hardening, creating bas-relief murals, or shaped charged cutting tools use the interplay between the Taylor wave loading by an explosive with the metal hardening and plasticity processes that occur under shock.

The application of the loading pulse produced by a high explosive to change the state in materials has been used to synthesise elements within the extreme pulse applied. For example, diamond can now be synthesised in the laboratory by loading high-carbon steels with explosive pulses and introducing a dislocation microstructure not found in the natural material that increases its toughness. Diamonds produced under HE load are industrial quality and small in size due to the brevity of the pulse, yet the opportunity to tap synthesis routes through martensitic changes probed by HE is an exciting possibility for the future since it is possible that dynamic processes may be harnessed to capture and maintain theoretical limits in strength, hardness or some other material property in a controlled manner.

Yet with great benefits come accidents that accompany such a powerful source of energy. The quest for understanding of the safe limits within which explosives must be contained has occupied societies since their discovery. In the military environment, sappers have been prone to accident since tunnelling led to the all too frequent dangers of the protagonist left 'hoist by his own petard'. Yet it was in the mining industry that the largest dangers were addressed. William Bickford developed the miner's safety fuse in 1831, creating a composite in which black powder was wound in with a flammable rope outer and which burnt at a known rate. Accidents in nitroglycerine production and transport in the mid-nineteenth century led Nobel to the development of dynamite. Indeed the concept of engineering a reactive composite with an energetic stabilised in an inert binder is key to a series of later developments leading to the polymer-bonded materials of today.

The largest accidental explosion devastated Halifax Nova Scotia on Thursday, 6 December 1917 when the SS *Mont-Blanc*, loaded with wartime explosives, collided with the Norwegian SS *Imo*. The detonation killed 2000 people in the near vicinity with over 9000 injured in the wider district. The air shock levelled surrounding buildings and structures along the adjacent shore, snapped trees, bent railings and drove a tsunami in the harbour. It ranks as one of the largest conventional explosive events, in a similar class to nuclear explosions but only 2.9 kilotons (kt) compared with the 21 kt of Fat Man which destroyed Nagasaki in 1945. In the interwar period a number of accidents showed dangers in the use of other materials not used in munitions. The Oppau explosion in Mannheim occurred on 21 September 1921 when a silo storing 4500 tons of ammonium sulphate and ammonium nitrate detonated killing 561 and injuring 1500 with effects felt 20 miles away. At that time the term 'critical diameter' was not understood but realisation soon hit that a poor explosive can build reaction to full detonation if the required inertial confinement is present. In 1926, lightning strike detonated 1500 tons in a US Naval store, which led to the establishment of quantity distance rules to minimise risk to personnel by fixing closest occupation.

The Second World War brought unprecedented production, and with that an unprecedented frequency of accidents in the use and storage of explosives. One of the largest blasts occurred at RAF Fauld near Hanbury in Staffordshire. On the morning of the 27 November 1944, incorrect removal of a detonator from a shell detonated 3500 tons of munitions stored in old mine-workings 35 m below the surface. The shock was felt in Birmingham and heard in London and Weston-super-Mare, 190 kilometres away.

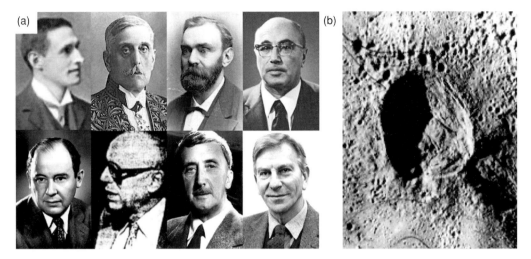

Figure 8.2 (a) Rogue's gallery of pioneers in energetic materials physics. Left to right, top to bottom: D. L. Chapman; J. C. E. Jouguet; A. B. Nobel; Y. B. Zeldovich; J. L. Von Neumann; W. Döring; F. P. Bowden; and A. D. Yoffe. (b) Explosion at RAF Fauld 1944. The crater was caused by the collapse of a mine storing 3500 tons of HE and is 250 m in diameter. It can still be seen today.

The dam of a reservoir close by was breached, venting a tidal wave that destroyed a factory at the mine entrance. In all 80 people died, and as the surface collapsed inward a crater 250 m across and 30 m deep was formed which is still visible today (Figure 8.2(b)).

Since the Second World War, the largest explosives accidents have involved ammonium nitrate storage, transportation or production. In July 1947 a ship loaded with 3300 tons caught fire in port in Brest and even through it was towed out of the harbour, the explosion caused 29 deaths and serious damage. This followed an incident in April of that year when a fire was detected in a cargo ship in Texas City; 2600 tons detonated setting fire to a second vessel 250 m away which contained a further 1000 tons of sulphur and 1000 of ammonium nitrate which burned for 16 hours before exploding the next day. The first of these events drove an air shock which knocked two planes flying 500 m above out of the sky. In all 567 died in this disaster and 5000 were injured. The explosion drove a tsunami 4.5 m high and flung the anchor of a second ship over a mile from its starting point. These disasters and their effects introduced a series of major changes to regulations for the storage and handling of energetic materials but also for volatile chemicals that can burn under pressure. Over the last decades major accidents have occurred across the world in chemical plants where explosion has occurred: Flixborough (1974; explosion in caprolactum plant resulted in 28 deaths and 89 injuries); Novisibirsk (1979; chemical factory 300 deaths); and Chernobyl (1986; explosion and fire in the graphite core of one of four reactors released radioactive material that spread over part of the Soviet Union, Eastern Europe, Scandinavia and later Western Europe, claimed 31 dead; total casualties unknown) to pick just three of the largest. This litany of accidents shows that

Table 8.2 Some selected properties of energetics referred to throughout this chapter. Detonation velocity varies with density and for some materials an empirical fit exists for its dependence of the form $D = a + b\rho$, where ρ is the density of the explosive considered. Velocity of detonation (VoD), Chapman–Jouget pressure (P_{CJ}), heat of detonation (ΔH)

Explosive	Density (g cm^{-3})	VoD (mm μs^{-1})	P_{CJ} (GPa)	ΔH (MJ kg^{-1})	a	b
CL20	2.00	9.40	42.0	6.94		
HMX	1.89	9.11	39.0	5.61		
NG	1.59	7.65	25.3	6.20		
NH$_4$NO$_3$	1.73	2.66				
PETN	1.77	8.26	33.5	6.40	1.82	3.70
RDX	1.77	8.70	33.8	5.72	2.56	3.47
TATB	1.88	7.76	29.1	4.52	0.343	3.47
TETRYL	1.71	7.85	26.0	4.77		
TNT	1.63	6.93	21.0	5.40	1.67	3.34
Pb(N$_3$)$_2$	2.90	5.20	20.4	1.93		

there have been major incidents several times a decade up to the present time and in each case localised transient temperature at some ignition point has caused explosion if energetic materials were confined so that pressure could build. The violence of the events leaves investigators understanding the causes of the incidents to resemble a detective story of modern accident. Forensic explosives investigation can use physical indicators such as crater size, or local sensors such as cracked man-hole covers, blown out windows, bent lamp-posts or road signs as well as chemical analysis of samples from the site to deduce the source of ignition, the magnitude of the blast and the composition of the energetic that destroyed a site.

The properties of selected energetic materials covered in this chapter are given in Table 8.2. The approximate heat of detonation per mole can be determined from heats of formation of reactants and products of detonation through

$$\Delta H_{explosion} = \Delta H_f(\text{detonation products}) - \Delta H_f(\text{explosive}). \tag{8.2}$$

8.2   Classification of energetics

All reaction in energetic materials starts with burning or deflagration within a solid and may transit to detonation; a reactive shock wave. Propellants and pyrotechnics deflagrate whilst primary and secondary high explosives detonate. The detonation of high explosives moves metal in hydrodynamic plastic flow, even welding plates to one another, yet propellant reaction creates lower amplitude and longer duration pulses which can drive rocket motors to lift structures, fire bolts in emergency, inflate air bags when triggered, cut motors, open valves or switch a range of safety systems quickly and permanently to ensure a threat is neutralised *in extremis*. Further, new devices employ explosively driven magnetic field compressors, whilst energetics are vital in seismology,

Table 8.3 Energy-releasing processes

	Pressure (GPa)	Rate (g s^{-1})	Power density (W cm^{-2})
Acetylene flame	1	1	10^2
Propellant	10^3	10^3	10^6
Detonating HE	10^5	10^6	10^{10}

and even in the civil nuclear engineering of the plants of the future. Finally, pyrotechnics react to give light, sound or heat for pleasure or effect. Fireworks are now the staple of public celebration as well of course as a reminder of attempted infamy in the mother of parliaments. These three terms, high explosive (HE), propellants and pyrotechnics, group the major classes of composition to be considered in what follows. Formally, high explosives are energetic materials that detonate, meaning that a shock wave travels through them supported by chemical reaction behind the front. A propellant is a material that burns, or deflagrates, producing high-pressure gas which can be used to propel rockets or projectiles, drive machinery or fill some emergency constraint like an air bag. Finally, a pyrotechnic deflagrates to ignite propellants or to produce particular effects: delays, heat, light, gas, smoke and/or sound. Collectively propellants and pyrotechnics constitute the class of low explosives. A high explosive defines its unique range of properties by the power it generates rather than the energy it liberates and thus the speed of the processes is key to the use of the energetic material. Since flames burn subsonically, waves from adjacent boundaries can change the flame speed. However, since a detonation wave is a shock, the operation of a HE will be independent of its surroundings and reproducible temporally and in effect. This is the key difference between low and high explosives and the main advantage in engineering with detonating explosives. This results from the definition of a detonation wave: an inert shock front driven by rapid exothermic reaction close behind.

Table 8.3 shows some typical output metrics for burning hydrocarbons, propellants or detonating explosive to illustrate these points. Clearly the power density differentiates the fuel from the low- and high-explosive materials in each case and it is the fact that the HE contains the oxidiser within the molecule that leads to extra power gained from the jump in reaction rate as one moves from deflagration to detonation. Before passing on it is worth defining historical terms that differentiate classes of detonable high explosives. A *primary* HE is one that detonates easily on application of a small mechanical or electrical stimulus but with small output. A *secondary* HE is one that requires a shock wave to enter it in order to initiate detonation. In such a scheme, *tertiary* explosives are those so insensitive to shock that they cannot be reliably initiated by primaries and thus require an intermediate booster to increase the shock strength before they can detonate. This defines the triggering sequence of explosives that culminates in the detonation of the last element in the explosive train. It is not until a detonator initiates a booster attached to the high explosive that a trigger can reproducibly detonate the main charge (see Figure 8.3(a)). Constructing a firing chain using this classification allows safe use of hazardous materials in engineering and this concept serves as the basis for safe working practices with HEs.

The chemistry of these processes concerns oxidisation of a fuel. For gunpowder that process relies on mixing three compounds and produces solids which absorb the output heat, as well as hot gases which may expand and do work. A similar exercise may be undergone to look at the reaction of three other compounds from Figure 8.1: nitroglycerine (NG; the active explosive in dynamite), trinitrotoluene (TNT) and RDX.

The three reactions are

$$\text{NG: } 4C_3H_5N_3O_9 \rightarrow 12CO_2 + 10H_2O + 6N_2 + O_2,$$
$$\text{TNT: } 2C_7H_5N_3O_6 \rightarrow 7CO + 5H_2O + 3N_2 + \textbf{7C}, \qquad (8.3)$$
$$\text{RDX: } C_3H_6N_6O_6 \rightarrow 3CO + 5H_2O + 3N_2.$$

It can be seen that both TNT and RDX have insufficient oxygen to fully oxidise the fuel while NG has excess; indeed TNT produces solid carbon (in bold). They are thus said to be oxygen deficient and indeed TNT gives a carbon residue after detonation. An indication of the fuel consumption within a molecule is given by the oxygen balance (Ω), defined to be the percentage by weight of oxygen, positive or negative, remaining after explosion assuming that all the carbon and hydrogen atoms in the explosive are converted into CO_2 and H_2O. For instance

$$C_aH_bN_cO_d \text{ gives } \Omega = (d - 2a - b/2) \times 1600/M, \qquad (8.4)$$

where M is molecular mass. This gives an oxygen balance of $+3.5\%$ for nitroglycerine, $+20\%$ for ammonium nitrate, -74% for TNT and -22% for RDX. This approach is particularly useful for making explosive mixtures where the blending ratio is adjusted to give $\Omega = 0$. For example, ammonium nitrate/TNT (80:20) and cellulose/NG (8:92; otherwise known as blasting gelatine) have $\Omega = 0$.

8.2.1 Microstructures of engineering energetics

Figure 8.1 has shown the molecular structure of several energetic materials, but just as the response of metals under dynamic loading is more dependent on the mesoscale structure of the metal than its atomic number, so the response of the explosives used in the field (dubbed engineering energetics here) is fixed by processes occurring on the microstructural scale not just on the chemistry of the molecular conformation. There are a series of core threats that particularly insensitive munitions must meet for qualification. These are slow and fast cook-off (two heating rates in a fuel fire), bullet/fragment impact, sympathetic detonation of a charge placed close by and shaped charge attack (penetration by a metal jet). A range of energetic composites must meet these threats and thus it is useful to briefly overview the microstructural form of modern engineering energetic materials. The propellant mixtures such as black powder that developed from the earliest times were of variable quality and consisted of a packed mixture of crystals of varying

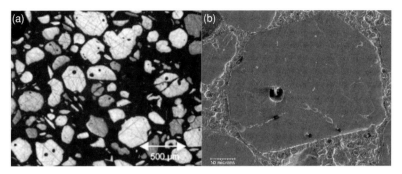

Figure 8.3 Micrographs of an RDX:HTPB, 88:12 research composition at two magnifications.

morphology prepared in different manners for insertion into guns or excavated holes to render some effect. However, when the molecular explosives of the mid-nineteenth century appeared, more control could be applied. The production of the compounds as melts or slurries allowed them to be poured into moulds or shells for use in the field. Another alternative was to pack the crystalline powder into containers in various manners and consequently different degrees of precision. Cast and pressed formulations were typical in military systems up to the 1970s and in legacy munitions. Slurries and prills of commercial materials are still commonly used for quarrying and mining up to the present day.

Cast HEs, poured molten into metal shells and then slowly cooled and stored for many years in open stores were prone to several damage mechanisms through their lives. Firstly differential cooling could leave cracks within the crystalline insert at the end of the production process. Secondly cyclical heating and cooling in storage between day and night and potentially in geographically hot and cold regions ($-50\,^{\circ}$C to $50\,^{\circ}$C cycle) could also fragment the filling with differential thermal conduction in the components. These effects can be summarised in the term *ageing*, which includes a range of mechanisms that reduce the performance of the charge on the one hand, but also render it more sensitive to accidental ignition on the other.

Since the 1980s there has been a great effort to replace cast or pressed explosives with a filling that is more tolerant to thermal fluctuations and low-intensity mechanical loads and jolts. The result is the plastic bonded explosive (PBX). An explosive composite is engineered which binds energetic crystals within a matrix which is generally inert but may be engineered to be energetic itself. Clearly the packing of the crystals now makes it impossible to reach full density of the filler energetic and this results in attendant loss in performance in the formulation. However, by careful selection of bimodal mixes of different crystal sizes, the packing can be adjusted to maximise performance and by this means materials over 95% of the theoretical density of the pure energetic can be produced. Micrographs of a research composition of RDX in an HTPB binder in the proportions of 88:12 by weight are shown in Figure 8.3(a); large, predominantly rounded, equiaxed crystals of diameter c. 300 μm, and smaller, more angular crystals whose long axis is a maximum of c. 100 μm. An initially bimodal set of crystals has been mixed and yet the form of the section shows that they are damaged in the production process with sharp edges rounded. Indeed there are regions where the RDX appears entirely absent. These regions at higher magnification show a large volume of debris of less than 10 μm size entrained within, making the binder essentially reactive and carrying a volume of RDX within it. There is also evidence of fracture with damaged crystals penetrated by the HTPB during material processing. Indeed a large proportion of the crystals show cylindrical tunnels through them or voids within (Figure 8.3(b)). These are solvent bubbles from the RDX crystal growth and are present in large numbers and at a length scale of tens of μm within. The large fraction of micron dimension grains entrained in the rubber filler is clear to see. Modern PBXs look to minimise risk by using more insensitive explosives such as TATB to replace RDX and HMX in the PBX. These modern compositions are known as insensitive high explosives (IHEs) for this reason.

RDX is one of a range of explosives found in systems used throughout the world. The intergranular friction and compaction that occur in loose or pressed powders and entrained volume of porosity that is inevitably introduced in older engineered systems is clear. But cast compositions present different issues in terms of cooling cracks produced during processing or thermal cycling during life. Finally, composite explosives have a mesoscale defect distribution within the crystals and with entrained particles in the binder. All of these features at their respective length scales are of a size that map well to the kinetics of the rate-limiting steps of chemical reaction; thermal conduction limitations allow temperature build-up and mechanical localisation from accidental driving impulses all contribute to accidental ignition. These ingredients make designing with energetic materials for both safety and performance fascinating and complex.

The explosives stored at present include cast and filled stockpile munitions, propellants and pyrotechnics. The newer systems contain PBXs and the move at present is to design and use IHE formulations for the next generation. There are several groups of indicators applied to explosives and most of these quantities are efforts to quantify hazard. One is sensitivity, principally to one of the three identified groups of problem stimuli: thermal, mechanical and electrical. A second includes thermodynamic quantities including the heat and temperature of explosion but also standard data such as density, toxicity, volatility etc. relevant when considering for material storage. Attempts to classify explosives performance have taken different routes to assess output and effect. These include empirical measures such as *brisance*, which refers to the ability to shatter bomb casings and the like, and also to the rapidity with which the pulse delivers its peak power. There are, however, quantitative terms that will be referred to below such as the CJ pressure, detonation velocity and so on. The goal of modern safety and licensing is to reduce the cost and risk of fielding an explosive for general use. The aim in a modern system is to use increased physical understanding to assess any risk, particularly using validated models for energetic behaviour in hazard scenarios and these tenets are laid in place, agreed internationally and set down in UN licensing standards. The regulations cover the following issues: storage; ageing; sympathetic detonation as a result of an accidental initiation close by; tests to prove the explosive meets IHE standards; resistance to heating (in particular performance in a fire); and mechanical insult by impact such as hitting by stray bullets. In each of these areas there exists a series of agreed qualification tests used to license the material for production and use.

This work will not attempt to include the historical hazard and safety tests and their analysis; the reader is referred to other texts which discuss these in more depth. This book has concentrated on the behaviour of materials in extremes and it is the performance of energetics or energetic composites that will occupy this chapter. As with the other classes of material, this text will concentrate on close to fully dense explosives, highlighting the chemical differences that lead to varying properties. Similarly performance at ambient temperatures occupies the majority of their uses and so is the basis for the treatment here. There are other more specialist texts that cover these issues in greater depth and the reader is referred to these at the end of the chapter. The core materials issues addressed here include the equations of state of unreacted and reacted materials, strength of explosives and more specifically their failure, chemical reaction when there is mechanical localisation, and the response of aged material to impulsive loads.

8.3   Reaction pathways

An energetic material may exhibit a series of responses according to the stimulus it receives. Stored at varying temperature and humidity over years, a chemical will naturally slowly decompose and this process is called *ageing*. If an explosive is heated unconfined in an area open to the atmosphere, ignition will lead to flame propagation that may spread across the explosive in milliseconds. Conversely it may detonate, driving a shock wave that will propagate through a confined sample in microseconds. All of these are possible in the same material; the contrasting responses are due to the different loading impulses that the explosive is subject to and the varying boundary conditions that deformation and reaction mechanisms operate within. However, when an explosive ignites, it is a direct thermal fluctuation that has caused reaction to start and variations in other state parameters trigger a sequence of mechanisms that result in associated thermal affects. In what follows, pathways to detonation will be described. Some specific terminology will assist so that, in what follows, the start of chemical reaction which leads to burning or deflagration will be termed *ignition*, whilst that which leads to the formation of a steady detonation wave will be dubbed *initiation*.

There are two final stages a burning energetic may achieve after ignition: a burn may transit to a violent reaction (BVR) or some route may lead to full detonation. The sensitivity of a HE to BVR is controlled by a series of factors including its porosity, its damaged state, charge geometry and the chemical nature of the HE, but most particularly by its confinement. Transits to detonation are historically ascribed to one of three routes; all of them transitions from an initial state to a propagating, reaction-driven shock.

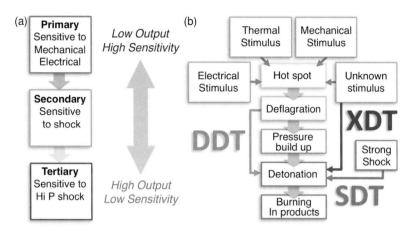

Figure 8.4 Reaction pathways: (a) the explosive train; (b) transitions to detonation. XDT = unknown to detonation, DDT = deflagration to detonation, SDT = shock to detonation transition.

The formal mathematical derivation of conditions in a detonating wave has already been presented in Section 2.12 and the reader is recommended to revisit that section to understand the concepts and terminology used in what follows. The first process involves the transit of an inert shock to a reactive front which is known as shock to detonation transition (SDT). The second process starts as a stable deflagration which accelerates to a detonation wave and is known as a deflagration to detonation transition (DDT). The final process is a catch-all for the many undefined cases where detonation resulted but forensic efforts could not tie down the cause and this is known as an unknown (X =?) to detonation transition (XDT). The stages in the three classifications are illustrated in Figure 8.4(b). In the general case a route can be tracked to detonation in a particular explosive. There are a known suite of mechanisms that have a mechanical, electrical or thermal excursion as their source. All of these lead to a local volume element that is heated to a critical threshold temperature for a time that allows reaction to propagate and lead to a stable burn. This local volume element was dubbed the *hot spot*. Using this idea, the shock to detonation transition can be ascribed to a heterogeneous agglomeration of hot spots excited by the inert shock at the front of the wave to react over the reaction zone to completion at the Chapman–Jouguet (CJ) plane at the rear where the flow becomes sonic. This process takes place over a microsecond and the burn time out from individual nuclei is of the order of nanoseconds. The deflagration to detonation transition takes place over milliseconds with confined burning at sites increasing in rate as the pressure builds until a compressed plug is driven into the packed material and SDT occurs. Of course there may be other unidentified mechanisms by which burning may begin and they may be lumped into the catch-all X to detonation transition mechanism discussed above. XDT may take seconds or longer since the events which establish burn are undefined. However, once the burn is established and confinement exists then DDT may occur as above.

Thus critical boundary conditions for local burning to transit to more violent and rapid chemistry are key to understanding processes leading to deflagration and detonation and these will be discussed in more detail below. It is confinement of the products and the pressure build up accelerating reaction kinetics that allows DDT to occur. In the case of SDT, burning that starts behind the front under compression establishes a pressure sufficient to drive the wave forward against releases from free surfaces. The transit from a purely inert shock to a steady reactive wave is an unsteady part of the flow that requires explanation for different systems and forms interesting variations in initiation discussed below. In the following pages the schematic path (of Figure 8.4(b)) will be followed through from the first moments at which a temperature excursion occurs within an energetic solid, to the final state when it has transited to full detonation. Of course boundary conditions within the flow may not confine the material sufficiently, or at some point the energy liberated from a formulation may not be sufficient to balance losses from other causes and reaction will continue as a burn or will decelerate and die. The following will consider the most energetic routes as limits on the range of behaviours found in observed response.

8.4   Ignition and initiation

If a local mass is heated at a faster rate than losses by conduction, convection or radiation can dissipate energy then the temperature will rise in that volume element to a point at which reaction thresholds can be exceeded and ignition occurs. The stimulus may be purely thermal, a laser focused on a region for instance, or may be a discharge path where current flows resistively, heating a conducting volume, for example, in an arc discharge through an explosive. Alternatively plastic work may be dissipated in localised flow within a solid that may cause the heating of a local region. In all cases these conditions will correspond to a trade off between dynamic heating and cooling by conduction. This reflects the discussions in earlier chapters where wave equilibration of stress states again gave way in certain geometries and circumstance to localisation of flow and the same heating that accompanies slip in shear or cavity collapse in compression in this case reaches an ignition threshold to start burning. Thus temperature rises occur in shock heating, viscous flow or from shear localisation (resulting from friction or from intrusion for instance) whilst cooling accompanies release, expansion, dissipation or radiation, and this trade off determines onset of reaction in energetic materials.

In the case of mechanical deformation, inelastic flow mechanisms localise heating and result in ignitions which lead to combustion. The amplitude and duration of the mechanical impulse drives the rate of the process and determines the temperature rise that results, but since loading samples a statistical population of defects, it also determines the number of propagating reaction sites that result. Thus a spectrum of global responses from a violent reaction to full detonation may result. The scale of the reaction site will also determine behaviour. Defects exist at the atomic scale (nm), at the microscale in metagrains (assemblages of atoms at the nm scale), at the mesoscale, at the dimension of constituent crystals (μm) and clearly also at the component or structural scale at which observation is made. A series of experiments for propagating hot spots defines the length and timescales for local ignition to occur under particular ambient pressure states, and these define the physical constraints on ignition for the material of interest.

8.4.1   Hot spots

The temperature for thermal ignition of an energetic material under ambient conditions defines a figure for comparison between various materials. The primary explosives lead and silver azide ignite at 345 °C and 275 °C, respectively, whilst the secondaries PETN, NG, RDX, HMX and TNT ignite at 190 °C, 200 °C, 260 °C, 280 °C and 300 °C. Finally, the insensitive composition TATB ignites at 384 °C. For completeness, ammonium nitrate ignites at 200 °C and black powder between 300 and 400 °C. Before discussing dynamic thermal loads, it is important to recognise the process of cook-off, in which an energetic material decomposes violently after being subject to a temperature exceeding critical thresholds in the material as a result, for instance, of fire. Bowden and Yoffe (1952/1985) were the first to point out that the amount of work dissipated by dynamic processes is insufficient to result in bulk heating that will ignite explosives. Working in parallel, Tabor (Bowdon and Tabor, 1950/2001) showed that plastic deformation at surfaces under frictional load was found at local contact points between the rubbing solids which connected the surfaces. As discussed previously, friction occurs at the component scale with localisation at the mesoscale that heats significant volumes of energetic in shearing bands. Thus, the total energy must be divided into local volumes with temperatures high enough to exceed mean thresholds and these localised plastic regions in rubbing solids must be mirrored as reaction nuclei within explosives. Thus energy deposition is localised at inhomogeneities and burning spreads as the locus of local thermal ignition points, which combine into a reaction zone as time progresses. The constraints on these local zones are defined by their size and by the magnitude and time duration of the temperature excursion. A critical hot spot is one expanding from a volume and at a temperature just sufficient to cause a burn to propagate against these dissipative effects. In their work on a variety of explosives, Bowden and Yoffe were able to home in on a range of physical attributes that could be ascribed to a critical hot spot (these are summarised in Table 8.4).

Clearly different chemical and physical microstructures will place a range around dimensions and temperatures, yet the order of each number fixes the dimension and filters the kinetics of mechanisms that will lead to propagating reaction within an explosive. The hot spot is a mesoscale entity focusing attention on the state of the microstructure at this level. The PBX microstructure of Figure 8.3 shows a defect population precisely at the dimensions required to produce critical hot spots. As to the operating mechanisms for ignition, the timescale for which temperature must be maintained makes the phenomena of interest in seeding such structures slower than wave related mechanisms. A series of headings for conducted experiments with notional results has been presented in the table. In the following paragraphs each will be expanded for further comment.

Table 8.4 Notional conditions for a critical hot spot in condensed-phase explosives

Physical property/mechanism	Range
Minimum size	1 μm
Minimum time duration	10 μs
Minimum temperature	700 K
Adiabatic shear	Mesoscale for ms
Void collapse	Flow for 10s μs
Friction	Mesoscale for ms
Electrical	Heating, if current level high enough
Fracture	Not in crystal; possibly with grits
Fission tracks	On lattice scale
Dislocation pile ups	< 1 μm, flow stresses of explosive crystals low
Lattice thermal fluctuations	On lattice scale

Adiabatic shear localises plastic work by deformation of a crystal in compression as a prelude to structural deformation of components. The reader is referred back to the discussion of shear bands in Chapter 5 for a complete discussion. A band of high plastic flow is established when work done in the zone (which leads to thermal softening of the region) exceeds the rate at which thermal conduction can make the temperature field homogeneous through the surrounding bulk material. The size of the band, and the time taken to develop it, are derived from models equating the plastic work and thermal softening and if temperature can build over significant flow times to exceed ignition thresholds then burning will commence. Shear bands are developed to sizes of 10 μm or greater and over times of more than 10 μs. Thus loading impulses must have pulse lengths exceeding these values if reaction is to occur. When it does so, the start of burning can be seen in deformation bands and their intersection where plastic work is localised (see the discoloured heat-sensitive emulsion recovered from HMX impacted in a drop tower shown in Figure 8.5(a)). The details of shear banding have been described earlier and analyses are to first order. Yet shear is the vital mechanism that ignites energetic materials in accident scenarios since the kinetics of band formation couple to rate-limiting steps in reaction and for this reason alone the phenomenon deserves more effort to fully address its consequences. In particular, friction, an integrated set of multiple deformation mechanisms, has shear at local asperities or around confined grits at its core. Local inhomogeneities on surfaces onto which explosives may fall in storage depots or in loading or transport represent major hazards that must be limited to avoid accident in the field.

Mesoscale voids exist within crystals and at grain boundaries in the composite ener-getics that are used in the field (for example, see Figure 8.3). Voids or bubbles will collapse under compressive loads and heat the vapour or gas contained within them adiabatically to high temperature. Voids will be microns or greater in size and if they are swept by a shock front they will collapse asymmetrically, with one wall accelerated and driven towards its downstream fellow, inverting to form a jet of material that impacts and isolates a toroidal vapour pocket as the jet drives into the downstream explosive. The

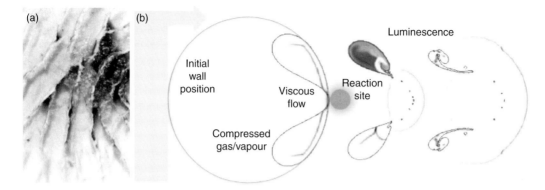

Figure 8.5 Hot spot ignition: (a) shear banding in HMX; (b) shock-induced cavity collapse in high explosive. Source: (a) illustration from Field, J. E., Bourne, N. K., Palmer, S. J. P. and Walley, S. M. (1992) Hot-spot ignition mechanisms for explosives and propellants, *Phil. Trans. R. Soc. Lond. A*, **339**: 269–283.

impact pressure and temperature induced in the bubble contents drive reaction in the material, whilst the compressing gas heats the contained vapour to extremely high temperatures. Figure 8.5(b) shows the initial bubble size and three snapshots of the collapse of a spherical cavity driven by a shock moving from the left-hand side. A jet on the involuted wall impacts the downstream explosive and starts reaction in the first frame whilst strong shocks reverberate and heat vapour in the toroid compressed in the flow. The shock in the next frame isolates high temperatures (white is 10 000 K) which strips electrons and causes luminescence from the trapped gas, whilst all the time thermal conduction heats the surrounding explosive and fuels the reaction that propagates out from the reacting hot spot. Shocked explosives are thus sensitised by the presence of entrained cavities. The principal ignition mechanism is the heating on jet impact, which ensures local turbulent flow, and thus viscous heating that also maintains the elevated temperature field for a significant time. Finally, entrained gas or vapour will increase effects further. The importance of this mechanism is reflected in the fact that commercial explosives (AN emulsions) are sensitised for detonation with the addition of glass micro-balloons to the mixture before it is poured down the bore-hole prior to accurately timed and reproducible detonation.

The spatial extent, pressure achieved and time duration of the hot spot achieved by a collapsed void is favoured by convergent walls in arbitrary geometries (such as wedges) since the jetting that results will be faster. Further, such a site will actually be favoured by non-spherical voids since the number of critical hot spots is a function of volume, duration and temperature achieved. However, at the continuum, it is the statistical distribution of propagating burn sites that will determine DDT for low-amplitude loading or the transit to full detonation in the case of an inert shock to detonation transition.

Electrical hazards for high explosives stem from ignition from high-temperature channels confined with the bulk of a material when current flow causes Joule heating of a resistive path that exceeds the ignition threshold in a material. Such current flow might originate from lightning strike on a storage building or container or from accidental sparks from charged workers or electrical equipment nearby. Primary explosives in particular are extremely sensitive to spark discharge and working arrangements for explosive should thus ensure any handling is done while properly earthed. Similarly, explosive stores and working laboratories subject to discharge take precautions against such discharges in the design of the workplace and working practices.

It may be assumed that fracture of an explosive crystal might localise plastic work and cause reaction to start quickly. However, firstly the process zone at the crack tip is too small whilst secondly the crack tip sweeps through the bulk of the crystal at high speed so that the temperature is not confined to a high value in a volume element for a sufficient length of time. Finally, a single crystal liberates a small amount of energy creating the free surface and has a high thermal conductivity thus cooling any heated site quickly after its formation. Thus fracture through a crystal will cause a subcritical hot spot that will not propagate into the rest of the bulk material. It is possible, however, that tough polymers in a PBX may fracture more slowly, liberating three orders of magnitude more surface energy, have low thermal conductivity and exhibit slowly moving crack tips. In this case a high-temperature polymer heating an adjacent crystal directly may cause ignition. However, fracture has little claim to be considered as a mechanism of primary importance.

Temperature excursions at lower length scales are of course possible and will always occur and yet these are subcritical since they induce temperature sources of dimension much less than a micron held for a time less than a microsecond. Such features will include the submicron dislocation structures discussed earlier in the deformation of crystals or tracks from elementary particles bombarding the explosive from radiation sources. Thus no thermal fluctuations around defects at the lattice dimensions meet the criteria required for localisation and so no scale below the mesoscale needs consideration.

8.4.2   Thermal explosion

Burning from a hot spot can be described as a series of reactions taking place in zones above an explosive's surface. It was known early on that burning depends on surface area and proceeds perpendicular to surfaces (Piobert, 1839; see Fickett and Davis, 2000) but a simple description to describe the kinetics of the process came at the end of the century with the burning rate law (Berthelot and Vieille, 1882). This states that

$$r = \beta P^{\alpha}, \tag{8.5}$$

where α = burning rate coefficient; β = pressure index (typically 0.9 for gun propellants) and r = burning rate (typically few mm s^{-1} in direction perpendicular to the burning surface). Although this law is of simple form and empirical, it has been tested for over a century and describes well the behaviour of burning explosives and thus it is used to describe deflagration processes to the present day. Its most prominent feature is the dependence of the rate upon pressure. This ensures that once burning starts in a confined energetic material the pressure and thus the rate of reaction will build, leading to burning that will inevitably accelerate unless waves from boundaries release pressure before its completion. This model of chemical reaction in a burning solid was expressed analytically in a simple thermal explosion theory assuming Arrhenius kinetics (Frank-Kamenetskii, 1969). He expressed the heating of a slab of material of specific heat c_v to heat generated in a reaction Q and lost by thermal conduction (conductivity κ) in a time-dependent heat conduction equation with a reaction term having an Arrhenius temperature dependence thus

$$c_v \rho \frac{\partial T}{\partial t} = \kappa \nabla^2 T + Q A e^{-E/R(T - T_o)}. \tag{8.6}$$

Assuming that there comes a point where heat generation exceeds thermal losses, one may calculate a time to thermal explosion, τ:

$$\tau = \frac{c_v R T^2}{A Q E} e^{E/RT}, \tag{8.7}$$

Equation (8.7) shows that the reaction runs away after a small time, dependent critically on the temperature generated at the hot spot.

In the case of hazard from insults to an explosive, the matrix or edge of grains are a boundary condition on the hot spot within the energetic crystal and provide the surface where cooling occurs. Conversely within a shocked PBX, the hot spot sits within a cool energetic volume with hot polymer around its boundaries. Thus the boundary conditions determine the thermal conduction into the grains and their size will thus determine the time to thermal explosion (Eq. (8.7)).

This suite of reactions, starting at a distribution of hot spots within the volume, consume reactants by burning out to neighbouring explosive between the sites and so consumes larger volumes of material, with an ever greater reacting surface area, as the progressive burn fronts grow. Product regions coalesce and if the volume is uniformly seeded the burn enters into a regime where the amount of material left to consume begins to decline – a regressive burning phase. If the volume is behind a reacting front then a new population of hot spots are entrained into the flow as it progresses (Figure 8.6). Imagine sitting on a shock front sweeping through a PBX and seeding a population of hot spots reacting behind the front and now burning at the rates consistent with the high pressures behind the Von Neumann spike (defined in Section 2.12). The specific area of the available reacting explosive rises and falls through a progressive then a regressive burn behind the front. All reaction has ceased at the CJ plane where the flow becomes sonic in product gases so that the degree of reaction is one on this plane. The maximum reaction rate occurs when the progressive burn finally ceases, which is at the centre of the reaction zone just where ZND theory predicts the maximum reaction rate to be. If the wave is at a free surface, the pressure is released and critical hot spots are less violent and fewer in number; the front is not driven and so is curved in this region with two-dimensional flow.

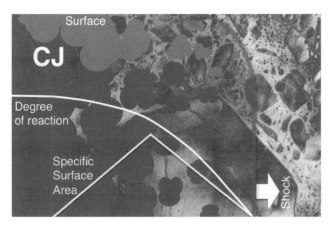

Figure 8.6 Reaction zone in the detonation wave. Burning sites behind an inert shock front showing progressive and regressive burn and degree of reaction which follow the forms shown. Reaction is complete at the CJ zone at the rear where the flow becomes sonic. Release from the upper surface quenches reaction and the shock pressure drops in this region, slowing the wave.

8.4.3   The burning rate at pressure in HMX

Energetic materials have reaction timescales determined by impulse transmission through the bulk and these include elastic and plastic waves as well as burn fronts that are induced by localised heating. The rate at which these fronts can travel is a function of the compression in the material determined by the pressure in the driving impulse. Recent work has shown that there is a threshold above which there is an increase in burning rate as confining pressure increases towards detonation.

One of the most studied high explosives is HMX. The material exists in four solid phases α, β, γ and δ whose properties and crystal structure are well documented. It has been found that the β and δ phases are the two most relevant to ignition problems. Of these the β phase is most stable and has the highest density (Figure 8.7(a)). However, the δ phase is long-lived, has a higher reaction rate, is more sensitive to ignition and dominates behaviour in the low-pressure regime when explosives are damaged. There is a 6.7% volume expansion on transition from the β to the δ phase and this exacerbates crystal fracture in composites that are exposed to heating under ambient conditions such as in the cook-off hazard tests. The sensitivity of δ HMX is comparable to the most sensitive secondary – PETN. However, above c. 200 MPa the material is found in the denser β phase and it is this that dominates in the shock states as detonation is approached. Thus HMX offers a candidate of high pedigree to assess burn rates in energetic materials confined at high pressure.

Recent compelling work measuring burn rate directly was conducted by Zaug and colleagues (see Esposito *et al.*, 2003) where two grain sizes (a pressed pellet of greater than and one less than 10 μm average grain size) were compressed in diamond anvil cells and ignited on one face with the appearance of the burn at the other recorded. The results from this work are summarised in Figure 8.7(b). The plot shows several features of note. Firstly there is transition to higher burning rate at pressures greater than

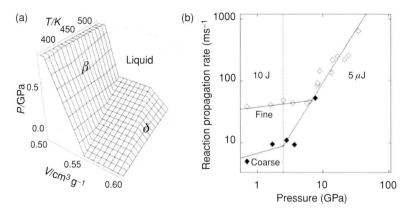

Figure 8.7 (a) Phase diagram for the principal stable phases of HMX in the low-pressure burning regime, β and δ. (b) Reaction propagation rate in HMX crystalline presses confined within a diamond anvil cell. The vertical dotted line indicates the pressure at which an inert shock becomes overdriven in HMX (the Weak Shock Limit; WSL). Source: (b) after Esposito, A. P., Farber, D. L., Reaugh, J. E. and Zaug, J. M. (2003) Reaction propagation rates in HMX at high pressure, *Propell. Explos. Pyrot.*, **28**: 83–88.

10 GPa regardless of the grain size of the HE. Secondly there is a much lower transition from slow to fast burning rate in the coarse-grained material than the fine-grained. It should be noted that these samples represent a small number of single crystals in a target of the order of 50 μm in depth. They will thus constitute a few grains in the coarse case but more in the fine. Thus the available surface areas for burning will be different in the two cases. The dotted line shows the calculated pressure at which defects saturate as compression increases loading down the easy direction in the crystal. Above this value the coarse-grained HMX deflagrates at higher speed. The fine-grained target accelerates to this greater value at a pressure a few times higher. Here there is a more isotropic mix of crystals with larger surface area but the transit still occurs to the same higher rate. This defect saturation makes combustion through the grain possible rather than surface Vieille burn discussed earlier. Indeed in these experiments, the laser pulse energy required to ignite one face of the explosive was two million times less in the compressed targets with higher burn rates than in the lower pressure targets with slower ones.

8.4.4   Shock to detonation – initiation

After the previous discussion it is possible to visualise a steady detonation wave with mesoscale burning sites growing behind an inert shock front and completing the reaction at the CJ zone to the rear of the inert shock travelling forward into the explosive. This view must tie to observations at the continuum scale and several features of shocked flow are known that can illuminate behaviour. To correlate with simple one-dimensional theory, the simplest system to consider is that of a rigid piston entering an explosive and

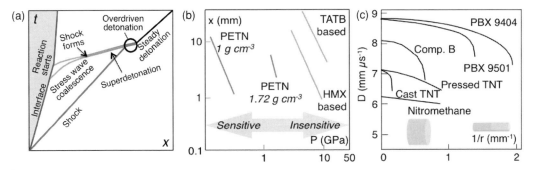

Figure 8.8 Shock to detonation: (a) schematic SDT; (b) pop plots showing log x vs. log P; (c) VoD vs. critical diameter (r) test data for detonation extinction.

driving an inert shock into the medium. A schematic x–t diagram showing the features of the flow is shown in Figure 8.8(a).

The detonation front has an inert shock (the Von Neumann spike) that develops until reaction completes at the rear of the detonation zone, the CJ plane. In this position, the pressure in the explosive may be approximately determined by

$$P_{CJ} = \frac{\rho_0 D^2}{4},\tag{8.8}$$

where D is the detonation velocity and ρ_0 the initial density of the explosive.

The compression front raises the pressure and temperature in the shocked flow beyond the ignition temperature of the medium and thermal explosion occurs. There is of course an induction time for this to occur, which means that the initial shock has run ahead some distance before reaction starts. From this time onwards further hot spots build reaction and this population of pressure generators drive compression waves forward into an ever-densifying material. This pressure build up sends strengthening waves ahead, which will converge to form a shock front ahead of the piston. This shock can form a superdetonation which eventually overtakes its inert precursor, and subsequently decays from an overdriven state to a steady detonation front travelling at the expected VoD for the medium. This model may apply to homogeneous media like liquid systems and indeed this is termed the homogeneous SDT model. However, in the heterogeneous solid explosives described here, where defects exist in the unshocked medium, wave amplitude will grow and activate larger volumes of hot spots as a convective burn reaches the shock front from the rear. Thus no overdriven stage is generally observed; rather there is a smooth growth behind the inert shock which transits to full detonation after some run distance. Of course the flow behind the shock front is subsonic so that effects from boundaries can communicate behind the growing shock. Thus there is observed dependence on particle size, pressed density, additives, etc. in this region.

A detonation wave travels supersonically and thus its velocity and performance are in principle independent of geometry and sample preparation, since information on boundaries should not reach the high-pressure region on the central axis of the reacting cylinder. Thus, knowing the jump conditions and CJ theory, it is only necessary to calibrate a limited number of detonation parameters, and that can be done by simply launching a shock and measuring a steady detonation velocity for the material. Such experiments are the staple for characterising ideal detonation properties. A suitable impulse is launched from a booster down a stick of explosives and, by using suitable gaps, the input pressure P of the initial shock can be varied as a parameter. There is an unsteady region as seen above but the charge soon settles to a steady detonation which can be recorded with sensors or streak photography as the bright reaction front breaks out of the side of the cylinder. The distance travelled before the detonation wave is steady and light can be recorded from the sample surface; this is called the run distance x. Alfonse Popolato (1972) conducted a systematic experimental study to determine the run distance to detonation for many explosives. He found that the higher the initial shock pressure, the more prompt the detonation became, with the value at zero run the threshold pressure for prompt detonation. Plots of the logarithm of run distance to detonation break out versus the logarithm of pressure in the inert pulse applied became known as Pop plots, and these give useful data on the kinetics operating in the reaction build up to detonation. Further, there are criteria for whether an inert shock will run to detonation at all; clearly it must be of the right amplitude, but critically it must also be of the correct duration, reflecting the magnitude of the impulse imparted.

Since the rate of work per unit area in a shock is Pu_p, the critical energy fluence E (energy per unit area of shock) for reaction can be shown to be

$$E = \frac{P^2 \tau}{\rho_0 U}, \tag{8.9}$$

where P is the shock pressure entering, τ the pulse duration and $\rho_0 U$ is the dynamic impedance of the unreacted explosive.

Figure 8.8(b) shows data for various compositions. It can be immediately seen that PETN, which is a sensitive secondary high explosive, has a short run length whilst TATB, which is an insensitive composition, has a much longer run distance required to reach detonation after loading from an inert shock. Thus since TATB is less sensitive than HMX it has become the prototype for the class of insensitive high explosives (IHEs) which will replace conventional materials in the medium term. Finally, two densities of pressed PETN powder are shown with the lower density pressing (with a greater number and larger size of voids within the compact) more sensitive than the higher density material where potential hot spot sites have been pressed out of the same material. Further, there are other features of the stimulus that affect the run distance measured in experiment. The results present Pop plots for a long pulse. However, if the pulse were short, release would catch and at some point begin to attenuate the stimulus. Further, the lateral extent of the shock holding the HE in a one-dimensional state will have an analogous effect. Finally, the geometry of the impactor and the flow that result have an effect on the critical pressure recorded. Thus this simple test to measure the velocity of detonation (VoD) of a sample energetic yields data which suggest that the simple one-dimensional picture borrowed from CJ or ZND theory for the detonation zone is very far from the case. Further, the temperature in the bulk explosive corresponding to the pressure for break out of full detonation is in all cases less than the measured ignition temperature of the energetic; proof that there is a population of local hot spots fuelling the steady front as shown in the sketch in Figure 8.6.

A second result comes from a further variant on cylinder tests firing propagating detonation waves into charges with a bare surface and reducing their diameter until the detonation wave eventually slows below the calculated value and stops propagating altogether. For all explosives there is some critical cylinder cross-section at which the detonation region at the centre is no longer confined. This has a failure radius, r, or a *critical diameter*, $2r$. This behaviour can be seen in Figure 8.8(c) which shows VoD changing in various explosives as the diameter of the cylinder is adjusted. As it narrows the detonation velocity drops and the front becomes more curved until eventually the shock is completely extinguished by lateral release. The critical diameter is found to depend on a whole series of properties such as density, structure, confinement, and even initial temperature. This is because the detonation front cannot be supported by gas pressure from chemical reaction since it is vented by lateral release as the volume element sampled by the confining pressures from reaction shrinks below a critical value which reflects the small concentration of burning sites progressing reaction.

Of course steady-state theories such as Chapman–Jouguet and ZND are one-dimensional and do not attempt to view details of mechanisms within the combustion zone. However, in a stable detonation there is flow at the mesoscale and transverse waves propagate in the reaction zone. Even detonations in explosive gas mixtures show cell structures from these effects that can be seen on the recovered tubes confining them. Here too it is the release from free surfaces that saps energy from the reaction zone and that leads to curvature of the front and even galloping or spinning waves near failure. Thus detonation is a two-(and at the mesoscale three-) dimensional process and needs to be considered in this manner to describe even simple tests.

In a mesoscale picture of the reaction zone in a granular explosive (such as that of Figure 8.6), release at the surface reduces pressure which slows the inert shock front and the burning rate of grains and thus the pressure (and the energy liberated) to drive the front. It takes larger grains longer to burn than their smaller neighbours. Indeed it is found that critical diameter varies with grain size, and that explosive containing larger grains require greater diameters to support detonation. Cast material has a lower concentration of defects to start burning from, whilst requiring heating of large volumes with smaller surface areas to burn; the figure shows that cast TNT requires greater radius cylinders to sustain detonation compared with pressed ones. Thus the key processes at work in the reaction zone are ignition of a hot spot in a suitable site, and its combustion in the high-pressure environment. Thermodynamic states at the continuum are the result of statistical populations of burning sites at the mesoscale and modern descriptions of key processes must address these localised sites to describe the observed behaviours in the reaction zone behind the shock front. However, the burning rate at these elevated pressures is key in determining the energy liberation behind the front and this will be returned to below. In a propagating wave there is a dynamic trade off between the pressure generation by chemical reaction and its release by outward flow from free surfaces. Thus any confinement alters that balance to sustain detonation or to make it easier to achieve.

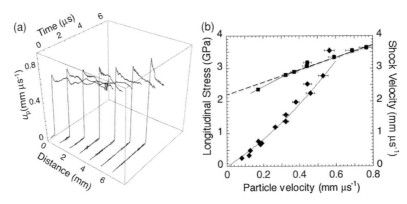

Figure 8.9 Experiments to characterise SDT: (a) particle velocity histories for the impact of a 10 mm alumina flyer at 680 m s⁻¹; (b) unreacted Hugoniot including shock velocity and stress–particle velocity data. Source: reprinted with permission from Bourne, N. K. and Milne, A. M. (2004) Shock to detonation transition in a plastic bonded explosive, *J. Appl. Phys.*, **95**(5): 2379–2385. Copyright 2004 American Institute of Physics.

Figure 8.9(a) shows an example of a recent study on the shock to detonation transition. An impacting flyer plate launches a one-dimensional shock into an explosive and particle velocity gauges introduced into the flow give a snapshot of the developing reaction in the front as the inert pulse transits to detonation. This gives detail in the reaction zone not seen in streak images of the luminous wave front. In this experiment an inert shock pulse of magnitude 3 GPa was input into the explosive whose microstructure was discussed earlier (Figure 8.3) and the experiment here was designed not to run for long enough to be affected by lateral release from free surfaces. There is a build up of particle velocity at this level not seen at lower stresses, indicating the onset of reaction and propagation of compression pulses forward to strengthen the shock. The release, seen as a reduction in particle velocity after 2 μs, is not apparent after the shock has run for 8 mm into the target. By the end of the experiment, the shock is on the point of steepening to reach full detonation.

To fully describe the observed behaviour, it is necessary to conduct a series of experiments to determine the Hugoniot of the unreacted explosive (Figure 8.9(b)). One also needs to know the equation of state of the product gases which can be recovered from the chemistry assuming the reactions complete to produce the expected range of stable products. However, the key step is in understanding the path the explosive takes to go from the unreacted to the fully reacted state (from the Von Neumann spike to the CJ zone in full detonation), and a model describing the grain burning was used in this case to simulate the developing kinetics observed in the experiment in this case. The grain distribution in the composite was already known from microscopy (Figure 8.3) and one interesting observation from the model was that in the few microsecond pulse introduced by this flyer, burning grains bigger than 10 μm in the simulation did not affect the pulse form before release since the burning of larger grains was too slow to contribute to the gas build up. Thus hot spots in the smaller grains built the shock to detonation and the processes occurring in the run distance depend on this

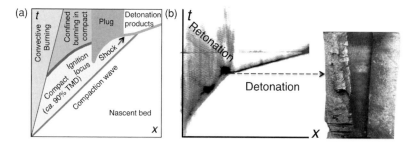

Figure 8.10 (a) Schematic DDT; (b) DDT in tube of CP pressed to 67% of theoretical density. Source: after Luebcke, P. E., Dickson, P. M. and Field, J. E. (1995) An experimental study of the deflagration-to-detonation transition in granular secondary explosives, *Proc. R. Soc. A.*, **448**: 439–448.

population rather than their larger surrounding counterparts. There comes a scale at which grains within an inert matrix must be regarded as a reactive continuum with its own properties and this work suggests that this is a volume element roughly ten grains in size. Similar concepts applied to energetics may be used to probe reaction and failure.

8.4.5   Deflagration to detonation

The deflagration to detonation transition is a reaction path key in addressing the concerns for safety in handling explosives and understanding accidents and their causes. It describes the transit of a local burn to full detonation; a process that takes approximately a millisecond in contrast to the microsecond in the SDT process. DDT is observed primarily in pressed granular material and investigations of the mechanisms for its operation have focused on behaviours found in porous beds. Indeed, when compacts are pressed close to their single crystal density it is found that the formulation becomes incapable of reliable initiation except by the strongest stimuli; this phenomenon is called *dead pressing*.

A schematic overview of operating processes is sketched in Figure 8.10(a). A burn starts thus generating gas that transmits heat from the flame region to adjacent surfaces by conduction. Burning of this sort is slow, at a nominal speed of c. 0.01 m s^{-1}, and proceeds by heat conduction from the gas flame region to the surface and to a lesser extent radiation transport from the gas to the solid. As the pressure builds a compaction wave travels out from the initial burning site compressing the compact to around 90% of full density, which triggers hot spots at defects such as cracks or voids and allows the burning front to accelerate. Such convective burning involves heat transfer augmented by mass flow and travels at a speed of c. 10 m s^{-1} that nucleates further hot spots ahead and leads to a combustion wave travelling at hundreds of m s^{-1} into the bed. Fast reaction behind this front sends waves forward into the compressed bed and when they coalesce, further densification forms an impermeable plug. This is accelerated by the high pressures generated by reaction behind it as a piston, which shocks the explosive and eventually detonates downstream by SDT. Thus a burn transits to detonation over a run distance under conditions of confinement and on detonation a rearward wave, travelling back into the plug and called a *retonation* is formed. The time these processes take and the precise velocities of each step depend upon the density of the explosive, the degree of the confinement, and the physical morphology of the crystals. The whole is a conglomeration of mechanisms: pressure-dependent burning, compaction and SDT into a porous bed. All of these together test understanding of processes and their kinetics to the full and, if they are to be described, a suite of connected models must track each step correctly. However, in each of these steps, the key need for strong confinement is clear and the lesson for avoidance of such incidents is to prevent pressure build up at the earliest possible stage.

Figure 8.10(b) shows the results of DDT experiment on an explosive bed (CP) packed into a metal tube. The streak photograph (in negative), taken down a slit window set down the length of the tube, is shown to the left with the metal confinement recovered after detonation has occurred to the right. The burning front appears dark (shown in negative in this picture). The slow build up of reaction is mirrored in the slight expansion of the metal cylinder bore in the early stages. At later times it can be seen to be larger still after the point at which detonation has occurred. The walls themselves are also smoother (and slightly reflective after surface melting) beyond this position. The streak shows the build up of reaction from the hot wire igniter at the end of the tube. The transit to the much faster (and hotter) detonation is clear to see, but the rearward retonation is less so. Clearly the compaction wave which runs ahead of the bright front is not luminous, yet its presence can be inferred from the streak record which shows the build up of burning to the break out of full detonation. Such an experiment contains a compressed bed of material yet such compacts may be divided into two classes according to their morphology. In open beds, pores are joined with flow space to form an interconnected network. The second class do not have connected pores and are closed cell pressings containing voids embedded in a reactive matrix. When confinement prevents gas permeability the compact becomes dead pressed and hot spots must now be formed from collapsing gas spaces within the dense energetic matrix.

Although each energetic material has different physical properties and of course a number of different possible chemical pathways, there are common links between these transits to detonation which have been outlined above. However, there are special cases where materials show a different chemical response or loading impulses that present a different set of boundary conditions. One such is ignition by direct laser irradiation onto a target surface. Here, at first sight, the loading is purely thermal if the wavelength is matched to a bond frequency so that it might be absorbed at least. A surface skin layer is heated for the 10 ns of the pulse and then conducts into the bulk of the explosive. Vaporised material confines the surface reaction site and this spreads into the bulk as discussed above.

Figure 8.11 shows a reacting column of PETN. A Nd-YAG laser (wavelength 1.06 μm and of pulse length 10 ns) irradiates PETN of pressed densities 1.3 g cm^{-3} at three increasing power densities, 0.62, 0.73 and 1.6 GW cm^{-2}. The reaction spreading from the surface was photographed using a streak camera with the slit across a diameter of

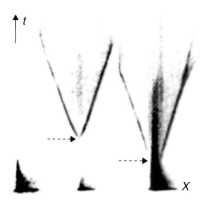

Figure 8.11 Laser initiation of PETN. SDT in tube of PETN pulsed with a laser of wavelength 1.06 μm. Streak photographs (in negative) of an explosive–laser pulse interaction surface; 1.06 μm laser at PETN of density 0.9 g cm^{-3} and at three different power densities. An arrow indicates transit to detonation (time running up the page). In the first (to the left) only surface plasma is recorded, indicated by the dark region. In the centre record, a plasma is generated, followed by a delay and finally a detonation front is recorded. The third shows laser plasma and a detonation front, but this time with a reduced delay to detonation. Source: reproduced from Bourne, N. K. (2001a) On the laser ignition and initiation of explosives, *Proc. R. Soc. Lond. A.*, **457**: 1401–1426.

the column and irradiation in the centre. The results are shown in Figure 8.11. At the lowest power density, the incident laser light and that from chemical reaction occurring during the time for which the pulse was present can be seen. Clearly, there is no sustained reaction that can continue once the laser source has been removed at this level. When the power is increased to 0.73 GW cm^{-2}, the light flash from the laser may still be seen; however, there is a self-sustaining reaction which breaks out in light emission after c. 400 ns. Clearly, reaction is building in the stage before detonation (and light emission) but the physical mechanisms operating after the ignition are not clear. However, once there is full detonation occurring, the streak record of the waves shows a constant velocity, indicating that it is travelling into a region at the expected detonation velocity, not accelerating or decelerating as it travels from a higher to a lower density region.

At the highest power of 1.6 GW cm^{-2}, the delay to detonation is reduced, but increasing power density beyond this value does not decrease it below this fixed induction time. The mechanisms operating, like the DDT experiment shown earlier, suggest SDT in the final stages, and further there is always a delay before the detonation breaks out, suggesting some burn process necessary to build to full detonation. Yet the vaporisation of surface layers also drives a shock pulse of nanoseconds for a short distance, although the pulse is not driven after 10 ns. Thus after the delays required to ensure localisation before the flow phase begins, this series of operating mechanisms interact to give steady detonation; to understand the sequence in which they operate, coupling loading impulse to operating kinetics is key to explaining the response. This message comes down through all of the scenarios this book has considered and remains the goal of understanding and then optimising dynamic extreme-loading technologies.

8.5   Approaches in describing growth of reaction

In order to model energetic materials there are several approaches that can be used to describe the reaction leading to detonation within continuum hydrocodes. The simplest option is to describe explosives with a model assuming CJ detonation. There the only requirement is the product equation of state and the velocity of the detonation (VoD). The latter may be calculated using a physics- and chemistry-based thermochemical code (such as CHEETAH for solid explosives) that can predict the performance of ideal and non-ideal high explosives and explosive formulations. Armed with this information and knowing the position at which the detonation initiates, energy can be released into mesh elements based on the distance from the ignition point at each time step. This method is known as programmed burn. In the case of geometries in which line of sight distance is inappropriate to calculate burn times it is possible to use a Huygens construction instead and this method is known as Huygens burn.

Deficiencies of a simple 1D theory have led to more complex models which include burning within the detonation zone, take into account grain size and allow some treatment of free surfaces and these can model effects such as critical diameter behaviours with some success. These models formulate an expression for reaction rate in the material as a function of some variable in the flow and are fit to experimental results to fix scaling parameters. There are two classes of model that have attempted this. The first expresses the reaction rate as a function of a local hydrodynamic variable behind the shock (often pressure) and with the equation of state and equations of motion calculates a new state at the next time step (Lee and Tarver, 1980). In the second the reaction rate is assumed to be a function of the local shock strength and the time that a particular position has experienced load and solves again with the thermodynamic and flow descriptors to update the state. In this case the reaction rate is expressed as a function of a shock parameter which does not grow behind the shock and the entropy of the solid component of the inert explosive has been used successfully (James and Lambourn, 2001).

By these means it is increasingly possible to model non-ideal behaviours in explosives with continuum hydrocodes although there is still need to fit many of the observed behaviours to a suite of experiments since mesoscale response where burning occurs cannot be represented on modern computers in the same manner as problems arise in the inert case with localisation of deformation at lower length scale features.

8.6   The response of energetic materials to dynamic compression

In previous sections the response of materials has been described in terms of processes occurring within the microstructure as a shock wave transitions through the material, focussing on mesoscale volume elements which contain a representative population of defects. In the case of crystalline materials defects, those that triggered deformation were dominated by those at the mesoscale. Their response to a compression pulse triggers

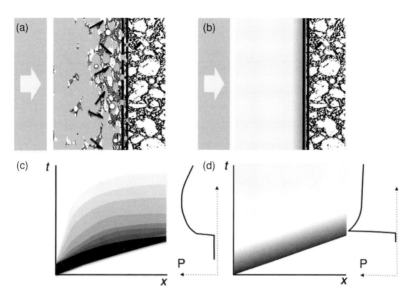

Figure 8.12 Mesoscale response in a reacting PBX. The reaction starts in (a) at a series of discrete sites which are dominated by voids that collapse to ignite material. The grey levels represent degree of reaction. A representation of degree of reaction takes black to be zero and light grey to be one (complete). In (b) the inert shock does plastic work in the front and starts reaction which has completed by a plane at the rear. In (c) and (d) x–t diagrams for the schematics are shown with schematic pulse histories. In (c) the black unreacted region has reaction triggered within which spreads as plastic work generates more heat as deformation proceeds. The peak pressure is behind the inert front and moves towards it over time as the wave encounters more explosive and further burning occurs. In (d) all plastic deformation is within the shock front and reaction completes at the CJ plane in full detonation. The pressure rises to the Von Neumann spike and decays to the pressure in the products as the reaction completes.

energy deposition at these sites since when loaded under compression the defects respond in three phases: a localisation regime where the instabilities establish slip planes; a flow phase where it equilibrates; and an interaction phase where flow planes lock. With these concepts it was possible to explain the observed features of deformation in inert materials. In the case of energetic solids there is the possibility of triggering chemical reaction that starts during the localisation phase of the response. However, the kinetics of the rate-limiting step filter the defects of relevance to ignition to those that are greater than micron dimension (at the mesoscale) as observed in the discussion of hot spots above. In the case of low-amplitude impulses clearly dislocations are nucleated and slip occurs within HE crystals as in metals. However, the energy deposition rate at the dislocation (the microscale) cannot initiate reaction. In suitable geometries and with sensitive and brittle primary explosives, fracture may localise flow and ignite the most sensitive of energetic materials. This is observed in primaries such as lead azide.

Figure 8.12 is a schematic of a PBX microstructure loaded by a 1D shock. Figure 8.12(a) and (c) show low-amplitude shocks in the explosive whilst Figure 8.12(b) and (d) show the same material in full detonation. At lower amplitude in secondary explosives it is mesoscale defects that control response. The first stage of compression (localisation) reduces the volume by pore collapse starting slip or inducing shear bands on the component scale to allow the material to flow and accommodate the strain applied by the loading pulse. This plastic deformation may increase local temperature rise above ignition thresholds in the grains and local burning sites result, which consumes material, increases pressure with the generated products and thus loads their neighbouring grains to burn at a higher rate again as time progresses. As the pulse amplitude rises, the defect density follows, and temperature increases as reaction propagates. At a critical pressure, the Weak Shock Limit (WSL) is reached in the inert explosive and the material flows within the shock front. Beyond this stress there is homogeneous heating and flow so that at the WSL the burning rate moves to its faster value and full detonation ensues in the reaction zone. When burning is confined within the shock at detonation pressures there is full defect activation and as seen earlier in diamond anvil cells, a lower impetus is required to nucleate burn than in the lower pressure (weak shock) regime where strength is still a factor. Thus the *WSL* in condensed phase *HE* is its detonation pressure.

The correlation by Grady discussed previously has also been applied and shows shock width decreasing with pressure in explosives as other solids. As pressure increases so also does the chemical reaction rate. However, clearly in an unconfined charge a steady state is reached in which work done in the inert rise of the front leads to completion of chemical reaction at the rear to stabilise a detonation front and this is balanced with release processes at interfaces which transmit transverse waves into the bulk (Figure 8.12(b) and (d)). This steady detonation state can be overdriven if pressure does not release laterally so that VoD is a quantity that can vary based on confinement conditions in the charge as observed.

Whereas response to an impulse was determined in the weak shock regime by defect spacing there is a second length scale at work in energetic materials. This is defined as the rate-limiting time step for reaction that corresponds to a volume of material that must remain under load before chemistry can complete. Thus there are always thresholds in thickness (volume) that must be swept if burning is to be established; overdriving a shock on a surface with a laser pulse for instance has some induction time before homogeneous detonation is observed.

The discussion above has indicated that in the detonation regime, the microstructure can be regarded as homogeneous down to the microscale. However, clearly there is a boundary in the weak shock regime below which the flaw distribution within the material is key and loading paths occur that include localisation before ignition can occur. In this region the flaws activated are further apart and the lower pressures result in slower burn rates. Reaction fronts thus spread and coalesce behind the deformation front and reaction completes behind the front (Figure 8.12(a) and (c)). The onset of reaction in (a) is shown starting at a series of discrete sites which are dominated by voids that collapse to ignite material. The light grey regions in the figure represent areas reacting and as one moves back into the flow these coalesce (Figure 8.12(a)). The schematic *x–t* diagram beneath (Figure 8.12(c)) shows a compressed, black, unreacted region in which the travelling shock triggers reaction which spreads as plastic work generates further heat and triggers more reaction as the deformation proceeds. The peak pressure builds behind the inert front as reaction progresses and moves towards it over time as the wave encounters more explosive and consequently further burning occurs.

In an explosive the flaws of importance are those that couple with the rate-limiting reaction kinetics of the HE of interest. The F number introduced in Chapter 1 can be used to filter mechanisms of importance from those that will not complete. Here the pulse length must be compared with the time required for a burning site to establish. In the discussion earlier the critical mechanisms for ignition were shown to be cavity or void collapse at the mesoscale and shear within structural elements at the meso- or structural scales. The localisation of deformation transits to a population of burning sites and the plastic work done in the front in this weak shock regime is in the thermal TdS component of the energy. Different loading paths couple the rate at which work is done against the heat liberated as reaction progresses. The simplest to consider are those where there are multiple, subcritical, stepped one-dimensional shocks where heat conduction is not important. HE crystals are complex unit cells, with high Peierls stress, so that there is little strengthening between shocks and the response of a stepped shock in this regime is thus almost equivalent to the sum of the two thermal components of the total energy (delayed in time by the step), which scale with the response of a single shock to the higher stress. This is a result that James has noted and has shown that in the weak shock regime entropy is the key parameter to track reaction rate (James and Lambourn, 2001).

In previous classes of material a microstructural unit (MSU) has been constructed that includes the defects within deforming solids. In the case of explosives it is more appropriate to include a volume containing a high density of reaction sites. Workers in the field have used concepts like this to treat continuum HE response for many years so details will not be developed further here. However, it should be noted again that the density of reaction sites (critical hot spots) reaches a maximum when the shock becomes overdriven, beyond which the response is essentially homogeneous.

8.7 Explosives engineering

A detonating high explosive can be used to reproducibly drive metals or fail brittle solids into fragments, and to give a flavour of design with such capability it is instructive to view interactions of a detonating high explosive and an inert contact plate (Figure 8.13(a)). The detonation wave is a travelling front for which the rear of the shock behind the thin reaction zone is at the CJ pressure for the explosive. When it acts directly on the plate it takes the material to the CJ state itself and releases into the products behind the CJ plane. This state is assigned a state 1 in the P–u_p diagram of Figure 8.13(c) and shown on the wave diagram of Figure 8.13(b). The shock then rebounds from the rear of the plate and releases it back to the ambient state (point 2) accelerating it from rest in the process. At this point brittle solids may fragment under the impulse applied as seen in previous chapters, particularly if the shock is divergent and there is an in-plane strain. Assuming this is not the case, however, the release returns to interact once more with the loading surface and acquires state 3 on the release isentrope of

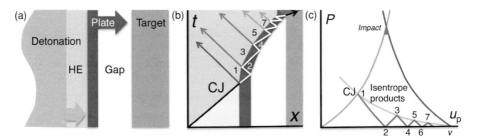

Figure 8.13 Explosive acceleration of inert materials. (a) Detonation wave interacts with an inert plate. The pulse reflects and accelerates the plate. (b) Stress rings down in the plate and accelerates the plate to impact onto a target. (c) States achieved plotted in P–u_p space. The plate accelerates to an impact speed and then hits the target and induces a pressure pulse.

the explosive. The returning compression accelerates the free surface once more taking the velocity to that of state 4, and the process continues until after multiple reflections during the flight the plate reaches a velocity v. On impact with the target, a high stress impact results which may be accessed by the intersection of the flyer and target Hugoniots as in Figure 8.13(c). It is clear from the figure that by this means the pressure achieved by impact is many times that from direct contract of the HE with the plate. Of course the pulse from a high explosive is a Taylor wave, and the complex release structure formed, and the dynamic interactions with the CJ state in the products, make the full analysis of the acceleration profile of the surface motion geometry and material dependent.

However, in the Second World War Gurney (1943) addressed the problem of detonating explosive shells with a simple analysis that predicted the terminal velocity of fragments thrown by an accidental detonation in an inert casing. He equated the energy liberated by the expanding explosive products with the kinetic energy of the propelled shell fragments and arrived at a prediction of the terminal velocity achieved which can be applied to various geometries with suitable integration of the defining equations. The continuum analysis assumes symmetrical (radial) flow and equilibration of the shock in the driven inert confinement. If the mass of case projected is M, the mass of charge is C and the chemical energy of the explosive converted to drive them is E, then the terminal velocity of the fragments v is given by

$$v = \sqrt{2E} \left(\frac{M}{C} + \alpha \right)^{-1/2}, \tag{8.10}$$

where $\alpha = \dfrac{1}{3}$ for a sandwich, $\alpha = \dfrac{1}{2}$ for cylinders and $\alpha = \dfrac{3}{5}$ for spheres. This analysis has proved remarkably accurate given its simple assumptions and can be applied with relevant integrations to other geometries. It has been found to fit best to behaviour when $M/C > 0.1$ in tests done to assess its success. An empirical aspect of this analysis is the fraction of the chemical energy that is converted into kinetic energy of fragments. This term hides many of the losses in the system for interactions of the material and explosive. However, for a range of military energetics it has been calibrated and lies between a half and three quarters of the enthalpy of detonation. Relevant data can be found in the literature. Further, some materials fragment early in the process which means they will not benefit from the full drive since gas will flow between the surfaces after separation. In this case the fragments are found to gain about 80% of the expected throw velocity.

The loading of metal plates to push them at speed has many engineering applications. A detonation wave meeting a shaped, normally metal, liner accelerates it to focus the explosive energy into the propulsion of an accelerated inert. One well-known application concerns the driving of conical or V-shaped metal sections to form an axial or planar jet of material which cuts a deep thin cavity or slices through structural elements. Such jets can typically travel from c. 7 to up to 15 km s^{-1}. Linear-shaped charges are used to cut large structures (to cut the structural supports of oil rigs in Britain's North Sea, for instance) or to separate multi-stage rockets. The flow of metal jets is interesting since the material is at high pressures and subject to very large strains; conditions for which the standard derivation of constitutive models for a material is invalid. Thus there is still work necessary to fully describe the flow and break-up of these jets as they extend into media and fail their target materials. However, there are further interesting effects that result from the interaction of the diverging wave with the metal microstructure.

The discussion of reaction zones in detonation above has highlighted the limits of ideal detonation theory which was in any case derived initially from observations made in the gases. In most cases detonation is in some manner not ideal. Equally there are occasions where one aims for non-ideal detonation to ensure a particular response. One of these situations is the wish to extend the loading pulse produced by an explosive. In this case it is necessary to ensure that there is some reaction behind the sonic plane. A high CJ pressure and large shock is favoured if one wishes to push metal (as seen above) but if one wishes later time energy release then changes in formulation are necessary. In this case one aims to maximise the potential for useful work output (*heave*). One means of doing this is in the addition of metal powder to a formulation, which reacts at later time. Addition of aluminium to an explosive allows late-time reaction with water, for example, and this enhances the reaction possible in underwater explosions where there is a wish to create large gas bubbles.

When bubbles are created underwater, the large gas bubble experiences a series of forces which act after the expansion phase is complete to collapse the cavity. Obviously the displaced volume creates upthrust since the cavity is buoyant. If the bubble is in free water it can migrate upward in this phase but when inertia takes over the walls will accelerate inward and collapse the volume. If there are no constrictions to flow, the bubble will collapse symmetrically. However, if the walls are close to boundaries, whether they be the seabed, the surface or some structure in the water, flow is favoured from the surface away from the flow constriction and a collapse of the cavity will occur by a jet towards the wall. Thus a ship or submarine loaded by such a pulse feels a concentrated impulse which can cause the structure to whip under the load, breaking it under the large bubble oscillations that occur close to it.

Jet impact at any scale damages impacted targets through the high pressure on impact and high temperatures in the collapsing bubble. Figure 8.14 shows the impact of a vertical water jet onto a transparent PMMA surface. Since the impact is symmetrical, a

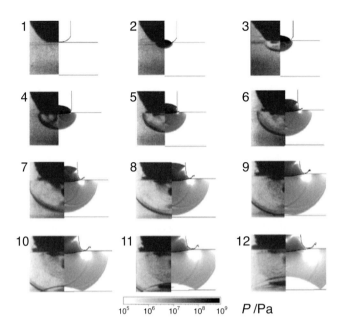

Figure 8.14 Impact of an underwater jet from a collapsing bubble. Impact viewed side on with rear lighting, interframe time 200 ns; exposure time per frame 40 ns. Right-hand side of each frame shows hydrocode simulation with contours of pressure. Source: reproduced from Bourne, N. K. (2005b) On impacting liquid jets and drops onto polymethylmethacrylate targets, *Proc. R. Soc. Lond. A.*, **461**: 1129–1145.

high-speed sequence of pictures and a series of simulations are shown side by side in the figure. The jet impacts and shocks a target and for a brief period, until release waves have come in from the jet edge, loads the target at a high shock pressure. The divergent shock travels out in this period compressing the structure underneath. A spreading cylindrical release front interacts on the jet axis and loads the structure in tension whilst the jet itself, now released, flows inward still exerting a reduced Bernoulli pressure on the surface. In Figure 1.8 a metal rod was shown collapsing onto a hard target and flowing over its surface. The analogies between these two scenarios are clear and these connections will be picked up in the final chapter.

Bubbles and porosity act to concentrate flow and nucleate damage at all scales in materials. A collapsing cloud of bubbles may pit and erode a blade surface if a cavitation cloud lies over a surface. Equally a large explosive bubble may break the back of a ship if suitably placed nearby. Finally, a cavity within an energetic may react under the high temperatures and pressures as it collapses under a shock load from a detonator. Thus a cavity or porosity within an energetic material can focus energy from a travelling pulse to create transient impulses which can fail or ignite it in the flow in the weak shock regime. In explosives their presence must be controlled by design to sensitise if required or be eliminated if a hot spot ignition might cause safety problems in the field.

Applying explosives in engineering requires coupling of the expanding products at high pressure to other media in contact or close by. Fixing geometry and controlling interfaces is key to efficient operation of any device, but equally reproducible firing trains must be constructed alongside these to ensure safe and secure operation. Energetic materials offer a series of solutions for many processes and better understanding of their operation and use can only see these expand in the future.

8.8   Final comments

The present chapter has reviewed the response of condensed-phase energetic materials to insults ranging from low-amplitude mechanical insults to shocks driving detonation. History has developed explosives technologies to safely harness the power that they deliver and understanding their behaviour under mechanical load has resulted in safer and more effective systems for engineering today. To control properties, most modern formulations are composites which bind energetic crystals within a matrix that is generally inert but may be engineered to react. Thus primary defects lie at the length scale of the composite and like engineering ceramics consist primarily of porosity of micron dimensions. Collapse of pores or the shear developed around them as the material strains convert mechanical impulses into temperature rises and thus these defects localise deformation and ignite the composite, presenting a safety issue for the materials and in their use. History has shown that major accidents have generally resulted from a local temperature rise that has spread and grown to detonation while the material was confined. There are a number of routes of varying severity that accidental ignition may take. In a less severe instance a burn transits to a violent reaction (BVR). Incidents such as low-speed impact that might result from dropping components, friction with interfaces, or fragments impacting and penetrating the charge come under this class of response.

More intense stimuli can lead to incidents that end in full detonation of the explosives. The deflagration to detonation transition is a hazard that occurs particularly for pressed charges where pore space allows combustion waves to transit burn through to full detonation. Munitions caught in fuel fires or subject to extended hot cycling can lead to such events. If the impulse supplied is sufficiently violent then the explosive can transit an inert shock to full detonation through a shock to detonation transition. In all of these cases it is defects that cause hot spots which react to generate products at pressure so increasing the burn rate in the material. It is the confinement of that pressure that determines whether the material transits to detonation or burns in a violent manner.

These physical pathways show that the onset and growth of reaction is fully three-dimensional at the mesoscale in the low-pressure regime. When mechanical insults light hot spots at defects the same constraints on simple one-dimensional descriptions of performance apply as with the simple application of purely the conservation relations in the weak shock regime in inert materials. Energetics are for the most part crystal composites that respond to mesoscale stimuli on microsecond timescales.

As compression increases, however, a boundary is reached at which defect concentration becomes saturated and the burning may be regarded as homogeneous at the microscale.

Again the engineering of practical systems involving production materials relies on controlling the defects introduced both to ensure that the product achieves the performance designed for, but also that accidental loading does not perturb its operation. It is the understanding of the scale and behaviour of these defects in practical situations that allows design for the future using this class of materials, noting the behaviours at different length and timescales that they will exhibit.

8.9   Selected reading

Bailey, A. and Murray, S. G. (1989) *Explosives, Propellants and Pyrotechnics*. London: Brassey's.

Cooper, P. W. (1997) *Explosives Engineering*. New York: Wiley.

Davis, W. C. (1987) The detonation of explosives, *Sci. Am.*, 256(5): 106.

Fickett, W. and Davis, W. C. (2000) *Detonation: Theory and Experiment*. Mineola, NY: Courier Dover Publications. Reprint, originally published 1979.

Field, J. E., Bourne, N. K., Palmer, S. J. P. and Walley, S. M. (1992) Hot-spot ignition mechanisms for explosives and propellants, *Phil. Trans. R. Soc. Lond. A.*, 339: 269–283.

James, H. R. and Lambourn, B. D. (2001) A continuum-based reaction growth model for the shock initiation of explosives, *Propell. Explos. Pyrot.*, 26: 246–256.

Lee, E. L. and Tarver, C. M. (1980) A phenomenological model of shock initiation in heterogeneous explosives, *Phys. Fluids*, 23: 2362.

9 Asteroid impact

9.1 Introduction

Matter in the universe exists in a series of states; three that are well known from standard experience, solid, liquid and gas, and the plasma state in which gas is ionised. Other more esoteric states are possible but the four above occupy the main thrust of this book. This volume has confined itself to pressures in the range up to a megabar and temperatures below 10 000 K in which solids exhibit strength that is based upon the interaction of valence electrons. Beyond a critical energy density bonding is determined by further interactions of inner orbital electrons and concepts from the ambient cannot be extended.

Equally the forces acting on matter applicable to this work are electrostatic or gravitational. Electrostatic forces may act over great distance but as length scale increases there is sufficient matter that substances behave as neutral. The long-range attractions at the microscale are due to Van de Waals' forces that might operate over distances of the order of 10 nm between polymer chains and are important in binding matter at the mesoscale. Components on a scale of centimetres are naturally held under gravity in stacks or compressed under lateral forces by some restraint. The strength of such an interface in tension is determined by that of the pin or joint that constrains the interface between the two components. At this scale, flow occurs by hinging around pivots under load or by slip along the fracture line with frictional heating at the interface. At the planetary scale forces are gravitational and slip occurs down faults that allow flow under shear.

9.2 Structural scales in condensed matter

The geometrical and compositional constituents of condensed-phase materials are described by a series of terms. These are summarised in Table 9.1. The length scales of relevance to the concerns of this book have been discussed previously but are summarised for completeness here. Each scale consists of an assembly of elements. Each of these elements is a coherent unit which when combined together constitutes the assemblage considered. Elements may have different properties from each other and may be combined in different ways according to how they bond with one another. Clearly there are boundaries between them and at a given scale these boundaries represent defects at that length scale from which failure of the interfaces may propagate. Thus a length scale

Table 9.1 Length scales, defects and boundaries in compression of condensed matter

Scales	Micro	Meso	Macro	
Elements	Atoms	Grains	Components	Structures
Structural form	Crystalline, amorphous	Polycrystalline and potentially textured	Formed or natural morphology	Assemblage of components
Defects	Point/line defects	Voids, inclusions, cracks, shear bands	Element boundary	Module boundary
Bonding	Electrostatic (bond)	Electrostatic: weak due to increased displacement Amorphous metal, VdeW polymer Composite joint: brittle	Joint: weld, diffusion boundary, compact	Joint: weld, diffusion boundary, compact
Localisation	Dislocation	Shear band Fracture	Hinge, shear	Hinge, shear
Structural transformation	PT/twinning	Pore collapse Densification	Packing of components	Collapse of structure
Flow mechanism	Dislocation slip	Frictional surface slip Grain plasticity	Rubbing of fracture planes Slip or rotation at shear bands (hinging)	Hinging and shear of components around line or planar joints

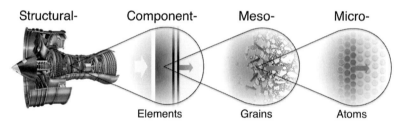

Figure 9.1 Structural scales in condensed matter. A structure at the macroscale is composed of components jointed (generally by man) in some manner. A volume of one of the components contains grains that respond in a three-dimensional way to any load. This scale is called the mesoscale. Each grain is composed of atoms stacked in a particular manner that defines the microscale. The number of degrees of freedom increases as one crosses to the scale beneath.

consists of an assemblage of elements and defects with boundaries that join the structure and fix its strength.

The macroscale contains components which may be composed of different materials and which together form the elements found in a structure (Figure 9.1). However, a component has a microstructure which consists of a series of grains that are further made of assemblies of atoms. Grains and their organisation and deformation are considered at the mesoscale where their conformation and bonding defines the microstructure. Grains themselves contain constituents at the microscale which itself is a packing of unit cells composed of atoms which are considered at the atomic scale. Thus a typical element at the macroscale contains components that are formed of mesoscale grains, and these have particular packing determined at the microscale. The microscale itself consists of unit cells of repeating atoms that are defined at the atomic scale.

Between each scale is some transition boundary (a welded joint, a grain boundary or a dislocation plane, for instance) and this offers an impedance mismatch between regions that within have common properties. These interfaces define zones of bonding between elements at that scale and defects are primarily concentrated at these locations within the component. The strength of components in an assemblage is greater than the strength of the joint and it is defects at these scales which define the limits of performance in structures under load.

The assemblage, whether it be at the atomic or the structural scale, is formed by constituents and some volume of defects that determine the assemblage's geometrical packing and bonding to one another. This controls the average strength of the assemblage integrated over a volume element that contains a representative number of constituents and defects such that each element has the same properties as the next. Deformation time considered at a particular scale has two parts: a first in which such a microstructural unit is swept by a loading pulse and equilibrates the strain state within it; and a second steady deformation where the strain rate in the unit is constant and the stress state has equilibrated. As the stress amplitude in the load increases, the number of activated defects increases from the ambient state and the MSU thus successively decreases in size.

9.3   Regimes of compression

This book has considered a range of compression that has spanned the onset of yield in materials under load up to the point at which the bonding ceases to involve merely valence electrons and the assumptions of solid mechanics become ill-defined. This threshold has been termed the *finis extremis* (FE) throughout the text. The easiest means of studying behaviours at the highest pressures is to apply the load dynamically, but the states may clearly exist statically as well. The compression is applied by a stress and induces a strain in a material and, whilst strength may be defined in similar fashion, continuum mechanics can be applied to the region defined. When strength can be defined, any loading can be decomposed into a hydrodynamic and a shear component and the behaviour described as stress increases. This defines the region of interest for this book. The regimes and thresholds are summarised in Table 9.2.

Within the elastic range the strength remains constant and dynamically an elastic pulse may travel through the medium whilst the material returns to its original dimensions after the load is released. In shock loading the elastic limit is called the Hugoniot elastic limit (HEL) where plasticity processes are initiated at the largest flaws in the bulk of the material. Above the yield surface, however, plastic processes occur and the compression may be decomposed into two regimes. An elastic wave still propagates ahead but higher stresses drive a plastic front through the material as defects are activated and deformation proceeds within the solid. The elastic deformation is unsteady in the initial stages as

Table 9.2 Loading states in condensed matter. The ranges are separated by thresholds whose names and approximate values are displayed

State	Threshold stress/ pressure (order of magnitude)	Name of stress/ pressure range
Plasma		Super-extreme
Ionisation threshold	10 TPa +	
WDM (HEDP)		
Finis extremis	100 GPa	
Strong shock		Extreme
Weak shock limit	50 GPa	
Weak shock regime		
Yield surface	0.1 GPa	
Elastic range		Elastic

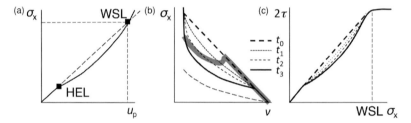

Figure 9.2 Behaviours in the weak shock regime. (a) The WSL defined as the point at which the shock becomes overdriven. (b) Hugoniot in *PV* space showing transits to the steady state at different times. The heavy line is a typical WS loading path. (c) Strength as a function of stress from the elastic to steady state at the WSL.

deformation processes take time to complete and the precursor decays until the flow steadies over some run distance. In this regime the plasticity processes collapsing large flaws take time that delays the compression front behind the elastic wave. Flow in the material down slip lines activated at defects may interact after some time and may thus harden the compressed assemblage, making it stronger than the ambient structure was before load. As the driving stress is increased, however, more and more defects are activated and compressed within the front, which approaches the elastic wave until a threshold is reached at which all defects have been accessed within the material and compression occurs within a single front. At this compression there is no remaining defect volume and the strength of the assemblage saturates. This level is defined to be the Weak Shock Limit (WSL). Below this limit, behaviour is conventionally dubbed the weak shock regime (Figure 9.2(a)).

In this regime the material moves over time towards a state in which it has achieved its fully compressed density within the shock front. This process is quickened as the amplitude of the pulse increases. Of course the impulse supplied must be sufficient in duration that the defects may be closed and plastic processes are able to work on the

Table 9.3 WSL and theoretical strength for metals and a glass

Material	Density (g cm^{-3})	Particle velocity (mm µs^{-1})	P_{WSL} (GPa)	σ_{Th} (GPa)	$\frac{1}{3}P_{WSL}$ (GPa)
W	19.20	0.96	96.1	27.0	32.0
Ta	16.60	0.62	42.4	11.0	14.1
Ni	8.87	0.83	42.4	13.8	14.1
Cu	8.93	0.55	23.4	7.7	7.7
Al6061	2.70	0.78	13.5	4.3	4.5
Al1100	2.71	0.75	12.9	4.3	4.3
SL glass	2.49	3.90	14.7	4.7	4.9

solid. Compression of polymers for example must compress the structure before atomic interactions can begin. The compression thus always commences with some localisation of the flow at the defect sites under load. Flow takes place over some time driven by the kinetics of the deformation mechanisms operating and ends at the finish of the pulse or at the completion of the process (Figure 9.2(b)). Kinetics is key: as seen earlier the Peierls stress in metals slows slip in BCC metals compared with their FCC analogues.

In the weak shock region the mechanical response can have an elastic component that propagates through undisturbed material ahead of the slower plastic front behind. The pressure pulse activates more and more defects at the meso- and microscale until a stress limit is reached at which all have been closed. This is the WSL in the material under load (Figure 9.2(c)). At this point porosity has been eliminated from the microstructure and the hydrodynamic pressure overcomes the strength of the material. A single shock pulse propagates, within which all inelastic processes occur. The shock overdrive point itself can be recovered assuming continued elastic response and the shock equation of state for the material, and examples are shown in Table 9.3 for a range of metals and a glass. The theoretical strength for a material has been calculated from the early 1900s and assumes an ideally packed microstructure that is rarely found in reality (except in the thinnest whiskers of material) since defects dominate response. It was shown earlier that the fraction of the shear modulus $\mu/2\pi$ is close to more rigorous calculations of the quantity. If indeed the WSL corresponds to the point at which the pressure P exceeds the theoretical strength σ_{Th} then one would expect

$$\sigma_{Th} = 1/3 P_{WSL} \tag{9.1}$$

at this point and the table confirms that this is the case. The agreement is good given the approximations in the calculation. The WSL for glass is taken as the point at which fracture occurs within the shock front. This was observed in experiments showing delayed failure and can also be recovered from calculating pressure on the Hugoniot (Figure 6.11). Polymers are not included in the table since their WSL occurs when the Van de Waals' bonding is overcome at the 20–30 GPa limit where fully hydrodynamic behaviour is observed.

The elastic strength of the material defines the point at which defects are first activated and inelastic processes can take place. At small strains (or in dynamic experiments small times) that defect population is confined to be at lower length scales and the value is higher and approachs the theoretical strength with the shortest pulses. In the glass above, for instance, the theoretical strength is close to the HEL of the material observed in shock experiments. At the upper limit of the weak shock regime is the WSL where the pressure overcomes material strength. Between these thresholds lies a region where material behaviour is dominated by the inelastic flow mechanisms that accommodate the strain in a pulse. These include slip and fracture to close the defects present but also martensitic phase transformation and twinning which operate within this regime. All of these processes expend plastic work in the flow and increase the temperature behind the front, but it is only at the WSL (and above) that all processes are complete within it. Thus explosives show run to detonation and brittle solids exhibit failure waves behind the initial front as the inelastic processes operate. Most of the plastic work done in this region has been expended as thermal energy and this dominates reaction processes where ignition occurs.

Above the WSL is the strong shock regime where the strength is at its maximum, and when strength is probed with some reload it is found to decrease with pressure only due to the thermal heating that occurs. In this region the simple shock sketched in the early chapters operates with material behaviour homogeneous and transitions thermodynamic. The shock front contains all operating processes with the flow behind steady and the state well-defined. The front itself obeys the Swegle–Grady laws with the energy and the time of the processes operating determining its width. Since times are short there is no mixing, so that whilst phases behave homogeneously, diffusion does not occur and release results in a microstructure with inclusions in the same positions within the material.

With increasing shock strength comes further heating in the front and greater compression of the homogeneous material. As the atoms are forced closer together, the point is reached at which inner shell electrons become involved in the bonding and the localisation of electron density changes nature from that observed at lower amplitudes. Thus there are two thresholds within which behaviours occur: the WSL and the *finis extremis*. Whereas they are transitions occurring at different thresholds in different materials there are indications that may be seen in bulk behaviour. Figure 9.3 shows all available Hugoniot data for the two metals FCC Cu and BCC W. The deviation between the data and the hydrodynamic curve is plotted for a range up to 400 GPa. There is large experimental scatter in the data suite taken from a range of sources going back to the 1960s with different pedigrees of the metals. However, there are trends that can be seen in the data. There is an indicator path annotated on the W curve where the behaviour is spread over the pressure range. In the weak shock regime one would expect the deviation between the curves to alter simply because the strength of the material increases over this region (Figure 9.2(c)). In the strong shock regime, the two curves would be expected to be displaced from one another by a constant offset due to the strength, which decreases with shock strength as the temperature increases, and this appears largely to be true. However, beyond the second threshold the curves diverge markedly and the Hugoniot (despite

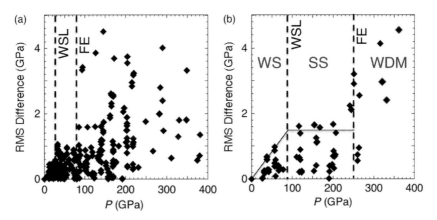

Figure 9.3 Offset of the Hugoniot from the hydrodynamic curve for FCC and BCC metals, showing the Weak Shock Limit (WSL) and *finis extremis* (FE) for (a) copper and (b) tungsten. WS, SS and WDM indicate the weak shock, strong shock and warm dense matter regimes, respectively. The solid line on the tungsten data indicates the limit of the envelope of scatter up to the FE.

the scatter in the data) is clearly following a very different path since other forces are operating to determine the equation of state. The FCC and the BCC metals selected show different threshold levels but similar behaviour and the transit to the WDM regime is *c.* 100 GPa for copper (around the WSL for tungsten), whereas it is 300 GPa for tungsten (Figure 9.3).

As stress increases further, compression forces nuclei closer resulting in overlap of inner electron shells and the bonding becomes localised in a very different manner from that at lower stresses. The material at this stage is compressed and heated by virtue of the routes taken to achieve this state and is thus dubbed warm dense matter (WDM). Further compression finally crosses the ionisation threshold and a dense plasma is created. These last two super-extreme states are beyond the limits that this book will consider and these states cannot be described by solid mechanics but rather cross the threshold into high-energy-density physics.

9.4 Phases of loading

The processes discussed through this book all result from an impulse that sweeps a volume and interacts with microstructure at that scale with results at those beneath. Most processes observed at the engineering scale commence at the macroscale, excite local plasticity at the mesoscale, and lead to slip at the microscale. As the amplitude of the incident impulse increases, the response occurs at lower length scales increasingly quickly until, at the point where the stress level reaches the *finis extremis*, the response is homogeneous at the microscale since the bond strength is overcome in the flow. As seen above, the behaviour in the weak shock regime is more interesting and consists of the initiation and subsequent operation of inelastic flow processes in sequential time periods. This behaviour is described below.

There are three phases in the deformation of matter that operate over particular timescales in each of the length-scale domains described in the previous section:

- *localisation* – flow surfaces are created by nucleation and deformation from available defects;
- *flow* – planes of material slip past one another to allow the material to accommodate the applied strain;
- *interaction* – flow is halted or arrested by the meeting of slip surfaces or by their termination at boundaries.

Each of these follow one another chronologically as the dynamic process proceeds.

At each scale there is a communication speed that defines the response of the element. At the microscale it is that of acoustic phonons which communicate mechanical insult through the bonds within the grain. Statistical mechanics can recover bulk properties from such a description. At the mesoscale it is waves which travel between regions of different impedance to equilibrate the stress state within the component or get driven from moving surfaces. At the structural scale the stress and strain field is equilibrated through the structure and the applied strain acts to collapse it under load.

The *modi operandi* for the three phases of response are indicated in Tables 9.4–Tables 9.6. Inelastic flow begins by creation of the means to deform the solid by localisation at the length scale beneath. In the weak shock regime, flow is initiated at flaws of one type or another in the material under load. At the mesoscale, grains are crystals or amorphous regions and contain atomic (or larger) defects, or alternatively regions that contain second-phase atoms. Locked-in strains, due to inclusions or previous work done on the crystals during processing, once activated under load may initiate propagation of dislocations in crystalline grains or begin strain localisation in amorphous regions. Finally, fracture may be initiated at strain concentrations. At the mesoscale are pores – at grain boundary triple points, for instance – or inclusions introduced by particular processing methods (texture). On collapse both these sites and those of second-phase inclusions will concentrate local plasticity, collapsing the microstructure to accommodate the applied strain and defining an inhomogeneous mesoscale 3D strain field. Before any steady flow can occur waves must equilibrate the stress states in the component so that it can enter a constant strain rate flow phase, as discussed below.

At the component scale the compression must be accommodated by bulk homogeneous flow or localisation at shear bands forming within the component. In brittle materials, local tensile strains at surfaces or pores initiate fracture in the bulk which can travel through the component. If it has already some degree of comminution then static frictional forces must be overcome prior to sliding at these surfaces. Finally, at the structural scale localisation can form hinges, or failure of joints between interfaces can proceed again to steady slip under the right circumstances. In every case the stress and strain fields are not uniform at any scale and the deformation processes at this early time are unsteady and inhomogeneous (Table 9.4).

Table 9.4 Sites of initiation and mechanistic routes in the localisation phase of dynamic compressive inelastic response

Scale	Sites of initiation	Operating process
Micro	Flaws: Point defects and larger assemblages Strained regions in substructure	Initiation of slip by shedding dislocation Initiation of fracture at twin plane
Meso	Pores within microstructure or at grain boundaries Weakened or dry joints between grains in composite micro-structures Second-phase particles Grains with locked in strains	Onset of plasticity around closing pore volume Overcome interface bonding Inhomogeneous strain fields around particles drives waves to equilibrate stress states
Component	Localisation in shear band Initiation of fracture within the component Overcome static friction in fractured material	Slip at surfaces Strain build up and heating as wave propagation equalises strain field to allow steady shear strain to operate in band (dynamic friction)
Structural	Formation of hinge Failed jointing between components Overcome static friction at jointing surface	Localisation of shear deformation in hinge location(s) Ductile or brittle failure of joint Shear at asperities before steady rubbing at interface begins: friction

The principal period over which the imposed strain from an impact is accommodated by the material is within the flow phase of target response. At the microscale slip can occur along dislocations in grains whilst in amorphous solids, shear strain can accommodate flow with shear bands. If cracks have opened and the static friction has been overcome, steady rubbing can proceed at some dynamic rate with heating at the interface so that a defined slip interface is formed. At the component scale bulk plasticity barrels the geometry or alternatively, if flow has localised into bands, motion along shear trajectories or fracture planes accommodates the imposed strain. Finally, at the largest scale these processes allow slippage along existing faults or give rise to hinging at bands in the structure, resulting in crushing of the edifice or sliding along a fault plane (Table 9.5). Flow at all scales that involves motion at interfaces results in heating at the slip line and this is termed friction at the continuum. A similar term may be applied at the mesoscale, but the forces involved and the scale of the band on which it occurs are very different so that comparisons are qualitative. Applying friction as a term to describe motion at the microscale is not appropriate since there is little relative motion when volumes are considered at this scale. This is an example of a term that can be regarded as scale-specific, describing a means for energy dissipation during flow on a meso- or component-scale plane. It is thus best applied at the macroscale, where there is an established pedigree.

The flow process continues until some mechanism intervenes at a lower length scale generally to terminate the phase. At the microscale this may involve the interaction of dislocations with one another or with a boundary with some different impedance. Dislocations form structures as they interact with one another that on relaxation may

Table 9.5 Sites of initiation and mechanistic routes in the flow phase of dynamic compressive inelastic response

Scale	Means of operation	Operating process
Micro	Dislocations within crystals	Slip
	Shear bands in amorphous solids	
Meso	Cracks	Shear
	Shear bands	
Component	Bulk plasticity	Barrelling
	Accommodation of strain through motion along shear trajectories or fracture planes	
Structural	Hinging	Crushing
	Faulting	Sliding

Table 9.6 Sites of initiation and mechanistic routes in the Interaction phase of dynamic compressive inelastic response

Scale	Sites of termination of flow	Operating process
Micro	Dislocation interaction with other dislocations, grain boundaries or free surfaces	Pile up/release
	Formation of cell structures	
Meso	Crack interconnection	Comminution
	Failure from boundaries of different impedance	Hardening/softening
	Hardening/softening of elements within of microstructure	
Component	Locking of surfaces against boundaries	Densification
	No further shear strain possible on available slip planes	
Structural	Hinging of components against one another terminated	Locking
	Crushing of structure against boundaries	

annihilate line length or organise into cellular forms within the material. Cracks may also interact with particles or pores or alternatively meet each other and free new subparticles within the component. This process is called comminution of the material under load. Grains within a component may harden or soften according to their orientation or the nature of boundaries close by (Table 9.6). At the component scale, rotation and flow of grains within the substructure will eventually lock or release at boundaries. At this time no further shear strain is possible and the component has densified. Finally, a structure crushes to a more dense body in its collapse on a more rigid interface. Individual components lose degrees of freedom to rotate or deform and lock into a denser form. There is an analogy in structural response to the weak and strong shock regimes discussed earlier, but this time the crushing strain within the structure marks the transition in behaviour. It is the time and thus the strain that the driving piston loads for that determines the final density and when it reaches full density a new loading phase is begun.

These three loading phases accommodate the strain applied within a structure loaded by a compression impulse. The flow phase is steady and the stress state is in equilibrium with the target at a uniform strain rate. The localisation and interaction phases on the other hand are transient, have heterogeneous stress and strain states and no global defined strain rate since there is no uniform stress or strain state over the scale volume. The flow is unsteady in these phases. Within each regime processes are ductile or brittle according to the bond strength and the magnitude of the impulse supplied. Further, there are a few key quantities that limit response times and these are wave or phonon speeds within the materials of the target. For any experiment designed to investigate flow in materials under load, the strain rate at the scale of interest for the test must be constant before useful data can be extracted for further use.

9.5   Weak shock behaviour in metals, brittle solids, plastics and explosives

As matter is loaded to higher amplitudes the response of the material homogenises and some material characteristics become less prominent. The richest range of behaviours is found in the weak shock regime where defect structures dominate behaviour and their populations and geometry must be investigated and understood to explain response. The first phase will be densification at the largest pore volume in the structure. If there is significant porosity at the component scale as might be the case in foam or in a composite (glass microballoons are added to commercial explosives to sensitise formulations to shock for instance) then this must close before plasticity at lower length scales can begin. A planar spalled surface travels at twice the particle velocity behind the shock so that this defines an order of magnitude for a time of collapse. If the population has not closed before the impulse releases then later stages of compression cannot operate and the foam has absorbed the impulse. This behaviour was represented in the $P-\alpha$ model considered earlier.

Once the component has reached full density the three phases of response proceed sequentially as outlined above. Here the defect populations and the available flow mechanisms for metals, brittle solids, polymers and energetic materials determine the response in different material classes. Schematic responses for each class are shown in Figure 9.4. Of course each individual material will have its own unique character either in composition, texture or defect population. However, these straw men will serve as representatives of each to illustrate dynamic response. In each a one-dimensional pulse at the continuum (component-) scale is considered driven from left to right into a mesoscale composite. Localised flow surfaces are represented as zig-zag lines and porosity and grain boundaries are indicated in the microstructures. Of course in any real solid the impact surface has microstructure at the mesoscale on both the impactor or the target and this surface state will be important at lower scales representing a geometrical localisation zone which *must* be collapsed before flow may begin. However, here it will be assumed that the surface state has the properties of the bulk and that the flow phase may begin instantaneously.

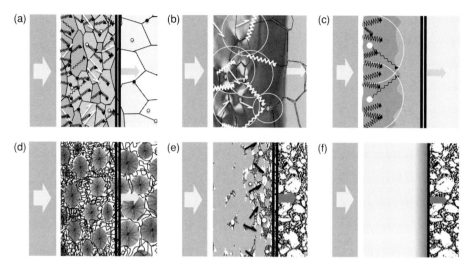

Figure 9.4 Schematic mesoscale weak shock response of (a) metals, (b) polycrystalline ceramics, (c) glasses, (d) thermoplastic, (e) PBX, (f) loading in strong shock.

Snapshots of the response of metals and crystalline solids are shown in Figure 9.4(a). The mesoscale structure consists of individual grains of particular stacked crystal phases, each with its own anisotropic nature and bonded with boundaries that have an intermediate disordered nature. There may be voids introduced within grains by processing or inclusions around which crystal growth has nucleated in the growth phase. At the microscale within the stacking there may be point defects or assemblages of lower density or merely regions where packing is disordered. Grain processing will mean that the crystals contain some population of line defects as well. All these features will result in a mesoscale texture which will determine the initial ambient properties at the component scale in the material.

The loading of the microstructure at the wave front will start the three compression phases at different locations behind the front. The motion of atoms around defects will localise flow to dislocation slip planes and compression can slip at these planes once they are established. The fastest processes in the crystal can proceed by a shuffle of atomic planes and this results in twinning or martensitic phase transformation within the crystals, introducing a new boundary between phases of different stacking and accommodating a fixed proportion of strain within the target. Once localisation has settled the grain to a steady stress state, plastic flow may occur within it equilibrating the stress and strain states across the assembly of crystals within the mesoscale structure around. These processes occur by averaging the effect of each defect with wave propagation at that scale. The MSU was introduced in Chapter 5 to aid consideration of representative elements and the schematics of Figure 9.4 are representations of these for each material class. A fixed population of defects was considered in the unit over which the stress state could equilibrate in a sweep across the structure. Clearly as stress amplitude increases the unit size reduces, until at the WSL it has shrunk to zero volume since bond strengths may be overcome. At stress levels in the weak shock regime the flow phase ends with dislocation tangling if their density is high enough or alternatively with interaction with grain boundaries. At the component scale wave reflection from free surfaces then begins release that unloads the microstructure annihilating dislocations and relaxing tangled structures into cells.

A special case of such a microstructure is a polycrystalline ceramic where the grains are elements in which slip is difficult with high Peierls barriers (Figure 9.4(b)). The ultimate case of such a composite is boron carbide where the composition of individual regions at the microscale is not fixed. Flow in polycrystalline ceramics may occur by slip but there are fewer activated systems and crystals are anisotropic. Twinning is thus favoured to accommodate some part (but only a few percent) of the strain applied and further deformation must be by slip or fracture down suitable planes. Observed recovered microstructures show that cracks can propagate down either the twin planes formed or equally down the weakly bonded grain boundaries, which in these composite materials generally have some glassy phase present to bond crystals. Defects at the mesoscale include pores and unbound crystals which favour failure down these weaker paths. With localised slip planes, flow is by the motion of deformed crystals down fractured interfaces as compression occurs by plasticity within the crystals. Clearly there are different thresholds for crystal yield and grain boundary cleavage and this is reflected in the complex waveforms seen in these materials. In the weak shock regime failure occurs over time and flows the compression front at some distance approaching it as the WSL is reached. Multiple yield stresses for the individual phases in the ceramic are typical of these mesoscale composites but also of materials formed of layers at the component scale. The MSU for such assemblages must contain enough grains and defects to give a continuum average after integration over it and the timescale for sweeping the cell determines the unsteady flow phase with which such materials respond to impact.

Pure brittle solids with no slip must accommodate strain by fracture and then flow of comminutiae when the fragment size becomes small enough that further cracks cannot form (Figure 9.4(c)). Materials in this class include amorphous glasses and complex crystalline solids such as boron carbide. The weak shock behaviour in these materials is dominated by zones of high defect concentration. In glasses this is principally on the outer face of components. Within the microstructure are included processing bubbles and strain concentrations with fracture planes. However, their concentration is much lower than that found at the component outer faces in these strong solids. Fracture propagates in brittle solids from these flaws until they coalesce to form a failed region in the bulk. This marks the phase in response and for the times the MSU takes to stabilise, the stress state at the component scale is unsteady. As the stress amplitude in the initial impulse increases, the failure front approaches the elastic front and at the stress at which the WSL is reached, failure occurs within the front.

Polymer materials must collapse their packed chains against Van de Waals' forces before strong shock behaviour can be observed (Figure 9.4(d)). With the short length of pulses delivered by present shock loading platforms at these lower stress levels, the collapse of the chain structure has not occurred in these materials when compared to the response of engineering metals and brittle solids to a similar duration pulse. Overcoming these forces dominates the behaviour at the microscale. Strength can be seen to evolve from the impact face into the microstructure as strain is imparted from an impactor. Typically the response is highly anisotropic at the mesoscale with compression within spherulites in thermoplastics or within amorphous networks between these regions. Geometrical rotation and packing of the chains results in a rich phase diagram with which the strain from the impulse may be absorbed. It is only when these interchain forces have been overcome that the material can compress within a single front and this is the WSL at c. 20–30 GPa identified in Figure 7.12.

In an energetic composite such as a PBX (Figure 9.4(e)) the localisation at defects results in the ignition of sites and burning within the solid under load. Like failure behind a shock in a brittle glass, individual sites see localisation of the flow and this triggers chemical reaction to begin. The mesoscale burning zones interact and coalesce behind the front and the pressure generated feeds back into greater burning rates if confinement is present around the explosive. The flow is unsteady at these times but with suitable confinement strengthens the front until burning occurs to completion within the shock itself and detonation results. The WSL in the energetic crystals allows the pressure to break bonds and burn the material homogeneously within the front, but the detonation pressure is fixed by the chemical energy of the materials liberated within the CJ zone.

Figure 9.7(f) shows such a detonation wave with chemical decomposition complete to the rear of the reaction zone. However, the figure may be regarded as a schematic snapshot of all classes of solid in the strong shock regime. The front is strong enough to overcome the strength of the matter into which it progresses and deformation processes (and subsequent chemical reaction in the case of HE) are confined within that front. The flow behind it is homogeneous since all defects are activated, excepting regions that result from inhomogeneities of different elements. Further, the strains within shock fronts are always small so that mixing does not occur on microsecond timescales. Thus the geometrical configuration of second phases is preserved when release arrives to cool the flow and re-establish a microstructure in the solid. In this state the material has density gradients and is at high temperature but response is homogeneous and whilst the shock front does not have a steady stress state the flow behind has. Thus in the strong shock regime region the MSU does not have significance.

9.6   Response of materials to extreme dynamic loading

The last sections have sketched the length scales and amplitude ranges within which condensed matter responds to dynamic impulses. The length scales discussed, from the microscale to the structural scale, extend from nanometres to kilometres – a factor of 10^{12}. The scope of compression considered extends from the yield strength of materials ranging from of the order of 100 MPa to pressures beyond the *finis extremis* in the strongest materials at c. 1 TPa; a range of 10^7 GPa in pressure. Beyond this extreme

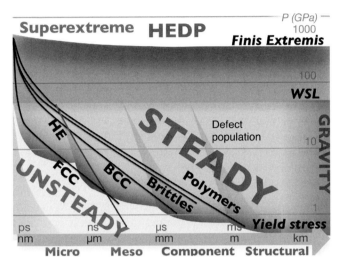

Figure 9.5 Schematic of the mechanical behaviours observed in condensed matter over the operational regimes of various deformation processes. The pressure/stress in the loading is plotted as a function of the length (converted to time via wave speed) of the target volume accessed.

regime lie superextreme states of warm dense matter and beyond highly confined plasma. Within this range lie the states of condensed matter accessible in the laboratory and describable using a mathematical framework based on strength formed through bonding where valence electrons are shared between atoms. This space is illustrated in Figure 9.5 and indicates a schematic of the mechanical behaviour observed in condensed matter over the operational regimes of deformation processes considered here. In this region temperature is omitted since general principles are addressed below, but the reader could extend these surfaces to cover temperatures to melt and beyond from what has been said previously in the text. The elastic region is light grey, the weak shock regime grey, and the strong shock regime is darker still. The *finis extremis* marks the upper limit of the extreme behaviour considered here and divides observations made in the solid mechanics regime from superextreme behaviour in the region described by high-energy-density physics (HEDP).

Each scale is divided from the next by a boundary at which an interface exists with the assemblage of elements at each scale: grain boundaries between the micro- and mesoscales; surfaces between the meso- and component scales; and joints between elements at the component and structural scales. At these interfaces are centred populations of defects that control behaviour in the weak shock regime, with that population disappearing above the WSL at *c.* 50 GPa. Darker regions are sketched in the figure at scale boundaries to indicate the regime of defect distributions for each scale. These disappear at the WSL.

At the lower limit in the microscale the yield strength of the solid is at the theoretical strength for the material in a defect-free volume and this is also the WSL. However, this reduces as defect populations are activated and the distance scale increases. It thus declines in a series of steps, dropping as each defect boundary is crossed.

The scale is displayed in length-scale units as well as time. The two are connected by the communication speed at that scale which determines the region of space swept by a front during that timescale; by phonons at the microscale, the wave speed at the meso- and component scales, and slower at the structural scale. Figure 9.5 is a schematic so this value is taken as several thousand m s^{-1} on average to represent gross behaviours.

When first loaded a material has the localisation phase to settle to more homogeneous flow within an MSU. As the pulse amplitude increases the flow steadies more rapidly as the kinetics for deformation processes are driven faster. Of course in impact loading there comes a scale at which the compression of a rough surface will dominate the process since surface layers have different properties to bulk, but in what follows material response will be assumed to dominate the transition from unsteady to steady flow. Each class of material, and subsets of that within, can accommodate compression by localisation at different rates. The fastest to steady are metals where dislocations nucleate and slip can begin quickly to transport the applied strain in compression across the material. FCC metals with lower Peierls stress and higher dislocation densities can steady faster than BCC materials with HCP structures limited by few systems. Brittle solids must localise fracture before grains can flow and their strength and the slower speeds make the transition from unsteady to steady flow slower than with metals. Polymers must collapse their structure and this is generally limited by the velocity and duration of the drive rather than wave transport before steady stress states are induced behind the front. There is thus a run distance for each class of material before a steady state occurs behind the front, and at the WSL this reduces to the width of the shock front in which all deformation processes run to completion before the steady flow behind occurs. In this region the time for completion of the processes decreases (Swegle–Grady) by the reciprocal of the stress level to the power four. Thus the figure has a series of curves for each class of material that represent the division in time and in run distance before the flow steadies and processes at that stress level have run to completion for a particular driving compression. The most varied and interesting responses lie in the weak shock regime where individual mechanisms have not completed and the effect of the existing defect states is a factor in the response that results. However, the boundary between steady and unsteady response has a strong influence on the models that describe material behaviour and the experiments that inform them for the derivation of mechanistic laws.

Figure 9.6 shows (a) the experimental and (b) the modelling platforms in the phase space discussed. The present range of devices can cover a large part of the region identified for investigating material behaviours. Lasers deliver the shortest pulses but reach the highest pressures in condensed matter research. Single- and two-stage launchers cover pressures up to the *finis extremis* at greater length scale. Explosive lenses can deliver pulses in a limited region around the WSL, whilst Hopkinson bars at intermediate strain rate and machines quasi-statically map the yield surface. The elastic and weak shock

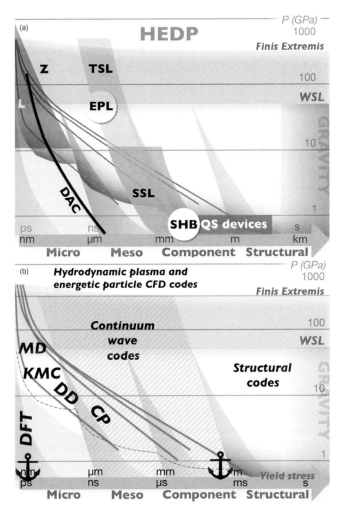

Figure 9.6 (a) Experimental and (b) modelling platforms for understanding extreme material response. These include lasers (L), Z pinch devices (Z), diamond anvil cell (DAC), loading by explosive plane wave lenses (EPL), single- and two-stage launchers (SSL, TSL), Hopkinson pressure bars (HPB) at intermediate rate and quasi-static loading machines (QS) at the continuum. The numerical techniques and numerical platforms used employ density functional theory (DFT), molecular dynamics (MD) or kinetic Monte-Carlo (KMC), dislocation dynamics (DD) or crystal plasticity (CP). Continuum wave and structural codes occupy the meso- and macroscale response. MD is tied to DFT (left anchor) and continuum codes to QS calibrations (right anchor).

regime are covered by devices operating on samples that are found at the component scale. The difficulties arise when extending experiments to investigate structural response close to the *finis extremis* where large enough volumes of material cannot be routinely tested. The results obtained from what is available, however, can be used to derive or validate models at the scales of interest using a series of numerical platforms.

The code platforms divide their effect at the micro-/mesoscale boundary with particle interaction codes at the microscale and continuum wave codes at the mesoscale and above. The molecular dynamics methods operate on an approximation to the wavefunction in an atom and neglect electronic contributions to the energy. On the other hand the continuum codes fit data at yield at the structural scale that is used to populate models for behaviour at the mesoscale or simulations up to the *finis extremis*. These codes can only qualitatively represent response in their present state unless fitted to results close to the application simulated and advance in this arena thus calls for better consideration of the defect populations at different scales in the weak shock regime. Further, one must reformulate consideration of the mechanisms that become important as one crosses scale boundaries in order that predictive power can be better ensured with these codes in the future. Simply mapping molecular dynamics onto larger and faster computer platforms cannot address the missing physics that is lost as one crosses defect populations at higher scale boundaries. Similarly engineering continuum codes cannot be applied at the microscale and in some cases not the mesoscale either as continuum fits neglect operating physics. At the microscale *ab initio* methods are needed to generate atomic potentials.

9.7   Response of components and structures to extreme dynamic loading

The engineer using materials to design structures must view their response at the continuum scale at which they operate. John and Bertram Hopkinson as early as the late nineteenth century realised there was a difference in the strength observed in static or 'dead' loading and in the dynamic response observed in impact. Orowan in the early twentieth century realised that materials were structures at different scales of observation and that discussion of response applies not to a homogeneous element but to an assembled structure. There are a series of material properties which the engineer much consider and describe in classifying their use and these have been incorporated into the developments in solid mechanics which continued through the last few centuries. Dynamic loading has several consequences which show particular behaviours and which must be considered. These have features in common with the behaviour at high energy densities but the practising engineer is concerned principally with material in the elastic state and impact engineering is dominated by the study of loading around the yield stress rather than effects at higher amplitudes. This section will not consider all of the observed consequences to yield as they occupy books on their own (see the bibliography). However, one at least must be highlighted before moving forward.

The most ubiquitous illustration of dynamic properties is shown in Figure 9.7. The yield stress of a material is determined in a series of experiments and plotted versus the continuum strain rate for that test. Such curves show upturns with a rapid increase in yield stress with faster loading rate. There are several other properties that metals in particular are observed to show that are also considered a particular feature of high-strain-rate loading. These include the transition from slip to twinning which appears

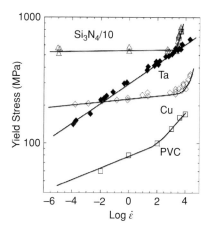

Figure 9.7 Yield stress vs. continuum strain rate for the ceramic Si_3N_4, metals Ta and Cu and polymer PVC. A threshold is reached beyond which yield stress increases rapidly. The trajectories for such behaviour are different for FCC and BCC metals. Such behaviour is generally called strain rate sensitivity (SRS) (see reviews such as Armstrong and Walley (2008) for more details). Sources: Ta data after Hoge and Mukherjee (1977); copper data after Follansbee *et al.* (1984); PVC data after Walley *et al.* (1989); and Si_3N_4 after Lankford (2005).

to occur at higher stresses. Also a transition from ductile to brittle behaviour as the strain rate increases (as well of course as the temperature reduces). Finally, shear banding occurs, as mentioned previously in Chapter 5.

It is to be expected that strength of a solid be different when the sample is free to expand laterally under compressive load on the one hand and inertially confined (at least for the timescale of the experiment) on the other. This has already been discussed in Chapter 2 and the inertial confinement increases yield stress by a factor given in Eq. (2.5). This depends upon the Poisson's ratio of the material and for metals, where that is close to a third, inertial confinement augments the yield stress by about 50%. In real metals the yield stress also increases across the lower rates where confinement is not important and exceeds this factor over the full range up to the highest rates so that clearly the metal response itself must additionally drive the process and this is immediately seen since FCC and BCC microstructures show different generic behaviours.

Clearly defect concentrations exist at different scales in metals. The yield surface is determined priniciply by that at the mesoscale since in fully dense materials this is where the largest defects lie. Thus plasticity and twinning are activated first from these sites. At higher strain rate the pulse has shorter rise time and hence sweeps less metal so that there comes a strain rate at which the mesoscale defects are not accessed and the rise to yield occurs within a grain. The strain at yield is of the order of 0.1% and given that grain size generally lies between 10 and 100 μm this limit occurs at $c.\ 10^4$–10^5 s^{-1}. The high defect density in FCC metals still allows time for the interaction phase to occur so that strengthening may occur. BCC metals, on the other hand, show a different response since dislocation densities are lower and Peierls stress higher. In this case dislocation interaction at boundaries is critical and the grain size dependence of yield stress (the Hall–Petch effect) is more prominent. As the strain rate increases, the loading time drops and less boundaries are sampled so that yield strength rises. Clearly there comes a point at which the pulse is so short that a dislocation cannot be driven to meet a grain boundary. In this case (laser pulses into metals with grain sizes of tens of μm, for example) discussion of Hall–Petch effects is not relevant since dislocations only travel a micron or less.

Other material classes have similar behaviours. One particular example is shown in the figure for an amorphous polymer PVC. Yield occurs here when the Van de Waals' bonding is overcome. The fast pulse reduces the degrees of freedom for accommodation of the strain and this leads to a higher yield stress. Put another way, with long load times the polymer can creep and accommodate the strain at a lower stress whilst at high strain rate the pulse is limited and higher stress is required. With other polymer microstructures there are several notes of caution for dynamic tests. Firstly, this is a measure of the onset of compression of a structure not at full density. Secondly, such polymers are generally anisotropic and a Hopkinson bar test for instance must explicitly recognise this in the deconvolution of the data. Whereas polymers densify when loaded quickly, with their response sensitive to the applied strain, brittle materials can only fracture. Thus increasing the strain rate in a pulse corresponds to decreased loading time that activates fewer flaws. In polycrystalline ceramics there is the possibility of limited plasticity (generally twinning) within grains and again once the rise to yield does not include the mesoscale, grain boundary flaw population, the yield strength rises. Such plasticity initiates cracking down twin boundaries as observed in alumina. An example is shown in the figure for silicon nitride where the yield stress of the ceramic has been scaled by a factor of ten so that it may be compared with the response of metals and polymers. This material has more uniform grain size than the metals shown and the rise thus occurs more sharply, reflecting the tighter mesoscale flaw distribution for the ceramic.

As seen earlier, the observed strength of a material approachs its theoretical value as the shock amplitude approaches the WSL. At this point the strength has saturated and the material behaves homogeneously. The shear stress required to move a dislocation results from interaction with an activation volume whose magnitude depends both upon the rise-time of the pulse and the temperature it is at. As the strain rate increases the flow stress climbs and the activation volume approaches that of the unit cell. Thus the WSL may be thought of as the level at which defects saturate the volume, at which all defects are activated by the pulse. In the regions of lower strain rate theory shows that the activation volume is three orders of magnitude higher thus including the grain boundary defect volume which lowers the required yield stress. Interestingly HCP metals form two groups, one behaving similar to FCC and the other to BCC metals. Those with their unit cell c/a ratio less than the ideal, such as titanium or the alloy Ti64, show a BCC-like response whilst magnesium, zinc and cadmium with c/a above this value show an FCC-like behaviour. It is the mesoscale flaw distribution that controls these responses and thus at these strain rates texture and processing have a recognisable effect on the observed response.

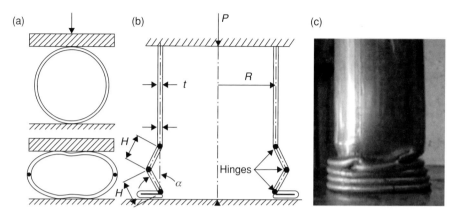

Figure 9.8 The plastic deformation of a cylinder crushed within a load frame: (a) radial; (b) axial crush; (c) crushed cylinder, with external diameter 51 mm and wall thickness 1.6 mm. Sources: after Calladine and English (1981), Reddy and Reid (1980).

All these features show strain rate to be a faithful indicator of loading behaviour when the stress state is steady. It reflects the extent of the material sampled during the rise of the pulse and thus when that distance becomes less than the size of defects, at most particularly the mesoscale flaw scale, the yield stress rises. However, strain rate is a continuum quantity and there comes a length scale in which flow is unsteady and inhomogeneous. At any scale where that is the case, strain rate is ill-defined as seen above since it is a measure of the conditions that exist at the end of the unsteady phase. This time will be the localisation phase of the structure or component that is loaded and it will relate to stress-state equilibration by wave interaction within it. When the operating plasticity mechanisms can run to completion within the rise of the pulse and the defect distributions of importance can be accessed and stress equilibrated within it, then the speed of the loading reaches a regime in which the time for which the pulse is applied is not important. This corresponds to the constant F (defined previously) becoming greater than one. In this regime it is merely the magnitude of the strain that is applied that determines response. However, slower processes operate that are generally not captured in tests within the laboratory since their effect is not important for the life time of structures in dynamic environments unless coupled with high temperatures for the components under load. These are diffusion processes and will be discussed later.

The onset of plasticity and localisation in FCC metals subject to impulses of increasing duration has been tracked through response at different length scales throughout the text. Consider the response of a structure constructed from aluminium (a tube) under the application of a dynamic load from a dropping mass. In this case the processes of plastic localisation and buckling occur over longer time and length scales than the shock regime but again follow similar behaviours driven by physical mechanisms in the material.

Figure 9.8 shows deformation of a cylinder oriented (a) radially and (b) axially to the loading direction. The case illustrated in (b) requires a phase establishing flow before plastic hinges are defined. It may be contrasted with the analogous case in (a) where the same cylinder is loaded radially, the plastic state is established within microseconds and the deformation occurs at a constant rate with strain accommodated by hinging. The latter flow is classified as type I behaviour for structural response and the former as type II. Type II behaviour illustrates the transition between stress-state-limited and structure-limited response, whilst in type I, structural response is attained very quickly and deformation occurs uniformly at lower stress values. Application of the concepts described earlier would define the relaxation time for acquisition of this state to be the time required to reach it and a load could be conceived in which $F > 1$ for radial loading (a) but $F < 1$ for a longitudinal loading (b). Further, one can see from this example that a relaxation time for a mechanism might equally be substituted for a relaxation strain which the drive acting on the component might apply.

The stress levels at later times required to fail the structure are now very much less than the theoretical strength of the material. In experimental and theoretical analyses, previous work has shown that the axial failure of cylinders was due to the yielding and folding of the cylinder into a series of segments with discrete geometry and features (Figure 9.8(b) and (c)). The cylinder yields at plastic hinges and folds at these to accommodate the applied load. This localisation into discrete zones allows strain to be accommodated by buckling, inversion and splitting when stress is applied to the cylinder. The accommodation of such large strain, however, has many practical uses since it acts as an efficient way to use deformation to absorb large amounts of energy in plastic work as the cylinder collapses.

In the first moments, material interfaces are not accessible by a volume element within the bulk and the strength is governed by an ever-greater accessed volume that evolves from atomic-scale effects to wave interactions between the mesoscale defect structure over the first microsecond. As wave propagation equilibrates the material state over the component volume, the global strain rate becomes defined for the deformation of a component. These times are sufficiently long for laboratory scale components that the structure responds to impulse applied not as a function of time but as a function of strain after loading is applied. The localisation phase of deformation is controlled by the equilibration time for the wave transmission through the elements of the structure and these are observed as inertial effects within the components. This phase defines differences in the time taken for a structure to respond to dynamic deformation, with the controlling mechanisms occurring in the earliest part of the compression. This is reflected in Perrone's rule, which states that the single level of strain rate which best characterises the process of deformation as a whole is that which occurs in the initial stages of deformation. Flow is achieved by a population of shear bands that develop and then allow shear to progress within the structural component. As the loading progresses, mechanical deformation proceeds to accommodate the imposed strain from the drive. Heat is dissipated in these bands but strain is accommodated at a constant rate within the structure.

At the scale of elements of the macrostructure, the strain must develop until mechanical deformation has equilibrated to the state observed in quasi-static loading. This is engineered through the action of inertial forces developed within the structure by rapid acceleration which parts of it experience during the impact. The flow phase can then occur when stress-state equilibration has occurred within individual loaded components. These mechanisms result in behaviours ascribed to two classes of structure. The dividing characteristic between these two groups is the difference in the times required to establish stress state stability in the loading direction (and uniform strain during the deformation). Mechanically this reflects the time (which is directly related to the strain) to localise shear zones at positions within components, followed by the growth of motions in resolved planes perpendicular to the loading direction, to accommodate the strain imposed by the load. Further, all of the lateral deformation must be compatible with the direction of the imposed loading impulse as it is absorbed by the target. In such a case it can be shown that the transverse acceleration of structural components is independent of the velocity of the impact and that a higher impact speed will produce a larger compressive plastic strain. At some time into the deformation the longitudinal compressive strain rate becomes zero and a second phase begins in which transverse bowing increases further by means of the rotation of plastic hinges until eventually the motion ceases. Finally, an interaction phase occurs as the cylinder supports load through its compressed volume and its strength again climbs.

The time for stress-state equilibration of a metal grain is tens of nanoseconds, that of grains of different orientation or similar scale defects ranging from hundreds of nanoseconds to a microsecond. The equilibration of the stress state within the structural component in the direction of the loading may take microseconds for a thin plate and hundreds of microseconds for a rod, and once the stress state has equilibrated and a strain rate has been defined for the component, transverse accelerations develop shear localisations in the structure and plastic hinges develop; these processes together take tens of milliseconds. Finally, the structure accommodates the strain over hundreds of milliseconds until the stress state is equilibrated with the incoming driver and the relative deformation ceases. The stress–strain response of the structure under dynamic load may be mapped to a stress–time one. At the earliest times and the smallest strains the strength is the theoretical maximum of the metal under load. After a microsecond it adopts the strength of the microstructural volume element sampled. When wave equilibration occurs the stress state in the component is uniaxial stress and yields at the flow stress of the metal. Finally, the structure defines its deformation mode with plastic hinges and the stress state required is at a lowered state until it reaches a maximum locking strain or the driving impulse is absorbed. The relaxation times for structural response to load are defined by the scale of the component and the relaxation times for the deformation process operating.

Salisbury Cathedral in Wiltshire has the tallest tower and spire in the UK. As with many of these great buildings it is a structure that has evolved over the centuries and from 1310 to 1330 building started to add the additional 6500 tonne edifice. The original design did not have supports sufficient to support the new structure and columns began to bend and indeed the tip of the tower shifted a metre out of alignment. There were even reports that the pillars creaked and moaned as the tower and spire were built higher. Buttresses, bracing arches and anchor irons were added over the centuries and the tower is now stable (Figure 9.9(a)). However, the supporting pillars at the corners of the spire

Figure 9.9 Salisbury cathedral. Left: buttress additions under the tallest (123 m) cathedral tower and spire in UK (6500 tonnes) added 1310–1330. Right: Purbeck marble pillars bent under stress during the build; support was added later.

are seen to bend inwards under the weight. The marble has crept under the applied stress with a relaxation time of centuries to bow, as one might expect of such a strut under the applied load. In fact this marble is not metamorphic and homogeneous but is a composite formed from densely packed shells of the fresh-water snail *viviparous* (Figure 9.9(b)). Thus this graphically illustrates that, given long enough times, large composite stuctures can deform plastically; literally in this case at a snail's pace. It is only that our observation time is too short to observe these effects over a human lifetime.

It is interesting to review the discussions of deformation considered through this book from the point of view, and using the terminology commonly used in engineering and solid mechanics. The previous chapters have described a series of mechanisms that operate during deformation in condensed matter, some on very short timescales driven by fast loading. None are instaneous; it is only possible to regard them as such when the relaxation time for the process is very much less than time of interest for the observation ($F \ll 1$). Most in some respect involve the closure of defects and movement of material to adjacent sites and application of energy to a system to increase transport to compress a material under the applied load. These features are also ascribed to diffusion so that a mechanical impulse may be regarded as a rate control in the same manner as temperature. A time-dependent deformation of this sort is normally regarded as creep and all processes that lead to plasticity in solids may be defined as such since they have a threshold for operation, an activation energy and a relaxation constant over which they operate. Thinking in this manner it is natural that loading rate and temperature have similar effects on yield stress since both control diffusion processes. Plasticity is often thought of as a mechanism that begins beyond a material's yield stress, whilst creep occurs at stresses below that level. However, in nested operating mechanisms the yield stress is not a fixed threshold as the volume sampled and the time over which observation takes place is changed.

The length scales in condensed matter have defect populations concentrated at their boundaries. Jointed components at the laboratory scale fix a further boundary (defect) within the macroscale at which new regimes of deformation are initiated as they fail under load. These scales have regimes of localisation, flow and interaction as deformation proceeds within each scale and particulary the flow behaviours have been described by analytical laws so that computation of response can be used in the design of both materials and structures. Applying these flow laws at different length scales is commonplace and scaling of response is a staple of engineering practice. However, scaling can only work where the physics is driven by the defect populations at that scale. New scales have different operating mechanisms and so such concepts cannot be applied. Thus applying atomistic molecular dynamics to engineering continuum problems fails, because potentials, approximate at the microscale, cannot describe the effects of defects at the meso- and component scales. It has been noted that smooth particle hydrodynamics has applied a particle scheme at the continuum with engineering potential and perhaps this is the best means of describing failure at the continuum.

The opening chapter introduced a series of natural events that resulted in extreme loading of materials both engineered and natural (for instance, Pompeii and the plug that seals the crater of Vesuvius). In the final section, impact on the Earth by the bolide that killed the dinosaurs will be reviewed through the understanding presented in the preceding chapters. Although the details of what precisely happened cannot be known, some idea of the magnitude and severity of the effects can be drawn to illuminate observations of the resulting impact.

9.8   Impact at Chicxulub

In any material where diffusion is a primary mechanism for creep, reduction of the mean free path of any component by reducing thermal relaxation to defect paths or freezing motion by loading states that only last a short time has a similar effect on the yield stress of a material. The correlation between temperature and strain rate in polymers is a clear example of this but similar effects occur in composites as well. The effects of dynamic loading and temperature upon material behaviour can be seen to be complementary and in some cases equivalent stimuli to reduce mechanical strength in a structure. Both extreme temperatures and pressures but also high-rate loading by impact or explosion are responsible for the morphology of planets in the solar system.

Throughout the universe, matter has agglomerated, and gravitational attraction has then operated, ensuring collisions between rocky bodies occur. In the early solar system,

Table 9.7 World impact craters with diameter > 50 km. An active group maintains the Earth Impact Database for the most up-to-date information (see http://www.passc.net/EarthImpactDatabase/index.html)

Crater name	State/province	Country	Diam. (km)	Age, min. (Ma)	Age, max. (Ma)
Vredefort	Free State	South Africa	300	2019	2027
Sudbury	Ontario	Canada	250	1847	1853
Chicxulub	Yucatan	Mexico	180	64.93	65.03
Manicouagan	Quebec	Canada	100	213	215
Popigai	Siberia	Russia	100	30	40
Acraman	South Australia	Australia	90	450	
Chesapeake Bay	Virginia	USA	85	34.9	36.1
Puchezh-Katunki	Nizhny Novorod Oblast	Russia	80	172	178
Morokweng	Kalahari	South Africa	70	144.2	145.8
Kara	Nenetsia	Russia	65	70	76
Beaverhead	Montana	USA	60	550	650
Tookoonooka	Queensland	Australia	55	123	133
Charlevoix	Quebec	Canada	54	342	372
Kara-Kul	Pamir mountains	Tajikistan	52		5
Siljan	Dalarna	Sweden	52	366.9	369.1

impact was the mechanism by which bodies of large mass cleared their orbits around the Sun. Yet throughout time objects on larger orbits in the vast Oort cloud, on paths that pull them towards the Sun, have entered the solar system. These comets have peppered the planets with impacts of varying magnitude and left the surfaces of all the objects, most now mapped by a continuing series of probes, sculpted by the marks of impactors across their surfaces. It is only on Earth, a planet whose surface is 70% covered by water and on a land reformed by active geology and cloaked under vegetation, that the signs are more difficult to see. Yet even here the increased sophistication of satellite imagery has revealed at least 170 impact structures with more hidden beneath the sea and almost all of those in the last 50 years (Table 9.7). Although the number of impacts is decreasing with time, there is at least one multi-kiloton impact somewhere on Earth each year and geological evidence, and simple statistics, inevitably means that there will be one of a size capable of affecting human survival at some time in the future.

9.8.1   Impact sites on Earth

There are a large number of extra-terrestrial objects capable of colliding with the Earth and incoming impactors may be lumped into one of two classes. Asteroids are porous rocky bodies found generally within the inner solar system, particularly the asteroid belt (between Mars and Jupiter). They comprise little total mass (less than 5% of the lunar mass), but nevertheless have the potential to collide if they stray into planetary orbits. The other class of bolides is comets, which may be regarded as primarily being composed of dirty ice (little other mass since their origin is from outside the solar system). Although their density is lower, their velocity is higher than asteroids. Potential impactors may be divided into two sets. A larger group comprising stray bolides that cause significant local effects and may even trigger global climatic perturbations, but are insufficient to induce significant permanent effects on the planet or on life upon it. A select further number comprise bolides of significant energy (rocky examples may be large enough to be described as planetessimals) capable of causing global effects upon the surface and atmosphere but of insufficient mass and velocity to perturb the Earth's orbit. One early impact lies in a class of its own since it defined all aspects of planetary geology, the appearance of an atmosphere and the subsequent evolution of life upon the planet; that was of a planetessimal on to the proto-earth about 4.45 billion years ago. One theory has the impactor the size of the planet Mars (about half the size and one tenth of mass of that of present-day Earth). In another a larger object gave a more glancing blow. In either case a portion of the ejecta from the impact zone was expelled into Earth orbit and lay in a disc of orbiting material for the time it took for gravity to cause condensation of the cloud into a defined object – the Moon. Advances in smooth particle hydrodynamics (SPH) modelling of such a process have allowed this theory to become accepted over the years since it was proposed and the computational schemes necessary to perform such a simulation with great mass flow in a rapidly evolving geometry were described in Chapter 2.

Cometry impact has been recently observed on an adjacent planet. In 1994 the comet Shoemaker-Levy 9 impacted Jupiter under the gaze of astronomers on Earth. It was the first comet observed to be orbiting Jupiter rather than the Sun, illustrating the significant role the massive planet plays in cleaning the inner solar system of many smaller bodies. The comet was split on its approach to the Sun into 21 pieces up to 2 km in diameter and with a mass of $c.$ 10^{13} kg at their largest (the mass of the Moon is 7×10^{22} kg) approaching at 60 km s^{-1}. The impact produced fireballs that at their greatest reached 25 000 K in a plume over 3000 km high and left darkened spots up to 12 000 km across in the clouds circulating the planet. Shock fronts were generated and observed travelling in the waveguides provided by the different density layers around the planet. Over 6 days, 12 discrete impacts tracked a line of dark scars across the light clouds and these were observed for many months to come, illustrating the valuable role Jupiter may have played protecting the development of life on Earth by shielding it from the impact of larger mass objects. The energy within such an impact was of the order of 10^{22} J, around 10^5 times greater than the largest nuclear explosion. The pressure generated may be simplistically estimated by assuming an ice comet impacting a liquid surface; applying hydrodynamic equations gives a pressure of the order of 1 TPa. Of course at these pressures the material is out of the strong shock regime and into super-extreme states where very different physics operate in the dense, hot matter produced at impact, so the equations of state are very different from those encountered in the weak and strong shock regimes. Nevertheless these figures give some indication of the states achieved in such impacts.

One bolide that did reach Earth, whether a comet travelling at 60 plus km s^{-1} from outside the solar system or an asteroid at 20–40 km s^{-1} from within, entered onto a collision course 65 Ma years ago and resulted in mass extinction events on the planet which have been documented in detail over the last 20 years. The impact occurred at a time when the Earth had active volcanism on a scale not seen today so that it had the effect of magnifying the effects of vast basaltic magma eruptions from the Deccan traps (Northern India). The joint effects of worldwide increase of nitrogen and sulphur dioxide levels in the atmosphere due to volcanism, coupled with the instantaneous injection of a lethal extra dose of further gases, with dust and tephera injected above the statosphere, dealt an unrecoverable blow to all creatures of mass greater than 25 kg; this impact led to an extinction event.

9.8.2   Mass extinctions

Since its appearance the number of species of life on Earth has expanded for hundreds of millions of years with periodic contractions at key times. The earliest single cells, preserved as fossils, date from around 3.5 billion years. At times, however, there have been cataclysmic extinctions involving large numbers of species and more than 99% have been lost as they failed to adapt to the prevalent environment. The rate of genetic mutation has also varied as a result of environmental factors such as radiation.

The foremost extinction (sometimes called the 'Great Dying') occurred at the end of the Permian (245 Ma ago). Earth's largest extinction killed 57% of all families and 83% of all genera. The best-known (but less destructive) extinction occurred at the end of the Cretaceous period and the beginning of the Paleogene (65.5 million years ago). This event (dubbed the K–Pg extinction, where the K may denote *Kreide*, German for chalk) saw 17% of all families, 50% of all genera and 75% of species wiped out; in particular larger animals over 25 kg. There has been much reference to this event as the KT extinction, where T denoted the Tertiary era of which the Paleogene and later the Neogene were subdivisions; however, K–Pg will be used in what follows. The event also marks the end of the Mesozoic Era and the beginning of the Cenozoic Era. Other major extinctions occurred in the late Triassic–Jurassic (200 Ma), the late Devonian–Carboniferous (360 Ma) and the late Ordovician–Silurian (444 Ma). In each of these events is it necessary to separate sudden from protracted mechanisms if one is to hypothesise impact as a possible mechanism. Of course many species declined slowly as a result of competition with other evolving forms, so gathering evidence for an instantaneous extinction in the fossil record is difficult since large numbers of relevant sediments need to be examined.

There are a variety of candidate mechanisms for mass extinctions, but any plausible one has to prevent recovery in a population to ensure it cannot adapt or manoeuvre sufficiently to survive and with the ingenuity and variety of life, that is frequently a difficult task. One class of mechanisms capable of this has to involve geographical change with no escape. The second involves habitat degradation. Certainly variation of sea level has caused problems for species when routes for migration are blocked or food sources are isolated. Such difficulties can also result from volcanic activity but this mechanism has other concomitant effects. Processes that release SO_2 and NO_2 degrade climate principally by causing heating or (by shielding the surface from light) global cooling. These effects kill life at the lowest levels and affect anything in the food chain above. In each of these mechanisms there are two features worthy of investigation: the first is the magnitude of the effect; the second concerns the speed of its imposition. Life may be adaptable, but not if the effect upon it is too fast for significant reaction. The fastest, global effects result from impact of an extraterrestrial body of around 10 km diameter or greater. The impact of the bolide responsible for the extinctions at the K–Pg boundary was estimated to have released 5×10^{23} J (10^{14} tons of TNT equivalent, which is 2 million times more powerful than the Tsar Bomba mentioned in Chapter 1). This places this event as 250 000 times more energetic than Krakatoa and that energy concentrated in a fraction of the time that the volcano liberated its energy. Nor would such an impact have caused merely mechanical effects, since there would also be an intense radiation pulse in the first moments. To confirm that such a collision occurred, geological evidence must be differentiable from other candidates and be spread as a thin layer to confirm a rapid event. Alternatively, artifacts must be found in one layer and not in later ones, with a sharp transition between the two. However, observables are scattered and the difficulty of identifying a boundary is increased if the sediments are very old and the number of individuals to find is small.

A leap in understanding was made by the Alvarez's (father and son) in 1980. They opened the way for a worldwide hunt with their new theory that the dispersion of ejecta from the impact of a comet or asteroid, travelling at tens of kilometres per second, was the source of the element iridium dispersed in a thin layer of silt deposited around the globe at the Cretaceous–Paleogene (K–Pg) boundary. Iridium is found at a concentration of 0.4 parts per billion (ppb) in rocks on Earth. However, meteorites recovered show much higher levels, recording 470 ppb in some cases. The K–Pg layer shows levels arount 6 ppb which is almost 20 times that possible from erosion of the Earth's crust alone. It was only later that this feature was connected with a crater several hundred kilometres across whose centre lies beneath a village called Chicxulub on the north coast of the Yucatan peninsular in Mexico (Figure 9.10). The crater was shown to be at the epicentre of a trail of evidence that radiated out from Mexico into the Caribbean and North America.

The data collected are compelling. There is a discrete layer of ejected dust and ash worldwide and, closer to the impact site, spherules of silicate glass at the K–Pg boundary all peppered with the tell-tale iridium content. Around the wider Americas there is not only evidence of damage of an impact origin, but also of global wildfires, cooling and acid rain. Thankfully, because of the Yucatan's geography (a shallow sea over a stable carbonate platform), the crater's interior morphology has been preserved. It is centred on the northern coast of the Yucatan, 1 km below ground level and accessible only through recovered cores and gravitational anomaly surveys. It is the youngest (and thus least altered) of the crater sites that correlate to extinction events, and is dated to 65.5 million years ago at the end of Mesozoic period (the K–Pg boundary). Other large

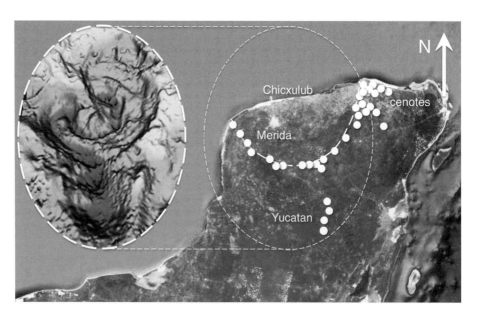

Figure 9.10 The Yucatan peninsular, Mexico. Inset: gravitational anomaly picture of an area showing the Chicxulub crater. The heavy dotted line and the dots showing cenote sties around the rim of the crater mark the outer trough. This ring is 180 km in diameter. The asymmetry in the inset gravitational anomaly picture indicates the NW path for the comet. Source: after NASA.

impact craters, Sudbury (Ontario, Canada) and Vredefort (South Africa), are nearly 2 billion years old, heavily eroded and tectonically modified.

Piecing together the evidence gives bald consequences for such a collision. Firstly, it is not clear if the impactor was a comet or an asteroid; comets are less dense and thus the body would have been larger, but since the bolide vaporises it is the energy it delivered that determines its effect. Nevertheless, a hundred million tons of impactor, travelling at between 20 and 70 kilometres per second, entered the Earth's atmosphere and careered down to the ground (Figure 9.10). It is most likely that the projectile was on a trajectory such that it impacted the earth at an angle; there is evidence in the gravitational anomaly scan for the impact site to suggest this. Such asymmetric signatures suggest a trajectory for the Chicxulub bolide from the southeast to the northwest at a 20–30° angle from the horizontal. Extinctions may thus, as a consequence of this, been most severe and catastrophic in the Northern Hemisphere.

An impacting bolide using simple considerations is most likey to enter the atmosphere at 45° (Figure 9.11). This results from motion of the Earth into a flux of bolides. Imagine a planet of radius r travelling into a flux of magnitude ϕ. The probability P of impacting the planet is $\pi r^2 \phi$. If the bolide travels at an angle θ to the normal to the surface then the chances of hitting an elemental ring of size dx, x from a diameter are $dP = 2\pi x \phi \, dx$. Since $x = r \sin \theta$,

$$dP = 2 \sin \theta \, \cos \theta \, d\theta, \tag{9.2}$$

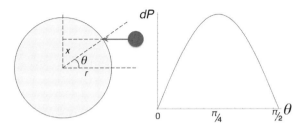

Figure 9.11 Bolide impact at most probable impact angle of $45°$ neglecting gravity.

where the incremental probability has been normalised by the total probability of impact. This gives an angle of $45°$ as the most probable with a normal or glancing impact of probability zero. This simple analysis neglects gravity, although adding in these effects does not change the result. The geophysical anomaly of the Chicxulub structure exhibits asymmetry. There is a central gravity high (running northwest) encircled by a horseshoe-shaped gravity low (Figure 9.10 insert). This corresponds to the 180 km diameter ring that disrupts the pre-existing gravity field.

Hitting the earth, it and the topmost layers of the impact site, immediately vaporised as temperature was instantaneously raised by the passage of the shock launched into the Earth and back through the asteroid. Part of the atmosphere was itself vented into space at impact. At that time, the Yucatan was a shallow sea so tsunamis were launched, travelling at around $500\ \mathrm{m\ s^{-1}}$ and with wave fronts 100 m high. Around 100 km^3 of rock fragments and melted silicate rock, transformed into liquid glass, sprayed upwards and broke into a mist of droplets which cooled and settled as minute spherical glass tektites found throughout the crater and across the ejecta beds though the USA and Caribbean. 10^{14} tons of vaporised comet rose to 100 km altitude and 10 to 20% of that did not fall back but remained at altitude. Further, the concentrations of energy in those few seconds at impact resulted in an intense electromagnetic pulse and, after the opacity once again allowed visible light to transmit as the temperature dropped, there was a brilliant flash of light; possibly the brightest ever seen on earth. The magnitude of the pressure in the divergent shock travelling outward from impact is reflected in quartz, shattered or shocked irreversibly into the high-pressure phase stishovite, with crystals fractured in a characteristic way. The highest pressures are indicated by material melted and then resolidified. At impact, pressures of the order of 1 TPa were created at ground zero and the divergent shock, flowing outward from the impact site, cracked the rocks nearby into small grains that allowed the layer beneath the impact site to flow as if a liquid. Like the tongue ejected upwards that follows a liquid drop falling onto a water surface, a plume of excavated material was ejected back into the air that spread dust and material into the upper atmosphere wherein it circulated for many months to come (see Figure 9.13 later).

The kinetic energy, E, at impact is simply given by the mass, m, and velocity, v, of the impactor, which is assumed to be a spherical impactor of density, ρ, and diameter, δ. Then

$$E = \frac{1}{2}mv^2 = \frac{\pi}{12}\rho d^3 v^2. \tag{9.3}$$

The Deep impact probe launched a 350 kg copper nose-cone which impacted at 10.3 km s^{-1}. This gave a kinetic energy of 1.96×10^{10} J which is equivalent 4.7 tons of TNT. Such an impact excavated a crater 100 m in diameter and up to 30 m deep. Note that 1 ton of TNT is equivalent to 4.184×10^9 J. Asteroids are typically of density 2.5–3 g cm^{-3} if rocky and 7.8 g cm^{-3} if iron. They travel in the range 10–20 km s^{-1}. In contrast comets are really icy dirt-balls of density 0.5–1.5 g cm^{-3}. However, they travel much faster – typically 30–70 km s^{-1}. The Chicxulub bolide has been estimated to be 10 km in diameter travelling at 10–70 km^{-2} and may have released an estimated 400 zetta-joules (4×10^{23} joules) of energy, equivalent to 100 teratons of TNT (10^{14} tons on impact). By contrast, the most powerful man-made explosive device ever detonated had a yield of 50 megatons, making the Chicxulub impact two million times more powerful. In relation to natural phenomena the largest known explosive volcanic eruption, which released approximately 10 zettajoules and created the La Garita Caldera, was more than an order of magnitude less powerful than the Chicxulub impact and released the energy over a much longer time. The pressure, p, may be estimated from the simple shock equation of state for a material of density ρ inducing a particle velocity in the flow of u_p, but this will not take into account the actual form of the Hugoniot under these super-extreme conditions:

$$p = \rho_0(c_0 + Su_p)u_p. \qquad (9.4)$$

Nevertheless the figure of 0.5–5 TPa indicates the regime that these events access. The temperature in this region can again be estimated using the methods described in Chapter 2. Here the temperature ΔT was given by

$$\Delta T = \frac{u_p^2}{2c_v}, \qquad (9.5)$$

which under these conditions indicates shock temperatures of the order of 10 000 K. Thus these pressures and temperatures are far in excess of the *finis extremis* and the material in this state is a hot dense plasma. The vaporised impactor and immediate impact site form a fireball which expands over the impact site. This will initially be opaque until expansion cools the vapour below the transparency temperature of air, which is *c*. 3000 K. At this time the opacity drops so that thermal radiation may escape primarily in the visible and infrared. At this time the fireball radius is of the order of ten times the impactor diameter; it will indeed appear as an extra sun to anything surrounding it. However, the curvature of the Earth limits the distance such a radiation source can affect, but within 1000 km the thermal flux will be high enough to spontaneously ignite vegetation.

The seismic affect of the impact will be felt at much greater distance. The Gutenberg–Richter equation may be used to estimate seismic magnitude M using

$$M = 0.67 \log E - 5.87, \qquad (9.6)$$

where E is the kinetic energy of the impactor. This gives a ground shock of magnitude over 10 for the impact. Such an earthquake would cause massive rock damage, with large rock masses projected or displaced, lines of sight distorted and trees and surface

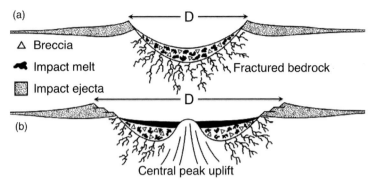

Figure 9.12 Crater morphologies: (a) simple; (b) complex. Source: after NASA.

features obliterated. It is estimated that around 10^{-4} of the impact energy was converted into seismic activity. The shock effects over those few vital seconds set the scene for the slower-scale events to come. Over the next tens of minutes, a flow of vapour, ejecta and melt, entrained in noxious gas, spread out from the site in a deadly wave. This low-angle ($<15°$) vapour phase would have expanded within a corridor that widened northward over west-central North America. This cloud will have preceded other effects and rapidly deposited a turbulent mixture of materials from the upper crust down to depths of 5–10 km.

The intense heat set firestorms raging in the immediate vicinity which spread around the world killing all animals of any size above ground in a lethal oven. And within the noxious gas atmosphere, the blast wave would have triggered chemical reaction forming acids of nitrogen and sulphur polluting the land and water. On the ocean surface, a tsunami was triggered whose 100 m wave amplitude battered the shores of adjacent lands, extending up to 20 km inland. At greater depth, the marine life left was decimated, with even the ammonite, which had survived previous mass extinctions, taken. On land, the ground would have risen as much as 100 m as far as 100 km away, causing a massive earthquake of magnitude 12 or 13. And after all these effects had happened, a crater remained 15–20 km deep and a minimum of around 180 km in diameter.

Crater morphology can take two forms; in the first, smaller impactors excavate simple depressions with concoidal form and raised rims (Figure 9.12(a)). As the crater size increases, loose debris in the walls slumps inward to partially coat the floor. In contrast complex craters have a central peak or ring, depending on their size (Figure 9.12(b)). These features are caused by the rebound of the surface material following the impact. The formation of a central jet and outward travelling ripples is well known when water drops impact liquid surfaces. Figure 9.13 shows the first moments of a drop impacting a surface as a schematic to the right with shock waves travelling into the water and back into the drop. Note the curvature of the drop makes the contact edge initially supersonic until material can release to form outward jets later into the impact. Three frames from a high-speed sequence of the drop impacting the water surface are shown. The dark, dyed drop can be seen just about to impact in the first picture. In the second,

Figure 9.13 Impact of a water drop on to a liquid surface. Left: schematic simulation; right: 4 mm dyed drop falling under gravity onto water with the appearance of a Worthington jet (Worthington and Cole, 1897; 1900). The frames were taken 0, 20, 115 ms (left to right) after the flash was triggered. Note the rebound in the third frame and the detachment of a darker drop from the jet. It is the same dyed volume of water as impacted. Source: after Cheny, J.-M. and Walters, K. (1999) Rheological influences on the splashing experiment, *J. Non-Newtonian Fluid Mech.*, **86**: 185–210.

20 ms later, the water drop has spread across a cavity formed in the surface and ejected a ring of droplets upward from the flow. This symmetric form resembles a coronet with a halo of fine ejecta and the image is often named as such. In the final frame the crater collapses on itself to eject a jet of material upward. In this case the jet itself sheds a drop at its apex and this is the same dyed volume that impacted. In the hydrodynamic flow between a comet there will be a critical velocity and dimension for which the pressures generated by the shocked material exceed the WSL of the target material. On the scale of the impactor the event is then analogous to that in the liquid case with dense fluid rebounding and ripples flowing outward in the compressed medium. As release and cooling occur the rock reforms its solid phase and marks of the trademark central jet and one or maybe more rings may be observed in the remaining rock (see Figure 9.10 for Chicxulub).

The critical diameter at which simple craters become complex is 3.2 km on Earth and slumping fills the floor with debris, making complex craters relatively shallow. As impactor energy increases further, the central peak has other rings surrounding it and one such can be seen in Figure 9.10. These effects depend critically on large volumes of material reaching pressures above the weak shock limit where they can behave hydrodynamically. In such circumstances the momentum of the incoming bolide is conserved and flow acts to rebound the projectile in a jet flowing upwards. The impact initially forms a transient cavity much deeper than the final forms observed. Imagine the kinetic energy of the bolide is converted into potential energy of the centre of mass of the volume excavated. Then the dependencies will involve a ratio of densities and a dependence on a term in the ratio v^2/g. To account for the forms of observed structures, an equation for the volume of a cavity excavated by a bolide can be estimated from that resulting from a nuclear explosion. A fit for the diameter of the transient cavity D'

Table 9.8 HELs for common rocks and typical pressure range for vulcanism

Rock	HEL (GPa)
Quartz	4.5–14.5
Feldspar	3.0
Olivine	9.0
Dolomite	0.3
Granodiorite	4.5
Granite	3.0
Volcanic processes	0.5–1.0

excavated is given by

$$D' = 1.61 \left(\frac{\rho_i}{\rho_t}\right)^{\frac{1}{3}} d^{0.78} v^{0.44} g^{-0.22} \sin^{\frac{1}{3}} \theta, \tag{9.7}$$

where ρ_i and ρ_t are the densities of impactor and target; d, v and θ are the diameter, velocity and angle of incidence of the bolide; and g is the acceleration due to gravity. For simple craters the collapse process results in a relation between transient and final diameters

$$D = 1.25 D', \tag{9.8}$$

whereas for craters larger than the critical size ($\Delta = 3.2$ km on Earth)

$$D = 1.17 \frac{D'^{1.13}}{\Delta^{0.13}}. \tag{9.9}$$

This gives a diameter of 180 km for the Chicxulub crater in good agreement with the measured value.

In the intense shocked zone around the impact site where matter was still condensed, silicates changed phase to coesite and then stishovite above c. 30 GPa. Outside this zone, pressures dropped and the rocks approached the HELs of the material. Some HELs for rocks are given in Table 9.8. As the shock wave transmitted out and dispersed, the pressure behind it released into the elastic regime and phase transitions partially reversed and heterogeneous microstructure returned. The expanding front put tensile stresses into the rock and cracked it into recognisable shatter cones showing the direction of the shock's travel.

Over the next months, the spread of the dust, and the formation of an acid atmosphere, put organisms into a critical state. The Earth was chemically hostile and the curtain of dust shielding the Earth's surface from the Sun prevented photosynthesis. For years the temperatures on Earth dropped below freezing, but after this time the tremendous energy injected by the impact and the greenhouse gases liberated into the atmosphere is estimated to have caused them to rise by 10 degrees or so for the next thousand years making large numbers of animals unable to survive on land. Whether it was the cold, heat, acid rain or their cumulative effect on parts of the food chain that finished individual species is uncertain, but all of these effects were integrated over the years that followed the impact. Nevertheless there were survivors including mammals that could escape the holocaust at the surface. Firstly, these animals were small and could burrow away from danger; underground was the best place to be as the fire storms swept across the land. Secondly, small animals have such numbers worldwide that reduction by even as much as 90% would still leave a viable population; a fact not true for large dinosaurs which were fewer in number.

Other mass extinctions over geological time might also be impact related but there is not enough evidence yet found to conclusively ascribe another to the visit of a comet or asteroid. Results have been reported at the Permian–Triassic boundary for an impact-related extinction but there are also large craters identified that did not affect life. Further, whilst there is no doubt that Chicxulub was formed by impact at the right time to have caused severe problems, there is still not unanimous agreement that it was the single cause for the K–Pg extinction. Other culprits have been identified, such as intense volcanism occurring at the same time in what is now India (the Deccan traps), which would also eject matter into the atmosphere and limit photosynthesis. However, evidence has been found there as well of an enclosed sedimentary layer with the visible K–Pg boundary sandwiched between volcanic deposits. There are even other contemporary impact craters, such as the multi-ringed Silverpit crater found at the bottom of the North Sea, a potential British culprit; the Boytsh crater in Ukraine; and the Shiva crater (off India, although its claim is not strong). Silverpit and Boytsh both date from 65 million years ago and perhaps this indicates that a series of impacts occurred at this time. In 1994 the comet Shoemaker Levy 9 broke up under gravity into smaller parts that followed the same trajectory onto the surface of Jupiter. Their separation from each other ensured that each fragment impacted on a different point on the planet's surface as Jupiter rotated, leaving a line of 21 impact sites over 4 days. Thus if Chicxulub were only one of several impactors, the global effects on Earth would be magnified and the times required for recovery would be much greater. Finally, some favour a radiation-induced extinction by virtue of the explosion of a nearby supernova. However, such an event would pollute sediments with the plutonium isotope Pu 244, but this has not yet been found in the K–Pg layer (so well populated with iridium).

Although much has been deduced concerning events in those far-off times, there is still need to understand details of the processes that occurred. In particular, there are parts of high-pressure science (akrology) not yet developed sufficiently to answer these questions. To truly track the interplay of physical, biological and chemical effects needs cross-disciplinary effort to tie down the complex interactions between them. For instance, models can be used to estimate the mass and nature of active gases released by the impact, but their interaction with the complex structural elements of our planet, and more importantly the ecosystems that exist upon them, is a great unknown. Even descriptions of mechanical properties, such as the constitutive models for geologic materials, are in a crude state at present. Similarly the detailed cause and effect of tsunamis, the physics of wave breaking, and the scale of impacts that would be necessary to cause global effects remain hypotheses. The division of processes is confused and so often localisation phenomena are omitted as perturbations rather than the key to defining response. Yet engineering observations such as Perrone's rule recognise the importance of considering all stages of deformation in a full analysis. If truly predictive computation is to be achieved, proper formulation of complete laws for all stages need to be formulated if responses are to be predicted. Finally, the *finis extremis* for iron is probably around 100 GPa. The core of the Earth is above this pressure and assumptions about its structure based upon extrapolations from materials science in the range accessible in the lab is again first order at present. Materials in super-extreme conditions represent a new forefront for understanding over the next years.

When the event discussed is long ago and the evidence required is so difficult to acquire, it is not surprising that alternative explanations exist for the extinction events observed. One has been touched on already, and that is the Deccan traps, which are the biggest volcanic feature covering a large part of India. This massive scale of volcanism occurred between 60 and 68 Ma and undoubtedly has had an affect on populations, which declined in number over this period. It is thus plausible that rather than volcanism and impact being alternatives, the impact pushed many species that were already in decline to destruction. Interestingly some have speculated that the event is connected with the hypothetical Shiva Crater, which is dated to have been formed at the same time as Chicxulub. Further, the effects of these mechanisms leading to rapid climate change give rise to similar effects which could have occurred without external stimuli at all. Other more gradual mechanisms include epidemic disease and even egg predation of land species by the rising mammals, which then became dominant after the extinction. Other extra-terrestrial candidates might include radiation from a nearby supernova or a theory tied to the observation of the period to extinction that an as yet unidentified binary twin star to the Earth (Nemesis), brings gravitationally bound bolides towards our planet every 26 million years as the system rotates. Rather more plausible is the observation that the major extinctions are of the order of 100 Ma apart which correlates with the transit of our solar system through the galactic plane. Our Sun is around 26 000 light years from the centre of the galaxy and takes 230 ± 10 million years to orbit. Moreover, the Sun orbits in a trajectory out of the main galactic plane. Thus twice a revolution it crosses it and feels increased gravitational forces, which might conceivably perturb objects, particularly those held loosely – such as at distance in the Oort cloud, for instance. The theory then wonders if a greater density of extra-terrestrial bodies comes towards our planet every 115 million years which might correlate with observed extinctions. Just as a note, our galaxy is within the galactic plane at present!

In summary, there is evidence of evolution reaching major extinction events over the period life has existed on earth. Of the five large extinctions, that at the Cretaceous–Paleogene boundary (65.5 Ma ago) has been hypothesised to result from the impact of an extra-terrestrial bolide with an energy 5×10^{23} J. The physical state at the impact site includes pressures in the TPa range, temperatures of the order of 10 000 K and a hot plasma source at the impact site. This launched electromagnetic and mechanical waves which radiated out into the surrounding media. Shock fronts changed mineral phases and, lowering in amplitude, fractured rock at greater distance. In the sea, tsunamis were driven which intruded deep into land they encountered. Finally, shock-heated gases started wildfires around the globe. Material and gases in the stratosphere blocked sunlight from the surface of the earth for years. Then greenhouse gases trapped in the atmosphere caused raging temperatures for a millennium to come.

The fossil evidence shows life retains a similar state for long periods (stasis) followed by rapid evolution and then abrupt appearance of a great diversity of new species. Certainly periodic extra-terrestrial impact will clear the way for new evolutionary opportunities unless the victims have enough control over their fate that they might defend their niches. Nevertheless, meteorites are falling to earth repeatedly with little warning; indeed an asteroid was discovered and reported early on 6 October 2008, which entered Earth's umbra on 7 October. Most of the mass was lost in an explosion at 37 km altitude but 4 kg of debris was nevertheless still recovered from the impact site in the Nubian Desert of northern Sudan. It was estimated to be c. 100 tonnes with a kinetic energy of 6×10^{12} J, much smaller than that at Chicxulub. Nevertheless if the impact had been over a major city damage would have occurred and if there were a warning it would have been extremely short. Further study and better measurements will hopefully better warn of impacts on Earth in the future.

9.9   Final comments

This book has described deformation under loading that takes condensed matter from a level just beyond the yield surface to the point at which electronic bonding becomes different in nature and strength assumes a different meaning. The loading can be fast and span a few atomic layers or so long that components of a structure deform according to the strain applied. In all cases as pressure mounts the strength depends upon the defects in the microstructure and the material approaches its theoretical strength as the pressure passes the Weak Shock Limit. All deformation expends work and while slip generates heat faster than it can conduct away the flow will not be isothermal. Processes at each scale can be divided into localisation, flow and interaction regimes. The scale of a localisation at a higher scale is that of a flow at the scale beneath: a shear band which is a slip line at the component scale consists of homogeneous flowing grains at the mesoscale. However, scales are defined and scale boundaries represent energy barriers to the microstructures formed at the scales beneath.

Deformation mechanisms themselves exist at each scale and have different terminology but similar effect. All can be described as a suite of lower-scale mechanisms integrated with an activation threshold and a relaxation constant at the scale of interest. The coupling of the pulse applied to excite the shear and the time constant for the process was defined by a dimensionless constant, the Freya, to judge which phase of deformation the material was undergoing. If the pulse length exceeded the time constant for the process then it was complete within the impulse and a process could operate within the deformation. Equally, at higher length scales, it was shown that the wave equilibration within structures was fast compared with the times over which strain could be applied and that in this case, the deformation switched from a dynamic process to one where stress response to strain was that observed.

The interplay between impulse length and response has probed mechanisms operating across vast time- and length scales. The properties of matter at the microscale with nanosecond relaxation times and below defines the subject matter of the physical sciences, those of the mesoscale at the micron frontier, materials science, whilst those at the component and structural scales, mechanical and civil engineering. Structures within the crust define geophysics at planetary scales, whilst those beyond that again enter planetary science and cosmology. The reader will now recognise common behaviours and common responses, and that a framework exists to describe phenomena at all of these scales and at compressions up to the point at which strength changes meaning. The common area of response is an overarching framework within the field of *akrology* – the science of the mechanisms operating, and the responses resulting in condensed matter loaded to extreme, mechanical states.

The behaviours observed are, as in many walks of science, based on a few simple principles, yet the complexity that results is painted across the wonders of nature. These few truths are as much philosophical as scientific. Every process has a starting point and none is instantaneous. Further, the homogeneous continuum deforming steadily has heterogeneous, localised regions of slip at the next scale down. No material at ambient conditions is defect free and homogeneous viewed at the correct magnification. On the other hand, looking at larger volumes and longer times allows description as a homogeneous flow with integrated properties; the mountains flowed to the sea, as Deborah said. Only when the compression is so extreme that bonding is overcome beyond the *finis extremis* does homogeneity become a reality and as pressure mounts beyond this point, structure and the unique properties of the elements homogenise too.

However, the high-energy-density physics of superextreme states is a topic for another book. In the extreme states described here, the variety and complexity of nature is encompassed within the remits of akrology, where opportunities to understand and engineer matter in extremes will expand and prosper in the years ahead.

9.10   Selected reading

Materials at high strain rates

Armstrong, R. W. and Walley, S. M. (2008) High strain rate properties of metals and alloys, *Int. Mater. Rev.*, 53: 105–128.

Follansbee, P. S., Regazzoni, G. and Kocks, U.F. (1984) The transition in drag-controlled deformation in copper at high strain rates, *Inst. Phys. Conf. Ser.*, 70: 71–80.

Hoge, G. and Mukherjee, A. K. (1977) The temperature and strain rate dependence of the flow stress of tantalum, *J. Mater. Sci.*, 12: 1666–1672.

Lankford Jr., J. (2005) The role of dynamic material properties in the performance of ceramic armour, *Int. J. Appl. Ceram. Tech.*, 1(3): 205–210.

Walley, S. M., Field, J. E., Pope, P. H. and Safford, N. A. (1989) A study of the rapid deformation behaviour of a range of polymers, *Phil. Trans. R. Soc. Lond. A.*, 328: 1–33.

Chicxulub impact: the KPg extinction

Collins, G. S., Melosh, H. J. and Marcus, R. A. (2005) Earth Impact Effects Program: a web-based computer program for calculating the regional environmental consequences of a meteoroid impact on Earth, *Meteorit. Planet. Sci.*, 40: 817.

Jenniskens, P., Shaddad, M. H., Numan, D. *et al.* (2009) The impact and recovery of asteroid 2008 TC3, *Nature*, 458: 485–488.

Pierazzo, E. and Melosh, H. J. (2000) Understanding oblique impacts from experiments, observations and modelling, *Annu. Rev. Earth Planet. Sci.*, 28: 141–167.

Powell, J. L. (1998) *Night Comes to the Cretaceous: Dinosaur Extinction and the Transformation of Modern Geology*. New York: W.H. Freeman.

Reiner, M. (1964) The Deborah Number, *Phys. Today*, 17(1): 62.

A   Relevant topics from materials science

In the text a range of problems encountered by materials under extreme conditions has been described. To understand them, knowledge of the response of structure at the microscale is necessary and this has been assembled in the ambient state by materials science. Chemical reaction is possible in some substances in the condensed phase, and these are described as energetic, but in general physical deformation precedes chemistry in loaded materials. A fundamental focus for this field will be to try and understand the nature of the strength of solids. It will become clear that this is a difficult objective since complex behaviour results from the two classes of process that define strain: that in which length or volume changes with constant shape and one in which the shape changes at constant volume.

In what follows the response of these will be followed through from the microstructure at the atomic level to their form at the continuum. The various materials classes – metals, brittle materials, polymers and composites of all three – will be looked at to highlight particular features of their behaviour which go towards defining how the macroscopic boundary conditions of the loading excite response from the individual atomic architectures. The framework to describe observations is materials physics and this will be summarised below to aid the reader. It is by no means complete and much more rigorous texts exist for the student of materials science; however, it serves to allow a reader from an alternative background access to the necessary concepts to make the comments elsewhere in the text more tractable.

A.1   Structures

A.1.1   Crystalline and amorphous

There are two classes of material commonly encountered: crystalline solids where atoms are at ordered positions; and amorphous substances where their location follows a distribution of interatomic spacings and where the structure is disordered. Crystalline materials are formed by cooling from a liquid where material is dissolved in the fluid. If forming from melt, small crystals nucleate from inhomogeneities, growing until they fuse to form a polycrystalline microstructure. A non-crystalline material with no long-range order is termed amorphous or glassy. Ionically bonded substances crystallise readily and are commonly encountered in nature as are covalent crystals such as diamond and silica. Polymers are found in amorphous and partially crystalline states where not only interatomic, but also Van der Waals' bonding is important.

The theory of cohesion underlying that of strength is defined by the electrostatic attractions between atoms: the valence electrons which form bonds below the *finis extremis* in condensed matter. There are different bonding types including covalent with Sigma and Pi bonding, ionic, and metallic and weaker bonds including hydrogen and Van der Waals' forces. Van der Waals' attractions exist because of correlations in the fluctuating polarisations of nearby electron distributions which result from quantum dynamics. The result of such interactions is a necessarily anisotropic force field acting over longer distances.

Covalent bonding results from the sharing of electrons in an orbital so that they are delocalised. However, there is a strong directionality associated with the bond that means that covalent solids are strong in compression since slip is difficult but weak in tension which results in them being brittle. Whereas a covalent bond delocalises electrons between atoms, an ionic bond occurs when an atom donates an electron to its neighbour, which results in distinct electrical charge on each. Ionic solids form three-dimensional crystalline microstructures determined by the charges of the ions and their sizes. In metals, a distinct bond between atoms does not exist. Rather the electrons are delocalised throughout the entire lattice and interact with the nuclei of the atoms within. In high-pressure environments covalent may transit to metallic bonding, for instance hydrogen in the core of Jupiter is in a metallic state. Crystalline microstructures will contain defects in stacking or both crystalline and amorphous ones might possess rogue second-phase particles introduced during growth or processing. The effect of these is to weaken the structure at localised points which can initiate slip failing the material.

A.1.2   Microstructures

The assembly of atoms or repeating units onto a lattice can be treated for the simplest packing, as stacking solid spheres. Consider the first layer being closest packed with a hexagonal coordination for each sphere and call it A. This layer contains two types of triangular void, allowing spheres from the next layer to occupy different sites either B or C. The next layer of spheres can thus occupy either B or C but not both. Similarly the next layer above a B layer can be either C or A and that above a C layer either A or B. The two most common structures that occur in nature are hexagonal close-packed (HCP), which is ABABAB . . . , and the cubic close-packed structure (CCP or FCC: face-centred cubic), which has packing ABCABC . . . . This is the most efficiently packed structure possible. However, nature adopts configurations according to other physical interactions and it is a complex matter to predict structures a priori.

A.1.3   Stacking faults

The FCC structure allows layers in the perfect stacking ABCABC . . . as seen above. A stacking fault is a one or two layer interruption to the stacking sequence, for example if

Table A.1 Packing density of various crystal systems

	Packing density %
Face-centred cubic (FCC)	74
Hexagonal close-packed (HCP)	74
Body-centred cubic (BCC)	68

the sequence ABC*AB*ABCAB . . . were found in an FCC structure then there would be a stacking fault at the position indicated. This is called an intrinsic stacking fault as a layer is missed at this position. Conversely, an extrinsic fault results from addition of a layer thus: ABCAB*A*CAB . . . . Stacking faults are important to dislocation dynamics in FCC metals and can be quantitatively assigned an energy barrier (called stacking fault energy, SFE), which modifies the ability of a dislocation in a crystal to glide onto an intersecting slip plane. The stacking fault results from the drive in FCC metals for dislocations to dissociate into separated partial dislocations which are themselves separated by a stacking fault. The separation between the partials, d, is determined by the lattice and the SFE of the fault (Σ) and can be shown to be

$$d = \frac{\mu a^2}{24\pi \Sigma},$$
(A.1)

where μ is the shear modulus and a is the lattice constant.

In shock loading dislocations are generated in the first few hundred picoseconds and these split in FCC metals into partials as the shock develops. The applied shear stress drives these through the lattice until they interact with each other, some boundary or an interstitial of some type. In order to travel further the fault must overcome the energy barrier so imposed and that can only occur if the partial dislocations recombine to overcome it. Recombination is easier if the partials are close and since this is inversely proportional to the SFE requires the latter to be high. Low-SFE materials will find it difficult to deform by slip and may adopt other mechanisms such as twinning. As an indication of these quantities, the stacking fault energy for silver is low (20 mJ m^{-2}), for copper it is intermediate (around 75 mJ m^{-2}), whilst those for aluminium and nickel are both higher still (around 200 mJ m^{-2}).

For a material with a lower number of valence electrons, the body-centred cubic (BCC) structure is frequently adopted, particularly at high temperatures. In the BCC packing, the material has no close-packed planes and therefore it is difficult for this structure to exhibit stacking faults (see Table A.1). Stacking faults have been considered, but not seen since a small shear to the lattice re-establishes the . . . ABAB . . . sequence.

The elements in the periodic table all display characteristic structures under ambient conditions and these have been shown across the periodic table in Figure 5.1. The packing in all of the materials considered depends critically on the geometry of the unit cell. One of the key factors that determines deformation in a close-packed lattice is the ratio of the distance parameters that define its form. In HCP crystals the relation that determines the distance between atoms (a) to the distance between planes (c) gives a measure of the

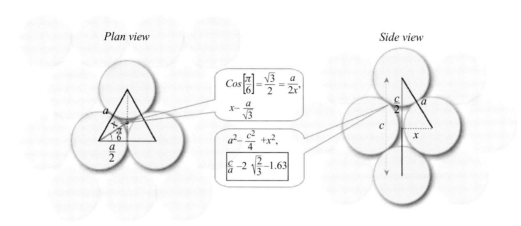

Figure A.1 Ideal HCP structure constructed from stacked spheres.

ideality of the packing for a particular material. This has an effect upon the accessible slip systems and thus a strong influence upon plastic deformation mechanisms.

The ideal axial ratio (c/a) for a HCP crystal structure can be calculated by considering non-interacting hard spheres packed in an HCP lattice. Figure A.1 shows a plan view of a plane within the structure spaced by interatomic distance a and with three atoms highlighted. To the right-hand side is a side-on view of the same lattice where further atoms are stacked in the plane above and below the interstice marked. The perpendiculars between the centres, defining the positions of the first plane of atoms to that through the second, fix a distance x which is related to the interatomic spacing as shown. Further, the distance, c, between similarly stacked planes (defining the cell), is related to that spacing so that a relation between c and a can be derived as shown, which in this ideal case is 1.63.

There are many materials that have the HCP crystal system, but the axial ratio is rarely ideal. In some HCP metals the ratio of these lattice parameters (c/a) is higher than ideal, such as cadmium which has an axial ratio of $c/a = 1.89$ or zinc where $c/a = 1.86$. In others it is less so, such as beryllium ($c/a = 1.57$) or titanium ($c/a = 1.59$). This non-ideal structure has implications for the ease of slip and group mechnical behaviour for pure HCP metals.

A.2   Strength of a crystalline material

It is frequently the goal of materials science to engineer substances in which strength is maximised so as to minimise the amount of material required and to produce longer-lasting components. It is possible to estimate the theoretical strength of a metal by assuming a particular bonding between the atoms and calculating the force required to break and then reform the bonds.

To illustrate the principles of such a calculation, let an approximate interatomic force law follow a sine function of form

$$\sigma = \sigma_0 \sin\left[\frac{2\pi x}{a}\right],$$

where σ_0 is the tensile strength of the solid. Hooke's law defines the slope of the σ–ε curve when ε is small to be E. It must translate a lattice spacing to break the bond and thus for small ε

$$\sigma = 2\pi\varepsilon\sigma_0 \text{ so that } \sigma_0 = \frac{E}{2\pi}. \qquad (A.2)$$

This value gives c. 20 GPa for copper, which is hundreds of times higher than a typical tensile strength of 50 MPa. More accurate force laws gives 0.05–$0.1E$ as a value but this still gives a value many times the actual strength. In the case of dislocation motion, a shear stress must act to move atoms on one plane an interatomic distance over those in the layer beneath. Again the moving atom sees a varying force that may be approximated as sinusoidal as it translates across the surface. Hooke's law connects the shear stress to the shear strain via the shear modulus, μ and an analogous argument for small ε gives the maximum shear strength τ_0 given by

$$\tau_0 = \frac{\mu}{2\pi}. \qquad (A.3)$$

Again substituting for copper gives a value of around 7 GPa for the force required to nucleate a dislocation. The reason that these calculations give bounding strengths that are so high is that defects dominate the deformation processes in these materials. Only when a material is formed in whiskers can strengths approach these theoretical limits.

What this indicates is that creating dislocations by homogeneous nucleation requires high stresses close to theoretical maxima required to break bonds and that whereas this may be possible in the strong shear region behind a propagating shock, it is energetically much more favourable to start slip at surfaces or interfaces in regions where steps or second-phase particles can concentrate stresses. Thus if the microstructure can take this easy route to accommodate the applied stress, it will do so, especially in structures where slip systems are limited.

A.3 Crystalline defects

Any ideal crystal consists of identical unit cells arranged on an ordered lattice. Further, it may form from a single element but could contain more in which case it will be classed a solid solution or an alloy. Solid solutions may be *substitutional*, where atoms sit on crystal sites normally occupied by the major component, or *interstitial* where they sit in the interstices. Nevertheless, a crystal consists of atoms or molecules positioned regularly in three dimensions. However, none is perfect since they are limited in size and always contain some irregularities in stacking called defects. These may involve a single atom added or removed from the array, a plane added or reoriented within the crystal, or a boundary within the structure. These are known respectively as point, line or planar defects. When material is missing, volume defects exist which represent key nucleation sites in dynamic loading.

A.3.1   Point defects

Point defects result from some change at a single atomic site within the crystalline array. These can take one of three forms: vacancies, interstitials or impurities. They result either during the growth of the crystal, or are imposed after by the effects of external stimuli such as radiation. An interstitial on the other hand is an atom interposed within the crystal onto a non-lattice site. Clearly it must be weakly bonded or have large atoms to accommodate other particles so that the host is generally covalently bonded or of high atomic number. Whereas interstitials are imposed within the lattice, an impurity is an atom that takes one of the regular occupants of a site and replaces it with another that is not normally in that position.

The concentration of point defects in a crystal is typically less than 1% of the atomic sites; however, extremely pure materials can now be grown. Nevertheless, despite their small volume fraction, they occupy a key role in controlling mechanical and electromagnetic properties within the solid.

A.3.2   Line defects

Dislocations are line discontinuities in a regular crystal structure. They are a type of topological defect, mathematically described as a soliton. Dislocations behave as stable particles in that they can be moved but maintain their identity as they do so. When two dislocations of opposite orientation are brought together, they may cancel each other out and this process is dubbed annihilation.

There are two basic types, edge and screw dislocations. An edge dislocation results from the insertion (or removal) of an extra half plane of atoms into the lattice with regions surrounding the linear apex region made of essentially perfect crystal. The insertion of the extra plane results in protrusions at the edge faces which are known as jogs. A screw dislocation changes the character of the atomic planes since they no longer exist separately from each other. They form a single surface, like a screw thread, a helix that translates from one end of the crystal to the other unzipping the lattice as it moves. Additionally it is possible for edge and screw dislocations to exist together in combinations of dislocation structures within the lattice.

Thus edge and screw dislocations form linear defect features that travel through the structure under an applied shear stress. The Burgers vector, denoted by b, represents the magnitude and direction of the lattice distortion of dislocation within a crystal. Preparation of the material, its environment and its history all determine the density of dislocations, nevertheless a representative indication of the distance between them in an annealed material is 500 nm. Of course, the planes on which the dislocations are to slip must be aligned to the direction of the impact loading. Clearly not all will be so and there will be two classes of defect in the flow. A *sessile* dislocation will be one for which the Burgers vector is not in the direction of slip so motion will not occur when impact occurs. On the other hand, *glissile* dislocations will move since their Burgers vector is in the direction of slip.

A.3.3   Planar defects

In addition to line defects, a discontinuity in a perfect crystal can be formed across a plane. The most easily identified planar defect is a grain boundary since grains contain defined crystal orientations on the micron to hundreds of microns length scale. Atoms within a grain boundary are generally in a partially ordered state, except in the case of twins when a perfect discontinuity is formed.

A.3.4   Twin boundaries

Twinning accommodates plasticity by moving atoms small distances in concert to allow microstructure to deliver macroscopic strain. The boundary separates crystal regions that are mirror images of each other and thus will be a single plane of atoms. Unlike a grain boundary, there is no region of disorder and the boundary atoms can be viewed as belonging to the crystal structures of either twin. Like a martensitic phase transformation these atomic shuffles are fast and accommodate strain in suitably orientated grains.

A.3.5   Microcracks

When small bodies impact or abrade surfaces a microcrack may form. These are of the order of 10 μm in size and tend to form at surfaces. They are thus found at grain boundaries and other regions of disorder. One of their major effects is to provide nucleation sites for fracture to fail the solid under high load.

A.3.6   Volume defects

Finally, volume (or bulk) defects complete a family increasing in dimension from point to line and then planar defects. Volume defects are three-dimensional voids requiring energy to form and possessing surface energy once created. Again a major effect of their presence is to provide sites from which failure can proceed through a loaded target. These defects are usually introduced during processing and fabrication operations. Processing may introduce pores, cracks or foreign particles and such inhomogeneities in the solid must inevitably alter its mechanical properties, weakening or strengthening it by nucleating tensile failure or hindering the movement of dislocations and thus increasing the strength.

Neither glasses nor polymers are homogeneous. Microstructure is not identical from region to region in a glass since filler metal concentrations and inclusions are in different density as one passes through the structure. Further, the bulk contains microcracks and bubbles entrained during the melt- and flow-casting processes in manufacture. Polymers are similarly complex. Thermoplastics typically consist of crystalline and amorphous regions. The crystalline regions (spherulites) can contain some of the defects described above. Polymer composites of this sort are called plastics.

The conclusion to be drawn from this section is that nature is far from homogeneous. All materials are defective to some extent in the ambient state and the continuum response observed when they are loaded is a consequence of these imperfections and their interaction over the time for which the load is present.

A.4   Plasticity

For small amplitude loads, the stress applied is proportional to the strain observed (Hooke's law) and the deformation that occurs is recovered when the loading is reversed. On the other hand, the processes involved in a shock front compress a material such that it has deformed to a state where irreversible flow has occurred, taking it to a strain that after unloading will result in a residual compression of the target. The mechanisms leading to this state are generally described as plasticity, a term that comes originally from metals, although it is also applied to the limits of elastic behaviour in other classes of material. To respond perfectly plastically implies that once the elastic limit has been reached, deformation can occur without any further increase in loading stress as the drive continues. In real materials, however, such conditions are rarely achieved since hardening, softening or viscous processes occur necessitating higher stresses to overcome the microstructural barriers that oppose deformation in the material.

Plasticity in metals is usually a consequence of defects driving slip within the microstructure. Thus dislocation nucleation, generation and propagation are at the heart of the processes that determine the ability of a metal to respond to rapid jumps in stress. The initial defect population within a material and its form is thus a key determinant of its behaviour under load. The dislocation density is a measure of how many dislocations are present in a quantity of a material. Since a dislocation is a line defect, this is defined as the total length of dislocation per unit volume (units m/m^3) or the number of dislocation lines intersecting a unit area. An annealed material has a dislocation density of approximately 10^{12} to 10^{14} m^{-2}, while cold-worked material has a density of approximately 10^{20} m^{-2}. This means that on average the separation between the dislocations in a real material is between 100 nm and 1 μm for the annealed material and 100 pm for a cold-worked one. Of course, upon impact the applied stress can only drive the glissile dislocations so that at any one time only a portion of the defect population will be in motion depending upon their Burgers vectors and whether the activation barrier (the Peierls stress) has been overcome, allowing slip to occur. However, as the stress amplitude increases, more and more are activated and flow.

In the weak shock regime, only the mobile portion of the dislocation density is free to move in the lattice, as some fraction is pinned at sites within the material which apply resistance to glide. Further, the dislocation motion is localised at kinks rather than resulting from some concerted motion of a linear defect which results in plastic flow at the continuum resulting from local strains at dislocation lines. All of these statements indicate that plasticity at the microscale is a heterogeneous process manifested in different manners at different length scales.

To determine the transport of strain through a target driven by the action of a population of mobile dislocations requires consideration of the motion of one. If a dislocation glides on a plane in a crystal lattice with area A, of width l, length of sample h, and volume $V = Ah$, then as it sweeps across the full area, the total shear strain is

$$\gamma = \frac{b}{h}.$$ (A.4)

Now consider incremental slip by $\Delta\gamma$ which results in a slip of ΔA. Then

$$\frac{\Delta\gamma}{\gamma} = \frac{\Delta A}{A} \Rightarrow \Delta\gamma = \frac{b}{h}\frac{\Delta A}{A}.$$ (A.5)

If the dislocation moves at average speed, v, then in time Δt it sweeps across an area $\Delta A = vl\Delta t$, and therefore

$$\Delta\gamma = \frac{blv\Delta t}{V},$$ (A.6)

which summing over n dislocations gives a total shear displacement of

$$\Delta\gamma = \frac{nblv\Delta t}{V} = \rho vb\Delta t,$$ (A.7)

where $\rho = \frac{nl}{V}$. Thus, the shear strain rate is

$$\dot{\gamma} = \rho vb,$$ (A.8)

and this may be related to the longitudinal strain rate through a factor M, the Taylor factor, which depends upon the direction of loading and the orientation of active slip planes for single crystal. Then

$$\dot{\varepsilon} = \frac{1}{M}\rho vb.$$ (A.9)

For a polycrystal $M = 3.1$ for FCC and 2.75 for BCC metals.

This relation is known as the Orowan equation and illustrates the relation of the various lattice properties to the transport properties. Applying it is more difficult since the density of the mobile defect population must be sampled over an area where it is constant and over a suitable time interval. Nevertheless, it puts bounds on material displacements in short time intervals characteristic of some of the processes discussed in transient loading.

Dislocation motion meets resistance from other lattice defects such as grain or twin boundaries, stacking faults, kink bands, inclusions and precipitates, free surfaces, point defects, ferromagnetic domain walls and so on. Additionally the lattice itself provides an equivalent drag force that resists dislocation motion due to mechanisms that transport momentum across the dislocation path. These processes in metals and ionic crystals may be due to phonons and electrons. In covalent crystals additional features of the bonding make dislocation mobility more difficult. These are due to restrictions in the motion of kinks because of the directional bonds found in these crystals that result in materials such as SiC being hard even when their purity is high.

A.4.1   Peierls stress

If a dislocation is to move, driven by the shear stress provided by the impulse, it must continually slip the extra planes of atoms introduced through the crystal within which it travels. This means that the atoms must move across a periodic potential defined by the stacking of the crystal that varies as they move from regions of greater electronic interaction to those where the forces are less. This periodic potential maps to a stress barrier that must be overcome if slip is to occur. Exceeding these intrinsic barriers in perfect crystals is only the first stage since dislocation mobility also depends on pinning mechanisms at imperfections within the crystal structure. Intrinsic barriers (called Peierls–Nabarro barriers) are periodic as befits the translation symmetry of the crystal. The Peierls stress, τ_{PN}, is the minimum stress required to move a dislocation one lattice site giving a maximum to the resistance force $\tau_{PN}b$ exerted by the crystal lattice, per unit length of the dislocation line (where b is the Burgers vector). An expression for the shear stress is given by

$$\tau_{PN} = \frac{2\mu}{1-\nu}e^{\frac{-2\pi\alpha}{\beta}}, \tag{A.10}$$

where μ is the shear modulus, ν is Poisson's ratio, $\alpha = \dfrac{\lambda}{1-\nu}$ (the dislocation width), λ is the interplanar spacing and β is the interatomic spacing. Since dislocation mobility is affected by the Peierls–Nabarro barriers, the mechanical properties of a material depend on such fundamental characteristics as the nature of the crystal structure and the character of chemical bonds between the atoms. The value of τ_{PN} varies in the range 10^{-5}–10^{-4} μ for FCC and HCP materials, 10^{-4}–10^{-3} μ for BCC metals, and 10^{-3}–10^{-1} μ for ceramics and semiconductors. The original calculations and analytical formulae found in texts simplify the form and structure of the dislocation. However, such calculations are routinely done using more advanced potentials giving accurately computed microstructures. Nevertheless, it is comforting that the original analytical treatments give good agreement with the potentials that exist today. Of course the number of slip systems available to the crystal will affect the response observed. The requirement that at least five independent systems are needed for plastic deformation is known as the von Mises Criterion. If less than five independent slip systems are available, the ductility is predicted to be low in the material since each grain will not be able to deform with the body and gaps will open, i.e. it will crack although even if a material has five or more independent systems, it may still be brittle (e.g. iridium).

A.4.2   Dislocation propagation velocity

Dislocation motion is not only governed by the driving stress levels but is also perturbed by electronic and thermo-mechanical effects that can be ascribed to phonon interactions. The net result of these is to slow the propagation of a dislocation by providing a drag upon its motion. The bonding in the solid has a strong influence upon the resulting behaviour. Phonon drag is thus most prevalent in metals and ionic solids.

As dislocations are driven faster by higher stress levels, different mechanisms control their velocity. For driving shear lower than the Peierls stress, thermal activation must control the velocity of the dislocation since this is the only means of surmounting the barrier. Above this level, velocity increases at a rate controlled by drag. At the highest levels the velocity becomes limited since the shear wave speed must limit the dislocation velocity.

A.4.3 Deformation twinning

The process of loading a solid gives rise to two deformation mechanisms which allow plasticity to occur to accommodate the applied strain. One is slip described above, but a second mechanism is the rearrangement of the lattice to form mechanical (or deformation) twins. A twin is a region of a crystal in which shear forms a reorientation of the lattice across a boundary into a mirror image across a twin plane. The need for a material to have five independent systems to deform by slip has been recognised for some time. However, in microstructures where slip is difficult, it is possible to accommodate plastic strain by twinning as well. Thus metals show a range of behaviours that reflect the number of available slip systems and the ease of exciting mechanical twinning modes. Those that possess five or more include ductile FCC and BCC metals. Some possess nearly five (e.g. HCP magnesium) but as the ability to slip is reduced, twinning modes offer an easier independent route to accommodate a limited strain. The geometry of the HCP cell determines the limitations on slip, so that the c/a ratio (described earlier) has a strong influence on the deformation mode. Thus in HCP metals the direction of the c-axis in the unit cell makes a crystal's deformation highly anisotropic, and so mechanical processing such as rolling or extrusion will preferentially align along particular directions. Thus the resulting response to an impulsive load will be different down directions that reflect the processing route of the metal.

In the discussion of the response of polycrystalline metals in Chapter 5 it was shown that the orientation of individual grains has an effect upon the observed response. This is particularly true of crystals with reduced symmetry which result in a limited number of slip systems. A term used to describe such a property at the mesoscale is preferred orientation or *texture* which describes the development of order in the grain orientations in a polycrystalline aggregate. Quasi-static and dynamic processes are of course dependent on the times available for rearrangement to occur. Quasi-statically, uniaxial compression produces preferred orientations that differ from those produced by simple or constrained tension. In compression, when deformation occurs by single slip, the active slip plane normally rotates toward the compression axis, whereas in tension it is the active slip direction that tends to align with the axis of the applied stress. In the compression of a polycrystalline aggregate, where multiple slip occurs, the direction of lattice rotation is opposite to that in tension. Thus, the stable end-orientations in tension and in compression are different. Texture then refers to the orientation, shape and size of the crystals in the polycrystalline material and it is its developing form that describes the compression processes or the effects of release after at the mesoscale. Thus it is the texture of microstructure that will have a key influence on the measured states achieved when pulses exceed c. 100 ns.

Both slip and twinning deform the lattice by shear on specific crystallographic planes and along defined directions. However, there are important differences between the two. In particular the magnitude of shear displacement is variable in the case of slip and fixed in the case of twinning. Further, the lattice rotation in slip is gradual whilst the kinetics of twinning is fast. Indeed the process occurs at speeds that give, for example, an audible tin cry in that metal when it is plastically deformed. Finally, the parent grain will spawn new subgrains with twin boundaries within each crystal. The shear strain resulting from deformation of each unit cell is fixed and may be calculated. It is $\dfrac{1}{\sqrt{2}}$ for FCC and BCC, and depends on the c/a ratio for HCP materials.

When a dynamic load is placed on a polycrystalline solid, it applies not just as a magnitude but a direction of loading on grains within the material, and so their alignment to the load is critical to the mechanisms activated. For slip, the critically resolved shear stress of that load down particular planes determines whether the dislocation can travel. Twinning requires some equivalent shear component down the possible slip planes if twins are to be formed. Thus whilst there must be some activation barrier overcome to reorient crystalline planes, in a microstructure that contains a distribution of crystals of different orientation, size and shape there is no specific stress barrier to overcome. Further, since twinning can by definition only supply a fixed strain to the deformation in a specified direction, such a varied response through such a heterogeneous assemblage of crystals can result in a maximum strain of c. 8% through twinning dependent upon the microstructure. Finally, twinning is a uniaxial deformation mechanism and in HCP metals, twinning modes are distinguished by their ability to produce either tensile or compressive strain along the c axis.

Deformation thus proceeds with twinning activated at the elastic limit until the microstructure is reoriented such that grains are suitably aligned with the applied load. However, once twinning has been exhausted, further plastic deformation must be achieved by slip in the daughter material that remains.

A.4.4   The Hall–Petch relation

The discussion above has concentrated principally upon the response of single crystals with perfect stacking. However, it has been mentioned that defects and their generation and storage are at the heart of understanding the effects of loading upon response. One of the classes of defects mentioned were grain boundaries, which provide an energy barrier across which dislocations cannot travel since the slip planes in the new grain are generally in a different direction to those in the parent and the boundary itself is disordered with respect to the parent grain. Also, impurities tend to segregate at boundaries, altering mechanical properties and thus the response of the polycrystal itself.

Both Hall and Petch in independent studies concluded that the yield strength σ_y could be related to the grain size through the following relation

$$\sigma_y = \sigma_0 + \frac{K}{\sqrt{d}}, \tag{A.11}$$

where K is a constant and d is the grain size. This was discussed in relation to dynamic pulses in Section 9.7.

As grain size in a target becomes smaller, the number and area of boundaries increases and impeding dislocation movement increases the threshold for the onset of plasticity and thus the yield strength of the material. However, experimentally the effect is active in the range from 1 μm grain size and larger. Here the yield strength decreases as the grain size increases. Below a micron, the effects of length and timescale, which will dominate the behaviour discussed below, result in a decreasing yield strength as grains approach 100 nm or less.

A.5 Brittle behaviour: crack generation and propagation

The nature of the bonding in solids determines the behaviour in compression in the manner described above. The limited number of slip systems that give rise to strength in covalent solids for instance produces brittle failure when the material is loaded in tension. Amorphous solids have no plasticity and so here only brittle failure is observed. It was Griffith (1921) who resolved the discrepancy between the theoretical tensile strength of brittle solids and their observed values. He determined that the cohesive strength was only reached at the tip of a sharp crack. With an analysis based on energy balance for crack propagation he was able to show that brittle and ductile yield had different criteria. Brittle materials show fracture by cleavage down weak crystallographic planes. Materials with an amorphous structure fail by fracture, with cracks propagating normal to the applied tension. Ductile fracture occurs in a material where plastic deformation may occur ahead of the crack tip.

In pure brittle materials Griffith realised that only the local stress at the crack tip need exceed the theoretical strength of the solid. Fracture occurs when more elastic energy is released by crack extension than is needed to create new surfaces. Then the required stress, σ_0, will be given by

$$\sigma_0 = \sqrt{\frac{2\gamma E}{c}}, \tag{A.12}$$

where γ is the fracture surface energy, E is Young's modulus and c is the length of the crack. Failure criteria can be formulated in this manner for both ductile and brittle yield. More recent formulations of purely brittle yield under compression account for cracks closing under compression and transmitted normal and tangential stresses. This alters failure conditions and modifies the criterion which considers friction on crack faces. Thus in compression one adopts a Tresca/von Mises yield criterion for ductile materials and a modified Griffith criterion in the brittle case. Materials that are typically considered in this category are hard ceramics. One indicator of this property is given by the ratio of the hardness to the Young's modulus, H/E, which is $c.$ 10^{-3} for metals, 10^{-2} for alloys, but 10^{-1} for ceramics showing the great strength of these materials. They show high lattice resistance to dislocation motion and this leads to little plasticity at the crack tip which is displayed as brittleness. The fracture toughness of ceramics is thus low, typically an order of magnitude less than metals. Thus polycrystalline ceramics are best thought of as composites. They are real three-phase structures comprising a limited slip (hard) phase, a brittle glassy binder and typically some porosity. The different phases exhibit different behaviours with plastic flow possible in the crystalline filler and brittle fracture in the intergranular material. Such materials are strong in compression (particularly if confined) but weak in tension.

A.6   Summary

There are several mechanisms and microstructural features that determine the response of condensed-phase inert materials to load. These are:

(i) elastic behaviour of the ideal crystal;
(ii) defect distribution in the undeformed target;
(iii) mechanisms for inelastic flow and strain hardening/softening;
(iv) mechanisms for structural development under dynamic shear;
(v) mechanisms for defect growth under dynamic tension.

Armed with this suite of known kinetics and deformation modes, one can begin to construct valid constitutive models for the yield behaviour of materials under shock.

B   Glossary

Activation energy The minimum amount of energy which must be supplied to a chemical system to initiate a chemical reaction.

Adiabatic A process for which there is no heat transfer between a system and its surroundings. An adiabatic process that is reversible is isentropic (of constant entropy).

Akrology The science of the mechanisms operating, and the responses resulting in condensed matter loaded to extreme, mechanical states. From Greek, *akros*, 'extreme, highest, topmost.'

Arrhenius equation A simple formula for the temperature dependence of the rate constant (k) of a process or for the rate of a chemical reaction. It embodies a threshold A and an activation energy E for the process: $k = Ae^{-E/RT}$.

Asteroid Naturally formed solid bodies that orbit the sun, have no atmosphere and no signs of gas or dust coming from them. Most are found in orbit between the orbits of Mars and Jupiter. Density c. 2.5–3.0 g cm^{-3}; potential impact velocity on Earth 10–20 km s^{-1}.

Binder A wax, resin or plastic used to aid consolidation of a powered energetic into a composition.

Bolide A catch-all term to describe an asteroid or comet that might impact on another body.

Booster Any component of an explosive train which is interposed between the initiator and the main charge.

Brisance The shattering effect exhibited by a detonating explosive.

Combustion An exothermic oxidation reaction producing flame, sparks or smoke. The oxidant may be part of the material as in a propellant, or oxygen from the atmosphere or other source.

Comet Small bodies of rock, iron and frozen water and gases that orbit the sun in elliptical orbits. As they get close to the sun the gas vaporizes leaving a tail of dust and debris. Density 0.5–1.5 g cm^{-3}; potential impact velocity on Earth 30–70 km s^{-1}.

Comminution The process by which solids are fragmented and free particles (comminutiae) are reduced in size, by crushing, grinding and other processes.

Conservation equations Expressions that equate the mass, momentum and energy across a steady wave or shock discontinuity (Eqs. (2.22)–(2.24)). Also known as the jump conditions or the Rankine–Hugoniot relations.

Constitutive relation An equation that approximates the response of a material to external mechanical stimuli. This equation is a description of the material and distinguishes one material from another. It is normally an engineering expression for the strength and is rate-dependent. It is combined with the jump conditions to yield the Hugoniot curve which is also material-dependent.

Contact discontinuity A spatial discontinuity in one of the dependent variables other than normal stress (or pressure) and particle velocity. Examples such as density, specific internal energy, or temperature are possible. The contact discontinuity may arise because material on either side of it has experienced a different loading history.

Cook-off The ignition of an enclosed explosive by the conduction of heat from an external source or internal reaction.

Deflagration A rapid burning in which convection plays an important role.

Equation of state A (P, V, T) surface that describes the hydrostatic compression states of a given material. It defines a surface in thermodynamic space on which all equilibrium states lie. In the shock wave loading of inert materials, the initial and final states are frequently assumed to lie on the equation of state surface, and this equation can be combined with the jump conditions to define the Hugoniot curve.

Eulerian coordinates The coordinate system in which the spatial position (x) and time (t) are the independent variables. The dependent variables are expressed as functions of x as material moves through space. Also known as laboratory coordinates when the reference frame is that of an observer.

Explosiveness The rate at which a particular explosive when exposed to a given stimulus gives up its energy and/or the degree to which it does so.

Failure radius/diameter The radius/diameter of a cylindrical charge below which it is impossible to propagate a steady detonation.

Finis extremis The pressure at which the physics binding materials to one another changes since the energy density applied in the pulse becomes of the order of that of a valence electron. Core electronic states are perturbed and the nature of strength changes from the lower pressure regime. It is at a few megabars in most materials and heralds the boundary with the warm dense matter (WDM) regime. It represents an upper limit to the regime in which solid mechanics may be applied.

Gruneisen constant (gamma) The ratio of the logarithmic derivatives of the Debye temperature to that of the volume. Used to calculate thermal changes through mechanical work done on a solid.

Hot spots Mesoscale regions of high temperature where mechanical work is localised during deformation.

Hugoniot curve A curve representing all possible final states that can be attained by a single shock wave passing into a given initial state. It may be expressed in terms of any two of the five variables: shock velocity, particle velocity, density (or specific volume), normal stress (or pressure), and specific internal energy. This curve is not the loading path in thermodynamic space but the locus of end states. In energetic materials the Hugoniot will evolve and different curves exist for each degree of reaction.

Hugoniot elastic limit (HEL) The yield stress of the material under shock loading in conditions of one-dimensional strain. The elastic precursor wave is not steady until twinning, phase transformation, fracture or slip processes are complete and this may take time to establish according to microstructure.

Impedance (shock) Defined as $Z = \rho_0 U_s$. Describes the ability of material to generate pressure under given loading conditions; generally a function of pressure.

Impedance (acoustic or elastic) Defined as $Z = \rho_0 c_L$. The shock impedance in the limit of an infinitesimal disturbance; independent of pressure.

Isotropic uniformity in all orientations; derived from the Greek *iso* (equal) and *tropos* (direction). *Anisotropy* describes materials where properties vary with direction.

Isentropic A reversible adiabatic process, in which there is no change in the entropy of the system.

Jump conditions Expressions for conservation of mass, momentum, and energy across a steady wave or shock discontinuity (Eqs. (2.22)–(2.24)). Also known as the conservation equations or the Rankine-Hugoniot relations.

Lagrangian coordinates The coordinate system in which the material position (h) and time (t) are the independent variables. The dependent variables are described as functions of a particle position within the material which had coordinate $x = h$ at time $t = 0$. Also known as material coordinates.

Microstructural unit (MSU) A volume element that contains of the order of a thousand defects within a compressed solid that shrinks in dimension with defect generation until vacancies are removed at the Weak Shock Limit (WSL).

Monroe effect A local concentration of shock energy which occurs when a wave emerges from a reentrant shape.

Particle velocity The velocity associated with a point attached to the material as it flows through space.

Peierls stress (the Peierls-Nabarro stress) The force needed to move a dislocation within a plane of atoms in the unit cell.

Precursor decay (elastic precursor decay) The decay of the Hugoniot elastic time in a solid with distance from the impact face. It is a phenomenon that results from unsteady flow in regions where the loading pulse rises more quickly than the deformation mechanisms can accommodate. The run distance represents the equilibration of the stress state behind the shock as the inelastic processes operate to accommodate the applied strain.

Rayleigh line A chord that connects the initial state of a material on its Hugoniot curve to the final state on the curve; most frequently drawn in the P–V plane.

Riemann invariant A constant which is independent of position on a release wave that propagates into a uniform state. A mathematical tool to aid in solution of the wave equations.

Release (or rarefaction) wave A wave that reduces the normal stress (or pressure) inside a material as it propagates; the mechanism by which a material returns to ambient pressure after being shocked (the state behind a position within the wave is at lower stress than the state in front of it). Also known as an unloading, expansion, release, relief or decompression wave.

Sensitiveness A measure of the relative ease with which an explosive may be ignited or initiated by a particular stimulus.

Shock velocity The velocity of the shock front as it passes through the material. In the limit it approaches the bulk sound speed of the substance.

Stacking fault energy (SFE) A stacking fault is an interruption in a crystalline stacking sequence which carries with it a fault energy (the SFE) which also determines its equilibrium width. High SFE reflects an environment in which dislocation glide is favoured and leads to rapid strengthening and equilibration behind the shock front. When SFE is low, materials will find it difficult to deform by slip and may adopt other mechanisms such as twinning. SFE is a good indicator of deformation behaviour in FCC materials; there are no stacking faults possible in BCC structures.

Stiochiometric mix A mixture of chemical reactants in which the quantities of each componenet are such that they are balanced and all materials react.

Taylor wave The pressure-release wave following a planar, steady detonation front.

Thermal explosion Explosion resulting from exothermic reaction in an explosive charge in a region where heat is liberated more rapidly that it can be conducted away.

Weak Shock Limit (WSL) The stress at which all vacancies are compressed and the pressure overcomes the theoretical strength of the solid. Overdrive of the plastic wave occurs taking material into the strong shock regime.

von Neumann spike The leading (high-pressure) point at the head of a detonation front before any products have been generated.

C   Elastic moduli in solid mechanics

For isotropic materials, only two independent elastic constants are needed to describe the stress–strain behaviour. The relationships amongst the five elastic constants are shown in the table below. In the table, E = Young's modulus, ν = Poisson's ratio, G or μ = shear modulus, K = bulk modulus and λ = first Lamé constant.

	E	ν	G	K	λ
E, ν	–	–	$\dfrac{E}{2(1+\nu)}$	$\dfrac{E}{3(1-2\nu)}$	$\dfrac{E\nu}{(1+\nu)(1-2\nu)}$
E, G	–	$\dfrac{E-2G}{2G}$	–	$\dfrac{EG}{3(3G-E)}$	$\dfrac{G(E-2G)}{3G-E}$
E, K	–	$\dfrac{3K-E}{6K}$	$\dfrac{3KE}{9K-E}$	–	$\dfrac{3K(3K-E)}{9K-E}$
E, λ^{*}	–	$\dfrac{2\lambda}{E+\lambda+R}$	$\dfrac{E-3\lambda+R}{4}$	$\dfrac{E+3\lambda+R}{6}$	–
ν, G	$2G(1-2\nu)$	–	–	$\dfrac{2G(1+\nu)}{3(1-2\nu)}$	$\dfrac{2G\nu}{1-2\nu}$
ν, K	$3K(1-2\nu)$	–	$\dfrac{3K(1-2\nu)}{2(1+\nu)}$	–	$\dfrac{3K\nu}{1+\nu}$
ν, λ	$\dfrac{\lambda(1+\nu)(1-2\nu)}{\nu}$	–	$\dfrac{\lambda(1-2\nu)}{2\nu}$	$\dfrac{\lambda(1+\nu)}{2\nu}$	–
G, K	$\dfrac{9KG}{3K+G}$	$\dfrac{3K-2G}{6K+2G}$	–	–	$\dfrac{3K-2G}{3}$
G, λ	$\dfrac{G(3\lambda+2G)}{\lambda+G}$	$\dfrac{\lambda}{2(\lambda+G)}$	–	$\dfrac{3\lambda+2G}{3}$	–
K, λ	$\dfrac{9K(K-\lambda)}{3K-\lambda}$	$\dfrac{\lambda}{3K-\lambda}$	$\dfrac{3}{2}(K-\lambda)$	–	–

* The factor $R = \sqrt{E^2 + 9\lambda^2 + 2E\lambda}$.

The second law of thermodynamics requires that E, G and $K > 0$ and that $1 < \nu < {}^1\!/_2$.

Waves travel in a medium with modulus K and with a wave speed, c_0, at

$$c_0 = \sqrt{\frac{K}{\rho}}, \tag{C.1}$$

where ρ is the density. In a medium contained within a semi-infinite half-space, the longitudinal wave speed, c_L, reaches a higher value than this

$$c_L = \sqrt{\frac{K + \frac{4}{3}\mu}{\rho}}, \tag{C.2}$$

where μ is the shear modulus. Finally, shear waves travel with a speed c_s, depending on the shear modulus and density thus

$$c_S = \sqrt{\frac{\mu}{\rho}}. \tag{C.3}$$

Clearly these quantities can be combined so that

$$c_L^2 = c_0^2 + \frac{4}{3}c_S^2. \tag{C.4}$$

A wave travels down a bar at a speed, c_{Rod} given by

$$c_{Rod} = \sqrt{\frac{E}{\rho}}, \tag{C.5}$$

where E is the Young's modulus of the bar material when it is in a uniaxial stress state. A group wave speed is a more appropriate term for such a loading pulse since each frequency travels at a different velocity and thus a step edge disperses with distance. The full analysis of this phenomenon is due to Pochhammer (1876) and Chree (1889) but Rayleigh produced a simpler solution where he defined a group velocity for such waves as

$$c_g = (1 - 3v^2\pi^2 [a/\Lambda]^2)c_{Rod}, \tag{C.6}$$

where a is the radius of the bar and Λ is the length of the pulse.

Waves on the surface of materials may take various forms depending upon the trajectory of particles entrained into them. The best known of these is the Rayleigh wave (a special case of the Stonely wave) where the particles take elliptical trajectories in the region near the surface of a solid. The analytic expression for the Rayleigh wave speed has a complex form but may be approximated as

$$c_R = \left(\frac{0.863 + 1.14v}{1 + v}\right)c_S \tag{C.7}$$

D   Shock relations and constants

Some common shock relations between flow variables are given below.
The standard linear relation for shock velocity is

$$U_s = c_0 + S u_p. \tag{D.1}$$

This calculation of pressure using the shock equation of state

$$p = \rho_0 (c_0 + S u_p) u_p. \tag{D.2}$$

Strain behind a shock is defined to be

$$\varepsilon = \frac{v_1 - v_0}{v_0} = \frac{\rho_0}{\rho_1} - 1, \tag{D.3}$$

So that

$$\rho_0 U_s = \rho_1 \left(u_p - U_s \right) \Rightarrow \varepsilon = \frac{u_p}{U_s}. \tag{D.4}$$

This allows rearrangement to give

$$U_s = \frac{c_0}{1 - S\varepsilon},$$

and

$$p = \rho_0 U_s u_p = \rho_0 U_s^2 \varepsilon$$

or

$$p = \frac{\rho_0 c_0^2 \varepsilon}{(1 - S\varepsilon)^2}. \tag{D.5}$$

Shear strength $\tau = \frac{1}{2}(\sigma_x - \sigma_y)$ for an isotropic material, $\sigma_x = \frac{v}{1-v}\sigma_y$ in the elastic region and

$$P = \frac{1}{3}(\sigma_x + \sigma_y + \sigma_z),$$

which at the HEL is

$$P = \frac{1}{3}\left(\frac{1+\nu}{1-\nu}\right)\sigma_{HEL}. \tag{D.6}$$

The longitudinal stress is given by

$$\sigma_x = P + \frac{4}{3}\tau. \tag{D.7}$$

Now the isothermal $\qquad K_T = \rho \left(\frac{\partial p}{\partial V}\right)_T,$

the adiabatic bulk moduli $\qquad K_s = \rho \left(\frac{\partial p}{\partial V}\right)_s,$

the specific heats $\qquad c_p$ and $c_v,$ $\tag{D.8}$

the Gruneisen coefficient, $\qquad \Gamma = \frac{1}{\rho}\left(\frac{\partial p}{\partial E}\right)_p,$

the volumetric thermal expansion coefficient $\quad \beta$

and the adiabatic sound speed $\qquad c = \sqrt{\frac{K_s}{\rho}}, \tag{D.9}$

combine together thus

$$\frac{K_T}{K_s} = \frac{c_p}{c_v} = 1 + \beta\Gamma T, \tag{D.10}$$

$$\rho\Gamma = \beta\frac{K_T}{c_v} = \beta\frac{K_s}{c_p} = \frac{\beta\rho c^2}{c_p}, \tag{D.11}$$

$$T dS = c_v \left[dT - \left(\frac{\Gamma T}{\rho}\right) d\rho\right], \tag{D.12}$$

$$T dS = c_p \left[dT - \left(\frac{\Gamma T}{K_s}\right) dp\right], \tag{D.13}$$

$$T dS = \frac{1}{\rho\Gamma}[dp - c^2 d\rho]. \tag{D.14}$$

Data for common materials

The data listed in Table D.1 are typically used to generate the equation of state for the materials listed in the strong shock regime but clearly there are variants between different pedigrees. To do so assumes a linear relation of the form $U_s = c_0 + Su_p$ (D.1), where ρ_0 is the ambient density, c_p the specific heat at constant pressure and Γ is Gruneisen's constant. This treatment breaks down beyond the *finis extremis* where more complex interatomic interactions occur.

There are several standard sources with a wider range of materials tabulated and the three most important are listed at the end of this appendix.

Table D.1 Useful parameters for common materials. ρ is the density, c_L and c_S are longitudinal and transverse wave speeds, c_0 is bulk sound speed and S the shock constant both fitted to $c_0 + S u_p$, c_p is the specific heat and Γ is Gruneisen gamma for the material

Material	ρ_0 (g cm^{-3})	c_L (mm μs^{-1})	c_S (mm μs^{-1})	c_0 (mm μs^{-1})	S	c_p (J g^{-1}K^{-1})	Γ
Ag	10.49	3.83	1.69	3.23	1.60	0.24	2.5
Au	19.24	3.37	1.20	3.06	1.57	0.13	3.1
Be	1.85	13.2	8.97	8.00	1.12	0.18	1.2
Bi	9.84	2.49	1.43	1.83	1.47	0.12	1.1
Ca	1.55	4.39	2.49	3.60	0.95	0.66	1.1
Cr	7.12	6.61	4.01	5.17	1.47	0.45	1.5
Cs	1.83	–	–	1.05	1.04	0.24	1.5
Cu	8.93	4.76	2.33	3.94	1.49	0.40	2.0
Fe[1]	7.85	5.94	3.26	3.57	1.92	0.45	1.8
Hg	13.54	1.45	–	1.49	2.05	0.14	3.0
K	0.86	–	–	1.97	1.18	0.76	1.4
Li	0.53	–	–	4.65	1.13	3.41	0.9
Mg	1.74	5.70	3.05	4.49	1.24	1.02	1.6
Mo	10.21	6.48	3.49	5.12	1.23	0.25	1.7
Na	0.97	–	–	2.58	1.24	1.23	1.3
Nb	8.59	5.06	2.10	4.44	1.21	0.27	1.7
Ni	8.87	5.79	3.13	4.60	1.44	0.44	2.0
Pb	11.35	2.25	0.89	2.05	1.46	0.13	2.8
Pd	11.99	4.68	2.33	3.95	1.59	0.24	2.5
Pt	21.42	4.15	1.72	3.60	1.54	0.13	2.9
Rb	1.53	–	–	1.13	1.27	0.36	1.9
Sn	7.29	3.43	1.77	2.61	1.49	0.22	2.3
Ta	16.65	4.14	2.03	3.41	1.20	0.14	1.8
U	18.95	3.50	2.13	2.49	2.20	0.12	2.1
W	19.22	5.22	2.89	4.03	1.24	0.13	1.8
Zn	7.14	4.07	2.35	3.01	1.58	0.39	2.1
KCl[1]	1.99	4.51	2.79	2.15	1.54	0.68	1.3
LiF	2.64	7.16	4.31	5.15	1.35	1.50	2.0
NaCl[2]	2.16	4.47	2.57	3.53	1.34	0.87	1.6
Al-2024	2.79	6.36	3.16	5.33	1.34	0.89	2.0
Al-6061	2.70	6.40	3.15	5.35	1.34	0.89	2.0
SS-304	7.90	5.71	3.09	4.57	1.49	0.44	2.2
Brass	8.45	4.70	2.30	3.73	1.43	0.38	2.0
Water	1.00	1.00	–	1.65	1.92	4.19	0.1
PTFE	2.15	1.23	0.41	1.84	1.71	1.02	0.6
PMMA	1.19	2.72	1.36	2.60	1.52	1.20	1.0
PE	0.92	2.46	1.01	2.90	1.48	2.30	1.6
PS	1.04	2.24	1.15	2.75	1.32	1.20	1.2

[1]Above phase transition; [2]below phase transition.

Standard sources

Bushman, A. V., Lomonosov, I. V. and Khishchenko, K. V. (2002) *Rusbank Shock Wave Database*, http://teos.ficp.ac.ru/rusbank/.

Marsh, S.P. (1980) *LASL Shock Hugoniot Data*. Berkeley, CA: University of California Press.

van Thiel, M. (1966) *Compendium of Shock Wave Data (2 vols plus suppl.)*. Livermore, CA: Lawrence Radiation Laboratory.

Bibliography

Starting in 1993, a series of edited collections has been published which review particular areas of shock wave and high compression phenomena in condensed matter. For completeness a list of these volumes is reproduced here.

Asay, J. R. and Shahinpoor, M. (eds.) (1993) *High-Pressure Shock Compression of Solids*. Shock Wave and High Pressure Phenomena. New York: Springer.

Davison, L. W., Grady, D. E. and Shahinpoor, M. (eds.) (1996) *High Pressure Shock Compression of Solids II: Dynamic Fracture and Fragmentation*. Shock Wave and High Pressure Phenomena. New York: Springer.

Davison, L. W. and Shahinpoor, M. (eds.) (1998) *High-Pressure Shock Compression of Solids III*. Shock Wave and High Pressure Phenomena. New York: Springer.

Davison, L. W., Horie, Y. and Shahinpoor, M. (eds.) (1997) *High-Pressure Shock Compression of Solids IV: Response of Highly Porous Solids to Shock Loading*. Shock Wave and High Pressure Phenomena. New York: Springer.

Davison, L. W., Horie, Y. and Sekine, T. (eds.) (2003) *High-Pressure Shock Compression of Solids V: Shock Chemistry with Applications to Meteorite Impacts*. Shock Wave and High Pressure Phenomena. New York: Springer.

Horie, Y., Davison, L. W. and Thadani, N. (eds.) (2003) *High-Pressure Shock Compression of Solids VI: Old Paradigms and New Challenges*. Shock Wave and High Pressure Phenomena. New York: Springer.

Fortov, V. E., Altshuler, L. V., Trunin, R. F. and Funtikov, A. I. (eds.) (2004) *High Pressure Shock Compression VII: Shock Waves and Extreme States of Matter*. Shock Wave and High Pressure Phenomena. New York: Springer.

Chhabildas, L. C., Davison, L. W. and Horie, Y. (eds.) (2005) *High-Pressure Shock Compression of Solids VIII: The Science and Technology of High-Velocity Impact*. Shock Wave and High Pressure Phenomena. New York: Springer.

This text has introduced concepts and quoted results from a range of publications. Each chapter has had some selected reading added for the interested reader to delve deeper. All of of these are brought together in the following list compiled alphabetically by author.

Angel, R. J., Bujak, M., Zhao, J., Gatta, G. D. and Jacobsen, S. D. (2007) Effective hydrostatic limits of pressure media for high-pressure crystallographic studies, *J. Appl. Cryst.*, 40: 26–32.

Armstrong, R. W. and Walley, S. M. (2008) High strain rate properties of metals and alloys, *Int. Mater. Rev.*, 53: 105–128.

Asay, J. R. and Lipkin, J. (1978) A self-consistent technique for estimating the dynamic yield strength of a shock-loaded material, *J. Appl. Phys.*, 49: 4242–4247.

Asay, J. R., Konrad, C. H., Hall, C. A. *et al.* (1999) Use of Z-pinch radiation sources for high-pressure shock wave studies, in *New Models and Numerical Codes for Shock Wave Processes in Condensed Media*, ed. Cameron, I. G. Aldermaston, Berkshire, UK: AWE Hunting Brae, pp. 287–297.

Asay, J. R., Ao, T., Vogler, T. J., Davis, J.-P. and Gray III, G. T. (2009) Yield strength of tantalum for shockless compression to 18 GPa, *J. Appl. Phys.*, 106: 073515.

ASM Metals Handbook, Vol. 8: Mechanical Testing and Evaluation (2000). Materials Park, OH: ASM International.

Bai, Y. L. and Dodd, B. (1992) *Adiabatic Shear Localization*. Oxford: Pergamon Press.

Bailey, A. and Murray, S. G. (1989) *Explosives, Propellants and Pyrotechnics*. London: Brassey's.

Bell, J. F. (1973) The experimental foundations of solid mechanics, in *Encyclopedia of Physics*, Vol. VIa. Berlin: Springer Verlag.

Benson, D. J. (1992) Computational methods in Lagrangian and Eulerian hydrocodes, *Comput. Meth. Appl. Mech. Eng.*, 99: 235–394.

Berthelot, M. and Vieille, P. (1882) Sur la vitesse de propagation des phenomenes explosifs dans les gaz, *C. R. Acad. Sci. Paris*, 94: 101–108.

Berthelot, M. (1883) Sur la force des matières explosives d'après la thermochimie. Gauthier-Villars.

Barker, L. M. and Hollenbach, R. E. (1970) Shock-wave studies of PMMA, fused silica, and sapphire, *J. Appl. Phys.*, 41: 4208–4226.

Bourne, N. K. (2001a) On the laser ignition and initiation of explosives, *Proc. R. Soc. Lond. A.*, 457(2010): 1401–1426.

Bourne, N. K. (2001b) The onset of damage in shocked alumina, *Proc. R. Soc. Lond. A.*, 457(2013): 2189–2205.

Bourne, N. K. (2004) Gas gun for dynamic loading of explosives, *Rev. Sci. Instrum.*, 75(1): 1–6.

Bourne, N. K. (2005a) On the impact and penetration of soda-lime glass, *Int. J. Imp. Engng.*, 32: 65–79.

Bourne, N. K. (2005b) On impacting liquid jets and drops onto polymethylmethacrylate targets, *Proc. R. Soc. Lond. A.*, 461: 1129–1145.

Bourne, N. K. (2005c) The shock response of float-glass laminates, *J. Appl. Phys.*, 98(6), 063515.

Bourne, N. K. (2006a) Impact on alumina. I. Response at the mesoscale, *Proc. R. Soc. Lond. A*, 462(2074): 3061–3080.

Bourne, N. K. (2006b) Impact on alumina. II. Linking the mesoscale to the continuum, *Proc. R. Soc. Lond. A.*, 462(2075): 3213–3231.

Bourne, N. K. (2008). The relation of failure under 1D shock to the ballistic performance of brittle materials, *Int. J. Imp. Engng.*, 35: 674–683.

Bourne, N. K. (2009) Shock and awe, *Physics World*, 22(1): 26–29.

Bourne, N. K. (2011) Materials' physics in extremes: akrology, *Metall. Mater. Trans. A.*, 42: 2975–2984.

Bourne, N. K. and Field, J. E. (1999) On the impact and penetration of transparent targets, *Proc. R. Soc. Lond. A.*, 455: 4169–4179.

Bourne, N. K. and Gray III, G. T. (2005a) Computational design of recovery experiments for ductile metals, *Proc. R. Soc. Lond. A.*, 460: 3297–3312.

Bourne, N. K. and Gray III, G. T. (2005b) Soft-recovery of shocked polymers and composites, *J. Phys D. Appl. Phys.*, 38: 3690–3694.

Bourne, N. K. and Millett, J. C. F. (2000) Shock-induced interfacial failure in glass laminates, *Proc. R. Soc. Lond. A.*, 456(2003): 2673–2688.

Bourne, N. K. and Millett, J. C. F. (2001) Decay of the elastic precursor in a filled glass, *J. Appl. Phys.*, 89(10): 5368–5371.

Bourne, N. K. and Milne, A. M. (2004) Shock to detonation transition in a plastic bonded explosive, *J. Appl. Phys.*, 95(5): 2379–2385.

Bourne, N. K., Rosenberg, Z. and Field, J. E. (1995) High-speed photography of compressive failure waves in glasses, *J. Appl. Phys.*, 78(6): 3736–3739.

Bourne, N. K., Millett, J. C. F., Rosenberg, Z. and Murray, N. H. (1998) On the shock induced failure of brittle solids, *J. Mech. Phys. Solids*, 46(10): 1887–1908.

Bourne, N. K., Millett, J. C. F. and Field, J. E. (1999) On the strength of shocked glasses, *Proc. R. Soc. Lond. A*, 455(1984): 1275–1282.

Bourne, N. K., Rosenberg, Z. and Field, J. E. (1999) Failure zones in polycrystalline aluminas, *Proc. R. Soc. Lond. A.*, 455(1984): 1267–1274.

Bourne, N. K., Millett, J. C. F., Chen, M., McCauley, J. W. and Dandekar, D. P. (2007) On the Hugoniot elastic limit in polycrystalline alumina, *J. Appl. Phys.*, 102: 073514.

Bourne, N. K., Millett, J. C. F., Brown, E. N and Gray III, G. T. (2007) The effect of halogenation on the shock properties of semi-crystalline thermo-plastics, *J. Appl. Phys.*, 102: 063510.

Bourne, N. K., Gray III, G. T. and Millett, J. C. F. (2009) On the shock response of cubic metals, *J. Appl. Phys.*, 106(9): 091301.

Bourne, N. K., Millett, J. C. F. and Gray III, G. T. (2009) On the shock compression of polycrystalline metals. *J. Mat. Sci.* 44(13): 3319–3343.

Bowden, F. P. and Tabor, D. (1950/2001) *The Friction and Lubrication of Solids.* Oxford: Oxford University Press. Originally published 1950, revised edition 2001.

Bowden, F. P. and Yoffe, A. D. (1952/1985) *Initiation and Growth of Explosion in Liquids and Solids.* Cambridge: Cambridge University Press. Originally published 1952, revised edition 1985.

Brannon, R. M., Wells, J. M. and Strack, O. E. (2007) Validating theories for brittle damage, *Metal. Mater. Trans. A*, 38: 2861–2868.

Bridgman, P. W. (1914) Two new modifications of phosphorus, *J. Am. Chem. Soc.* 36(7): 1344–1363.

Bridgman, P. W. (1952) *The Physics of High Pressure.* Bell and Sons, reprint.

Bringa, E. M., Rosolankova, K., Rudd, R. E. *et al.* (2006), Shock deformation of face-centred-cubic metals on subnanosecond timescales, *Nature Materials*, 5: 805–809.

Bushman, A. V., Lomonosov, I. V. and Khishchenko, K. V. (2002) *Rusbank Shock Wave Database*, http://teos.ficp.ac.ru/rusbank/.

Calladine, C. R. and English, R. W. (1984) Strain-rate and inertia effects in the collapse of two types of energy-absorbing structure, *Int. J. Mech. Sci.*, 26: 689–701.

Cao, B. Y., Meyers, M. A., Lassila, D. H. *et al.* (2005) Effect of shock compression method on the defect substructure in monocrystalline copper, *Mat. Sci. and Eng A.*, 409: 270–281.

Carter, W. J. and Marsh, S. P. (1995; republished by J.N. Fritz and S.A. Sheffield from a report put together in 1977), *Hugoniot Equation of State of Polymers.* Los Alamos, NM: Los Alamos National Laboratory.

Cerreta, E. K., Gray III, G. T., Hixson, R. S., Rigg, P. A. and Brown, D. W. (2005) The influence of interstitial oxygen and peak pressure on the shock loading behavior of zirconium, *Acta Materialia*, 53: 1751–1758.

Chapman, D.L. (1899) On the rate of explosion in gases, *Philos. Mag.*, series 5, 47: 90–104.

Chen, M. W., McCauley, J. W., Dandekar, D. P. and Bourne, N. K. (2006) Dynamic plasticity and failure of high-purity alumina under shock loading, *Nature Mater.*, 5: 614–618.

Cheny, J.-M. and Walters, K. (1999) Rheological influences on the splashing experiment, *J. Non-Newtonian Fluid Mech.*, 86: 185–210.

Chree, C. (1889) The equations of an isotropic elastic solid in polar and cylindrical coordinates, their solutions and applications, *Trans. Cambridge Philos. Soc. Math. Phys. Sci.* 14: 250.

Cochran, S. and Banner, D. (1977) Spall studies in uranium, *J. Appl. Phys.*, 48: 2729–2737.

Collins, G. S., Melosh, H. J. and Marcus, R. A. (2005) Earth Impact Effects Program: a web-based computer program for calculating the regional environmental consequences of a meteoroid impact on Earth, *Meteorit. Planet. Sci.*, 40: 817.

Collins, G. W., Da Silva, L. B., Celliers, P. *et al.* (1998) Measurements of the equation of state of deuterium at the fluid insulator–metal transition, *Science*, 281: 1178–1181.

Cooper, P. W. (1997) *Explosives Engineering*. New York: Wiley.

Courant, R. and Friedrichs, K. O. (1948) *Supersonic Flow and Shock Waves*. New York: Interscience.

Cox, B. N., Gao, H., Gross, D. and Rittel, D. (2005) Modern topics and challenges in dynamic fracture, *J. Mech. Phys. Solids*, 53: 565–596.

Curran, D. R., Seaman, L. and Shockey, D. A. (1977) Dynamic failure in solids, *Phys. Today*, 30: 46.

Curran, D. R., Seaman, L. and Shockey, D. A. (1987) *Dynamic Failure of Solids*. Amsterdam: North-Holland.

Da Silva, L. B., Celliers, P., Collins, G.W. *et al.* (1997) Absolute equation of state measurements on shocked liquid deuterium up to 200 GPa (2 Mbar), *Phys. Rev. Lett.*, 78: 483–486

Dai, L. H. and Bai, Y. L. (2008) Basic mechanical behaviors and mechanics of shear banding in BMGs, *Int. J. Imp. Eng.*, 35: 704–716

Davis, W. C. (1987) The detonation of explosives, *Sci. Am.*, 256(5): 106.

Davison, L. and Graham R. A. (1979) Shock compression of solids, *Phys. Rep.*, 55(4): 255–379.

DeCarli P. S. and Jamieson J. C. (1961) Formation of diamond by explosive shock, *Science*, 133: 1821–1822.

Döring, W. (1943) Über Detonationsvorgang in Gasen (On the detonation process in gases), *Annalen der Physik*, 43: 421–436.

Dolan, D. H. and Gupta, Y. M. (2004) Nanosecond freezing of water under multiple shock wave compression: optical transmission and imaging measurements, *J. Chem. Phys.*, 121: 9050–9057.

Dremin, A. N. and Adadurov, G. A. (1964) Behaviour of a glass at dynamic loading, *Fiz. Tverd. Tela*, 6(6): 1757–1764.

Duffy, J., Campbell, J. D. and Hawley, R. H. (1971) On the use of a torsional split Hopkinson bar to study rate effects in 1100-0 aluminum, *J. Appl. Mech.*, 38(1): 83–92.

Duffy, T. S. (2005) Synchrotron facilities and the study of the Earth's deep interior, *Rep. Prog. Phys.*, 68: 1811–1859.

Duffy, T. S. (2007) Strength of materials under static loading in the diamond anvil cell, in *Shock Compression of Condensed Matter 2007*, eds. M. D. Furnish, *et al.* Melville, NY: American Institute of Physics, pp. 639–644.

Duffy, T. S., Shen, G., Shu, J., Mao, H.-K., Hamley, R. J. and Singh, A. K. (1999) Elasticity, shear strength, and equation of state of molybdenum and gold from X-ray diffraction under nonhydrostatic compression to 24 GPa, *J. Appl. Phys.*, 86: 6729–6736.

Duvall, G. E. and Graham R. A. (1977) Phase transitions under shock-wave loading, *Rev. Mod. Phys.*, 49(3): 523–580.

Esposito, A. P., Farber, D. L., Reaugh, J. E. and Zaug, J. M. (2003) Reaction propagation rates in HMX at high pressure, *Propell. Explos. Pyrot.*, 28: 83–88.

Fickett, W. and Davis, W. C. (2000) *Detonation: Theory and Experiment.* Mineola, NY: Courier Dover Publications. Reprint, originally published 1979.

Field, J. E., Bourne, N. K., Palmer, S. J. P. and Walley, S. M. (1992) Hot-spot ignition mechanisms for explosives and propellants, *Phil. Trans. R. Soc. Lond. A*, 339: 269–283.

Follansbee, P. S., Regazzoni, G. and Kocks, U.F. (1984) The transition in drag-controlled deformation in copper at high strain rates, *Inst. Phys. Conf. Ser.*, 70: 71–80.

Fortov, V. E. (2011) *Extreme States of Matter on Earth and in the Cosmos.* Berlin: Springer.

Frank-Kamenetskii, D. A. (1969) *Diffusion and Heat Transfer in Chemical Kinetics*, 2nd edn., translated by J. P. Appleton. New York: Plenum Press.

Freund, L. B. (1990) *Dynamic Fracture Mechanics.* Cambridge Monographs on Mechanics and Applied Mathematics. Cambridge: Cambridge University Press.

French, B. M. (1998) *Traces of Catastrophe: A Handbook of Shock-Metamorphic Effects in Terrestrial Meteorite Impact Structures.* LPI Contribution No. 954. Houston, TX: Lunar and Planetary Institute.

Gama, B. A., Lopatnikov, S. L., Gillespie Jr., J. W. (2004) Hopkinson bar experimental technique: a critical review, *Appl. Mech. Rev.*, 57(4), 223–250.

Germann, T. C., Tanguy, D., Holian, B. L., Lomdahl, P.S., Mareschal, M. and Ravelo, R. (2004) Dislocation structure behind a shock front in fcc perfect crystals: Atomistic simulation results. *Metall. Mater. Trans. A*, 35: 2609–2615.

Gibbons, R. V. and Ahrens, T. J. (1971) Shock metamorphism of silicate glasses. *J. Geophys. Res.*, 76, p. 5489–5498.

Grady, D. E. (1997) Shock-wave compression of brittle solids, *Mech. Mater.*, 29: 181–203.

Grady, D. E. (2006) *Fragmentation of Rings and Shells: The Legacy of N.F. Mott.* New York: Springer.

Grady, D. E. (2007) The shock wave profile, in *Shock Compression of Condensed Matter 2007*, eds Furnish, M. D., Elert, M., Chau, R., Holmes, N. C. and Nguyen, J. Melville, NY: American Institute of Physics, pp. 3–11.

Grady, D. E. (2010) Structured shock waves and fourth power law, *J. Appl. Phys.*, 107: 013506.

Grady, D. E. and Kipp, M. E. (1993) Dynamic fracture and fragmentation, in *High-Pressure Shock Compression of Solids*, eds. J. R. Asay and M. Shahinpoor. New York: Springer-Verlag, pp. 265–322.

Gray III, G. T. (2000a) Classic split-Hopkinson pressure bar testing, in *ASM Handbook, Vol. 8: Mechanical Testing and Evaluation.* Materials Park, OH: ASM International, pp. 462–476.

Gray III, G. T. (2000b) Shock wave testing of ductile materials, in *ASM Handbook, Vol. 8: Mechanical Testing and Evaluation.* Materials Park, OH: ASM International, pp. 530–538.

Griffith, A. A. (1921) The phenomena of rupture and flow in solids, *Philos. Trans. R. Soc. London A*, 221: 163–198.

Gupta, Y. M., Winey, J. M., Trivedi, P. B. *et al.* (2009) Large elastic wave amplitude and attenuation in shocked pure aluminium, *J. Appl. Phys.*, 105: 036107.

Gurney, R. (1943) The initial velocities of fragments from bombs, shells and grenades. Report no. 405, Ballistic Research Laboratory, Aberdeen, MD, AII-36218.

Harding, J., Wood, E. O. and Campbell, J. D. (1960) Tensile testing of materials at impact rates of strain, *J. Mech. Engng Sci.*, 2, 88–96.

Hayes, D. B., (1974) Polymorphic phase transformation rates in shock-loaded potassium chloride, *J. Appl. Phys.*, 45: 1208–1217.

Hermann, W. (1969) Constitutive equation for the dynamic compaction of ductile porous materials, *J. Appl. Phys.*, 40: 2490–2499.

Hill, R. (1963) Elastic properties of reinforced solids: some theoretical principles, *J. Mech. Phys. Solids*, 11: 357–372.

Hoge, G. and Mukherjee, A. K. (1977) The temperature and strain rate dependence of the flow stress of tantalum, *J. Mater. Sci.*, 12: 1666–1672.

Holian, B. L. (1988) Modeling shock wave deformation via molecular dynamics, *Phys. Rev. A*, 37: 2562–2568.

Holtkamp, D. B., Clark, D. A., Ferm, E. N. *et al.* (2004) A survey of high explosive-induced damage and spall in selected metals using proton radiography, in *Proceedings of the Conference of the American Physical Society on Shock Compression of Condensed Matter*, Portland, OR, 20–25 July 2003. Melville, NY: American Institute of Physics, pp. 477–482.

Honeycombe, R. W. K. (1984) *The Plastic Deformation of Metals*, 2nd edition. London: Edward Arnold.

Hopkinson, B. (1914) A method of measuring the pressure produced in the detonation of high explosives or by the impact of bullets, *Proc. R. Soc. Lond. A.*, 89(612): 411–413.

Hu, J., Zhou, X., Dai, C., Tan, H. and Li, J. (2008) Shock-induced bct-bcc transition and melting of tin identified by sound velocity measurements, *J. Appl. Phys.*, 104: 083520.

Hugoniot, P-H. (1887a) Mémoire sur la propagation du mouvement dans un fluide indéfini, *J. Math. Pures Appl.* (4th series), 3: 477–492 and 4: 153–168.

Hugoniot, P-H. (1887b) Sur la propagation du mouvement dans les corps et spécialement dans les gaz parfaits (première partie), *J. l'École Polytech.*, 57: 3–97.

Hugoniot, P-H. (1889) Sur la propagation du mouvement dans les corps et spécialement dans les gaz parfaits (deuxième partie), *J. l'École Polytech.*, 58: 1–125.

Inglis, C. E. (1913) Stresses in a plate due to the presence of cracks and sharp corners, *Proc. Inst. Naval Arch.*, 55: 219–241.

Irwin, G. R. (1985) Fracture mechanics, in *ASM Metals Handbook, Vol. 8: Mechanical Testing and Evaluation* (2000). Materials Park, OH: ASM International, pp. 439–458.

James, H. R. and Lambourn, B. D. (2001) A continuum-based reaction growth model for the shock initiation of explosives, *Propell. Explos. Pyrot.*, 26: 246–256.

Jenniskens, P., Shaddad, M. H., Numan, D. *et al.* (2009) The impact and recovery of asteroid 2008 TC3, *Nature*, 458: 485–488.

Jouguet, E. (1906) Sur la propagation des réactions chimiques dans les gaz: les ondes de choc, *J. Math. Pures Appli.* (6ème Sér.), 2: 5–86.

Jouguet, E. (1917) *Mécanique des Explosifs*. Paris, France: Octave Doin.

Kalantar, D.H., Remington, B.A., Chandler, E.A. *et al.* (1999) High pressure solid state experiments on the Nova laser, *Int. J. Impact Engng.*, 23: 409–420.

Kanel, G. I., Rasorenov, S. V., Fortov, V. E. and Abasehov, M. M. (1991) The fracture of glass under high pressure impulsive loading, *High. Press. Res.*, 6: 225–232.

Kelly, A. (1973) *Strong Solids*, 2nd edition. Oxford: Clarendon Press.

Knudson, M. D., Hanson, D. L., Bailey, J. E., Hall, C. A., Asay, J. R. and Anderson, W. W. (2001) *Phys. Rev. Lett.*, 87: 225501.

Koller, D. D., Hixson, R. S., Gray III, G. T. *et al.* (2006) Explosively driven shock induced damage in OFHC Cu, in *Shock Compression of Condensed Matter 2006*, eds. Furnish, M. D., Elert, M. L., Russell, T. P. and White, C. T. Melville, NY: American Institute of Physics, pp. 599–602.

Kolsky, H. (1953) *Stress Waves in Solids*. Oxford: Clarendon Press.

Lankford Jr., J. (2005) The role of dynamic material properties in the performance of ceramic armour, *Int. J. Appl. Ceram. Tech.*, 1(3): 205–210.

Lawn, B. R. (1993) *Fracture of Brittle Solids*, 2nd edition. Cambridge: Cambridge University Press.

Lee, E. L. and Tarver, C. M. (1980) A phenomenological model of shock initiation in heterogeneous explosives, *Phys. Fluids*, 23: 2362.

Luebcke, P. E., Dickson, P. M. and Field, J. E. (1995) An experimental study of the deflagration-to-detonation transition in granular secondary explosives, *Proc. R. Soc. A.*, 448: 439–448.

MacLeod, S. G., Tegner, B. E., Cynn, H. *et al.* (2012) Experimental and theoretical study of Ti-6Al-4V to 220 GPa, *Phys. Rev. B*, 85: 224202.

Mader, C. L. (1979) *Numerical Modeling of Explosives and Propellants*. London: CRC Press, reprint.

Mallard, E. and Le Chatelier, H. (1881) On the propagation velocity of burning in gaseous explosive mixtures, *C. R. Acad. Sci. Paris*, 93: 145–148.

Malvern, L. E. (1969) *Introduction to the Mechanics of a Continuous Medium*. Englewood Cliffs, NJ: Prentice-Hall.

Marchand, A. and Duffy, J. (1988) An experimental study of the formation process of adiabatic shear bands in a structural steel, *J. Mech. Phys. Solids*, 36: 251–283.

Marsh, S. P. (1980) *LASL Shock Hugoniot Data*. Berkeley, CA: University of California Press.

McClintock, F. A. and Walsh, J. B. (1962), Friction on Griffith cracks in rocks under pressure, *Proc. 4th Natl. Congr. Appl. Mech.*, Berkeley, CA, pp. 1015–1021.

Merzkirch, W. (1993) *Flow Visualization*. London: Academic Press.

Meyers, M. A. (1994) *Dynamic Behavior of Materials*. Chichester: Wiley.

Meyers, M. A., Aimone, C. T. (1983) Dynamic fracture (spalling) of metals, *Prog. Mater. Sci.*: 1–96.

Millett, J. C. F., Bourne, N. K. and Rosenberg, Z. (1998) Observations of the Hugoniot curves for glasses as measured by embedded stress gauges, *J. Appl. Phys.*, 84(2): 739–741.

Millett, J. C. F., Bourne, N. K. and Stevens, G. S. (2006) Taylor impact of polyether ether ketone, *Int. J. Impact Eng.*, 32(7): 1086–1094.

Millett, J. C. F., Whiteman, G. and Bourne, N. K. (2009) Lateral stress and shear strength behind the shock front in three face centered cubic metals, *J. Appl. Phys.*, 105: 033515.

Millett, J. C. F., Bourne, N. K., Park, N. T., Whiteman, G. and Gray III, G. T. (2011) On the behaviour of body-centred cubic metals to one-dimensional shock loading, *J. Mater. Sci.*, 46: 3899–3906.

Mills, N. J. (1993) *Plastics: Microstructure and Engineering Applications*. London: Edward Arnold.

Minshall, S. (1955) *Phys. Rev.* 98: 271.

Mott, N. F. (1947) Fragmentation of shell cases, *Proc. Royal Soc.*, A189: 300–305.

Murray, N. H., Bourne, N. K. and Rosenberg, Z. (1998) The dynamic compressive strength of aluminas, *J. Appl. Phys.*, 84(9): 4866–4871.

Nesterenko, V. F. and Bondar, M. P. (1994) Localization of deformation in collapse of a thick walled cylinder, *Combus. Explos. Shock Waves*, 30(4): 500–509.

Neumann, J., von (1942) *Theory of Detonation Waves*, Office of Scientific Research and Development, Report 549, Ballistic Research Laboratory File No. X-122. Aberdeen Proving Ground, Maryland.

Odeshi, A. G., Bassim, M. N. and Al-Ameeri, S. (2006) Adiabatic shear bands in a high-strength low alloy steel, *Mater. Sci. Engng. A*, 419: 69–75.

Orowan, E. (1934) Zur Kristallplastizität III, *Z. Phys.*, 89: 634–659.

Pierazzo, E. and Melosh, H. J. (2000) Understanding oblique impacts from experiments, observations and modelling, *Annu. Rev. Earth Planet. Sci.*, 28: 141–167.

Pochhammer, L. (1876) Uber die fortpflanzungsgeschwindigkeiten kleiner schwingungen in einem unbegrenzten isotropen kreiscylinder, *J. Fur. Reine and Angewandte Math. (Crelle)*, 81: 324–336.

Popolato, A. (1972) in *Behaviour and Utilisation of Explosives in Engineering Design*, ed. R. L. Henderson. Albuquerque, NM: ASME.

Powell, J. L. (1998) *Night Comes to the Cretaceous: Dinosaur Extinction and the Transformation of Modern Geology*. New York: W.H. Freeman.

Raftenberg, M. N. (2001) A shear banding model for penetration calculations, *Int. J. Imp. Engng.*, 25(2): 123–146.

Rankine, W. J. M. (1870) On the thermodynamic theory of waves of finite longitudinal disturbances, *Phil. Trans. R. Soc. Lond.*, 160: 277–288.

Rayleigh, J. W. S. (1877) *The Theory of Sound*. London: Macmillan and Co.

Reddy, T. Y. and Reid, S. R. (1980) Phenomena associated with the crushing of metal tubes between rigid plates, *Int. J. Solids Struct.*, 16(6): 545–562.

Reiner, M. (1964) The Deborah Number, *Phys. Today*, 17(1): 62.

Remington, B. A., Bazan, G., Bringa, E. M. *et al.* (2006) Material dynamics under extreme conditions of pressure and strain rate, *Mat. Sci. Tech.*, 22: 474–488.

Rivas, J. M., Zurek, A. K., Thissell, W. R., Tonks, D. L. and Hixson, R. S. (2000) Quantitative description of damage evolution in ductile fracture of tantalum, *Metal. Mater. Trans. A*, 31: 845–851.

Romanchenko, V. I. and Stepanov, G. V. (1980) Dependence of the critical stresses on the loading time parameters during spall in copper, aluminum, and steel, *J. Appl. Mech. Tech. Phys.*, 21: 555–561.

Rosenberg Z. and Dekel, E. (2012) *Terminal Ballistics*. London: Springer.

Seigel, A. E. (1965) *The Theory of High Speed Guns*, AGARDograph 91, May.

Schön, E. (2004). *Asa-Tors hammare, Gudar och jättar i tro och tradition*. Värnamo: Fält & Hässler.

Smallman, R. E. and Bishop, R. J. (1999) *Modern Physical Metallurgy and Materials Engineering*, 6th edition. Oxford: Butterworth-Heinemann.

Stokes, Sir, George Gabriel (1851) On the alleged necessity for a new general equation in hydrodynamics, *Philos. Mag.*, I: 157–160, 393–394.

Swegle, J. W. and Grady, D. E. (1985) Shock viscosity and the prediction of shock wave rise times, *J. Appl. Phys.*, 58: 692.

Taylor, J. W. and Rice, M. H. (1963) Decay of the elastic precursor wave in Armco iron, *J. Appl. Phys*, 34: 364–371.

van Thiel, M. (1966) *Compendium of Shock Wave Data* (2 vols plus suppl.). Livermore, CA: Lawrence Radiation Laboratory.

Wadsworth, G. (2007) *Basic Research Needs for Materials under Extreme Environments*. Office of Science, US Department of Energy, available at http://science.energy.gov/bes/news-and-resources/reports/abstracts/#MUEE.

Walley, S. M. (2007) Shear localization: a historical review, *Metall. Mater. Trans. A*, 38(11): 2629–2654.

Walley, S. M., Field, J. E., Pope, P. H. and Safford, N. A. (1989) A study of the rapid deformation behaviour of a range of polymers, *Phil. Trans. R. Soc. Lond. A.*, 328: 1–33.

Weir, S. T., Mitchell, A. C. and Nellis, W. J. (1996) Metallization of fluid molecular hydrogen, *Phys. Rev. Lett.* 76: 1860.

Whitley, V. H., McGrane, S. D., Eakins, D. E., Bolme, C. A., Moore, D. S. and Bingert, J. F. (2011) The elastic-plastic response of aluminum films to ultrafast laser-generated shocks, *J. Appl. Phys.*, 109: 013505.

Worthington, A. M. and Cole, R. S. (1897) Impact with a liquid surface studied by the aid of instantaneous photography, *Phil. Trans. R. Soc. A.*, 189: 137–148.

Worthington, A. M. and Cole, R. S. (1900) Impact with a liquid surface studied by the aid of instantaneous photography; paper II, *Phil. Trans. R. Soc. A.*, 194: 175–199.

Wright, T. W. (2002) *The Physics and Mathematics of Adiabatic Shear Bands*. Cambridge: Cambridge University Press.

Xue, Q. and Gray III, G. T. (2006) Development of adiabatic shear bands in annealed 316L stainless steel: part I. Correlation between evolving microstructure and mechanical behaviour, *Metall. Mater. Trans. A.*, 37(8): 2435–2446.

Xue, Q., Nesterenko, V. F. and Meyers, M. A. (2003) Evaluation of the collapsing thick-walled cylinder technique for shear-band spacing, *Int. J. Impact Engng.*, 28: 257–280.

Zeldovitch, Y. B. (1940) On the theory of the propagation of detonation in gaseous systems, *Zhurnal Eksperimental'noi i Teoreticheskoi Fiziki (JETP)*, 10: 542–568.

Zeldovitch, Y. B. and Raizer Y. P. (2002) *Physics of Shock Waves and High-Temperature Hydrodynamic Phenomena*. Mineola, NY: Dover, reprint.

Zener, C. and Hollomon, J. H. (1944) Effect of strain rate upon plastic flow of steel, *J. Appl. Phys.*, 15: 22–32.

Zerilli, F. J. and Armstrong, R. W. (2007) A constitutive equation for the dynamic deformation behavior of polymers, *J. Mater. Sci.* 42: 4562–4574.

Zukas, J. (1990) *High Velocity Impact Dynamics*. New York: Wiley.

The Association Internationale pour L'Avancement de la Reserche et de la Technologie aux Hautes Pression, the International Association for the Advancement of High Pressure Science and Technology (AIRAPT) grew from the Gordon Research Conference on Research at High Pressures (GRCHP). It organises a conference, usually every 2 years, and papers presented are published in a special issue of the IOP Journal of Physics. Shock Compression of Condensed Matter is a topical group of the American Physical Society. It too holds a biennial conference and proceedings date back to 1981. In future it too will publish papers in an IOP journal. The Detonation symposium started in 1951 and is usually held every 4 years. It again has proceedings that include papers and questions posed by attendees at the meeting for each contribution. The dynamic behaviour of materials and structures (DYMAT) is a forum for applications of extreme loading such as crashworthiness in transport, terminal ballistics, blast effects and material processing and holds a conference every 3 years concentrating upon the engineering aspects of the area. There are proceedings for this meeting going back to 1985. These conferences together serve the spread from the physical science of materials under extremes to engineering in the field. Their proceedings serve as a repository of the work done in each area over the past half century. This book has focussed principally on the condensed state but there is a community that studies shock waves in gases. The International Symposia on Shock Waves go back to 1957 and again proceedings exist from the biennial conferences for the interested reader to access.

Index